EXPLORING THE SOLAR WIND

Edited by **Marian Lazar**

INTECHOPEN.COM

Exploring the Solar Wind

Edited by Marian Lazar

CBS, Edition **2017**

Published by InTech

Janeza Trdine 9, 51000 Rijeka, Croatia

Publishing Process Manager Danijela Duric

Technical Editor InTech DTP team

Cover Designer InTech Design Team

Additional hard copies can be obtained from orders@intechopen.com

Exploring the Solar Wind, Edited by Marian Lazar
p. cm.
ISBN 978-953-51-0339-4

INTECH

open science | open minds

Contents

Preface VII

Part 1 The Solar Wind: Overview of the Fundamentals 1

Chapter 1 **Solar Wind Laws Valid for any Phase of a Solar Cycle 3**
V.G. Eselevich

Chapter 2 **Solar Wind: Origin, Properties and Impact on Earth 29**
U.L. Visakh Kumar and P.J. Kurian

Part 2 The Solar Wind Elemental Compostition 47

Chapter 3 **Solar Wind Composition Associated with the Solar Activity 49**
X. Wang, B. Klecker and P. Wurz

Chapter 4 **Solar Wind and Solar System Matter After Mission Genesis 69**
Kurt Marti and Peter Bochsler

Chapter 5 **Measuring the Isotopic Composition of Solar Wind Noble Gases 93**
Alex Meshik, Charles Hohenberg, Olga Pravdivtseva and Donald Burnett

Chapter 6 **Solar Wind Noble Gases in Micrometeorites 121**
Takahito Osawa

Part 3 The Solar Wind Dynamics: From Large to Small Scales 141

Chapter 7 **Multifractal Turbulence in the Heliosphere 143**
Wiesław M. Macek

Chapter 8 **Field-Aligned Current Mechanisms of Prominence Destabilization 169**
Petko Nenovski

Contents

Chapter 9 **Small Scale Processes in the Solar Wind 195**
Antonella Greco, Francesco Valentini and Sergio Servidio

Chapter 10 **Kinetic Models of Solar Wind Electrons, Protons and Heavy Ions 221**
Viviane Pierrard

Chapter 11 **Suprathermal Particle Populations in the Solar Wind and Corona 241**
M. Lazar, R. Schlickeiser and S. Poedts

Part 4 The Solar Wind Magnetic Field Powered by the Sun 259

Chapter 12 **Impact of the Large-Scale Solar Magnetic Field on the Solar Corona and Solar Wind 261**
A.G. Tlatov and B.P. Filippov

Chapter 13 **Variability of Low Energy Cosmic Rays Near Earth 285**
Karel Kudela

Part 5 The Interaction of the Solar Wind with the Magnetosphere 315

Chapter 14 **Impact of Solar Wind on the Earth Magnetosphere: Recent Progress in the Modeling of Ring Current and Radiation Belts 317**
Natalia Buzulukova, Mei-Ching Fok and Alex Glocer

Chapter 15 **Ground-Based Monitoring of the Solar Wind Geoefficiency 337**
Oleg Troshichev

Chapter 16 **The Polar Cap PC Indices: Relations to Solar Wind and Global Disturbances 357**
Peter Stauning

Chapter 17 **Sudden Impulses in the Magnetosphere and at Ground 399**
U. Villante and M. Piersanti

Chapter 18 **Turbulence in the Magnetosheath and the Problem of Plasma Penetration Inside the Magnetosphere 417**
Elizaveta E. Antonova, Maria S. Pulinets, Maria O. Riazantseva, Svetlana S. Znatkova, Igor P. Kirpichev and Marina V. Stepanova

Chapter 19 **Solar Wind Sails 439**
Ikkoh Funaki and Hiroshi Yamakawa

Preface

The solar wind is a continuous outward stream of energetic charged particles from the Sun's hot corona. The high temperature in the solar corona measures more than one million degrees causing ionization of the hydrogen and formation of a hot plasma of protons and electrons. The solar plasma is so hot that it breaks free of the Sun's gravitational force and blows away from the surface in all directions giving rise to the solar wind. The intensity of the solar wind changes constantly, and when it gets stronger, we see more brighter aurora on Earth. Terrestrial magnetic field is compressed by the solar wind and distorted into a comet-shaped cavity known as the magnetosphere. The magnetosphere protects the Earth as it deflects the solar wind streams, which would otherwise blow the atmosphere away. However, the energetic solar flares and coronal mass ejections during times of an active Sun can drastically affect the solar wind and space weather conditions, and, implicitly, the advanced space technology we have become so dependent upon in our everyday lives. Understanding the changing solar wind and its effects on Earth and our life is therefore one of the most challenging tasks facing space scientists today, and many space exploration missions focus on the solar wind and its interactions with Earth.

This book consists of a selection of original papers of the leading scientists in the fields of Space and Planetary Physics, Solar and Space Plasma Physics with important contribu- tions to the theory, modeling and experimental techniques of the solar wind exploration. All chapters of this book were invited with the aim of providing a comprehensive view of the current knowledge of the solar wind formation and elemental composition, the interplane- tary dynamical evolution and acceleration of the charged plasma particles, and the guiding magnetic field that connects to the magnetospheric field lines and adjusts the effects of the solar wind on Earth.

The book is divided into five distinct sections: an introductive description of the solar wind properties and laws associated with different phases of the solar activity, and four key research topics with significant advances in the last decades. In the second section, the interested reader can find an extended analysis of the solar wind matter and elemental composition as measured in-situ by different spacecraft missions or from traces in microme- teorites. The third section is devoted to the solar wind dynamics ranging from the large-scale perturbations in the heliosphere to the small-scale kinetic processes of the wave-particle en- ergy dissipation. Magnetic reconnection is closely related to wave turbulence, which can be an efficient mechanism to dissipate magnetic energy into kinetic energy in small-scale, lo- calised processes. The fourth section highlights the role of the interplanetary magnetic field, which is powered by the Sun and extends through the corona further out in the solar wind. In the last section, four chapters report on the progress made in describing the solar wind interaction with the Earth's magnetosphere, focusing on principal geophysical effects as well as the wave turbulence and the problem of plasma penetration into the magnetosphere. The pressure exerted by the solar wind on the terrestrial magnetosphere has inspired a new and ambitious concept of propulsion for the so-called magnetic solar wind sails, which are the subject of the last chapter of our book.

It is necessary to point out that this book is not a monograph as it does not cover all aspects of the topic. Its purpose is to provide the means for interested readers to become familiar with the basic concepts as well as the recent progress in developing the observational techniques and theoretical models of the solar wind. I also am convinced that most of the research scientists actively working in this field will find in this book many new and interesting ideas.

Marian Lazar
Institute for Theoretical Physics,
Institute IV: Space and Astrophysics,
Ruhr-University, Bochum
Germany

Part 1

The Solar Wind – Overview of the Fundamentals

1

Solar Wind Laws Valid for any Phase of a Solar Cycle

V.G. Eselevich
Institute of Solar-Terrestrial Physics of Siberian Branch of Russian Academy of Sciences, Irkutsk Russia

1. Introduction

First, let us remind what a physical law is.

It is an empirically established, formulated strictly in words or mathematically, stable relation between repetitive phenomena and states of bodies and other material objects in the world around. Revealing physical regularities is a primary objective of physics. A physical law is considered valid if it has been proved by repeated experiments. A physical law is to be valid for a large number of objects; ideally, for all objects in the Universe. Obviously, the last requirement is especially difficult to test. We will, therefore, somewhat confine ourselves to the following comments:

a. We will lay down only SW physical laws, calling them simply "laws". Here, we will take into account that they meet the main above-stated requirements for physical laws.
b. Any law is fulfilled under ideal conditions, i.e., when its effect is not violated by outside influence. For instance, the Newton first law of motion may be tested only when the friction force is absent or tends to zero. Since SW conditions are often far from ideal, it is sometimes difficult to determine, lay down, and prove the existence of an SW physical law.
c. We will distinguish between the laws and their mechanisms of effect. For example, the law of universal gravitation is well known, but its mechanism is still unclear.
d. Obviously, the relevance of these laws is different. But all of them are of limited application. To illustrate, laws of simple mechanics are violated for relativistic velocities or superlarge masses of substance. The Ohm's law is valid only if there is current in the conductor. The SW laws are valid only for a hot ionised medium, etc.
e. It is good to keep in mind that a part of the SW laws defined below may later merge into one law. Time will show. As for now, considering the SW laws separately is reasonable, because in this way we can examine their mechanisms that are likely to be different.

Laying down SW laws actually implies that the "solar wind" subdiscipline of space science turns from multidirectional investigations and data collection into an independent branch of physics. This, based on established laws, provides a way to examine the SW behaviour in more complex situations, when it is under the effect of several factors at once, without resorting to statistical methods that are not capable of restoring the truth.

Laying down a law enables us to pose tasks of examining its mechanism as well as to discover new laws rather than repeating and rechecking well-known ones.

Knowing SW laws is of critical importance for developing a unified theory of SW that is practically absent now. The point is that SW obeys the diluted plasma dynamics laws with due regard to boundary conditions: on the one hand, it is the Sun; on the other, it is the galactic environment. The distance between the galactic environment and the Sun is $R \sim 2 \bullet 10^4 R_0$ (R_0 is the solar radius); the SW density decreases by law of $(R/R_0)^2$ (i.e., ~ $4 \bullet 10^8$ times). Thus, for SW at distances of order and less than the Earth's orbit ($R \approx 214R_0$), the infinity condition is simple: SW density tends to zero. However, the conditions on the Sun are totally determined by the experimentally established SW laws comprising such notions as coronal holes, bases of the coronal streamer belt, active regions, and magnetic tubes emerging from the solar convective zone - these are the sources of various SW on the Sun without knowledge of which it is impossible to impose boundary conditions there.

The sequence of the presentation is as follows: a brief wording of a law and then a reference to 2-4 first fundamental papers on this law according to their time priority (in some cases, more references will be given). They are in bold typed in the text, their authors are bold typed. For some laws we will explain their possible violations under the influence of other factors as well as possible problems associated with their implementation mechanisms.

I took the liberty of naming some SW laws, where considered it possible and important, after their discoverers, for example:

The Law of the Solar Wind (SW) Existence - the **Ponomarev-Parker Law;**
The Law of the Existence of Collisionless Shocks in the Diluted Plasma – the **Sagdeev Law**;
The Law of Two Mechanisms for Accelerating Solar Energetic Particles – the **Reams Law**.
The Law of the Relation between the Type-II Radio Emission and Collisionless Shocks - the **Zheleznyakov-Zaitsev Law**

2. Quasi-stationary solar wind laws

Law 1. "Of the solar wind (SW) existence": There is a diluted plasma stream - solar wind (SW) - from the Sun.

This law was theoretically substantiated in (**Ponomarev, 1957; Vsekhcvyatcky, et al., 1957; Parker, 1958**). They predicted the SW existence in the Earth's orbit based on the well-known high temperature of the coronal plasma that provided plasma acceleration due to pressure gradient forces.

The SW stream existence was confirmed by experiments at the Luna-2 and Luna-3 Automatic Interplanetary Stations (**Gringauz, et al., 1960**) and the Explorer-10 satellite (**Bonetti et al., 1963**).

However, Ponomarev and Parker failed to answer the question about the mechanism of the SW origin near the solar surface where the temperature is within 6000 degrees (i.e., how the plasma from the solar surface enters the corona). That is precisely why the Ponomarev-Parker law opened a new chapter in solar-terrestrial physics research that has been over half a century already.

Further investigations demonstrated that there are mostly three SW types (V.G. Eselevich, et al., 1990; Schwenn and Marsch, 1991; McComas et al, 2002): two quasi-stationary SW types with fairly long-lived sources on the Sun (over 24 hours, often weeks and even months): **the fast SW** (its maximum velocity V_M is 450-800 km/s) flowing out of coronal holes (CH), and **the slow SW** (its maximum velocity is 250-450 km/s) flowing out of the coronal streamer belt or chains (pseudostreamers). The third type is **the sporadic SW**. Its sources on the Sun exist less than 24 hours (flares, coronal mass ejections (CME), eruptive prominences).

The three SW types have different generation mechanisms that are still unclear. Therefore, their associated laws are laid down separately.

Law 2. "Fast SW": the sources of the fast SW on the Sun are coronal holes. The maximum SW velocity V_M in the Earth's orbit is related to the area (S) of a coronal hole, enclosed in the latitude range $\lambda = \pm 10°$ relative to the ecliptic plane (Fig. 1), by $V_M (S) = (426 \pm 5) + (80 \pm 2) \cdot S$ at $S \leq 5 \cdot 10^{10}$ km^2 and $V_M (S) \approx$ const $\approx$ 750-800 km/s at $S > 5 \cdot 10^{10}$ km^2.

This law was experimentally established in (**Nolte et al., 1976**), where six equatorial coronal holes were recorded in soft X-ray concurrently with time velocity profiles of fast SW streams in the Earth's orbit during ten Carrington rotations. It was verified by many subsequent investigations both for equatorial coronal holes and for extra equatorial ones, in particular:

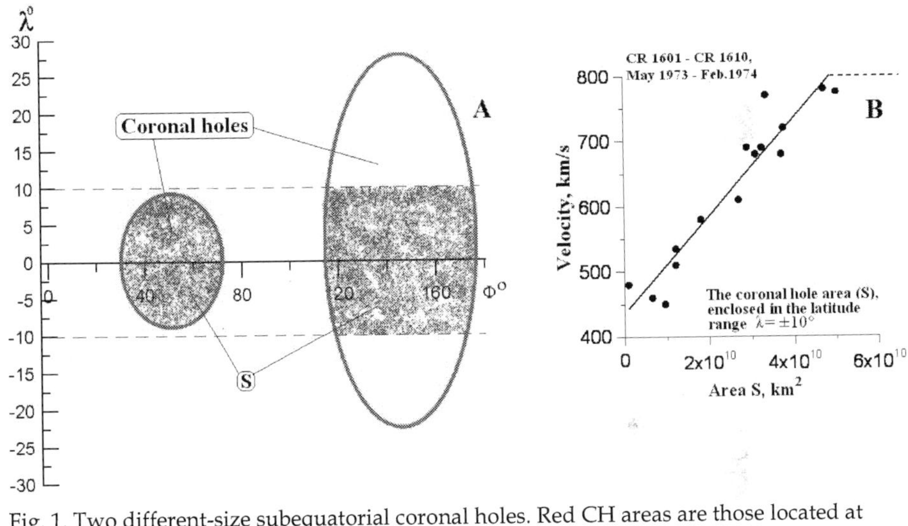

Fig. 1. Two different-size subequatorial coronal holes. Red CH areas are those located at latitudes λ within ±10° relative to the equatorial plane.

a. according to the Ulysses measurements, the maximum velocity V_M of the SW streams from the polar coronal holes, whose area $S > 5 \cdot 10^{10}$ km^2, was $V_M \approx$ const $\approx$750-800 km/s (Goldstein et al.,1996).
b. The dependence $V_M(S)$ on Law 2 was used to develop a method to compute the V(t) profile for the fast SW in the Earth's orbit from characteristics of any coronal holes (equatorial and off-equatorial) (V.G. Eselevich, , 1992 ; V.G. Eselevich, V. & M. V.

Eselevich, 2005). It provided a basis for the continuous website comprising the prediction of V(t) for the fast SW. The comparison between the predicted results at this website and experimental curves of V(t) over several years demonstrated high efficiency and validity of this method (Eselevich, et al., 2009).

c. Another independent method of testing Law 2 is the dependence of the superradial divergence "f" of magnetic field lines emanating from a coronal hole with maximum velocity V_M of the fast SW. This dependence was obtained in (V.G. Eselevich & Filippov, 1986; Wang, 1995). On its basis, another method to compute the V(t) profile for the fast SW in the Earth's orbit from characteristics of coronal holes (equatorial and off-equatorial) has been developed (Wang & Sheeley,1990; Arge & Pizzo, 2003). A website to predict V(t) profiles of fast SW streams in the Earth's orbit using this method (the V(f) dependence at the base of coronal holes) has been functioning continuously for many years. The method provides results in their reliability and validity close to the prediction method using the $V_M(S)$ dependence (Eselevich et al., 2009).

Since the value "f" is, in turn, a function of S (V.G. Eselevich & Filippov, 1986), the results of this method also support Law 2.

Law 3. "Streamer belts": the streamer belt with the slow SW in the Earth's orbit is recorded as areas with higher plasma density containing an odd number of the interplanetary magnetic field (IMF) sign changes or an IMF sector boundary.

Svalgaard et al. (1974) showed that the streamer belt separates areas with an opposite direction of the global magnetic field radial component on the solar surface. It means that at the base of the streamer belt there are magnetic field arcs along whose tops there goes a neutral line of the Sun's global magnetic field radial component (dashed curve in Fig. 2A). The intersections of the neutral line with the ecliptic plane (red horizontal line in Fig. 2A) are recorded in the Earth's orbit as sector boundaries of the interplanetary magnetic field (IMF) (arrow "sec" in Fig. 2B) (**Korzhov, 1977**).

All this was verified and developed in many subsequent studies (e.g., Gosling et al., 1981; Burlaga et al., 1981; Wilcox & Hundhausen, 1983; Hoeksema, 1984).

Law 4. "Streamer chains (or pseudostreamer)": Streamer chains with the slow SW in the Earth's orbit are recorded as areas with higher plasma density that contain an even number of IMF sign changes.

In (**V.G. Eselevich et al., 1999**) it was demonstrated that, except the streamer belt proper, there are its branches termed streamer chains. The chains in the white-light corona look like the belt itself - like areas with higher brightness. There is slow SW in them; its properties are approximately identical to those in the streamer belt. However, the chains differ from the belt in that they separate open magnetic field lines in the corona with identical magnetic polarity. Thus, the magnetic field structures, calculated in potential approximation, at the base of the chains have the form of double arches (in general case - an even number of arches), as opposed to the streamer belt where there are single arches at the base (an odd number of arches), see Fig. 2A. The properties of the streamer chains have been poorly studied so far; their name has not been established. So, in the very first paper (**V.G. Eselevich & Fainshtein, 1992**), they were termed "heliospheric current sheet without a neutral line" (HCS without NL); in (Zhao & Webb, 2003), "unipolar closed field region" (the streamer belt in that paper was termed "bipolar closed field region"). In the most recent investigations (Wang et al., 2007), they were termed pseudostreamers. In (Ivanov et al., 2002), manifestations of the chains in the heliosphere were designated as subsector boundaries. We will use the term "'streamer chains", and their manifestations in the Earth's orbit will be termed as subsector boundaries (arrow "subsec" in Fig. 2B).

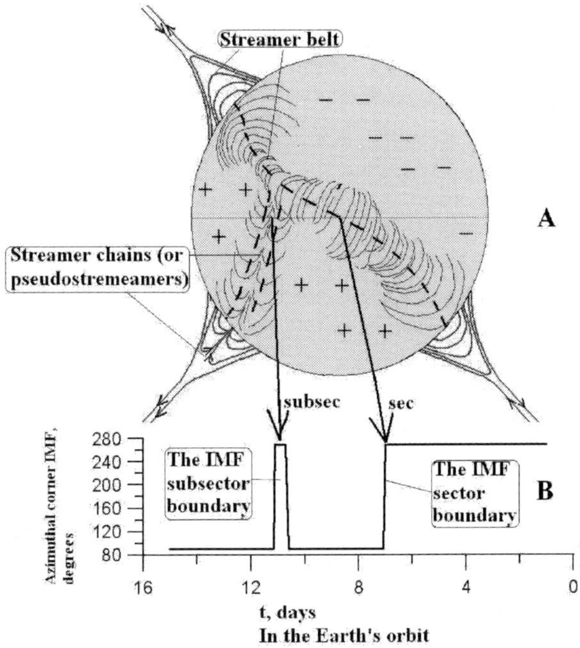

Fig. 2. A) The coronal streamer belt and chains separating, respectively, areas on the solar surface with opposite and equal direction of the Sun's global magnetic field radial component. The single dash is the neutral line (NL) of the magnetic field radial component passing through the tops of the magnetic field arcs at the base of the streamer belt. The double dash is two NLs along double magnetic field arcs at the base of the streamer chains. B) The IMF azimuthal angle distribution in the Earth's orbit on the solar surface. It corresponds to that in (A).

Law 5. "Interaction between fast and slow SWs" In the heliosphere, there is a region of collision between slow and fast SWs caused by solar rotation. Inside the region, slow and fast SW streams are separated by a thin surface termed interface.

It has been shown theoretically (**Dessler & Fejer, 1963; Hundhausen & Burlaga, 1975**) and experimentally (**Belcher & Davis, 1971; Burlaga, 1974**) that the radially propagating fast and slow SWs collide in the heliosphere (in the Earth's orbit, in particular) starting with $R>20R_0$ and on, owing to the solar rotation (the fast SW overtakes the slow one). Between them, at the fast SW front, develops a sharp boundary less than $\approx 4 \cdot 10^4$ km thick. It is termed interface. The longitudinal proton temperature and the radial and azimuthal SW velocities abruptly increase at the interface; the proton density abruptly decreases (Gosling et al.,

1978). Also, electron temperature, relative portion of alpha particles, alpha-particles velocity relative to protons (Gosling et al., 1978; Borrini et al., 1981), ratio of ion content O^{7+}/O^{6+} reflecting the coronal temperature, and Mg/O controlled by the FIP effect (Geiss et al., 1995) abruptly increase at the interface, while the flow of matter j = NV decreases. A valid parameter enabling separating the flows of these two types is an entropy in the form of $S = k \ln(T/N^{0.5})$ (Burton et al., 1999). Here, in the gas entropy formula, it is assumed that the polytropic index $\gamma = 1.5$. The well-defined difference in entropy between these two streams enables us to record the so-called trailing interface located at the trailing edge solar wind stream. The trailing interface separating the fast SW from the following slow SW differs from the interface at the front of the following fast SW and is likely to be somewhat thicker.

Thus, the time variation in the entropy allows to unambiguously separate any fast SW from the ambient slow SW (and vice versa). The sharp difference in the said parameters and, especially, in the entropy suggests that the genesis for these two types of SW streams is different.

Law 6. "Nonradialities of rays of the streamer belt and chains": Nonradiality of rays $\Delta\lambda$ of the streamer belt and chains depends on the latitude of λ_0 of their location near the Sun and peaks at $\lambda_0 \approx \pm 40°$.

The cross-section of the streamer belt in white light is a helmet-shaped base resting on the solar surface and extending upward as a radially oriented ray (solid curves in Fig. 3A). Inside the helmet, there may be loop structures of three types: I and II in Fig. 3A correspond to the streamer belt splitting up the regions of the radial global magnetic field component with opposite polarity (an odd number of loops under the helmet); type III corresponds to the streamer chains splitting up the regions with identical radial component polarity (an even number of loops). Type II is largely observed around the minimum and at the onset of an increase in solar activity at $\lambda_0 \approx 0°$. The symbol λ_0 denotes the latitude of the helmet base centre near the solar surface. The latitude of the helmet centre and, then, of the ray to which the helmet top transforms changes usually with distance away from the solar surface (dashed line in Fig.3 (I)). And only at R > 5Ro, the ray becomes radial, but its latitude (designated λ_E) may differ greatly from the initial latitude of λ_0 at the helmet base. The latitude change is an angle $\Delta\lambda$. A positive $\Delta\lambda$ corresponds to the equatorward deviation; a negative $\Delta\lambda$ corresponds to the poleward one. To exclude the necessity of considering the sign in Fig. 3B, we defined the deviation as: : $\Delta\lambda = |\lambda_0| - |\lambda_E|$ (i.e., equally for the Northern and Southern hemispheres).

The analysis of the measurements and the plot in Fig. 3 suggests that at R < 5Ro from the solar centre (**V.G.Eselevich & M.V. Eselevich, 2002**):

- the deviation of the higher brightness rays from the radial direction is equatorward for the latitude range up to ≈ ±60°, nearly identical in the Northern and Southern hemispheres (curve in Fig. 3B), and is slightly asymmetric relative to the axis $\lambda_0 \approx 0°$) when observed at the western and eastern limbs in the streamer belt and chains;
- the deviation value $\Delta\lambda$ unambiguously depends on the latitude of the ray λ_0 near the solar surface;
- the near-equatorial rays almost do not deviate from the radial direction ($\lambda_0 \approx 0°$) .

These conclusions were then confirmed in the investigations based on the extensive statistics for the complete solar cycle in (**Tlatov & Vasil'eva, 2009**).

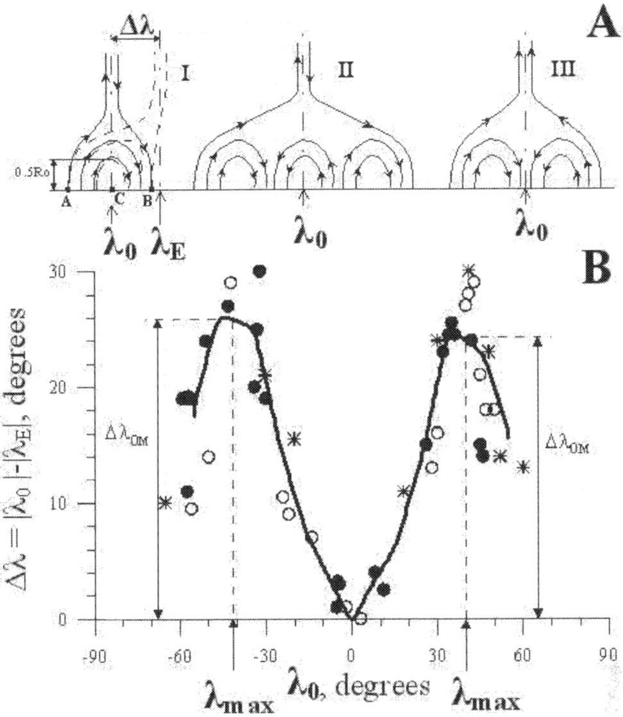

Fig. 3. A) The idealized magnetic field lines in the hamlet with a ray based on it: I and II in the streamer belt, III - in streamer chains. The dash in I indicates the pattern accounting for the streamer nonradiality effect. B) The dependence of the total angular deviation $\Delta\lambda$ on latitude λ_0 for 51 streamer belt brightness rays (black circles are the W limb; light circles, the E limb) and streamer chains (stars) over the period November 1996 through June 1998 as deduced from LASCO C1 and C2 data **(V.G. Eselevich & M.V. Eselevich, 2002)**.

The mechanism for the emergence of the ray nonradiality in the streamer belt and chains has been still unclear, but the law itself is the basis for testing any theory about the solar wind origin.

Law 7. "Of the streamer belt ray structure": The coronal streamer belt is a sequence of pairs of higher brightness rays (or two, closely spaced ray sets). Ray brightnesses in each pair may differ in general case. The neutral line of the radial component of the Sun's global magnetic field goes along the belt between the rays of each of these pairs.

The first experimental evidence for the existence of the coronal streamer belt regular ray structure was obtained in (**V.G. Eselevich & M.V. Eselevich, 1999**). Later, more detailed investigations carried out in (**V.G. Eselevich & M.V. Eselevich, 2006**) revealed that the spatial streamer belt structure has the form of two closely-spaced rows of higher brightness rays (magnetic tubes with SW plasma moving in them) separated by the neutral line of the global magnetic field radial component (Fig. 4a). Figure 4b shows the belt cross-section in the form of two rays enveloping the helmet on either side. The magnetic field direction (arrows and + - signs) in these rays is opposite. The pattern does not show the nonradiality of the rays in the streamer belt plane near the solar surface at R< 4-5Ro.

The double-ray streamer belt structure was considered as a result of the instability development. In the streamer belt type current systems, there is a proton "beam" relative to the main SW mass along the magnetic field (Schwenn & Marsch, 1991). In (**Gubchenko et al., 2004**), in the context of the kinetic approach, it was shown that the sequences of magnetic tube (ray) pairs analogous to those observed above may be formed along the belt due to exciting the "stratification modes" of oscillations. If it is true, then we deal with collective properties of diluted plasma that manifest themselves in forming cosmic-scale structures.

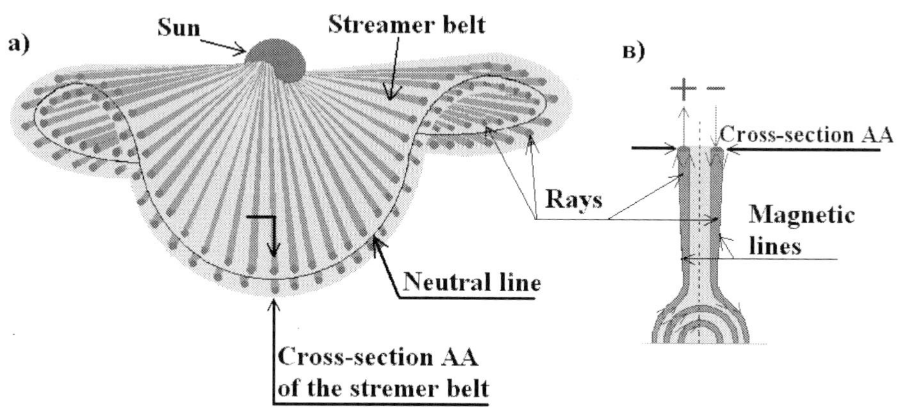

Fig. 4. The spatial ray structure of the coronal streamer belt (a); the streamer belt cross-section (AA) (b). In red rays of the top row of the streamer belt, the magnetic field is directed from the Sun (+); in green rays of the bottom row, to the Sun (-). The neutral line between rays (solid line).

We note that although the theoretically considered possible mechanism for the formation of the streamer belt ray structure yields the result qualitatively consistent with the experiment, the true cause of this very interesting phenomenon is still far from clear.

Law 8. "Of the heliospheric plasma sheet structure": The cross-section of the heliospheric plasma sheet (HPS) in the Earth's orbit generally takes the form of two density maxima of a characteristic size ≈2°-3° (in the heliospheric coordinate system) with a sector boundary between them. Such a structure is quasistationary (remains unchanged for nearly 24 hours). HPS is an extension of the coronal streamer belt structure (ray structure) into the heliosphere.

The streamer belt extension into the heliosphere is termed a heliospheric plasma sheet (HPS) (Winterhalter, et al., 1994) According to the findings of (Borrini, et al., 1981;V.G. Eselevich and Fainshtein, 1992), the quasistationary slow SW flowing into HPS in the Earth's orbit is characterised by the following parameters and features:

- a relatively low SW velocity V ≈ 250 - 450 km/s (the maximum velocity in the fast SW flowing out of coronal holes V ≈ 450 - 800 km/s);

- an enhanced plasma density with maximum values N_{max}>10 cm^{-3} (in the fast SW, Nmax <10 cm^{-3});
- anticorrelation of profiles of plasma density N(t) and of the magnetic field module B(t) on time scales of order of hours and more;
- a lower proton temperature Tp < 10^5 °K;
- one or several (an odd number) IMF sign reversals is the characteristic feature of the sector boundary or its structure.

The availability of all these signs is enough to unambiguously determine the heliospheric plasma sheet in the Earth's orbit.

According to (**Bavassano, et al., 1997**), the HPS cross-section is a narrow (with an angular size of ≈ 2° -3°) peak of plasma density with the built-in IMF sector boundary and is a sufficiently stable structure throughout the way from the Sun to the Earth (the pattern in Fig. 5A).

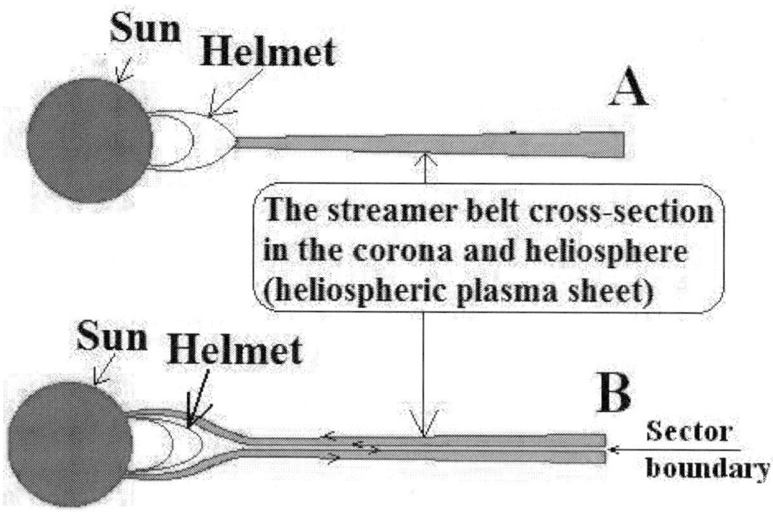

Fig. 5. The streamer belt cross-section structure in the corona and heliosphere (heliospheric plasma sheet) according to the results obtained in **(Bavassano, et al., 1997)** (A) and **(V.G. Eselevich & M.V. Eselevich, 2007b)** (B).

The HPS cross-section improved structure obtained in (**V.G. Eselevich, V. & M.V. Eselevich, 2007b**) proved to be slightly different from that in (**Bavassano, et al., 1997**) in the following characteristics:

a. The streamer belt cross-section in the corona and heliosphere is, in general case, two closely-spaced rays with identical or different values of density peaks, not one ray as it is assumed in (**Bavassano, et al., 1997**). The sector boundary is between the density peaks. One ray is observed, when the density peak of one ray is much *smaller* than that of the other (the pattern in Fig. 5B).
b. Rays do not start at the helmet top (like in the upper panel of Fig. 5A) but on the solar surface (Fig. 5B).

Mechanisms generating the slow SW in the streamer belt rays have been still unclear and are the subject for future research.

Laws 7 and 8 may later merge.

Law 9. "Of the heliospheric plasma sheet fractality": The fine structure of the heliospheric plasma sheet in the Earth's orbit is a sequence of nested magnetic tubes (fractality). Sizes of these tubes change by almost two orders of magnitude as they nest.

Analysing the data from the Wind and IMP-8 satellites has revealed that the slow SW in the heliospheric plasma sheet is a set of magnetic tubes containing plasma of an enhanced density (Nmax > 10 cm^{-3} in the Earth's orbit) that are the streamer belt ray structure extension into the heliosphere (**M.V. Eselevich & V.G. Eselevich, 2005**) (Fig. 6). Each tube has a fine structure in several spatial scales (fractality) from ≈ 1.5° -3° (in the Earth's orbit this equals to 2.7 -5.4 hours or (4-8)·10^6 km) to the minimum ≈ 0.03° -0.06°, i.e., angular sizes of nested tubes change by almost two orders of magnitude. In each spatial scale under observation, the magnetic tubes are diamagnetic (i.e., there is a diamagnetic (drift) current on their surface, decreasing the magnetic field inside the tube and increasing it outside). As this takes place, $\beta = 8\pi \cdot [N(Te + Tp)]/ B^2$ inside the tube is greater than β outside. In many cases, the total pressure $P = N(Te + Tp) + B^2/8\pi$ is practically constant both inside and outside the tubes in any of the above scales. The magnetic tubes are quasi-stationary structures. The drift (or diamagnetic) current at the tube boundaries is stable relative to the excitation of random oscillations in magnetised plasma.

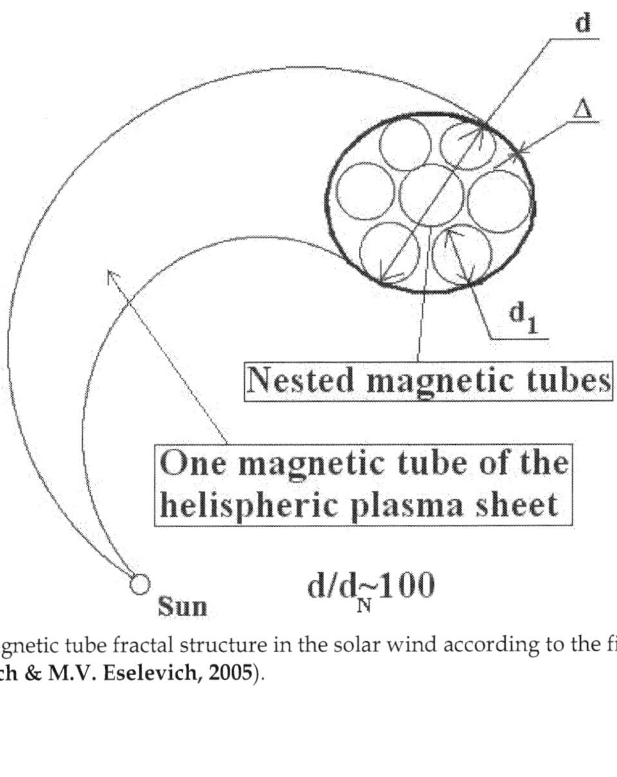

Fig. 6. The magnetic tube fractal structure in the solar wind according to the findings of (**V.G. Eselevich & M.V. Eselevich, 2005**).

The theory of possible evolution of such self-similar magnetic tubes (typical of fractal formations) in solar wind plasma was presented in (**Milovanov & Zelenyi, 1999**). However, no detailed comparison between the theoretical and experimental results has been made so far which is obviously necessary to understand the character of this interesting phenomena.

3. CME laws

Law 10. "Of the CME structure": The magnetic structure of a coronal mass ejection (CME) is a helical flux rope. In white-light images at a definite orientation to the sky plane, it can be seen as a bright frontal structure covering a cavity with a bright core.

It has been found that most CMEs with a big angular size (d > 30° - 50°) are helical flux ropes or tubes filled with plasma (**Krall et al., 2000**). This is supported by comparison between stereoscopic observation of CMEs with STEREO/SECCHI and calculations within the CME geometrical model in the form of a flux rope (**Thernisienet al., 2009**). According to (**Cremades & Bothmer, 2004**), axis orientation of the CME flux rope is nearly the same as the neutral line (NL) orientation near the CME source on the Sun or as the filament orientation along NL. The angle between NL and N-S direction on the Sun is denoted by γ. When observed in white light, "limb" CMEs in longitude $\Phi > 60°$, with high values $\gamma > 45°$, are of the simplest three-body form (**Illing & Hundhausen, 1985**): frontal structure (FS), region of a lowered density (cavity), and a bright core that is sometimes absent.

Law 11. "Of the generation mechanism for "gradual" CMEs": The generation mechanism for "gradual" CMEs is associated with the development of instability in the magnetic flux rope with its top in the corona and two bases in the photosphere.

"Gradual" CMEs (Sheeley et al., 1999; V.G. Eselevich & M.V. Eselevich, 2011) have the following peculiarities:

- the corona is the source of the leading edge of these CMEs at $1.2R_0<R<2.5R_0$ from the solar centre;
- CMEs start moving from the state of rest; i.e., the initial velocity $V_0 = 0$;
- the initial angular size in the state of rest $d_0 \approx 15° - 65°$.

At zero time, a gradual CME is an arch structure of helical flux ropes, filled with plasma, with two bases in the solar photosphere. In theoretical papers (**Krall et al., 2000; Kuznetsov Hood, 2000**), the eruption or the sudden motion of the arch structure of flux rope (localised in the solar corona) backward from the Sun is considered as a source of gradual CMEs. In (**Krall et al., 2000**), four specific drive mechanisms for the flux rope eruption forming CMEs are considered:

(1) flux injection, (2) footpoint twisting, (3) magnetic energy release, and (4) hot plasma injection.

In (**Kuznetsov & Hood, 2000**), no flux-rope equilibrium is caused by the increase in plasma pressure in the rope due to plasma heating. All these models show that eruption of the magnetic flux rope is possible in principle. However, only experimental investigation, being in close cooperation with theory, will throw light upon real causes of this process.

Laws 10 and 11 may later merge.

Law 12. "Of a CME initiation site": CMEs appear in bases of the streamer belt or chains.

Fig. 7a illustrates that there are almost no streamer chains (dashed curves) near the minimum phase. All CMEs (their positions and angular sizes are depicted by segments of vertical straight lines) appear near NL (solid curve) along the streamer belt (**Hundhausen, 1993)**. Number of streamer chains increases as solar activity grows. CMEs appear in bases of the streamer belt (near NL) or chains (dashed line) Fig. 7b,c,d (**V.G. Eselevich, 1995**).

Law 13. "Of a disturbed region in front of CME": Owing to the interaction with coronal plasma there is a disturbed region in front of CME.

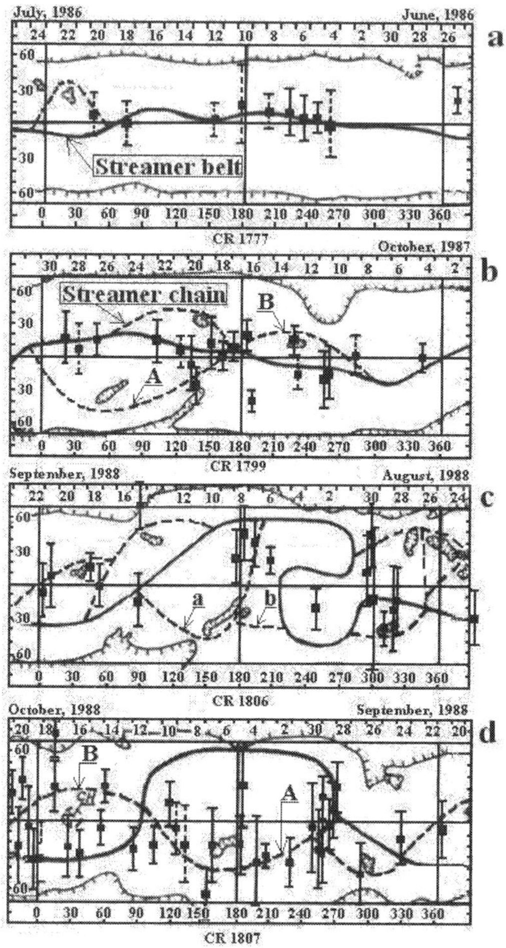

Fig. 7. Origin places of CME (vertical lines correspond to the CME angular size) relative to the streamer belt (solid curve is NL along the belt) and chains (dashed curve) for different Carrington rotations with an increase in solar activity from **(Eselevich, 1995)**.

The form of the frontal structure (FS) for the slow CME (its velocity relative to the undisturbed SW u < 700 km/s at $R<6R_0$) is close to the circle with radius "r" (shown dashed) centred at O (Fig.8A). This is confirmed by the coincidence between maxima of difference brightness distributions (see Fig. 8B) along two different directions (dashed lines 'a' and 'b' in Fig. 8A). For the slow SW the difference brightness profile is stretched in the CME propagation direction (Fig. 8B). This is a disturbed region arising from the interaction between CME and undisturbed SW (**M.V. Eselevich.& V.G. Eselevich, 2007a**). Examining the properties of the existing disturbed regions is important not only for understanding CME dynamics but also for identifying and studying the properties of the shock wave appearing in its front part at high velocities(u≥ 700 km/s) (see Law 15).

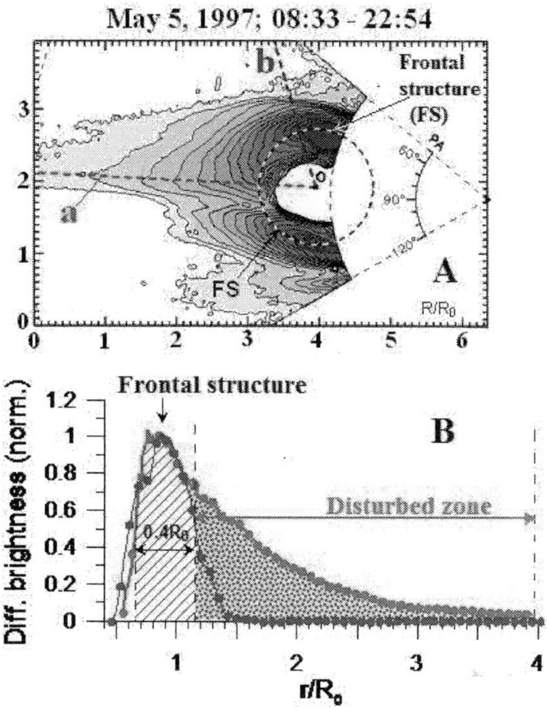

Fig. 8. (A) The difference brightness in the form of brightness isolines for the slow CME of 5 May 1997 (the velocity in reference to the undisturbed SW u ≈ 150 km/s). (B) The difference brightness profiles in the direction of two position angles (shown by dashed lines "a" (red) and "b"(blue) in (A). Value r is counted from the CME centre "O".

4. Shock wave problem. Laws of the CME-driven shock waves

4.1 Shock wave problem and its related law

First of all, let us divide this problem into two inequivalent components: collisional and collisionless shock waves.

Collisional shock waves. The waves are theoretically studied in gas (liquid) (Landau & Lifshitz, 1953) and plasma (Zeldovich & Riser, 1966). According to these studies, there are two main parameters of medium which are important for formation of the shock-wave discontinuity: velocity of sound (V_S) and mean free path (of gas or plasma) λ. It has been found experimentally (e.g., Korolev et al., 1978) that, as gas flow rate V exceeds value V_S, a shock wave discontinuity emerges where the Rankine-Hugoniot relations are valid. (This phenomenon is sometimes referred to as the "excess of velocity of sound"). As compared with gas, the structure of the shock front in plasma is complicated, since the scale where the ion heating takes place of the order of the mean free path for ions λ_i turns out different from the scale of heating for electrons $\lambda_e \sim (mi/me)^{1/2} \lambda_i$ (m_i and m_e are the ionic and electron masses, respectively) (Zeldovich & Riser, 1966). Experimental investigation into the structure of collisional shock front is, however, impossible because of small λ and λ_i in dense medium.

Collisionless shock waves. The situation gets worse in rarefied magnetised plasma which solar wind (SW) is. This can be explained by the fact that both parameters $\lambda_i=\lambda_p$ (λ_p is the mean free path of protons constituting SW) and V_S become, to a great extent, ambiguous for formation of the shock front, because λ_p in the Earth's orbit is of the order of the Sun-Earth distance. Apparently, the collisional shock wave with such a front thickness becomes meaningless. The second parameter (V_S) becomes indefinite, since V_S in magnetised plasma depends on the wave motion direction relative to the magnetic field direction. Fundamental theoretical works by R.Z. Sagdeev (review by Sagdeev, 1964) present the break in this deadlock. His research has shown that formation of the front with thickness $\delta << \lambda_p$ can be caused by collective processes in diluted plasma that are related to the development of an instability and its resulting plasma 'turbulisation'. As a consequence, the effective mean free path of protons dramatically decreases, being determined by the characteristic scale of the 'turbulence' $\delta_t << \lambda_p$. This scale plays the role of a new characteristic mean free path wherein the effective energy dissipation in the collisionless shock front may take place. So far, there has been no unified theory of front thickness in rarefied plasma that could explain various particular cases. There are numerous phenomena associated with collective processes.

Nevertheless, some limiting cases have not only been predicted theoretically (Sagdeev, 1964; Galeev and Sagdeev, 1966; Tidman, 1967) but also found in laboratory (Iskoldsky et al., 1964; Zagorodnikov et al., 1964; Paul et al., 1965; Alikhanov et al. 1968; Wong & Means, 1971; Volkov et al., 1974) and space experiments (Moreno et al., 1966; Olbert, 1968; Bame et al., 1979; Vaisberg et al., 1982). The comparison of the laboratory and satellite experiments has revealed a close agreement between them for certain collisionless shock fronts (V.G. Eselevich, 1983). Much experimental data on the structure of the near-Earth bow shock and interplanetary shock waves have been collected so far. There exists a possibility to analyse and interpret these data in order to deduce some experimental fundamental laws that will describe collective dissipation processes at the fronts of different collisionless shocks. Leaning on these laws, we will be able to elaborate a unified theory describing the front thickness in diluted plasma. However, these findings provide the basis for the law of collisionless shock existence given below.

Law 14. "Of the collisionless shock existence": The wave shocks with the front thickness being much smaller than the mean free path of ions and electrons may exist in rarefied plasma (The Sagdeev Law).

For some limiting cases, the collisionless shock has been predicted theoretically (**Sagdeev, 1964**). The existence of such waves has been proved both in laboratory (**Iskoldsky et al., 1964; Zagorodnikov et al., 1964; Paul et al., 1965**) and in the space plasma (**Moreno et al., 1966; Olbert, 1968**).

4.2 The CME-driven shock wave

The recent research into CME-driven shocks in the solar corona enabled us to deduce several new laws.

Law 15. "Of the formation of a shock in front of CME": A shock is formed in front of CME when its velocity relative to the surrounding coronal plasma exceeds the local Alfven one.

In the case of the fast CME (u ≥ 700 km/s), unlike in the case of the slow one (see Fig.8 B), the form of the difference brightness isolines is close to the frontal structure (FS) depicted by dashed circle in Fig. 9A. At the leading edge of the disturbed region in profile ΔP(R) (Fig. 9B),

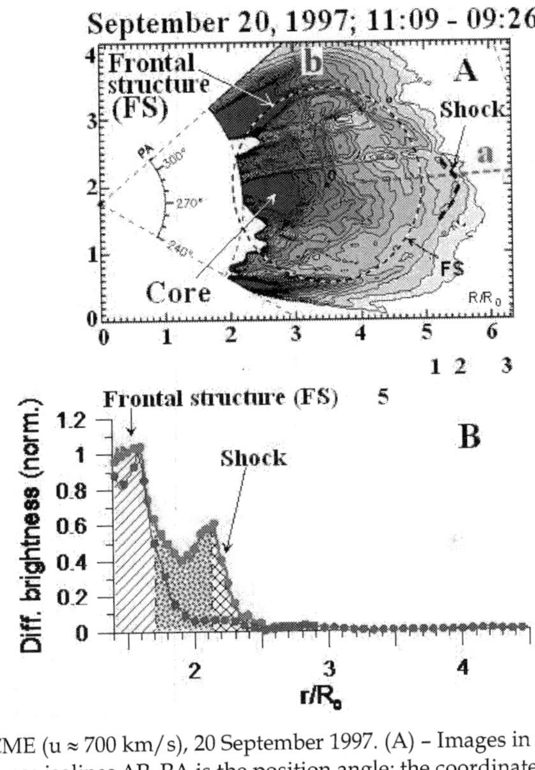

Fig. 9. The fast CME (u ≈ 700 km/s), 20 September 1997. (A) – Images in the form of difference brightness isolines ΔP, PA is the position angle; the coordinate axes are in units of R_0. (B) Difference brightness distributions with the distance r counted from the CME centre (point O) along two different sections "a"(red) and "b"(blue) whose directions are shown by the dashed lines in (A).

in the CME propagation direction (dashed straight line "a" in Fig.9A), there is a discontinuity (jump) with the scale of about 0.25 R_0 (inclined mesh). Fig.9A illustrates its position (segment of the heavy dashed curve).

The analysis (**M.V. Eselevich & V. G. Eselevich, 2008**) of dependence u(R) in Fig.10 allowed us to deduce the following law. When the CME propagation speed u, relative to surrounding coronal plasma, is lower than a certain critical speed u_C, there is a disturbed region extended along its propagation direction ahead of CME (these cases are highlighted by light marks). The formation of a shock ahead of the CME frontal structure in a certain vicinity relative to its propagation direction (events marked off by black marks) is determined by validity of the local inequality $u(R) > u_C \approx V_A(R)$ that can be true at different $R > 1.5R_0$ from the solar centre. Here, $V_A(R)$ is the local Alfven velocity of the slow SW in the streamer belt, calculated in (Mann et al., 1999) (green curve in Fig. 10). In the corona, V_A is approximately equal to the velocity of magnetic sound.

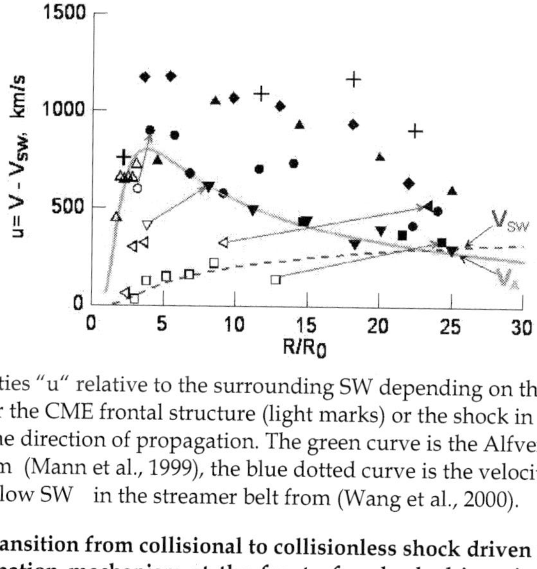

Fig. 10. The velocities "u" relative to the surrounding SW depending on the distance from the solar centre for the CME frontal structure (light marks) or the shock in front of CME (black marks) in the direction of propagation. The green curve is the Alfven velocity in the streamer belt from (Mann et al., 1999), the blue dotted curve is the velocity V_{SW} of the quasi-stationary, slow SW in the streamer belt from (Wang et al., 2000).

Law 16. "Of the transition from collisional to collisionless shock driven in front of CME": The energy dissipation mechanism at the front of a shock driven in front of CME at $R \leq 6R_0$ from the solar centre is collisional (R_0 is the solar radius). The transition from collisional to collisionless shock occurs at $R \geq 10R_0$.

According to (**M.V. Eselevich, 2010**), the front thickness δ_F of a CME-driven shock at $R \leq 6R_0$ increases with distance (the blue dashed curve in Fig. 11), remaining to be of order of the mean free path of protons λ_p (the two green dashed curves for coronal plasma temperature for $T = 10^6$K and $2 \cdot 10^6$K, respectively). This indicates at the collisional mechanism for energy dissipation at the shock front. At $R > 10\text{-}15R_0$, the formation of a new discontinuity having thickness $\delta_F^* \ll \lambda_p$ is observed at the shock front leading edge. The size of δ_F^* (within the measurement accuracy) does not vary with distance and is determined by the K spatial resolution of LASCO C3 ($K \approx 0.12R_0$) or STEREO/COR2 ($K \approx 0.03R_0$) in accordance with the data employed for these measurements. This implies that the real thickness is much

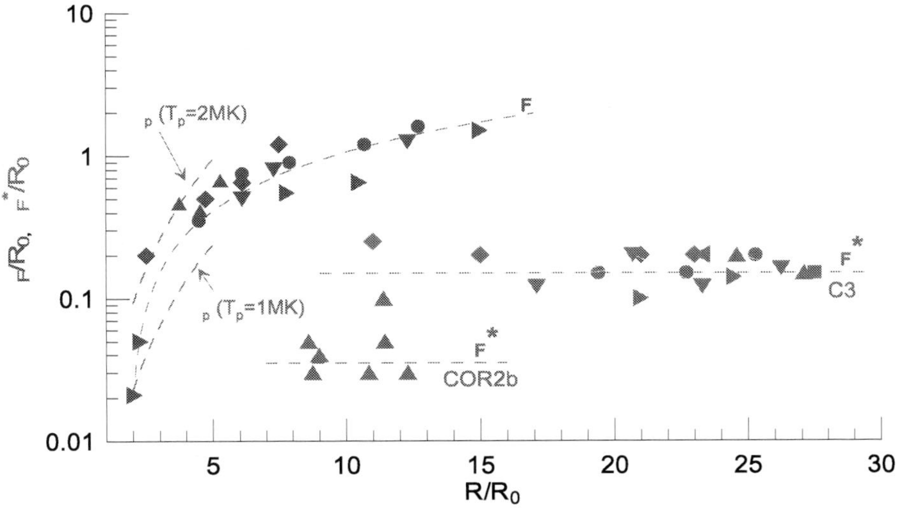

Fig. 11. The change in the CME-driven δ_F shock front thickness with distance R from the solar centre for seven different CMEs with high velocities. The calculated dependences: two green dashed curves show the mean free path of protons λp for two proton temperatures: T = 10^6 K and 2•10^6 K.. The blue dashed curve indicates the average thickness of the collisional shock front; the upper (red) and lower (violet) dashed lines stand for the average thickness of the collisionless shock front according to LASCO C3 and STEREO/COR2 data respectively from **(V.G. Eselevich, 2010)**.

less than the measured one (the image resolution is low), and the shock wave is apparently collisionless. To check this assumption, we have compared the dependence of the Alfven Mach number M_A on the shock wave strength ρ_2/ρ_1 with calculations within the ideal MHD for 10 shock waves (velocities being 800-2500 km/s) at the distance from $10R_0$ to $30R_0$ (**M.V. Eselevich & V.G. Eselevich, 2011**). As deduced from the comparison, the effective adiabatic index responsible for the processes at the front is within 2 to 5/3. This corresponds to the effective number of freedom degrees from 2 to 3 (Sagdeev, 1964). The similar dependence $M_A(\rho_2/\rho_1)$ has been obtained for the near-Earth bow shock and interplanetary collisionless shock waves. All these facts substantiate the assumption that the discontinuities under consideration, taking place in CME's leading edge at R≥ 10-15R_0, are really collisionless shock waves.

Law 17. "Of the blast shock driven by quite a powerful source of the sporadic SW (flares or CMEs)": A blast shock appears due to a pressure pulse resulting from quite a powerful flare or CME.

In the blast shock scenario (**Steinolfson et al., 1978**), the initial pressure pulse caused by a flare or a CME (**Uchida, 1968; Vrsnak & Lulic, 2000**) leads to excitation and propagation of a fast mode of the MHD wave in the corona. The mode transforms into a shock; the more powerful is the pressure pulse, the faster is the transformation. In the chromosphere, it has been first observed in the Hα line as the Moreton wave (**Moreton&Ramsey, 1960**); its manifestation in the corona is the so-called EIT wave (**Thompson et al., 1998**). The characteristic features distinguishing the blast shock from other types of disturbances and waves are: deceleration, broadening, and decrease in intensity of the profiles (**Warmuth et al., 2001**)

Law 18. "Of the existence of "foreshock" in front of the collisionless shock front": There is a region of an increased turbulence -"foreshock" - ahead of the front of collisionless bow and interplanetary shocks.

The experiments have shown that there is a region of an increased turbulence - "foreshock" - ahead of the near-Earth bow shock front (**Asbridge et al., 1968; Lin et al., 1974; Lee, 1982**) and the CME-driven shock (**Scholer et al., 1983; Lee, 1983**). Even though having different excitation mechanisms and sizes in the heliosphere, their shock front structures and "foreshock" characteristic features are the same. But their most important common feature is the diffuse plasma acceleration in the "foreshock" (**Desai and Burgess, 2008**).

In (Eastwood et al., 2005) presents a generalised pattern of the "foreshock" ahead of the near-Earth bow shock with its peculiarities and comments. Even though considerable successes have been achieved in developing the "foreshock" theory, many questions (the complete list is given in (**Desai and Burgess, 2008**) are still unanswered.

Law 19. "Of two mechanisms for solar energetic particle acceleration": There are two different classes and hence two different mechanisms for acceleration of solar energetic particles: Impulsive - particles are accelerated in flares and recorded at 1 A.U. in a narrow range of solar longitude angles. Gradual - particles are accelerated by CME-driven shocks and recorded in a wide range of solar longitudes (of about 200°).

Over the last thirty years, many papers have been written on impulsive and gradual events of solar energetic particles (SEP) (e.g., Cliver, et al., 1982; Kahler, et al., 1984; Mason et al., 1984; Cane et al., 1986, etc.); the papers have contributed greatly to the substantiation of this law. In our brief description, we will rely on the papers (**Reams, 1990; 1999**) presenting these two events in their pure form. Impulsive SEPs are driven by powerful solar flares in the western solar hemisphere. Having a small Larmor radius, they propagate along the Earth-related magnetic lines of force of IMF over a relatively narrow longitude range $\Delta\Phi \approx (20° - 40°)$. Their time profile has a narrow peak with a characteristic width of several hours (**Reams, 1999**). Gradual SEPs appear near the shock, ahead of CME, and are recorded over a wide range of longitudes ≈ 200°. Their time profile has a wider peak of several days (**Reams, 1999**).

According to [Desai and Burgess, 2008], these differences imply that mechanisms of collective particle acceleration in two events are not the same: impulsive ones are characterized by stochastic acceleration of coronal plasma heated during the flare; gradual ones feature diffuse plasma acceleration driven by the shock ahead of CME. In the case of gradual SEPs, plasma acceleration driven by the shock takes place at the front and in the "foreshock" region whose structure is similar to that of the "foreshock" ahead of the near-Earth bow shock (law 18). The mechanism for particle acceleration in flares is less well understood. In reality, impulsive and gradual SEPs are usually observed simultaneously. That is why laying down law 19 is important to study such complicated situations.

Law 20. "Of the relationship between the type-II radio emission and collisionless shocks": Type-II radio bursts are associated with processes of Rayleigh and Raman scattering of random, Langmuir electron oscillations occurring in the shock front and in the "foreshock" of collisionless shocks.

According to (**Zheleznyakov, 1965; Zaitsev, 1965**), type-II radio bursts can be associated with processes of Rayleigh and Raman scattering of random, Langmuir oscillations occurring in the front of collisionless laminar shocks. Due to the revealing of an increased turbulence region – "foreshock" - ahead of the front of the near-Earth bow shock (Asbridge et al., 1968; Lin et al., 1974; Lee, 1982) and interplanetary shock (Scholer et al., 1983; Lee, 1983), the Zheleznyakov-Zaitsev Law has turned out more universal, since the number of instabilities (and, consequently, of collisionless shock fronts) capable of exciting random Langmuir oscillations has increased. Indeed, it has been found that there are flows of energetic particles (electrons and ions) in the foreshock of the near-Earth bow shock (**Cairms et al., 1987**) and in interplanetary shocks (**Bale et al., 1999**); the flows move along the front of the undisturbed magnetic field. They are the most energetic part of heated plasma in the shock front. The collective process heating the front is of no importance. Due to the development of beam instability, electron flows in the "foreshock" excite electrostatic oscillations at the electron plasma frequency. As a result of Rayleigh and Raman scattering, these oscillations transform into the first and second harmonics of the type-II radio emission at the single and double electron plasma frequencies, respectively (**Kuncic et al., 2002**). This process is confirmed by direct observations of the simultaneous appearance of an increased level of electrostatic Langmuir oscillations ahead of the shock front and of type-II radio bursts at the same frequencies (**Bale et al., 1999**).

Laws 18, 19, and 20 may later merge.

5. Conclusion

1. This paper is the first attempt to lay down SW laws, using research results over the past 40 years. This needs to be done because
   - These laws enable further investigations into SW not only as a chaotically changing medium studied usually by statistical methods, but also as a quasiregular medium satisfying certain laws. This determines the choice of future investigation methods, largely non-statistical.
   - These laws allow us to study causes of possible SW behaviour deviations from the laws in more complex situations as well as to discover new laws.
2. The proposed list of the 20 SW laws is incomplete and it is to stand the test of time.
3. Particular attention should be given to five laws (14, 15, 16, 17, 18) dealing with shock waves: there is no unified theory of the front thickness in plasma for them that could explain various particular cases, though the laws are qualitatively understandable and physically meaningful. These five laws are most universal among all those listed above. But their mechanisms are still unknown. This line of investigation is very fruitful for both solar-terrestrial physics and plasma physics.
4. Priority of collisionless shocks over other most topical issues of solar-terrestrial physics was discussed by Sagdeev, R.Z. (Sagdeev, 2010) and Russell, C.T. (Russell, 2010) in their invited reports at COSPAR 2010.
5. Such analysis-generalization should also be conducted for the Sun (though it has been partially done in many monographs) as well as for the Earth's magnetosphere and ionosphere in their own right.
6. Laying down the SW laws actually implies that the space science "solar wind" subdiscipline turns from multidirectional investigations and data collection into an independent branch of physics.

6. Acknowledgments

I would like to express our profound gratitude to Corr. Member of RAS Viktor M. Grigoryev: the bulk of our research has been done in Solar Physics Department headed by him. I am also thankful to Academician of RAS Geliy A. Zherebtsov for his support and encouragement, enabling us to fight through every hardship when preparing this paper. I am especially grateful to Academician of USSR AS Roald Z. Sagdeev who discovered collisionless shock waves 50 years ago. His infrequent but extremely useful e-mails have contributed greatly to this chapter, allowing us to improve it dramatically.

I thank O.Kulish, K. Korzhova and Yuri Kaplunenko for the help in translation in the English.

The work was supported the Russian Foundation for Basic Research (Projects No. 09-02-00165a, No.10-02-00607-a).

7. References

Alikhanov, S.G.; Belan, V.G. & Sagdeev, R.Z. (1968). Non-linear ion-acoustic waves in plasma. *JETP Letters.*, Vol. 7, pp. 465.

Asbridge, J.R.; Bame, S.J. & Srong, I.B. (1968). Outward flow of protons from the earth's bow shock. *J.Geophys.Res*, Vol.73, pp. 777.

Arge, C. N. & Pizzo, V. J. (2003). Improvement in the prediction of solar wind conditions using near-real time solar magnetic field updates. *J. Geophys. Res.*, V. 105, No. A5, pp. 10465-10479.

Bonetti A.; Bridge H.S., Lazarus A.J., Lyon E.F., Rossi R. &, Scherb F. (1963). Explorer 10 plasma measurements. *J. Geophys. Res.*, Vol.68, pp. 4017-4063.

Belcher, J.W. & Davis, L. Jr. (1971). Large-amplitude Alfven waves in the interplanetary medium. 2, *J. Geophys. Res.*, Vol.76, pp. 3534-3563.

Bame, S.J.; Asbridge, J.R., Gosling, J.T., Halbig, M., Paschmann G., Scopke, N. & Rosenbauer, H. (1979). High temporal resolution observations of electron heating at the bow shock. *Space Sci. Rev.*, Vol.23, pp. 75-92.

Burlaga., L. F. (1974). Interplanetary stream interfaces, *J. Geophys. Res.*, Vol. 79, pp. 3717 - 3725.

Borrini G.; Wilcox, J. M., Gosling J. T., Bame S. J. & Feldman W. C. (1981). Solar wind helium and hydrogen structure near the heliospheric current sheet; a signal of coronal streamer at 1 AU. *J. Geophys. Res.* Vol.86. pp. 4565 -4573.

Burlaga, L.F.; Hundhausen, A.J. & Xue-pu Zhao. (1981). The coronal and interplanetary current sheet in early 1976. *J. Geophys. Res.*, Vol. 86, pp. 8893 - 8898.

Bavassano B.; Woo, R. & Bruno, R. (1997). Heliospheric plasma sheet and coronal streamers. Geophys. Res. Let., Vol.24, pp. 1655 - 1658.

Bale, S.D.; Reiner, M.J., Bougeret, J.-L., Kaiser, M.L., Kruker, S., Larson, D.E. & Lin, R.P. (1999). The source region of an interplanetary type II radio burst. *Geophys. Res. Lett.*, Vol. 26, No.11, pp. 1573 - 1576.

Burton, M.E; Neugebauer, M., Crooker, N. U., von Steiger, R. & Smith, E.J. (1999). Identification of trailing edge solar wind stream interface: A comparison of Ulysses plasma and composition measurements. J. Geophys. Res., Vol.104, No.A5, pp. 9925 - 9932.

Cliver, E.W.; Kahler, S.W., Shea, M.A. & Smart, D.F. (1982). Injection onsets of 2Gev protons, 1 MeV electrons in solar cosmic ray flare. *Astrophys. J.*, Vol.260, pp. 362-370, doi: 10.1086/160261.

Cane, H.V.; McGuire, R.E. & Rosenvinge, T.T. (1986). Two classes of solar energetic particle vents associated with impulsive and longduration soft X-ray flares. *Astrophys. J.*, Vol.301, pp. 448- 459, doi: 10.1086/163913.

Cairns, L.H. (1987). The electron distribution function upstream from the Earth's bow shock. *J. Geophys. Res.*, Vol.92, pp. 2315.

Cremandes, H. & Bothmer, V. (2004). On the three-dimensional configuration of coronal mass ejections. *Astronomy and Astrophysics*, Vol.422, pp. 307-332. DOI: 10.1051/0004-6361:20035776.

Dessler, A.J. &. Fejer, J.A. (1963). Interpretation of Kp- index and M-region geomagnetic storms. *Planet.Space Sci.*,Vol.11, pp. 505.

Desai, M.I. & Burgess, D. (2008). Particle acceleration at coronal mass ejection-driven interplanetary shocks and Earth's bow shock. *J.Geophys. Res*, Vol. 113, pp. A00B06, doi:10.1029/2008JA013219.

Eselevich, V.G. (1983). Bow shock structure from laboratory and satellite experimental results. *Planet. Spase Sci.*, Vol.34, No.11, pp. 1119-1132.

Eselevich, V. G. & Filippov, M.A. (1986). Study of the mechanism for solar wind formation. *Planet.Space Sci.*, Vol.34, No.11, pp. 1119-1132.

Eselevich, V.G.; Kaigorodov, A.P. & Fainshtein, V.G. (1990). Some peculiarities of solar plasma flows from coronal holes. *Planet. Space Si.* Vol. 38, No. 4, pp.459- 469.

Eselevich, V. G. (1992). Relationships of quasistationary solar wind flows with their sources on the Sun. *Solar Phys .*, Vol.137, pp. 179-197.

Eselevich, V.G. & Fainshtein, V.G. (1992). On the existence of the heliospheric current sheet without a neutral line (HCS without NL). *Planet. Space Sci.*, Vol.40, pp. 105 - 119.

Eselevich, V. G. (1995). New results on the site initiations of CMEs. *Geophys. Res. Let.*, Vol. 22 (20), pp. 2681 - 2684.

Eselevich, V.G. & Eselevich, M. V. (1999). An investigation of the fine ray structure of the coronal streamer belt using LASCO data. *Solar Phys.*, Vol.188, pp. 299 - 313.

Eselevich, V.G.; Rudenko, V.G. & Fainshtein, V.G. (1999). Study of the structure of streamer bels and chains in the Solar corona. *Solar Phys.*,Vol.188, pp. 277 - 297. Eselevich, V. & Eselevich, M. (2002). Study of the nonradial directional property of the rays of the streamer belt and chains in the solar corona. *Solar Phys.*, Vol.208, pp. 5 - 16.

Eselevich, V. G. & Eselevich M. V. (2005). Prediction of magnetospheric disturbances caused by a quasi-stationary solar wind. *Chin. Space Sci.*, Vol.,25 (5), pp. 374 -382.

Eastwood, O.3.; Lucek, E.A., Mazelle, C., Meziane, K., Narita, Y., Pickett, J. & Treumann, R.A. (2005). The foreshock. *Space Sci., Rev.*, Vol.118, pp. 41-94, doi: 10.1007/s11214-005-3824-3.

Eselevich, M. V. & Eselevich, V. G. (2005). Fractal. Structure of the heliospheric plasma sheet in the Earth's orbit. Geomagnetism and Aeronomy, Vol.45, No.3, pp. 326-336.

Eselevich, M.V. & Eselevich, V.G. (2006). The double structure of the coronal streamer belt. *Solar Phys.*, Vol.235, pp. 331 - 344.

Eselevich, M.V. & Eselevich, V.G. (2007a). First experimental studies a perturbrd zone preceding the front of a coronal mass ejection. *Astronomy Reports*, Vol.51, No. 111, pp. 947-954.

Eselevich, M.V. & Eselevich, V.G. (2007b). Streamer Belt in the Solar Corona and the Earth's Orbit. *Geomagnetism and Aeronomy*, Vol.47, No.3, pp. 291–298.

Eselevich, M.V. & Eselevich, V.G. (2008). On formation of a shock wave in front of a coronal mass ejection with velocity exceeding the critical one. *Geophys.Res.Let.*, Vol.35, pp. L22105.

Eselevich, V. G.; Fainshtein, V. G., Rudenko, G. V., Eselevich, M. V. & Kashapova, L. K. (2009). Forecasting the velocity of quasi-stationary solar wind and the intensity of geomagnetic disturbances produced by It. *Cosmic Research*, Vol.47, No.2, pp. 95–113.

Eselevich, M. V. (2010). Detecting the widths of shock fronts. Preceding coronal mass ejections. *Astronomy Reports*, Vol.54, No.2, pp. 173–183.

Eselevich, M.V. & Eselevich, V.G. (2011). Relations estimated at shock discontinuities txcited by coronal mass ejections. *Astronomy Reports*, Vol.55, No.4, pp. 359–373.

Eselevich, M.V. & Eselevich, V.G. (2011). On the mechanism for forming a sporadic solar wind, *Solar-Terrestrial Physics*. Issue17, pp. 127-136, RAS SB Publishers.

Gringauz K.I.; Bezrukikh V.V., Ozerov V.D. & Rybchinsky R.E. (1960). A study of interplanetary ionized gas, energetic electrons and corpuscular emission of the Sun, using three-electrode traps of charged particles aboard the second space rocket. *Reports of the USSR Academy of Sciences*, Vol.131, pp. 1301-1304.

Galeev, A. A. & Sagdeev, R.Z. (1966). *Lecture on the nonlinear theory of plasma*. pp. 38, Trieste, Italy.

Gosling, J. T.; Asbridge, J. R., Bame, S. J. & Feldman, W. C. (1978). Solar wind sreamer jnterfaces, *J. Geophys. Res.*, Vol.83, No.A4, pp. 1401 - 1412.

Gosling, J. T.; Borrini, G., Asbridge, J.R., Bame, S.J., Feldman, W.C. & Hansen, R.T. (1981). Coronal streamers in the solar wind at 1 a.u, *J. Geophys. Res.*, Vol.82, pp. 5438 - 5448.

Geiss J.; Gloeckler G. & von.Steiger, R. (1995). Origin of the solar wind composition data. *Space Science Reviews*, Vol.72, pp. 49-60.

Goldstein, B.E.; Neugebauer, M., Phillips, J.L., Bame, S., Goeling, J.T., McComas, D.,Wang, Y.-M., Sheeley, N.R., & Suess,S.T. (1996). Ulysses plasma parameters: latitudinal, radial, and temporal variations. *Astron.Astrophys.* Vol.316, pp. 296-303.

Gubchenko, V.M., Khodachenko, M.L., Biernat, H.K. , Zaitsev, V.V. & Rucker, H.O. (2004). On a plasma kinetic model of a 3D solar corona and solar wind at the heliospheric sheet, *Hvar Obs. Bull.*, Vol.28 (1), pp. 127.

Hundhausen, A.J. & Burlaga, L.F. (1975). A model for the origin of solar wind stream interfaces. *J. Geophys. Res.*, Vol.80, pp. 1845 - 1848.

Hoeksema J.T. (1984). *Structure and evolution of the large scale solar and heliospheric magnetic fields*. Ph. D. Diss. Stahford Univ.

Hundhausen, A.T. (1993). Sizes and locations of coronal mass ejections: SMM observations from 1980 and 1984 - 1989. *J.Geophys. Res* , Vol.98, pp. 13,177 - 13,200.

Iskoldsky A.M.; Kurtmullayev R.Kh., Nesterikhin Yu.E. & Ponomarenko A.G. (1964). Experiments in collisionless shock wave in plasma. *ZhETF*, Vol.47, No.2, pp. 774-776.

Illing, R.M. & Hundhausen, A.T. (1985). Disruption of a coronal streamer by an eruptive prominence and coronal mass ejection. *J. Geophys. Res.*, Vol.90, pp. 275 - 282.

Ivanov K.; Bothmer V., Cargill P.J., Kharshiladze A., Romashets E.P. & Veselovsky I.S. (2002). Subsector structure of the interplanetary space. *Proc. The Second Solar Cycle and Space Whether Euroconference*, pp. 317, Vicvo Equense (Italy).

Korzhov N P. (1977). Large-scale three-dimensional structure of the interplanetary magnetic field. *Solar Phys.*, Vol.55, pp. 505.

Korolev A.S.; Boshenyatov B.V., Druker I.G. & Zatoloka V.V. (1978). *Impulse tubes in aerodynamic studies*, pp. 5-80, Novosibirsk, "Nauka".

Kahler, S.W.; Sheeley, N.R.Jr., Howard, R.A., Michels, D.J., Koomen, M.J., McGuire, R.E., von Rosenvinge, T.T. & Reams, D.V. (1984). Associations between coronal mass ejections associated with impulsive solar energetic particle events. *J. Geophys. Res*, Vol.89, pp. 9683 - 9693, doi:10.1029/JA089iA11p09683.

Krall, J.; Chen, J. & Santoro, R. (2000). Drive mechanisms of erupting solar magnetic flux ropes. Astrophys. J., Vol.539, pp. 964-982.

Kuznetsov , V.D. & Hood, A.W. (2000). A phenomenological model of coronal mass ejection. Adv. Space Sci., Vol.26, No.3, pp. 539-542.

Kuncic, Z.; Cairns, I.H., Knock, S., & Robinson, P.A. (2002). A quantitative theory for terrestrial foreshock radio emission., *Geophys. Res. Lett.*, Vol.29, No.8, pp. 2-1, CiteID 1161, DOI 10.1029/2001GL014524.

Landau L.D. & Livshits E.M. (1953). *The mechanics of continuous media*. State Publishing House of Theoretical and Technical Literature, Moscow.

Lin, R.P.; Meng, C.I. & Anderson, K.A. (1974). 30-100kev protons upstream from the earth's bow shock. *J.Geophys. Res*, Vol.79, pp. 489 - 498.

Lee, M.A. (1982). Coupled hydromagnetic wave excitation and ion acceleration upstream of the Earth's bow shock. *J.Geophys. Res*, Vol.87, pp. 5063 - 5080.

Lee, M.A. (1983). Coupled hydromagnetic wave excitation and ion acceleration at interplanetary traveling shocks. *J.Geophys. Res.*, Vol. 88, No. A8, pp. 6109-6119.

Moreton, G. E. & Ramsey, H. E. (1960). Recent Observations of Dynamical Phenomena Associated with Solar Flares. *PASP*, Vol. 72, pp. 357.

Moreno, G.; Olbert, S. & Pai, L. (1966). Risultati di Imp-1 sul vento solare. *Quad. Ric. Sci.*, Vol.45, pp. 119.

Mason, G.M.; Gloecker, G. & Hovestadt, D. (1984). Temporal variations of nucleonic abundances in sllar flare energetic particle events. II - Evidence for large-scale shock acceleration. *Astrophys. J.*, Vol.280, pp. 902 - 916, doi: 10.1086/162066/261.

Mann, G.; Aurass, H., Klassen, A., Estel, C. & Thompson, B. J. (1999). Coronal Transient Waves and Coronal Shock Waves. In: Vial, J.-C., Kaldeich-Schumann, B. (eds.) *Proc. 8th SOHO Workshop Plasma Dynamics and Diagnostics in the Solar Transition Region and Corona*, pp. 477-481, Paris, France, 22-25 June 1999.

Milovanov A. V. & Zelenyi L. M. (1999). Fraction excititations as a driving mechanism for the self-organized dynamical structuring in the solar wind. *Astrophys. Space Science*, Vol.264, pp. 317 - 345.

McComas, D. J. ; Elliott, H. A.; von Steiger, R. (2002). Solar wind from high latitude CH at solar maximum. *Geophys. Res. Lett.* Vol.29(9), pp. 28-1, CiteID 1314, DOI 10.1029/2001GL013940

Nolte, J.T.; Kriger A.S., Timothy, A.F., Gold, R.E., Roelof, E.C., Vaina, G., Lazarus, A.J., Sullivan, J.D. & Mcintosh, P.S. (1976). Coronal holes as sources of solar wind. *Solar Phys.*, Vol.46, pp. 303-322.

Olbert, S. (1968). Summary of experimental results from MIT detector on Imp-1. *In Physics of Magnetosphere*, edited by R.L.Carovillano et al., p. 641, D.Reidel, Dordrecht, Netherland.

Ponomarev E.A. (1957). *On the theory of the solar corona.* Ph.D. Thesis in Physics and Mathematics. Kiev, Kiev University.

Parker E.N. (1958). Dynamics of interplanetary gas and magnetic fields. *Astrophys. J.*, Vol.128, pp. 664–675.

Sagdeev R.Z. (1964). Collective processes and shock waves in rarefied plasma, *Reviews of Plasma Physics*, Vol.2, pp. 20-80, M.: "Gosatomizdat".

Paul, J.W.H.; Holmes, l.S., Parkinson, M.J. & Sheffield, J. (1965). Experimantal observations on the structure of collisionless shock waves in a magnetized plasma. *Nature*, Vol.2, pp. 367-385.

Reams, D.V. (1990). Acceleration of energetic particles by shock waves from large solar flares, *Astrophys. J*, Vol.358, pp. L63 – L67.

Reams, D.V. (1999). Particle acceleration at the Sun and the heliosphere. *Space Sci. Rev.*, Vol.90, pp. 413 - 491.

Russell, C. T. (2010). Advances in understanding the plasma physics of the solar wind: contributions from STEREO. *Theses of the report COSPAR 2010*. D33- 0002-10.

Svalgaard, L.J.; Wilcox, W. & Duvall, T.L. (1974). A model combining the solar magnetic field. *Solar Phys.*, Vol.37, pp. 157 - 172.

Steinolfson, R. S.; Wu, S. T., Dryer, M. & Tanberg-Hanssen, E. (1978). Magnetohydrodynamic models of coronal transients in the meridional plane. I – The effect of the magnetic field, *Astrophys. J.*, Vol.225, pp. 259 - 274.

Scholer, M.; Ipavich, F.M. , Gloecker, G. & Hovestadt, D. (1983) Acceleration of low-energy protons and alpha particles at interplanetary shock waves . *J. Geophys. Res.*, Vol.88, pp. 1977 - 1988.

Schwenn, R. & Marsch, E. (1991). *Physics of the inner heliosphere v. I and v. II*, Springer Verlag, pp. 185, Berlin Heidelberg,.

Sheeley, N.R.Jr.; Walter, H., Wang, Y.-M. & Howard, R.A. (1999). Continuous tracking of coronal outflows: Two kinds of coronal mass ejections, *J.Geophys. Res.* Vol.104, pp. 24739 - 24768.

Sagdeev, R. Z. (2010). The role of space as an open physics lab in enriching of plasma science. *Theses of the report COSPAR 2010.* (D33-0001- 10)

Tidman, D.A. (1967). Turbulent shock waves in plasma. *Phys. Fluids*, Vol.10, pp. 547-568.

Tompson, B.J.; Plunkett, S. P., Gurman, J. B., Newmark, J. S., St. Cyr, O. C.. & Michels, D. J. (1998). SOHO/EIT observations of an Earth-directed coronal mass ejection on May 1997, *Geophys. Res. Lett.*, Vol.25, pp. 2465 - 2468.

Tlatov, A.G. & Vasil'eva, V.V. (2009). The non-radial propagation of coronal streamers in minimum activity epoch. *Proceedings of the International Astronomical Union*, Vol.5, pp. 292.

Thernisien A., Vourlidas, A. and Howard, R.A. (2009). Forward modeling TEREO/SECCHI data. *Sol. Phys.*, Vol.256, pp. 111- 130.

Uchida, Y. (1968). Propagation of hydromagnetic disturbances in the solar corona and Moreton's wave phenomenon, *Solar Phys.*, Vol.4, pp. 30.

Vsekhsvyatsky S.K.; Ponomarev E.A., Nikolsky G.M. & Cherednichenko V.I. (1957). *On corpuscular emission. "Physics of solar corpuscular streams and their impact on the Earth's upper atmosphere"*, M., Publishing House of the USSR Academy of Sciences.

Volkov O.L.; Eselevich V.G., Kichigin G.N. & Paperny V.L.(1974). Turbulent shock waves in rarefied nonmagnetised plasma. *JETP*, Vol.67, pp. 1689-1692.

Vaisberg O.L.; Galeev A.A., Klimov S.I., Nozdrachev M.N., Omelchenko A.N. & Sagdeev R.Z. (1982). Study of energy dissipation mechanisms in collisionless shocks with high Mach numbers with the help of measurement data aboard the 'Prognoz-8' satellite. *JETP Letters*, Vol.35, pp. 25 .

Vrsnak, B., & Lulic.S. (2000). Formation of coronal MHD shock waves - II. The Pressure Pulse Mechanism, *Solar Phys.*, Vol.196, pp. 181.

Wong, A.Y. & Means, R.W.(1971). Evolution of turbulent electrostatic shock. *Phys. Rev. Letters*, Vol.27, No.15, pp. 973-976.

Wilcox, John M. & Hundhausen, A.T. (1983). Comparison of heliospheric current sheet structure obtained from potential magnetic field computations and from observed polarization coronal brightness. *J. Geophys. Res.*, Vol.88, pp. 8095 - 8086.

Wang Y.-M. & Sheeley, N. R. Jr. (1990). Solar wind speed and coronal flux-tube expansion *Astrophys. J.*, Vol.355, pp. 727-732.

Winterhalter D.; Smith E. J., Burton M. E. & Murphy N. (1994). The heliospheric plasma sheet. *J. Geophys. Res.*, Vol.99, pp. 6667 - 6680.

Wang, Y.-M. (1995). Empirical relationship between the magnetic field and the mass and energy flux in the source regions of the solar wind. *Astrophys. J.*,Vol.449, pp. L157-L160.

Wang, Y.-M.; Sheeley, N. R., Socker, D. G., Howard, R. A. & Rich, N. B.(2000). The dynamical nature of coronal streamers. *J. Geophys. Res.*, Vol.105, No.A11, pp. 25133-25142, DOI:10.1029/2000JA000149

Warmuth, A.; Vrsnak, B., Aurass, H.& Hanslmeier. (2001). Evolution of EIT/Hα Moreton waves, *Astrophys. J.*, Vol.560, pp. L105.

Wang, Y.M.; Sheeley, N.R. & Rich. N.B. (2007). Coronal pseudostreamers. *Astrophys. J.*, Vol. 685, pp. 1340 - 1348.

Zagorodnikov, S.P.; Rudakov, L.I., Smolkin, G.E. & Sholin.,G.V. (1964). Observation of shock waves in collisionless plasma. *JETP*, Vol.47, No.5, pp. 1770-1720.

Zheleznyakov, V. (1965). On the genesis of solar radio bursts in a metre wave range. *Astron. Journal [in Russian]*, Vol.XLII, pp. 244 – 252.

Zaitsev, V.V. (1965). On the theory of type-II solar radio bursts. *Astron. Journal [in Russian]*, Vol. XLII, pp. 740 - 748.

Zeldovich, Ya. B. & Raiser, Yu.P. (1966). *Physics of shock waves and high-temperature hydrodynamic phenomena [in Russian]*, pp. 398 - 406, Publishing House 'Nauka', Moscow.

Zhao, X. P. & Webb, D.F. (2003). Source regions and storm effectiveness of frontside full halo coronal mass ejections. *J.Geophys. Res.*, Vol..108, No.A6, pp. SSH4-1, CiteID 1234, DOI 10.1029/2002JA009606.

2

Solar Wind: Origin, Properties and Impact on Earth

U.L. Visakh Kumar [1] and P.J. Kurian[2]
Physics Research Centre, St. Berchmans' College, Chanaganacherry, Kerala India

1. Introduction

In visible light, the sun appears as an isolated and perfectly shaped, disc like object in the sky. In the deepest interior of the sun hydrogen nuclei steadily fuse together to form helium nucleus and thereby release large amounts of heat that slowly leak in to the solar surface. The temperature of the dense core of the sun is a few million degrees. Towards the solar surface, the temperature gradually decreases and until it reaches its minimum value of about 4300 K. The sun and all other stars consist of plasma :the gas is so hot ,that it is ionized such that it can easily conduct electric currents and generate and carry magnetic fields. The outer solar atmosphere contains several distinct layers with qualitatively different properties. The photosphere ($T \cong 6000$ K) marks the boundary between the convection zone below and the chromosphere which is surrounded by corona.

When the intense light of the solar disk is shielded during the eclipse of the sun, a faint halo with a thread like structure and a form that is reminiscent of a crown becomes visible to the naked eye. This so called corona appears to extend into space over many solar radii. A rapid transition to the hot corona occurs approximately 0. 003 R_0 above the photosphere where R_0 is the solar radius. The solar corona consists of tenuous plasma that is highly structured by the strong magnetic field that finds its origin in the solar interior. From the corona coronal emission occurs due to highly ionized ions. This lead to the conclusion that the coronal plasma must be extremely hot ; for iron to be so highly ionized, the temperature of the ambient plasma must be a few million degrees.

The coronal temperature ($T \geq 10^6$K) thus turned out to be exceeding the photospheric temperature by almost a factor thousand. This indicates the presence of a physical process that actively heats the solar corona. The flow of energy through the solar atmosphere and the heating of the Sun's outer regions are still not fully understood. Apart from the magnetically closed coronal regions, part of the solar corona consist of open regions, above a height of 0. $1R_0$ where the magnetic field lines are not reentrant on the solar surface, but extend in to space.

The temperature of the coronal holes is one or two million degrees and this remains so over many solar radii into space so that the plasma is accelerated and escapes from the gravitational field of the sun to form the solar wind with an average flow velocity of

400km/s at 1 AU,the earth-sun distance. The entire system of sun, corona and solar wind constitutes the heliosphere, where the planets with their own magnetic fields appear as small islands in the stream.

2. Origin of solar wind

The basic types of solar wind are closely associated with the structure and the activity of the coronal magnetic field that changes over the solar cycle. The most fundamental aspect related to the magnetic field in the Sun is the working of the solar dynamo in the convection zone bellow the photosphere. It generates the magnetic field we observe on the solar surface including the corona. The corona is highly structured by the magnetic field of the sun. The base of the corona is a continually replenished ensemble of closed magnetic loops and open flux tubes, but above a height of 0. 1R_0, the open field lines begin to dominate. As the result of varying boundary conditions in the corona, three basic types of solar wind occur: Fast streams from large coronal holes (CHs); slow streams from small CHs and active regions (ARs), and from the boundary layers of coronal streamers; and the variable transient flows such as coronal mass ejections (CMEs), often associated with eruptive prominences, or plasmoids stemming from the top of streamers, and other ejections from ARs driven through magnetic flux emergence and reconnection(Marsch ,2006).

The steady solar wind consists of two major components: fast, tenuous, and uniform flows from large CHs, and slow, dense, and variable flows from small CHs, (Arge *et al.* , 2003) often from near the boundaries between closed and open coronal fields. The origin of fast streams seems clear, but the sources of the slow solar wind remain less obvious. The helium abundance is 3. 6% in high speed wind very constant in time and almost identical for all streams. Whereas in slow wind the abundance is only 2. 5% and is highly variable. The angular momentum carried away by the solar wind from the rotating sun is almost completely contained in the solar flow. This indicates that the fast wind starts from regions close to the solar rotation axis, while the slow wind is released only beyond 30 R_0.

CME's from the Sun are spontaneous expulsions of $\approx 10^6$ K blobs of coronal plasma which carries up to ten billions of kilograms of mass, ejected at speeds ranging from a few hundred km/s to as much as 2000 km/s. Concerning their occurrence rate, the CMEs tend to accumulate around maximum solar activity, when the corona is highly magnetically structured and of multi-polar nature. The constraints on the plasma are even more extreme for a transient CME than for a steady solar wind stream, as the CME plasma density is often much higher, and its flow speed may easily reach a multiple of the average solar-wind speed(Marsch,2006). Lusamma Joseph & P J Kurian (2010)finds elliptical distribution of CME speeds which indicates that magnetic field has a greater role in the dynamics of CME.

2.1 Solar wind from funnels in coronal holes

The coronal funnels are expanding magnetic field structures rooted in the magnetic network lanes. Using images and Doppler maps from the Solar Ultra Violet measurements of Emitted Radiation(SUMER)spectrometer and magnetograms delivered by the Michelson Doppler Imager (MDI)on the Space based solar and Heiospheric Observatory(SOHO) of ESA and NASA, a Chinese-German team of scientists have observed solar wind flows coming from funnel shaped magnetic fields which are anchored in the lanes of the magnetic network near the surface of the Sun. The fast solar wind seems to originate in coronal funnels with a flow speed of about 10 km/s at a height of 20,000 km above the photosphere(Tu *et al.* ,2005).

Just bellow the surface of the sun, there are large convection cells which are associated with magnetic fields. By magnetoconvection, they become concentrated in the network lanes, where the funnel necks are anchored. The plasma, while still being confined in small loops, is bought by convection to the funnels and released there ,like a bucket of water is emptied in to an open water channel. For CH of the period of *Skylab* observations of the Sun (February 1973 - March 1974) it was demonstrated that there was rather strong positive correlation between the area S_{CH} of a coronal hole recorded in the X-ray range of coronal emission and maximum velocity V_M (at the Earth's orbit) of the fast solar wind stream flowing out of it(Nolte, J. T *et al.* ,1976).

2.2 Solar wind from active regions

The ARs near solar maximum were clearly identified as the source regions of slow solar wind. For example, Liewer *et al.* (2003) investigated the magnetic topology of several ARs in connection with EUV and X-ray images. Synoptic coronal maps were employed for mapping the inferred sources of the solar wind from the magnetic source surface down to the photosphere. In most cases, a dark lane, as it is familiar for the small CHs, was seen in the EUV images, thus suggesting an open magnetic field. The in-situ composition data of the solar wind associated with these regions indicates high freezing-in temperatures of the heavy ions, a result that is consistent with the inference that the AR indeed is a genuine source of the solar wind.

3. Coronal heating and acceleration of solar wind

There is heating everywhere above the solar photosphere. Chromospheric heating occurs immediately above the photosphere where the plasma is mostly neutral. The plasma density is high enough for many collisions to occur. Thus, non magnetic mechanisms such as acoustic wave dissipation tend to be considered as the dominant source of energy deposition (Narain and Ulmschneider, 1990) but magnetic effects still may be important (Goodman,2000).

Base coronal heating "turns on" abruptly about 0. 003 R_0 above the photosphere and seems to extend out several tenths of a solar radius. Parker(1991)discussed the separation of heating mechanisms between the coronal base ($r \approx 1$-1. $5R_0$) and extended radial distances beyond the sonic point ($r \cong 2$-$5R_0$). In the coronal base there exists strong downward heat conduction generated by the sharp temperature gradient. The continually replenished "junkyard"of closed loops and open funnels at the coronal base (Dowdy *et al.* ,1986) evolves in to a relatively uniform flux expansion in the extended corona. In this region, the magnetic energy is probably dissipated as heat by Coulomb collisions (via, e. g. ,viscosity, thermal conductivity, ion-neutral friction, or electrical resistivity).

Extended corona is the region where the primary solar wind acceleration occurs. The vast majority of proposed physical processes involve the transfer of energy from propagating magnetic fluctuations(waves, shocks, or turbulence to the particles.)The ultimate source of energy must be solar in origin ,and thus it must some how propagate out to the distances where the heating occurs(Tu and Marsch,1995). At distances greater than 2 to 3 R_0, the proton temperature gradient is noticeably shallower than that expected from pure adiabatic expansion (Barnes *et al.* ,1995), indicating gradual heating of the collisionless plasma.

Near the distant termination shock, where the solar wind meets the interstellar medium, heating may occur when neutral interstellar atoms enter the heliosphere and become ionized, forming a beam or ring-like velocity distribution that is unstable to the generation of MHD waves (Zank *et al.* ,1999).

The dual questions of how the solar corona is heated and how the solar wind expands and accelerates have been considered together primarily by Hollweg(Hollweg 1986; Hollweg & Johnson 1988) since the solar wind is an outward extension of the corona. Thus, it is at least possible that similar physical processes are at work in both regions. Various mechanisms have been proposed in an effort to understand the heating of the solar corona and the acceleration of the solar wind. Unfortunately, there is no consensus among researchers about the physical mechanism(s) for coronal heating and for solar wind acceleration, even today.

The first calculation yielding a supersonic wind was performed by Parker in 1958 and was soon confirmed by in situ observations. The flow energy for the solar wind must come ultimately from what is provided at the base of the wind, where the flow speed is very small. Hence the asymptotic flow speed V_{sw}, at a very large distance where the flow kinetic energy dominates all other forms of energy, is constrained by the energy available as

$$\frac{V_{SW}^2}{2} \approx \frac{5K_BT_0}{m_p} - \frac{M_0G}{r_0} + \frac{Q_0}{n_0 m_p V_0} \tag{1}$$

The terms on the right-hand side of (1) are respectively:

- the enthalpy per unit mass, due to both the protons and the electrons,
- the gravitational binding energy per unit mass,
- the heat flux per unit mass flux,

At the base of the wind; the initial bulk kinetic energy has been neglected, as well as the asymptotic enthalpy and heat flux terms. With a coronal temperature of 2×10^6 K, the radius $r_0 \approx 7 \times 10^8$ m, and the solar mass $M_0 \approx 2 \times 10^{30}$ kg, the enthalpy provides only 0.8×10^{11} J kg^{-1},whereas the gravitational binding energy amounts to 2×10^{11} J kg^{-1}. Hence the available enthalpy is far from sufficient to lift the medium out of the Sun's gravitational well, so that the heat flux plays a key role. The heat is transported by the electrons, since they have a much greater thermal speed than the protons. With a coronal temperature of 2×10^6 K, the heat flux at the base of the wind provides about 2×10^{11} J kg^{-1}, which just balances the binding gravitational energy. The remaining enthalpy term yields a terminal velocity of a few hundred km s^{-1}, so that enough energy seems available to drive the wind. This result, however, is very sensitive to the temperature since the heat flux varies as $T^{7/2}$: with a temperature only 15% smaller, the right-hand side of (1) becomes negative.

Besides this, the wind which is the most stable, is the fastest and fills most of the heliosphere, comes from the coldest regions of the corona, where the electron thermal temperature (which determines the conductivity) is not significantly higher than 10^6 K. With such a temperature, the thermal conductivity falls short by roughly one order of magnitude of that required to drive even a low-speed wind. Hence the electron driven models were soon recognized to be insufficient to drive the high speed streams.

In 1942, by studying the mutual interaction between conducting fluid motion and electromagnetic fields Hannes Alfv´en discovered a new mode of waves, that later on were named as Alfv´en waves. An Alfv´en wave propagating in a plasma is a traveling oscillation of the ions and the magnetic field. The ion mass density provides the inertia and the magnetic field line tension provides the restoring force. The wave vector can either propagate in the parallel direction of the magnetic field or at oblique incidence. The waves efficiently carry energy and momentum along the magnetic field lines. The Alfv´en waves were identified in the solar wind by means of spacecraft measurements in late 60's. An early mention of the radiation pressure of Alfv´en waves can be found in Bretherton & Garrett (1969).

Alazraki & Couturier (1971) and Belcher (1971) inaugurated the concept of the wave-driven wind by noting that the waves exert a 'wave pressure' $-\nabla\langle\delta B^2\rangle/8\pi$ on the wind where B is magnetic field, the prefix δ denotes a fluctuation, and the angle brackets denote a time-average.

Alfv´en waves can travel a long distance to contribute not only to coronal heating but to the solar wind acceleration. The Alfven waves are excited by steady transverse motions of the field lines of the photosphere while they can also be produced by continual reconnection above the photosphere. Tomczyk et al. (2007) reported the detection of Alfv´en waves in images of the solar corona with the Coronal Multi-Channel Polarimeter instrument at the National Solar Observatory, New Mexico.

With heating and wave pressure, the wave-driven models were able to explain the high-speeds and hot protons observed in the fast wind in interplanetary space (e. g. , Hollweg 1978). These wave-driven models generally succeeded in explaining solar wind data far from the Sun, but they failed close to the Sun. The spacecraft gave us new coronal hole density data, which verified previous evidence that the density declines very rapidly with increasing r (Guhathakurta & Holzer 1994, Guhathakurta & Fisher 1998). That requires the flow speed to increase very rapidly with r. The wave-driven models could not achieve such rapid accelerations. The reason is simply that, close to the Sun, the wave pressure is small compared to other terms in the momentum balance.

The Ultraviolet Coronagraph Spectrometer (UVCS) aboard the *Solar and Heliospheric Observatory* (*SOHO*), launched in 1995, has been the first space borne instrument able to constrain ion temperature anisotropies and differential outflow speeds in the acceleration region of the wind. UVCS measured O^{5+} perpendicular temperatures exceeding 3×10^8 K at a height of 2 R_0. Temperatures for both O^{5+} and Mg^{9+} are significantly greater than mass-proportional when compared to hydrogen, and outflow speeds for O^{5+} may exceed those of hydrogen by as much as a factor of two. These results are similar in character to the *in situ* data, but they imply more extreme departures from thermodynamic equilibrium in the corona.

Because of the perpendicular nature of the heating, and because of the velocity distribution anisotropies for positive ions in the coronal holes, UVCS observations have led to a resurgence of interest in models of coronal ion cyclotron resonance. Wave-particle interactions, such as ion-cyclotron resonance, are considered now as the principal mechanism for heating of coronal holes, and ultimately driving the fast solar wind (Hollweg 2006; Cranmer 2002, 2004). The current understanding is that the solar wind is mainly driven by the pressure of hot protons, so the heating in coronal holes goes more into protons than electrons, because it is conveyed by the ion-cyclotron resonance rather than by currents, which is different from the DC heating models generally applied in the lower corona. By the late 1970s, various data were suggesting the importance of the ion-cyclotron resonance far from the Sun. But we still do not know the exact source of the high-frequency resonant waves(Hollweg, 2006).

Ion cyclotron waves (ICWs) are left-hand circularly polarized waves. They have been observed in a variety of space environments, including those upstream of and those within planetary magnetospheres (e. g. , Russell & Blanco-Cano 2007). These waves are often caused by newly created ions, accelerated by the electric field of a magnetized plasma flowing through a neutral gas from which the ions were produced (Gary, 1991; Huddleston & Johnstone ,1992).

Tomczyk *et al.* (2007) detected Alfven waves by using Coronal Multi-Channel Polarimeter (CoMP) at the National Solar Observatory, New Mexico. Observations showed the existence of upward propagating waves with phase velocity 1-4 × 10^6m/s. They concluded that the waves are too weak to heat the solar corona and added that the unresolved Alfven waves may carry enough energy to heat the corona.

Recent reports have claimed that the Alfv´en waves observed in the low solar atmosphere can provide an energy flux sufficient to heat the corona (De Pontieu, 2007; Jess *et al.* , 2009), but Alfv´en waves, which are linearly polarized waves at a much lower frequency than ion gyrofrequencies, do not directly interact with the core ions. They need an intermediary process to convert this energy flux to a form that can heat the coronal ions efficiently. One possible energy transfer is the production and subsequent damping of ICWs (e. g. , Cranmer,2000, 2004; Cowee *et al.* , 2007; Hollweg, 2008). The parallel wave numbers of the global resonant MHD mode are too low to directly provide ion heating through collisionless damping. At the same time, the global mode is characterized by small perpendicular length scales and thus by relatively strong currents, which can excite the ion cyclotron waves (Markvoskii,2001).

Energy flux density of the ICW is given as (e. g. , Banerjee *et al.* , 1998),

$$F_W = \sqrt{\frac{\rho}{4\pi}}\langle \delta v^2 \rangle B \quad \text{erg cm}^{-2}\,\text{s}^{-1} \tag{2}$$

The wave amplitude at heights 120'' off the solar limb is about, $\langle \delta V^2 \rangle = 2 \times (43.\ 9\ \text{kms}^{-1})^2$. Adopting the values for B = 5 G and $N_e = 4.\ 8 \times 10^{13}\ \text{m}^{-3}$ at r = 1. 25 R_0, Banerjee et al. (1998) found the wave flux density as $F_W = 4.\ 9 \times 10^5$ ergcm^{-2} s^{-1} which is high enough for the ion cyclotron resonance (ICR) process to be a good candidate for heating the coronal hole.

Less understood is the mechanism of the generation of the ion cyclotron waves in coronal holes. Generation of resonant ICW may be possible by stochastic magnetic foot point motions, magnetic reconnections and MHD filamentation instabilities or from MHD turbulent cascade. This latter mechanism is supposed to be the dominant one producing ICW that heat the coronal hole plasma and accelerate the solar wind particles (Cranmer, 2000). It is possible that waves with higher frequencies and wave numbers occur throughout the corona because of a turbulent cascade starting from MHD scales (Hollweg 1986; Hollweg & Johnson 1988).

The ion cyclotron waves are generated by a plasma microinstability that is driven by current fluctuations of lower frequency MHD waves. The current required to excite the instability is consistent with the spatial scales suggested by the observations and reasonably large magnetic field fluctuations. (Markvoskii, 2001& Vinas, Wong, & Klimas ,2000). According to another scenario, ion cyclotron waves are launched at the coronal base by reconnection events (Axford & McKenzie ,1992, Tu & Marsch ,1997).

Although ICWs cannot be remotely observed in the corona, the Solar and Heliospheric Observatory (SOHO) observations of ultraviolet emissions have been used to infer the presence of highly anisotropic heavy-ion distributions with strong mass dependent heating in the corona (e. g. , Kohl J. L. 1998; Cranmer S. R. 1999; Antonucci *et al.* , 2000). The data from STEREO A (2007 July 26–August 2) and STEREO B (2007 July 25–August 1) revealed that 246 ICW events appear discretely in the solar wind with variable durations. Unlike Alfv´en turbulence that often appears in the fast wind, or the whistler waves and mirror-mode waves ICWs were not generated by shocks and they are more often when the interplanetary magnetic field (IMF) is nearly radial(Jian. K,L. *et al.* ,2009).

It is well established that the ion cyclotron waves can provide ion heating and acceleration in good agreement with the observations such as ion temperature anisotropy (Marsch 1991; Kohl J,L. , 1998), faster outflow of heavier ions (Li et al. 1998; Cranmer S,R. , 1999), and higher temperatures of heavy ions compared to protons (Neugebauer 1981, 1992; Kohl J,L. ,1998, 1999). However many discrepancies have to be solved in the scenario of ICW heating mechanism in lower solar corona. Even though this mechanism is found successful in heavy ions , a model for proton cyclotron resonance heating in the lower corona needs to be proposed and verified.

4. Solar wind parameters near 1 AU and their interdependence

Solar wind, usually originating from the lower corona of the Sun, travels outwards through the heliosphere. Beyond the Alfv´en critical layer of about 10 R_0, the solar wind flows with an approximate terminal velocity up to 1AU. The bulk expansion of solar wind continues to accelerate until it is beyond around 8 R_0 and the acceleration is virtually completed by 10R_0 (B¨ohm- Vitense,1989). SOHO and interplanetary scintillation results show that the fast wind reaches its terminal speed by 10 R_0, and has already been accelerated. VLBA and EISCAT measurements show that solar wind velocity reaches a maximum value at about 10R_0 and it attains a terminal velocity at 10R_0 (Harmon *et al.* ,2005). However, the solar wind turbulence plays a major role in the post critical journey of the solar wind.

As the solar wind moves outwards, velocity and temperature remain coherent, whereas density does not (Richardson,1996). Various parameters, such as solar wind velocity, proton density, proton temperature and mean magnetic field, fluctuate in this scenario.

The interdependence among the solar wind parameters, namely solar wind velocity, proton temperature, proton density and mean magnetic field in the solar wind has been explained by a Multiple Linear Regression (MLR) model(Shollykutty John and P. J. Kurian[1],2009) . The model was proposed for the prediction of the solar wind velocity (response variable) based on the explanatory variables proton density, proton temperature and mean magnetic field in the solar wind, collected from the ACE satellite data during January 1998 - May 2006.

Solar wind velocity is the combined effect of the inertial uniform speed along with electron plasma wave velocity which is manifested by Langmuir waves, thermal velocity represented by ion acoustic waves and the Alfv´en velocity due to magnetic field.

$$V_{SW} = V_0 + \sqrt{\frac{4\pi}{k^2 n}\frac{e^2}{m_e} n(t)} + \sqrt{\frac{\gamma}{M}\frac{K}{T_0} T(t)} + \frac{B(t)}{\sqrt{4\pi\rho}} \quad (3)$$

where V_0 is the uniform speed due to inertia, k is the wave vector, n is the number density, B is the mean magnetic field, T is the ion temperature, K is the Boltzmann constant and M is the proton mass. The proposed MLR model is,

$$Y = \beta_0 + \beta_1 X_1 + \beta_2 X_2 + \beta_3 X_3$$

The consolidated data was fitted as,

$$V_{SW}(t) = 354.907 + 0.0216 X_1(t) + 0.00117 X_2(t) - 2.3925 X_3(t) \quad (4)$$

The driving potential in the solar wind can be expressed as ,

$$\phi = \phi_0 \exp(\frac{-x}{\lambda_D}) + \mathrm{K}(Te + \gamma Ti) - (\frac{B^2}{8\pi}) \quad (5)$$

The analysis revealed that the velocity and temperature are coherent in all cases, and the effect of temperature on velocity is also statistically significant. However, the proton density has an adverse effect on the solar wind velocity in major cases for the respective period.

There exists non-linear relationship between solar wind velocity and proton density and the variation is in an inverse manner (Fig 1). Solar wind speed increases with proton temperature and the relation is linear(Fig 2). The relation between solar wind velocity and

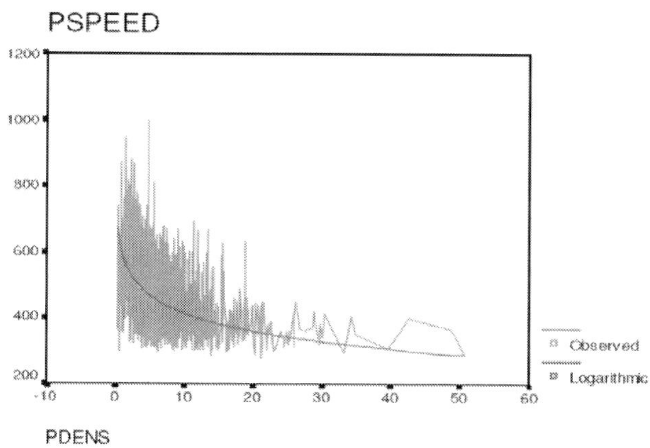

Fig. 1. Variation of solar wind proton speed (velocity) with proton density (for observed and fitted values). The abscissa is in cm-3 and the ordinate is in km s-1.

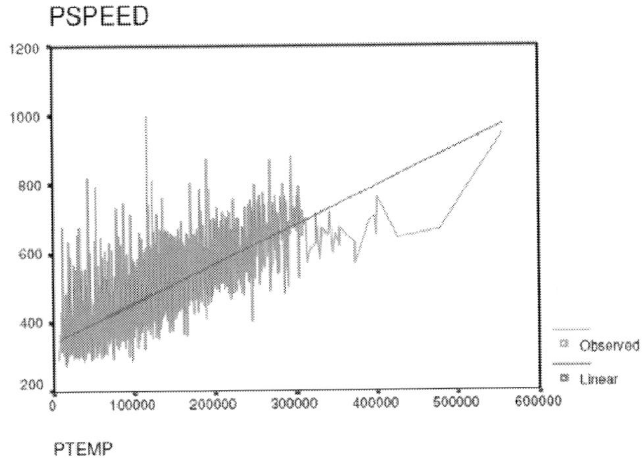

Fig. 2. Variation of solar wind proton speed (velocity) with proton temperature (for observed and fitted values). The abscissa is in Kelvin and ordinate is in km s^{-1}.

magnetic field is also non linear(Fig 3). Hence the study reveals that there is a significant correlation between solar wind velocity with the parameters proton density, proton temperature and average magnetic field in the solar wind.

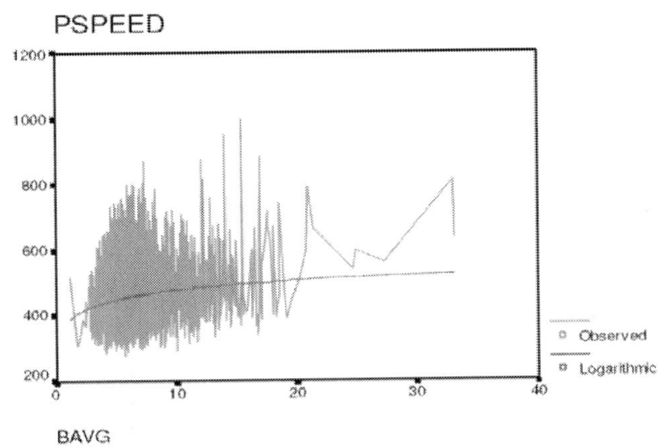

Fig. 3. Variation of solar wind proton speed (velocity) with *B* avg. (for observed and fitted values). The abscissa is in nano Tesla and the ordinate is in km s^{-1}.

Burlaga,L. F. ,(1993) has done a detailed analysis of the solar wind data obtained from various spacecrafts and he found some signatures of chaos (multifractals, intermittence and turbulence) in the solar wind. Buti (1996) showed that the chaotic fields generated in the solar wind can lead to anomalously large plasma heating and acceleration.

Shollykutty John&P. J Kurian[2](2009)proved the existence of deterministic chaos in the solar wind flow by analyzing the solar wind data of daily average values of solar wind velocity, density and temperature from January 1998 to October 2006 from ACE spacecraft measured *in situ* in the heliosphere at 1 AU using techniques of time series analysis. After calculating the natural logarithm of the correlation sum $C_m(\varepsilon)$ vs. ln ε for various embedding dimensions, they plot the slope of the curves for various embedding dimensions as in figures 4 a-c. The slope for which saturation occurs is the correlation dimension D_2 of the attractor. The attractor dimension for the velocity profile was 7. 84 bits/ and the

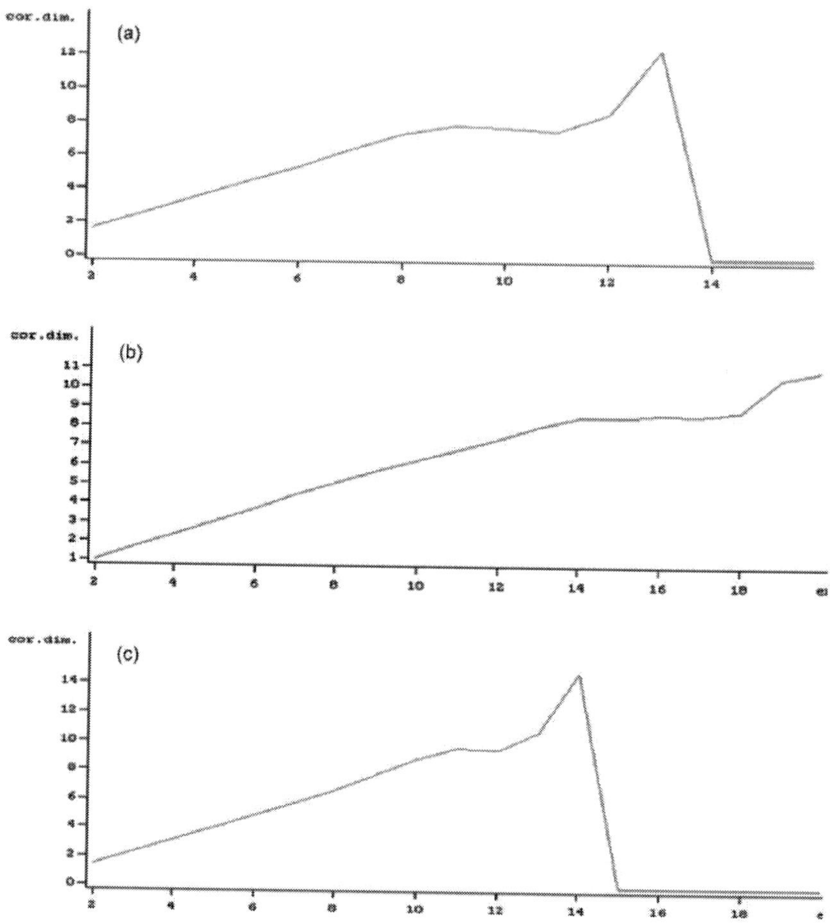

Fig. 4. (a) Graph showing the variations of correlation dimension with embedding dimension for velocity profile. (b) Graph showing the variations of correlation dimension with embedding dimension for density profile. (c) Graph showing the variations of correlation dimension with embedding dimension for temperature profile.

corresponding embedding dimension was 9. A positive value for λ_{max} which is equal to +0. 349 for the velocity data shows chaotic behaviour. The Kolmogorov entropy for this case is 0. 37.

Extending this idea of time series analysis to unfiltered density and temperature profiles to calculate the attractor dimensions, for the density profiles the attractor dimension is obtained as 8. 54 bits for the embedding dimension 14. The calculated LLE and Kolmogorov entropies are 0. 4938 and 0. 55. This shows that the density profiles also is chaotic. In the case of temperature profile, the correlation dimension was 9. 67 bits for an embedding dimension 11. The LLE and Kolmogorov entropy are 0. 403 and 0. 47. These results show that it forms another chaotic attractor.

The chaotic behavior is caused by the superposition of more than two modes of oscillation and is due to strong nonlinear coupling between them. At a distance of 1 AU the terrestrial magnetospheric fluctuations give rise to interaction between solar wind particles and the waves associated with them such as low frequency Alfven waves which leads to nonlinear behavior and chaos.

5. Interaction with Earth's magnetosphere

The Earth has an internal dipole magnetic moment of 8×10^{15} Tm^3 that produces a magnetic field strength at the equator on the Earth's surface of about 30,000 nT, and at 10 Earth radii (R_E) of about 30 nT(Russell,2000). This dipole moment is created by a magnetic dynamo deep inside the Earth in the fluid, electrically conducting core.

The magnetosphere is the region around a planet that is influenced by that planet's magnetic field. Earth's magnetic field is similar in overall structure to the field of a gigantic bar magnet and completely surrounds our planet. The magnetic field lines, run from south to north. Earth's magnetosphere contains two doughnut-shaped zones of high-energy charged particles, one located about 3000 km and the other 20,000 km above Earth's surface. These zones are named as the Van Allen Belts.

The particles that make up the Van Allen belts originate in the solar wind. When electrically charged particles(mainly electrons and protons)from the solar wind enters in to Earth's surface, the magnetic field exerts a force on them and can become trapped by Earth's magnetism causing the particle to spiral around the magnetic field lines and they accumulate into the Van Allen belts.

The positions at which, the field lines intersect the atmosphere, particles from the Van Allen belts often escape from the magnetosphere near Earth's north and south magnetic poles. Their collisions with air molecules create a spectacular light show called an aurora. This colorful display results when atmospheric molecules, excited upon collision with the charged particles, fall back to their ground states and emit visible light. Many different colors are produced because each type of atom or molecule can take one of several possible paths as it returns to its ground state. Aurorae are most brilliant at high latitudes, especially inside the Arctic and Antarctic circles. In the north, the spectacle is called the aurora borealis, or Northern Lights. In the south, it is called the aurora australis, or Southern Lights.

The MHD disturbances of three types propagate in this magnetized solar wind plasma. The fast mode wave compresses the magnetic field and plasma; the intermediate mode wave

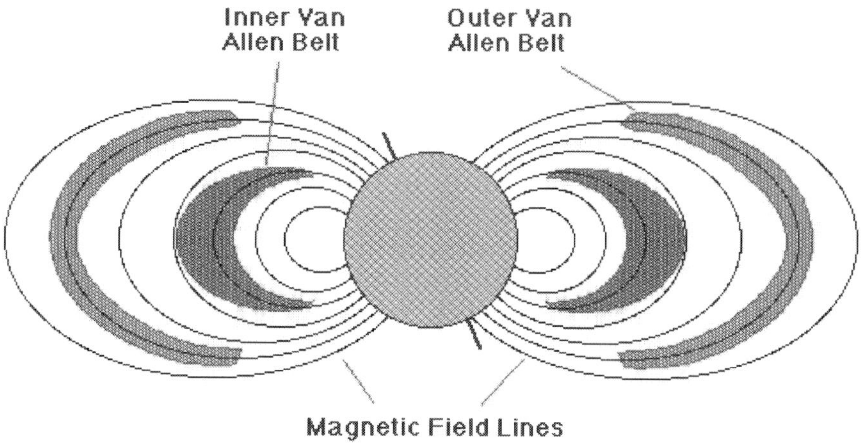

Fig. 5. Van Allen belts

bends the flow and magnetic field, but does not compress it; and the slow mode wave rarefies the field while it compresses the plasma and vice versa.

When the solar wind reaches the earth it causes change in the magnetic field topology, resulting in magnetic merging or magnetic reconnection. Magnetic reconnection at the day side magnetopause is the principal mechanism of energy transfer from the solar wind to the Earth's magnetosphere-this was first proposed by Dungey in 1961. According to Dungey when the frozen-in -condition is relaxed, the field will diffuse relative to the plasma in the magnetopause, allowing the interplanetary and terrestrial field lines to connect through the boundary . This process is termed as magnetic reconnection. The distended loops of open magnetic flux formed by the reconnection exert a magnetic tension force that accelerates the plasma in the boundary north and south away from the site where reconnection takes place, thus causing the open tubes to contract over the magnetopause towards the poles.

The open tubes are then carried downstream by the magneto sheath flow, and stretched in to a long cylindrical tail. Eventually, the open tubes close again by reconnection in the centre of the tail. This process forms distended closed flux tubes on one side of the reconnection cite, which contract back towards the earth and eventually flow to the dayside where the process can repeat. On the other side 'disconnected' field lines accelerate the tail plasma back in to the solar wind.

6. Geomagnetic storms

The Sun is the source of severe space weather. The U. S. National Oceanic and Atmospheric Administration (NOAA), categorizes space weather into three types, which each have their own measurement scales: geomagnetic storms, solar radiation storms, and radio blackouts. Large, violent eruptions of plasma and magnetic fields from the Sun's corona, known as coronal mass ejections (CMEs), are the origin of geomagnetic storms (National Academy of Sciences [NAS], 2008),while solar radiation storms and radio blackouts are caused by solar

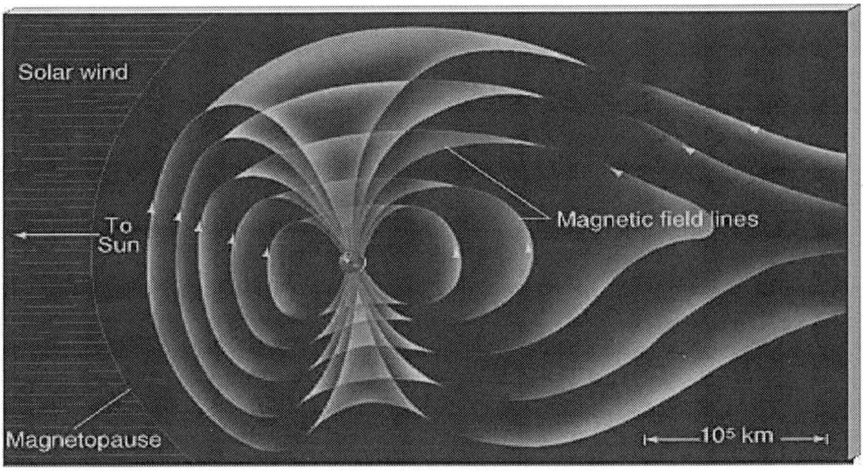

Fig. 6. Distortion in the earth's magnetosphere by the solar wind.

flares. CME shock waves create solar energetic particles (SEPs), which are high-energy particles consisting of electrons and coronal and solar wind ions (mainly protons). When CMEs head towards the Earth, these geomagnetic storms create disturbances that affect the Earth's magnetic field. It takes approximately two to three days after a CME launches from the Sun for a geomagnetic storm to reach Earth and to affect the Earth's geomagnetic field (NERC, 1990).

Geomagnetic storms have the potential to cause damage across the globe with a single event (Schieb,P. A&Gibson,A,2011). In the past, geomagnetic storms have disrupted space-based assets as well as terrestrial assets such as electric power transmission networks. The TEC of the Earth's ionosphere increases during a geomagnetic storm, which increases the density of the ionosphere and leads to signal propagation delays to and from satellites (Gubbins, et al. , 2007). Geomagnetically Induced Currents (GICs)were produced due to the fluctuation in the earth's geomagnetic field caused by the storms. These GICs can flow through power transmission grids (as well as pipelines and undersea cables) and lead to power system problems (Kappenman , 2000). Extra-high-voltage (EHV) transformers and transmission lines—built to increase the reliability of electric power systems in cases of terrestrial hazards—are particularly vulnerable to geomagnetically induced currents (GICs)with recent estimates stating that 300 large EHV transformers would be vulnerable to GICs in the United States (NAS, 2008).

The storm begins when the interplanetary shock wave reached the magnetosphere and compression occurs rather suddenly. After the sudden commencement of the storm there is a period of few hours of calmness. This period is the initial phase. After this the major phase of the storm begins which is called the main phase. The main field strength drops abruptly nearly 50 to 100 gammas below normal. During this period there may be positive and negative fluctuations of short duration. Finally in the recovery phase, the magnetic field strength of the earth returns in a some what irregular manner to a quiescent value. The recovery phase requires one or two days provided no other disturbances occur in the mean time.

There are several scales used to measure the severity of geomagnetic storms. The K and A_k indices are used to categorise the intensity of geomagnetic storms. A severe geomagnetic storm is categorized using K values ranging from 7 to 9 and A_k values ranging from 100-400(Molinski *et al.* ,2000). More severe storms are expressed with higher negative-value D_{st} indices. A severe geomagnetic storm is defined as any event with a D_{st} of less than -500 nanoTeslas (nT). In addition, geomagnetic storm intensity is frequently described in terms of positive nanoTeslas per minute (nTs/min). The 2003 geomagnetic storm 1 peaked at -410nT. No recorded geomagnetic storm since 1932 has exceeded -760 nT (Cliver and Svalgaard, 2004).

Countries located in northern latitudes, such as Canada, the United States, and the Scandinavian nations, are extremely vulnerable to geomagnetic storms. Power systems located in the northern regions of the North American continent are extremely vulnerable because of their proximity to the Earth's magnetic north pole (Kappenman *et al.* , 1990). Although higher geographic latitudes are more susceptible to geomagnetic storm activity than lower regions, damage from GICs have been witnessed in countries in lower latitudes, such as South Africa (Koen and Guant, no date) and Japan (Thomson , 2009). In addition to mapping out regions based on geological conductivity to predict GIC distribution, a more influential factor on GICs involves changes in the Earth's magnetic field (Thomson,2009). Together with ground conductivity, these magnetic field changes can generate electric fields which move GICs throughout electrical grids (Kappenman , 2000). GICs also are driven by currents from the earth's magnetosphere and ionosphere.

Because at all latitudes GIC movements are strongly correlated with the rate of change over time of the Earth's magnetic field, the only way to anticipate GIC movements would be to predict magnetic field movements, but predicting changes in the magnetic field is presently very difficult to do (Thomson,2009). In 2009,twin NASA spacecraft have provided scientists with their first view of the speed, trajectory, and three-dimensional shape of coronal mass ejections, or CMEs. This new capability will dramatically enhance scientists' ability to predict if and how these solar tsunamis could affect Earth.

7. Summary and conclusions

The solar corona consists of tenuous plasma that is highly structured by the strong magnetic field that finds its origin in the solar interior. As the result of varying boundary conditions in the corona, three basic types of solar wind occur: Fast streams from large coronal holes (CHs); slow streams from small CHs and active regions (ARs), and from the boundary layers of coronal streamers; and the variable transient flows such as coronal mass ejections (CMEs). The steady solar wind consists of two major components: fast, tenuous, and uniform flows from large CHs, and slow, dense, and variable flows from small CHs, often from near the boundaries between closed and open coronal fields. The origin of fast streams seems clear, but the sources of the slow solar wind remain less obvious.

The flow of energy through the solar atmosphere and the heating of the Sun's outer regions are still not fully understood. There is no consensus among researchers about the physical mechanism(s) for coronal heating and for solar wind acceleration, even today. The electron driven model is very sensitive to the temperature since the heat flux varies as $T^{7/2}$ but the wind which is the most stable, is the fastest and fills most of the heliosphere, comes from the coldest regions of the corona. At such a temperature, the thermal conductivity falls short by roughly one order of magnitude of that required to drive even a low-speed wind. Hence the electron driven models were soon recognized to be insufficient to drive the high speed streams.

With heating and wave pressure, the wave-driven models were able to explain the high-speeds and hot protons observed in the fast wind in interplanetary space . These wave-driven models generally succeeded in explaining solar wind data far from the Sun, but they failed close to the Sun. The reason is simply that, close to the Sun, the wave pressure is small compared to other terms in the momentum balance. Because of the perpendicular nature of the heating, and because of the velocity distribution anisotropies for positive ions in the coronal holes, UVCS observations have led to a resurgence of interest in models of coronal ion cyclotron resonance.

Wave-particle interactions, such as ion-cyclotron resonance, are considered now as the principal mechanism for heating of coronal holes, and ultimately driving the fast solar wind. But we still do not know the exact source of the high-frequency resonant waves.

At 1 AU there exists a significant correlation between solar wind velocity with the parameters proton density, proton temperature and average magnetic field in the solar wind. The terrestrial magnetospheric fluctuations give rise to interaction between solar wind particles and the waves associated with them such as low frequency Alfven waves which lead to nonlinear behavior and chaos at this distance.

When the solar wind reaches the earth it causes change in the magnetic field topology, resulting in magnetic merging or magnetic reconnection. The collision of the CME with the Earth excites a geomagnetic storm. As a natural event whose effect causes economic and technological hazards, geomagnetic storms require both domestic and international policy driven actions.

8. References

Alazraki, G. , Couturier, P. 1971, A&A, 13, 380.

Alfv´en, H. 1942, Nature, 150, 405.

Antonucci, E. , Dodero, M. A. , & Giordano, S. 2000, Sol. Phys. , 197, 115.

Arge, C. N. , Harvey, K. L. , Hudson, H. S. & Kahler, S. W. 2003, in: M. Velli, R. Bruno & F. Malara(eds.), Solar Wind Ten, AIP Conf. Proc. , Vol. 679, Melville, New York, USA, p. 202.

Aschwanden,M,J. (2008), J. Astrophys. Astr. 29, 3–16.

Axford, W. I. , & McKenzie J. F. 1992, in COSPAR Colloq. 3, Solar Wind Seven, ed. E. Marsch & R. Schwenn (Oxford: Pergamon).

Banerjee, D. , Teriaca, L. , Doyle, J. G. 1998, A&A, 339, 208.

Banerjee, D. , P´erez-Su´arez, D. , & Doyle, J. G. 2009, A&A, 501, L15.

Barnes, A. , Gazis, P. R. , and Phillips, J. L. ,1995, Geophys. Res. Letters. 22, 3309.

B¨ohm-Vitense, E. 1989, Introduction to stellar astrophysics. Vol. 2, ed.

B¨ohm-Vitense, E. (Cambridge: Cambridge University Press).

Bretherton, F. P. , Garrett, C. J. R. 1969, Proc. Roy. Soc. A, 302, 529.

Burlaga,L. F. , 1991, Geophys. Res. Lett. 18, 1651.

Burlaga,L. F. ,1993, Astrophys. J. 407, 347.
Buti,B. ,1996, Astrophys. Space Sci. 243, 33.
Cowee, M. M. , Winske, D. , Russell, C. T. , & Strangeway, R. J. 2007, Geophys. Res. Lett. , 34, L02113.
Cranmer, S. R. 1999, ApJ, 511, 481
Cranmer, S. R. 2000, ApJ, 532, 1197.
Cranmer, S. R. 2001, Proceedings of the 14th Topical Conference on Radio Frequency Power in Plasmas, May 7–9, Oxnard, California, AIP Press.
Cranmer, S. R. 2002, Space Sci. Revs. , 101, 229.
Cranmer, S. R. 2004, In: Proceedings of the SOHO 15 Workshop - Coronal Heating (eds). Walsh, R. W. , Ireland, J. , Danesy, D. , Fleck,B. , European Space Agency, Paris, p. 154.
Dowdy, J. F. , Jr. , Rabin, D. , & Moore, R. L. ,1986, Solar Physics, 105, 35.
Dungey,J. W. ,1961,Phys. Rev. Lett. , 6, 47.
Dungey, J. W. 1963. , in Geophysics: The Earth's Environment, edited by C. Dewitt, J. Hieblot, and A. Lebeau, pp. 505-550, Gordon and Breach, New York.
De Pontieu, B. 2007, Science, 318, 1574.
Eselevich,V. G. 2009,Cosmic Research, , 47, 95–113.
Gary, S. P. 1991, Space Sci. Rev. , 56, 373
Goodman,M. L. ,2000, Astrophys. J. 533, 501–522.
Guhathakurta, M. , Holzer, T. E. 1994, ApJ, 426, 782.
Guhathakurta, M. , Fisher, R. 1998, ApJ, 499, L215.
Gubbins, David, Emilio Herrero-Bervera, Encyclopedia of Geomagnetism and Paleomagnetism, Springer, The Netherlands, 2007.
Harmon, J. K. , & Coles, W. A. 2005, J. Geophys. Res. , 110, A03101.
Hollweg, J. V. , 1978,Rev. Geophys. Space Phys. 16, 689–720.
Hollweg, J. V. 1986, J. Geophys. Res. , 91, 4111.
Hollweg, J. V. , & Johnson, W. J. 1988, J. Geophys. Res. , 93, 9547.
Hollweg, J. V. 2006, J. Geophys. Res. , 111, A12106, doi:10. 1029/2006JA011917.
Hollweg, J. V. 2008, JA&A, 29, 217.
Huddleston, D. E. , & Johnstone, A. D. 1992, J. Geophys. Res. , 97, 12217.
Jess, D. B. , Mathioudakis, M. , Erd´elyi, R. , Crockett, P. J. , Keenan, F. P. , &Christian, D. J. 2009, Science, 323, 1582.
Jian. K,L. et al. ,(2009). ,Astrophys. J,701,L105.
John,S. K. ,&Kurian,P. J[1]. ,2009,Research in Astron. Astrophys. , 9, 485.
John,S. K. ,&Kurian,P. J[2]. ,2009,Pramana,72,743.
Joseph,L. & P J Kurian (2010) Journal of Physics: Conference Series ,208.
Kappenman, John G. and Vernon D. Albertson (1990), —Bracing for the Geomagnetic Storms: As Solar Activity Moves Toward an 11-Year Peak, Utility Engineers Are Girding for the Effects of Massive Magnetic Disturbances, IEEE Spectrum 1990.
Kappenman, John G. (2000), —Advanced Geomagnetic Storm Forecasting: A Risk Management Tool for Electric Power System Operations, IEEE Transactions On Plasma Science 28:6.
Koen, J. and C. T. Gaunt, Geomagnetically Induced Currents At Mid-Latitudes, Department of Electrical Engineering, University of Cape Town, South Africa.
Kohl, J. L. 1998, ApJ, 501, L127.

Kohl, J. L. 1999, ApJ, 510, L59.

Li, X. , Habbal, S. R. , Kohl, J. L. , & Noci, G. 1998, ApJ, 501, L133.

Liewer, C. P. , Neugebauer, M. & Zurbuchen, T. 2003, in: M. Velli, R. Bruno & F. Malara (eds.),Solar Wind Ten, AIP Conf. Proc. , Vol. 679, Melville,New York, USA, p. 51.

MarkovskiiI,S. A. , 2001, Astrophys J. 557:337È342.

Marsch, E. 1991, in Physics of Inner Heliosphere, vol. 2: Particles, Waves,Turbulence, ed. R. Schwenn & E. Marsch (New York: Springer), 45.

Marsch,E. ,2006,Origin and evolution of the solar wind. Solar Activity and its Magnetic Origin Proceedings IAU Symposium No. 233, V. Bothmer & A. A. Hady, eds.

Narain, U. , and Ulmschneider, P. ,1990, Space Science Reviews 54, 377.

NAS (National Academy of Sciences) (2008), Severe Space Weather Events—Understanding Societal and Economic Impacts Workshop Report, National Academies Press, Washington, D. C.

NAS (2009), Severe Space Weather Events—Understanding Societal and Economic Impacts: A Workshop Report - Extended Summary, National Academies Press, Washington, D. C.

NERC (North American Electric Reliability Corporation) (1990), March 13, 1989 Geomagnetic Disturbance, NERC, Princeton, NJ.

Neugebauer, M. 1981, Fundam. Cosmic Phys. , 7, 131.

Neugebauer, M,1992, in Solar Wind Seven, ed. E. Marsch & R. Schwenn (Tarrytown: Pergamon), 69.

NOAA (National Oceanic and Atmospheric Administration), ―Customer Services, National Weather Service Space Weather Prediction Center, http://www. swpc. noaa. gov/Services/index. html, accessed 20 September 2010.

Nolte, J. T. , Kriger, A. S. , Timothy, A. F. , et al. , 1976, Solar Phys. , vol. 46, pp. 303.

Parker, E. N. , 1958, Astrophys. J. , 128, 664.

Parker, E. N. , 1991,Astrophys. J. 372, 719.

Richardson, J. D. 1996, in Physics of Space Plasmas, ed. T. Chang & J. R. Jasperse (MIT Space Plasma Grouppublications).

Roberts, D. A. , & Miller, J. A. 1998, Geophys. Res. Lett. , 25, 607.

Russell, C. T. , & Blanco-Cano, X. 2007, J. Atmos. Sol. -Terr. Phys. , 69, 1723.

Russell, C. T. ,2000,Plasma Science,28(6), 1818 - 1830.

Schieb,P. A & Gibson. P. ,2011, OECD/IFP Futures Project on "Future Global Shocks", CENTRA Technology, Inc. , on behalf of Office of Risk Management and Analysis,United States Department of Homeland Security.

Thomson, A. W. P. (2009), ―Present Day Challenges In Understanding The Geomagnetic Hazard To National Power Grids, Advances in Space Research 45.

Tu, C. -Y. , & Marsch, E. ,1995, Space Science Reviews 73, 1–210.

Tu, C. -Y. , & Marsch, E. 1997, Sol. Phys. , 171, 363.

Tu, C. -Y. , Zhou, C. , Marsch, E. , Xia, L. -D. , Zhao, L. , Wang, J. -X. & Wilhelm, K. 2005, Science,308, 519.

Tu, C. -Y. , Zhou, C. , Marsch, E. , Wilhelm, K. , Zhao, L. , Xia, L. -D. & Wang, J. -X. 2005,Astrophys. J. 624, L133.

Tomczyk, S. , McIntosh, S. W. , Keil, S. L. et al. 2007, Science, 317, 1192.

Vernet,N. M. ,(1999), Eur. J. Phys. 20 , 167–176.

Vidotto,A,A& Pereira,J. V. ,2009,astro-ph. SR.

Vinas, A. F. ,Wong, H. K. , & Klimas, A. J. 2000, ApJ, 528, 509.

Zank, G. P. , Matthaeus, W. H. , Smith, C. W. , and Oughton, S. , 1999,"Heating of the Solar Wind beyond1 AU by Turbulent Dissipation," in Solar Wind Nine, edited by S. R. Habbal et al. , AIP Conference Proceedings 471, New York, 523.

Part 2

The Solar Wind Elemental Composition

3

Solar Wind Composition Associated with the Solar Activity

X. Wang[1], B. Klecker[2] and P. Wurz[3]
[1]*Key Laboratory of Solar Activity, The National Astronomical Observatories, CAS, Chaoyang District, Beijing*
[2]*Max-Planck-Institut für Extraterrestrische, Physik, Garching*
[3]*Physikalisches Institut, Universität Bern, Bern*
[1]*P.R. China*
[2]*Germany*
[3]*Switzerland*

1. Introduction

The studies of elemental abundance of solar wind ions allows to address by open questions in several major fields of research: solar physics, heliospheric and planetary physics, and astrophysics and cosmology. This chapter is intended to provide the current status of knowledge about the solar wind composition mainly in relation to the solar physics with the emphasis on the effects of solar magnetic field.

The composition of the solar wind is mainly determined by the composition of the source material at the solar surface, and then modified by plasma processes in the solar atmosphere, operating in the transition region and in the inner corona. In recent decades, attention in composition studies has shifted from its early models toward differences in chemical fractionation as well as considerable fine-structure in the region above the solar surface. In-situ measurements of the solar wind composition give a unique opportunity to obtain information on the isotopic and elemental abundances of the Sun (e.g. Bochsler, 1998).

The magnetic field on the surface and in the atmosphere of the Sun is considered by many to play a significant role in the plasma processes, which is reflected in composition changes. The magnetic field on the solar surface includes two components: open magnetic flux, which opens into the heliosphere to form the heliospheric magnetic field (also called the interplanetary magnetic field); and closed magnetic flux, in the form of loops attached at both ends to the solar surface. The open magnetic flux controls many of the important processes in the solar corona. Reames (1999) argues that the interaction of loops with open flux is essential for an impulsive solar particle event, i.e., magnetic field reconnection causing the re-distribution of the magnetic field in the loop, the transfer of magnetic energy to the local plasma, and the escape of energetic particles. The interaction and reconnection between open flux and coronal loops releases matter and energy from the closed onto open field lines, which may add to the energetisation of the solar wind. As a result, open flux is broadly distributed on the Sun (Fisk et al. 1999; Fisk 2003; Fisk & Zurbuchen 2006). The open flux also exhibits the reversal in polarity of the magnetic field over the Sun. The polarity of coronal mass ejection (CME) footpoints tends to follow a pattern similar to the Hale cycle of sunspot polarity. Repeated CME eruptions and subsequent reconnection will result in latitudinal transport of open flux and reverse the coronal fields (Owens et al. 2007). Understanding how the open flux of the solar surface behaves, how it is transported and distributed, is important for understanding the heliospheric plasma flow and the interplanetary magnetic field. The distribution of open flux can thus also be a sign of solar activity.

Solar wind charge states are indicative of the coronal electron temperature when assuming local thermodynamic equilibrium between the electrons and ions (Arnaud & Raymond 1992; Bryans et al. 2006). Each charge state pair freezes-in at a different altitude, where the coronal expansion time scale overcomes its ionization/recombination time scale. Recent studies of charge states in fast streams found that the inferred electron temperature given by the *in-situ* observed charge states are higher than those derived from the spectroscopic measurements of the electron temperature. To resolve this discrepancy, two assumptions were adopted: one is assuming a non-Maxwellian velocity distribution for electrons with a suprathermal tail in the near-Sun region (Aellig et al. 1999; Esser & Edgar 2000), the other is introducing the differential flow speed between the adjacent charge states of the same element with the assumption of the Maxwellian velocity distribution for the solar wind particles (Ko et al. 1997; Esser & Edgar 2001). However, Chen et al. (2003) found that the differential flow speed has no significant impact on the charge state distribution of most of the heavy ions. The only way to resolve this issue is to introduce a non-Maxwellian electron velocity distribution. As we know, the fast solar wind is accelerated by ion cyclotron waves possibly generated by the interaction and reconnection between open flux and small scale closed loops. Once ions are perpendicularly heated by ion cyclotron waves and execute large gyro-orbits, density gradients in the flow can excite lower hybrid waves through which electrons can then be heated in the parallel direction (Laming & Lepri 2007). A weak temperature gradient can lead to the development of non-Maxwellian suprathermal tails on electron velocity distributions, invalidating the Spitzer-Harm theory (Scudder & Olbert 1983). Therefore, solar magnetic field fluctuations might be a reason to cause the non-Maxwellian velocity distribution of electrons in the fast solar wind. But things would be different in the slow solar wind because slow solar wind plasma is believed to accumulate in closed loops for hours to days before being set free into the heliosphere. The ions in the loops execute sufficient heating and their velocity distributions become almost isotropic. The conditions for exciting lower hybrid waves might not be satisfied because of the isotropic velocity distribution. Thus the charge state distributions in the slow solar wind would display different signatures from that in fast solar wind. In Section 2, we check the signatures of the solar magnetic activity on the charge states of heavy ions (Fe, Si, Mg, Ne, O, C) in the slow and fast solar wind using *ACE* solar wind data, the "current sheet source surface" (CSSS) model of the corona, and *SOHO* MDI data during the 23rd solar cycle.

It is well established that the relative abundances of elements in the solar corona, solar wind, and solar energetic particles are similar (Meyer 1985). However, when compared to the photosphere, the particle populations do show an elemental fractionation that is organized by the first ionization potential (FIP). This elemental fractionation, the so called FIP effect is widely discussed in the literature, and there exist several models that give account of it, for example Marsch et al. (1995), Arge & Mullan (1998), Schwadron et al. (1999), and Laming (2004). These models even differ in terms of the underlying physical processes (Hénoux 1998), and at present there is no consensus about the actual mechanism responsible for the FIP fractionation happening in the solar atmosphere. Moreover, when looking in detail at the various particle populations they do show some differences in different domains, as demonstrated for Ca (Wurz et al. 2003a) and later for several other heavy elements (Giammanco et al. ApJ 2007). In particular, energetic particles often show severe fractionation in their elemental composition. On the other hand, low-speed CMEs are the most difficult to identify within the ambient low-speed solar wind (Reisenfeld et al. 2003). In many aspects, CME-related flow reflects most closely the low speed solar wind, except for the general enhancement of He. It is still an open question whether the CME-related solar wind needs to be considered as an independent type of flow, or whether a low-speed solar wind reflects a composition, produced by a multitude of small-scale CMEs, that dissolve in the inner corona to form a plasma stream, while only the large-scale CMEs preserve their plasma signatures out to distances of in-situ spacecraft observations and remain identifiable as independent events (Bochsler 2007). For example, the elemental composition during the passage of the January 6, 1997 CME was found to be different from the interstream, i.e., slow solar wind, and from coronal hole, i.e., fast solar wind, observed before and after it, respectively, with a mass-dependent element fractionation (Wurz et al. 1998). Moreover, in a study of CME plasma composition in the velocity range from 390 to 520 km/s a strong deviation in composition of heavy ions with respect to slow solar wind was found for each CME (Wurz et al. 2003b). Schwadron et al. (1999) demonstrated that wave heating can account for both a mass-independent fractionation and a bias of low-FIP elements in active region loops but not on continuously open field lines. Their model placed several constrains on the rate at which waves isotropicalize and thereby heat the species distributions, e.g. the spatial scale of the discontinuity must be smaller than the gyroradius of protons.

In the following we will examine what is the role of the solar magnetic fields and their temporal evolution on the FIP fractionation? In examining the physical processes that can account for the mass-dependent element fractionation, we will focus on the waves associated with an-isotropic ion distributions that would act more effectively on heavy ions than on protons. In the following we will include both quasi-stationary and intermittent solar wind and give the attention to the abundance variations associated with the solar magnetic effects.

The evolution of the open magnetic flux on the solar surface reflects the changing properties of the solar activity. The solar wind associated with the distribution of open flux is expected also to contain the information of solar activity. One fact is that large abundance enhancements are observed in the solar corona over open magnetic field structures such as polar plumes (Feldman 1992; Sheeley 1996), and in active regions surrounding a sunspot with diverging magnetic field lines (Doschek 1983). Another factor is based on the recent models on impulsive events, a significant fraction of heavy elements that reside on the actively flaring flux rope is energized, the resonant interaction operates mainly on heavy elements with charge states increasing systematically with energy (Möbius et al. 2003; Klecker et al. 2006; Kartavykh et al. 2007). We are therefore motivated to explore the effect of solar magnetic activity on the abundance variations of heavy ions in the solar wind plasma.

In Section 3, we compare the elemental abundance ratios Fe/O, Si/O, Mg/O, Ne/O, and C/O over the 23rd solar cycle, where attention is being given to the varying fraction of open magnetic flux on the visible side of the solar disc.

2. Effects of solar magnetic activity on the charge states of minor ions of solar wind

We extrapolate the photospheric magnetic field into a global heliospheric field by using the Current Sheet Source Surface (CSSS) model (Zhao & Hoeksema 1995). This model uses Bogdan and Low's solution (Bogdan & Low 1986) for a magnetostatic equilibrium to calculate the effect of large-scale horizontal currents flowing in the inner corona and, by introducing the cusp surface and the source surface, uses Schatten's technique (Schatten 1971) to calculate the effects of the coronal and heliospheric current sheets and volume currents. These currents maintain the total pressure balance between regions of high and low plasma density. To model the effects of volume and sheet currents on the coronal magnetic field, we divide the solar atmosphere into three parts, separated by two spherical surfaces, i.e., the cusp surface and the source surface (see Figure 1 of Zhao & Hoeksema 1995). The inner sphere, called the cusp surface, is located approximately at the height of the cusp points of coronal streamers. Above the cusp surface the coronal magnetic field is open everywhere. The outer sphere, called the source surface, is located near the reference height identified in the Parker's model above which the radially directed solar wind totally controls the magnetic field (Parker 1958). The cusp point is not easily defined and probably varies from place to place. However, the estimates of the height of cusp points from different experiments range from below 1.5 solar radii to above 3 solar radii (Zhao & Hoeksema 1995). For instance, typical coronal streamers in the K-corona are approximately radial structures extending beyond 1.5 - 2.0 solar radii. This implies a wide height range for the interactions of open and close field lines to happen. Recently, Laming and Lepri (2007) pointed out that any heating mechanism for electrons between 1.5 and 3 solar radii needs to explain the discrepancy between the SUMMER measurements of coronal electron temperatures and electron temperatures derived from Ulysses/SWICS charge state data of heavy ions in the fast solar wind. If most of the cusp points are located within 1.5 - 3 solar radii, it implies that waves generated by reconnections of open and closed magnetic flux will further power the solar wind beyond the point where it can be observed by SUMER, i.e., ions are heated by ion cyclotron resonant Alfvén waves and part of the ion energy then leaks to electrons through a collisionless process (Laming 2004). This would provide a reasonable explanation to the charge state issue in the fast solar wind (Laming & Lepri 2007).

Alternatively, the potential field-source surface (PFSS) model can be used to model the coronal magnetic field from the observed photospheric magnetic field. The difference between the models is that the PFSS model is without currents and the CSSS model is with current sheet-currents. There are two essential advantages of the CSSS model over the potential field-source (PFSS) model: first, in the CSSS model the field lines are open but not necessarily radial at the cusp surface which includes the effects of streamer current sheets; second, the source surface in the CSSS model is placed near the Alfvén critical point. *In-situ* observations of the heliospheric magnetic field should be compared with the magnetic neutral line near the Alfvén critical point. The radial component of the heliospheric magnetic field is latitude-independent, as detected by Ulysses (Smith & Balogh 1995), and can be taken as uniform on a spherical surface above 5 solar radii (Suess & Smith 1996). However, the magnetic field distribution on the source surface obtained using the PFSS model is not uniform, which does not agree with the Ulysses observations of the heliospheric magnetic field (Poduval & Zhao 2004). In addition, the CSSS model shows better prediction of solar wind and interplanetary magnetic field (IMF) polarity and intensity measured near the Earth's orbit than the PFSS model. The correlation coefficient between the observed interplanetary magnetic field and the calculated 27-day averages is 0.89 for the CSSS model, which is better than that of the PFSS model, which is 0.77 (Zhao & Hoeksema 1995). To obtain a uniform magnetic field on the source surface, we set the source surface at 15 solar radii (Zhao, Hoeksema, & Rich 2002) and the optimum cusp surface is determined by matching trial calculations of Carrington rotation-averaged open magnetic flux with *in-situ* solar wind speed.

SoHO/MDI daily magnetic field synoptic data are used to obtain the daily proportion of open magnetic flux on the front side (i.e., the Earth-ward side) of the photosphere (Fig. 1a) during the 23rd solar cycle. We define the open magnetic flux fraction *alpha = opnf/(opnf+clsf)*,

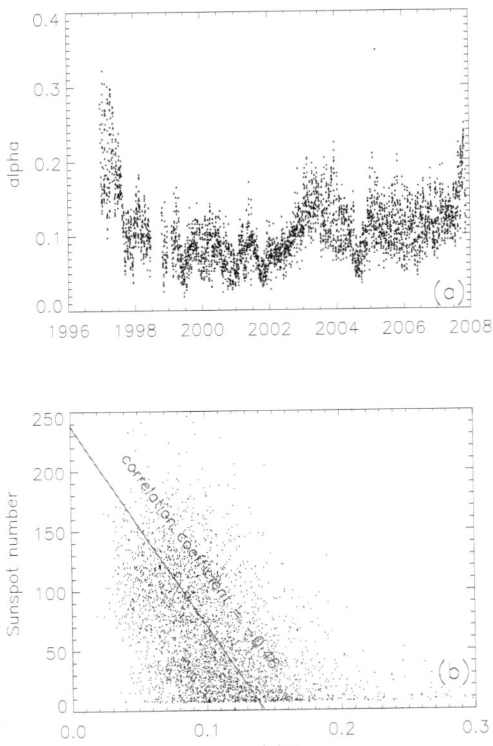

Fig. 1. (a) The fraction of the open magnetic flux on the front side of photosphere - *alpha* (for the definition of *alpha* see main body of the text) as a function of time during solar cycle 23rd. (b) The correlation of the daily fraction of open flux with the daily sunspot number.

where *opnf* indicates open magnetic flux and *clsf* close magnetic flux, respectively, on the front side of the photosphere. The correlation of the open magnetic flux fraction *alpha* with the sunspot number is displayed in Fig. 1b. Thus, we can use the parameter *alpha* to describe the intensity of solar activity in this paper, where a small *alpha* corresponds to a high solar activity and vice versa..

The observed solar wind is traced back to the source surface in the corona along the Archimedian spiral assuming little radial acceleration (constant speed) and pure radial flow, neglecting interaction between fast and slow solar wind streams. That is, the heliographic latitude at the source surface is the same as that of *in-situ* observed point. The Carrington longitude at the source surface is shifted to the west according to the daily values of *in-situ* observed solar wind speed (Neugebauer et al. 2002). Here ACE/SWEPAM daily average solar wind data are used to infer the longitude shift. Once the shifted longitude is obtained, we get the time of day when the solar wind comes out of the source surface, and we further get the daily fraction of open magnetic flux on the front side of photosphere of that day. By this mapping technique, we can associate the *in-situ* observed solar wind with low, middle, and high solar magnetic activity, i.e., $alpha>0.14$, $0.075<alpha<0.14$, *and* $alpha<0.075$, respectively.

We do not use the inverse relation between flux tube expansion factor and solar wind speed, because the solar wind will be traced back to a region bounded within a narrow range of longitude that sensitively depends on the mapping speed used, but rather emphasize the influence of the background magnetic field on the properties of solar wind, i.e., the fraction of the open flux on the front side of the photosphere.

Note that the solar wind speed profile obtained at 1 AU is the result of the interaction between solar wind streams of different speeds as they propagate outwards since the stream-stream interactions are inevitable. Our constant speed assumption would inevitably introduce the longitude shift error and cause a possible error in the mapping of solar wind with solar activity. However, we use the daily synoptic MDI data for the mapping that weakens the influence of the longitude shift error to some degree. For instance, a longitude error of less than 13 degree would not change the mapping result.

We analyzed ACE/SWICS charge state distribution data of heavy ions, from Fe, Si, Mg, Ne, O, to C, from the years 1998 to 2007. We used the criterion $O^{+7}/O^{+6} < 0.8$ to separate interplanetary coronal mass ejections (ICMEs) from the quasi-stationary solar wind, which is based on the results by Richardson and Cane (2004). The corresponding charge state distributions of Fe and Mg of the solar wind for six different speed ranges are compiled in Fig. 2. The wide range of charge states of the measured distribution is due to a mix of sources in the solar wind. As the solar wind speed is increasing, two opposite trends are identified: for iron, a charge state peak shifts from $Q=9$ to $Q=10$ with a tail extending to $Q=20$; for magnesium, the charge state peak shifts from $Q=10$ to $Q=9$ with a tail extending to $Q=5$. Opposite trends for Fe and Mg may be due to the effects of resonant acceleration at high altitude in the corona, where the magnetic effects dominate and preferentially accelerate species with lower charge state (that is with higher m/q) in the slow solar wind. The slow solar wind plasma is believed to accumulate in closed loops in the solar atmosphere for hours to days before being released into the heliosphere by magnetic field reconnection of closed with open flux. The heating will occur in the loops of the corona by the interaction of ions with MHD turbulence, i.e., magnetic fluctuations. If ions are heated

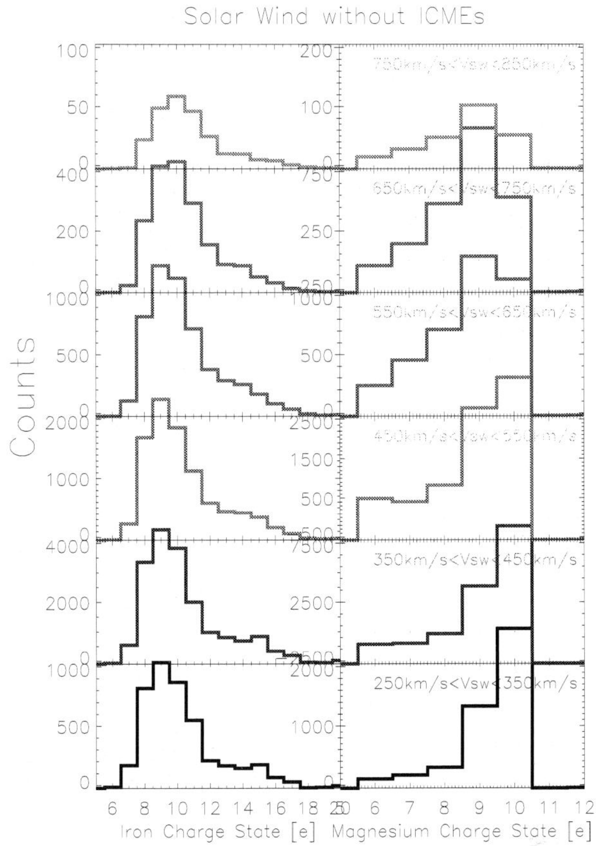

Fig. 2. Charge state distributions of Fe and Mg of the solar wind for six solar wind speed bins ranging from 250 km/s to 850 km/s during solar cycle 23rd.

by magnetic fluctuations, which have a power-law spectrum $P(\omega) \propto \omega^{-\lambda}$, the ions with higher *m/q* will be heated more strongly. Therefore, we see lower charge states of Fe ions in the slow solar wind. However, the *m/q* for Mg ions is nearly half of that for Fe ions, which implies less power at the Mg-resonance. As a result, the resonant heating at high altitudes of the corona is much less efficient for Mg ions than for Fe ions. So the charge distribution of Mg, dominated by Mg^{10+} over a wide range of electron temperatures, still keeps the information of the source temperature of the inner corona, with a different trend for the Fe charge states.

To explore the magnetic effects on the *m/q* distribution, we calculated the mean charge states within different solar wind speed bins as a function of *m/q*, which are displayed in Fig. 3a. For comparison, the dependence on the first ionization potential (FIP) is given in Fig. 3b. We find that the fractions of the high charge states ($Q_{Fe} > +10$, $Q_{Mg} > +7$, $Q_{Si} > +7$, $Q_{Ne} > +7$, $Q_O > +6$, $Q_C > +5$) increase with the solar wind speed when the *m/q* is above 3; below this

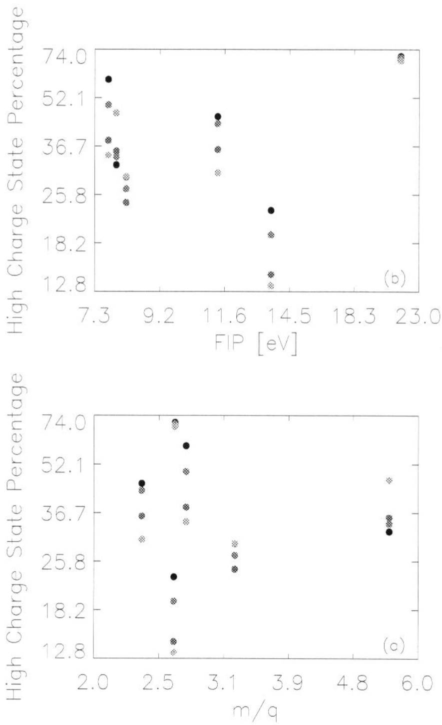

Fig. 3. (a) Mean charge states for four speed bins as a function of m/q (black: 250–350 km/s, green: 350–450 km/s, red: 450–550 km/s, yellow: 650–750 km/s); (b) Mean charge state for four speed bins as a function of FIP.

value, the fractions decrease with solar wind speed. No significant variation is found for Ne (m/q=2.63). The resonant heating of ions by the magnetic fluctuation with a power-law spectrum will preferentially select species with lower charge state (i.e., with higher m/q) in the slow solar wind, which leads to the higher fractions of lower charge states in slow solar wind. On the other hand, ions are perpendicularly heated by ion cyclotron resonant Alfvén waves in the fast solar wind, electrons would be heated as well through the lower hybrid waves excited by the density gradients in the flow (Laming & Lepri 2007). The increased electron temperature then further ionizes the plasma and leads to a higher ionization charge states in the fast streams. Once heated by lower hybrid waves, the electron distributions would depart from a Maxwellian velocity distribution (Laming & Lepri 2007). This is consistent with the theoretical assumption of non-Maxwellian velocity distribution for electrons to solve the discrepancy on coronal electron temperatures associated with in fast streams (Aellig et al. 1999; Esser & Edgar, 2000). In comparing the ionic charge states for different types of solar wind, Ko et al. (1999) found that the charge states of C and O for low latitude fast solar wind (V_{SW} > 500 km/s) are higher than those for south polar fast wind with the speed (V_{SW} > 700 km/s), but the charge states of Si and Fe for low latitude fast wind are lower than those for south polar wind (see Figure 5 of their paper). This result does not violate the m/q-dependent response of the charge states to the magnetic effects in the fast solar wind, although the authors attributed the difference to either lower ion velocity or higher electron density toward lower latitude, rather than the electron temperature. They also compared the charge states for low latitude slow solar wind to those of south polar solar wind. But they did not find lower charge states for Fe in the slow solar wind (see Figure 6 of their paper) as we found in our observations. The discrepancy is likely due to either the latitude-dependence of the charge states in the fast solar wind or the possibility that the solar activity dependence of the charge states is more significant in low latitude than that in high latitude. At least near solar minimum this would be the case. Further investigation of this issue needs a combined data set observed from low latitude to high latitude.

The mean charge states for all the six heavy elements are compiled in Fig. 4, separated into groups of different solar magnetic activity, in which the black, green, red points correspond to the low, middle, high solar activity, respectively, as defined above. The speed intervals correspond to the ones indicated in Fig. 2. The error bar shows the 1-sigma statistical error of the mean charge state. When the solar activity is high, the error bars are large in the fast solar wind. However, for solar wind speeds below 700 km/s, the mean charge states still reveal a significant overall variation of the charge state distributions with solar activity. The yellow points in Fig. 4 correspond to the mean charge states, averaged over all three solar activity cases with a half speed interval of Fig. 2. When the solar wind speed is above 550 km/s, we find that the mean charge states depend significantly on the solar activity. At lower solar wind speeds, no significant solar activity dependence is found. Also, when plotted as a function of solar wind speed, the mean charge states for iron display a trend to increase with solar wind speed, and on the contrary, for magnesium display a trend to decrease with speed. For the other four ions (Si, Ne, O, C) this trend changes from negative speed dependence to positive speed dependence at the point near 675 km/s.

In conclusion, we present several interesting findings in this section: first, we observe a dependence of the charge state distribution of heavy ions with solar activity. This dependence is more important in the fast solar wind than that in the slow solar wind; second, iron is different from other species in that it displays lower charge states in slow wind than in fast wind; third, the fractions of the high charge states for Fe and Si ($Q_{Fe} > +10$, $Q_{Si} > +7$) increase with the solar wind speed, while for the species with lower *m/q*, the fractions of the high charge states ($Q_{Mg} > +7$, $Q_O > +6$, $Q_C > +5$) decrease with the solar wind speed.

3. Solar wind elemental abundances related to the Sun's open magnetic flux

Using the CSSS extrapolation method introduced in Section 2, we analyzed the elemental abundance ratios Fe/O, Mg/O, Si/O, Ne/O, C/O, and He/O as measured by ACE/SWICS, with attention given to the fraction of the open magnetic flux. In Fig. 5 we compiled the 3D plots for the charge states and elemental abundances versus solar wind speed for different *alpha* value. Each data point corresponds to a two-hour average. This survey covers the data from DOY 36, 1998 to DOY 110, 2007 of the ACE mission. The quality flags in the ACE data we used are quality flag = 0 and 1.

Mean abundance ratios and the charge state ratios relative to oxygen for the six solar wind speed bins are compiled in Fig. 6. Considering the ratios in Fig. 6, a systematic

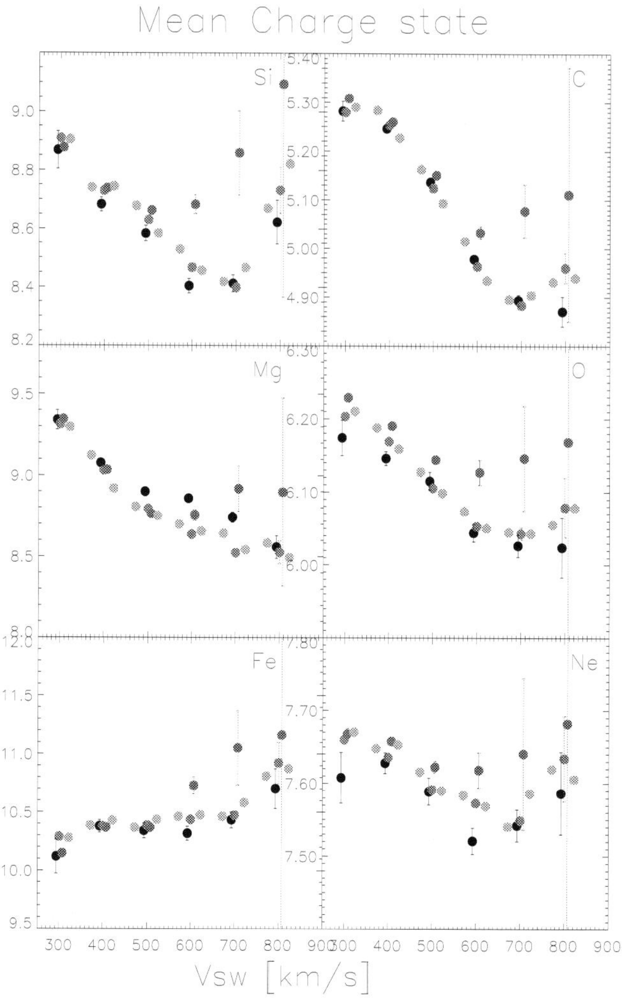

Fig. 4. Mean charge states for Fe, Si, Mg, Ne, O, C, underlying different solar magnetic activity (black - low activity, green - middle activity, red - high activity). Yellow points indicate the mean charge states of all solar wind bins without ICMEs.

instrumental error would only be a second order effect, the shown uncertainties are the statistical 1-sigma error. We find that, for the low-FIP elements Fe and Si, the charge state ratios (Q_O / Q_{Fe}, and Q_O / Q_{Si}) in the high solar magnetic activity bin (*alpha* < 0.07) are lower than the ratios in the higher activity bins when the solar wind velocity is between 550 km/s and 750 km/s, and they tend to decrease with solar wind speed. Although this dependence is not obvious in another low-FIP element Mg, the ratio Q_O / Q_{Mg} still shows some decreases in the high solar magnetic activity bin (*alpha* < 0.07) when the solar wind velocity is between 650

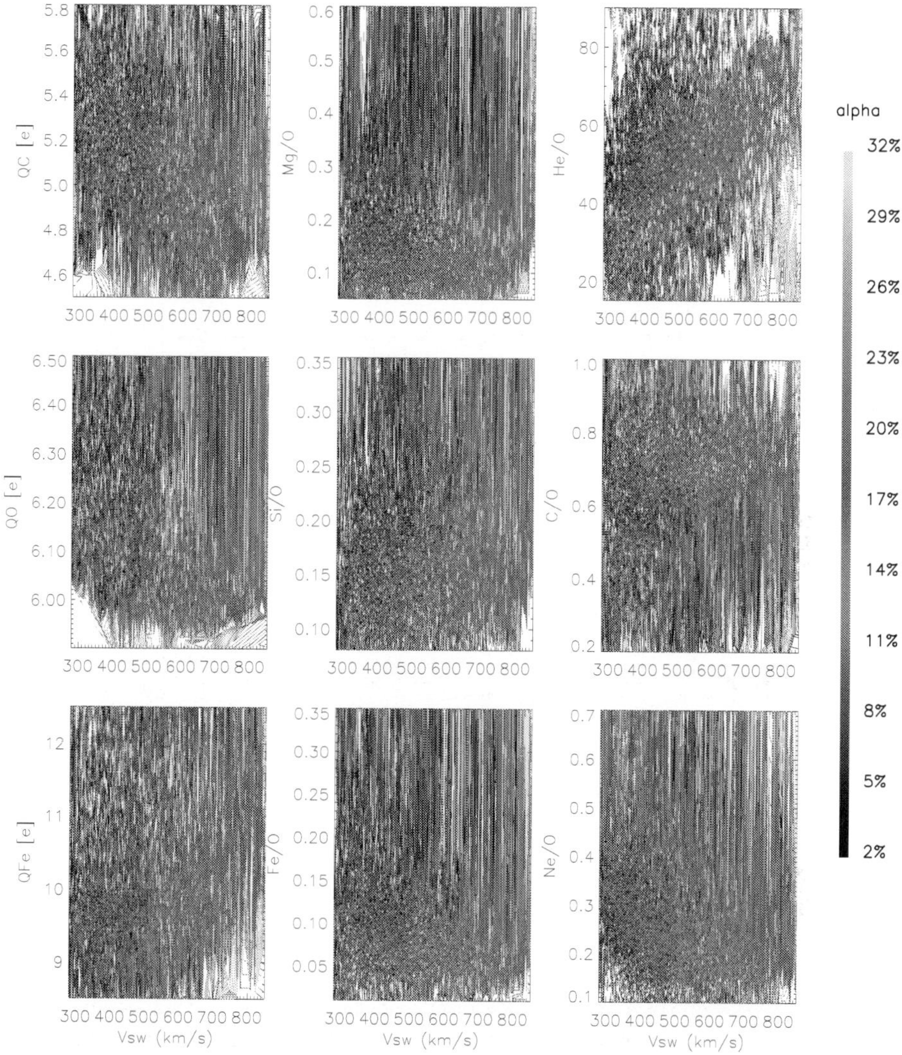

Fig. 5. 3D plots for the charge states and elemental abundance ratios versus solar wind velocity with different *alpha*. (*alpha* represents the fraction of the open magnetic flux on the visible side of the solar disc). Q_{Fe}, Q_O and Q_C represent the charge states of iron, oxygen and carbon, respectively. Each point corresponds to a two-hour average, and the survey covers the ACE mission from DOY 36, 1998 to DOY 110, 2007.

km/s and 750 km/s. For the high-FIP element neon, the ratio Q_O / Q_{Ne} in the the high solar magnetic activity bin (*alpha* < 0.07) is found to be higher than the ratios lower activity bins when the solar wind velocity is between 450 km/s and 650 km/s. For the mid-FIP element

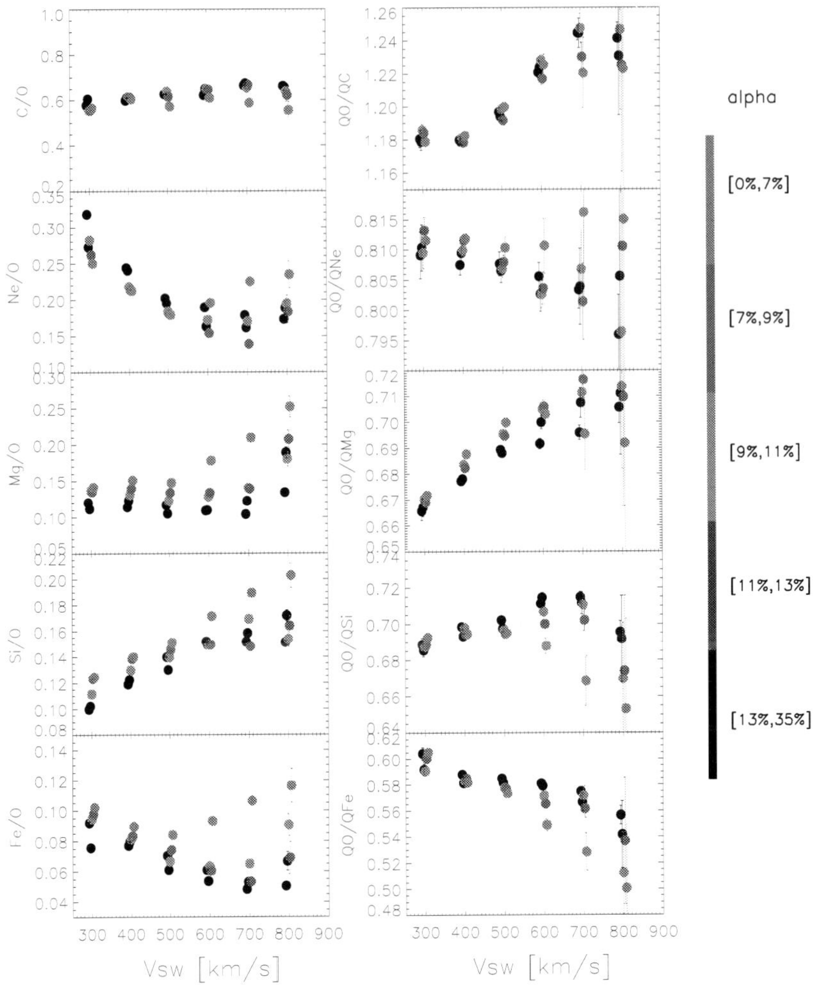

Fig. 6. Elemental abundance ratios and charge state ratios (relative to oxygen) versus solar wind speed for different *alpha* intervals. The error bars in the plot represent the statistical 1 sigma error of the mean.

carbon, the ratio Q_O / Q_C does not show a significant variation between different activity bins.

In summary, the variation of charge state ratios with solar magnetic activity seems to be correlated with FIP. In order to show this correlation more clearly, we compiled in Fig. 7 the variation of charge state ratios in different solar magnetic activity bins as a function of FIP for the velocity bin 650km/s < V_{SW} < 750km/s. The charge ratio, CR, indicates the ratio Q_O / Q_X. It is clear that the ratio CR in the high solar magnetic activity bin (*alpha* < 0.07) to

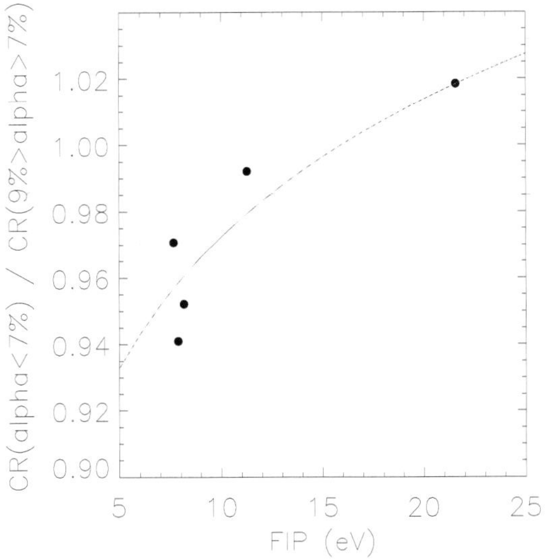

Fig. 7. Variation of charge state ratios in different *alpha* bins as a function of FIP for the velocity bin 650km/s < V_{SW} < 750km/s, where CR indicates the charge state ratio Q_O/Q_X. The axis of ordinate is the ratio of CR in *alpha* < 0.07 bin to CR in 0.07 < *alpha* < 0.09 bin, which increases from 0.94 to 1.02 as FIP is increasing. The red line shows a least-square fit to the data.

CR in the middle solar magnetic activity bin (0.07 < *alpha* < 0.09) bin exhibits a FIP dependence. We can fit this dependence with the empirical function y = 0.847*FIP**0.06, which is marked as red line Fig. 7. This dependence not only points to a non-negligible role of the configuration of background magnetic field in determining the charge states, indicating the importance of wave-particle interactions, it also suggests an important role of solar magnetic effects in the elemental fractionation of the solar wind.

We may note that both the charge state ratios and the abundance ratios in the slow solar wind do not show much variation between different solar magnetic activity bins. In the high solar magnetic activity bin (*alpha* < 0.07), the overall trend of the abundance ratios in the fast solar wind is clear: the abundance ratios rise with the solar wind velocity if the mass is above that of oxygen. When including a 10% to 20% measurement error for the ACE abundance data, the variation of C/O with solar magnetic activity is lost and leads to a constant value of C/O for all solar wind speeds. The ratio Fe/O basically decreases with solar wind speed for low solar magnetic activity (i.e., when *alpha* is relatively high), which agrees with the results of previous work (e.g. Wurz et al. 1999; Aellig et al. 1999). The difference occurs for increasing solar magnetic activity, i.e., *alpha* is decreasing. As we know, CMEs, which are associated with high solar activity (Yashiro et al., 2004), can have large increases in the abundances of heavy elements, especially iron (Wurz et al., 1998, 2003b). These increases in elemental abundance are in addition to the enrichments of the FIP effect, where a factor 4 to 10 in enrichment for iron in the CME plasma is possible in individual events (Wurz et al., 1998; Wurz et al. 2003b). We show the distribution of the three types of solar wind and the fraction of CMEs in the five solar magnetic activity bins and six velocity bins in Fig. 8. For high solar magnetic activity (*alpha* < 0.07), the fraction of the CMEs increases from 1% to 71% as the solar wind speed increases from 300 km/s to 800 km/s. If we make the statistical link between high solar wind speeds with the high fraction of CMEs during active times we can explain the enrichment of heavy ions for the fast solar wind in the low *alpha* bin. Statistically speaking, the correlation between solar activity and CMEs is well established (e.g. Hildner et al. 1976; Kahler 1992; Wang, Y. M. et al. 1998; Zhou et al. 2003; Gopalswamy et al. 2003; Yashiro et al. 2004).

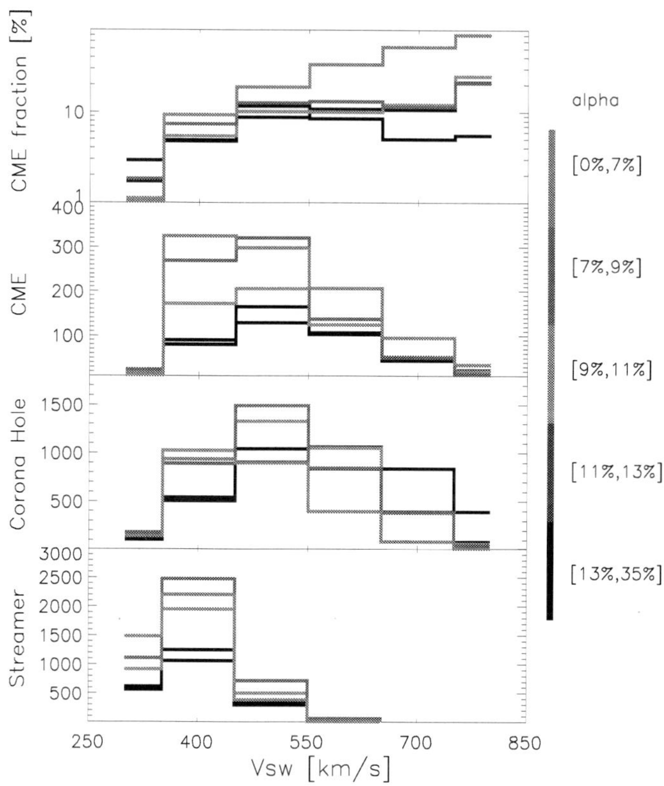

Fig. 8. Distribution of the three types of solar wind and the fraction of CMEs in the five *alpha* bins and six velocity bins.

However, Si/O and Mg/O are different from Fe/O in that their abundance ratios do not decrease with solar wind speed even in the high *alpha* bins, i.e., for low solar activity. We may note that this analysis is a statistical work resulting from the samples including all types of solar wind, and results might be different from for a more restrictive data selection customary in typical abundance studies. The role of the magnetic effect cannot be neglected in considering this difference. Is it possible that the magnetic effect plays a more prominent role for Si and Mg than for Fe in the low solar magnetic activity bins? Our recent statistical study shows that iron displays higher charge states in fast wind than in slow wind (Wang et al. 2008), but for silicon and magnesium ions, the mean charge states basically decrease with the velocity (V_{SW} < 750 km/s) in the low solar magnetic activity bins (Wang et al. 2008). If the low abundance ratio of Fe/O in the fast wind when *alpha* is above 0.07 can be partly attributed to the high charge states of Fe in fast wind, we can understand the absence of a decrease in Si/O and Mg/O for fast wind in the low solar magnetic activity bins (more power resonating with Mg and Si than with O because of their lower charge states). In other words, due to wave-particle interactions, the increased ratio of charge states (Q_O / Q_{Si} or Q_O / Q_{Mg}) would result in a constant abundance ratio (Si/O or Mg/O) in the fast solar wind.

The more prominent role of solar magnetic effects in fast solar wind may be due to the higher electron temperature powered by the MHD waves at the relative high altitude in the solar corona, i.e., at altitudes of 1.5–3 solar radii. Assuming that the fast solar wind is accelerated by ion cyclotron waves generated by the interaction and reconnection between open flux and small-scale closed loops at this height, ions are perpendicularly heated by the ion cyclotron waves and execute large gyro-orbits. Then density gradients in the flow can excite lower hybrid waves through which electrons then can be heated in the parallel direction (Laming & Lepri 2007). Once heated by lower hybrid waves, the increased electron temperature would push the elemental ionisation to higher charge states. But things would be different in slow solar wind because the slow solar wind is believed to accumulate in closed loops for hours to days before it is released into the heliosphere (e.g. Uzzo et al. 2003). The ions experience sufficient heating in the loops and their velocity distributions become almost isotropic. The conditions for exciting lower hybrid waves might not be satisfied because of the disappearance of an-isotropic velocity distributions. Therefore, the charge states and the abundance ratios in slow and fast solar wind respond differently to the solar magnetic effects.

We also study helium, another high-FIP element. Helium does not show the similar trend as neon because He cannot be regarded as minor ion in the solar wind plasma and its fractionation may be associated with the bulk solar wind acceleration (see discussion in Aellig et al. 2001; Kasper et al. 2007).

In Fig. 9 we give an overview of the measured solar wind abundances underlying different solar magnetic activity bins. For comparison, we plotted the elemental abundances as a function of both FIP (right column) and mass (left column). The observed solar wind abundance ratios (relative to oxygen), compared to the photospheric abundance ratios, show the well-known FIP pattern. The abundance ratios as function of FIP can be grouped in three plateaus: a low-FIP plateau for elements with FIPs < 10 eV which are the elements Mg, Fe, and Si showing the largest enrichments, separated by a step at about 10 eV until about 20 eV for the relative high-FIP elements including C and O with enrichments around 1. We mark the high-FIP element neon separately because it behaves different from C and O. The high-FIP element He often shows an abundance ratio of 0.5 (Aellig et al., 2001), i.e., it is depleted with respect to its photospheric abundance. The error bar marked on the low-FIP plateau represents the standard deviation of the average height of the low-FIP element group (Fe, Si, Mg).

In contrast to low-FIP (Mg, Fe, Si) and high-FIP elements (Ne, He), no obvious variations with solar wind speed are found for the two high-FIP elements carbon and oxygen (oxygen

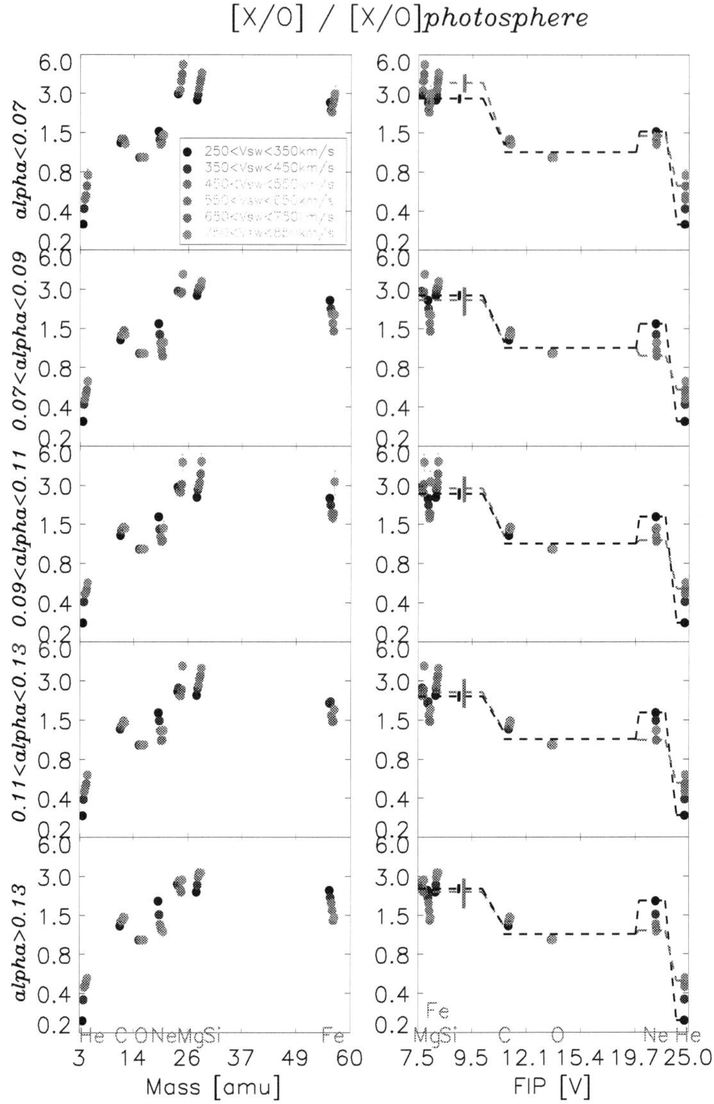

Fig. 9. Elemental abundance ratios (relative to oxygen) compared to photospheric ratios with different *alpha*, plotted as a function of FIP and mass. The FIP fractionation pattern basically shows three plateaus separated by a step at 10 eV between the low-FIP elements (Mg, Fe, Si) and mid-FIP elements (C and O), Ne is marked separately from C and O because of its different behaviour, another step is observed between the high-FIP element Ne and He. The error bar marked on the low-FIP plateau represents the standard deviation of the average height of the low-FIP element group (Fe, Si, Mg).

is used as reference). Carbon is slightly enriched relative to O, but this enrichment does not change significantly with the solar wind speed (see also Fig. 6). Therefore, the mid-FIP element plateau is plotted as a single plateau height that does not distinguish the fast wind from the slow wind. The average plateau height of low-FIP element group (Fe, Si, Mg) shows visible enhancement in the fast solar wind for high solar magnetic activity (*alpha* < 0.07), whereas it has no obvious variation in slow solar wind in all *alpha* bins. This suggests that mechanisms enhancing the upward transport of low-FIP elements should work equally well in low-speed CMEs and the ambient low-speed solar wind. The average enrichment factor for the low-FIP group (e.g. Mg, Fe, Si) relative to photospheric values is between 1.64 and 2.47 for low solar magnetic activity (*alpha* > 0.14), and it ranges from 2.78 to 4.15 when *alpha* is below 0.07 (large solar activity). For the high-FIP element Ne, the plateau height for the low-speed solar wind is always higher than that for high-speed solar wind until *alpha* is below 0.07, where they are close to each other. The plateau heights for Ne between different velocity bins can vary by as much as a factor of two. This is in contrast to the result in quasi-stationary solar wind, which is found to be constant and independent of solar wind type (e.g. von Steiger 1995; Geiss et al. 1994). But it agrees with the spectroscopically determined Ne/O ratios in active regions and flares (McKenzie & Feldman 1992) and it essentially agrees with the photospheric value (Anders & Grevesse 1989). In 1992, Widing and Feldman demonstrated that the ratio Mg/Ne in their spectroscopical observations can vary by as much as an order of magnitude, and its variation is correlated with the magnetic field morphology at the observed site on the solar surface. The largest Mg/Ne values are found in polar plumes and diverging magnetic fields (Widing & Feldman 1992). Although their report induced some arguments on effects of electron temperature or FIP (e.g. Doschek & Laming 2000), in view of our finding that the dependence of the charge state ratios with solar magnetic activity is correlated with FIP, it is quite possible that the effects of electron temperature and FIP effects are supplementing each other for the elemental fractionation, for example, through redistribution of open magnetic flux on the solar surface, and excitation of ion cyclotron waves. Ions that are perpendicularly heated by ion cyclotron waves further excite lower hybrid waves through which electrons are heated (Laming & Lepri 2007). Increased electron temperatures further influence the charge states of ions. Through wave-particle interactions, the modification of charge states would be finally reflected in elemental abundances.

4. Conclusions

In this paper, we present the influence of solar magnetic field activity on the charge state distributions and elemental abundances of heavy ions in the solar wind.

First, we observe a dependence of the charge state distribution of heavy ions with solar activity. This dependence is more important in the fast solar wind than that in the slow solar wind.

Second, iron is different from other species in that it displays lower charge states in slow wind than in fast solar wind.

Third, we find that the enrichment of low-FIP elements with respect to the photospheric values is around 2 when the fraction of the open magnetic flux on the visible side of the solar disc is above 0.14. The elemental fractionation of low-FIP elements rises to 2.78–4.15

when the fraction of the open flux is below 0.07, i.e., for high solar activity. These values are close to the enrichments of low-FIP elements with respect to the photospheric values in quasi-stationary fast- and slow solar wind, which suggests that the intrinsic mechanisms for the fractionation of stationary solar wind and intermittent solar wind might be similar.

Fourth, in the high solar magnetic activity (*alpha* < 0.07) case, the enrichment of heavy elements for the fast solar wind is well correlated with the high occurrence rate of CMEs in the solar wind. Both the charge state ratios and the elemental abundance ratios show less dependence on *alpha* in slow solar wind than in fast solar wind.

Fifth, The dependence of the charge state ratios on *alpha* is correlated with FIP, which implies an important role of solar magnetic field activity in the elemental fractionation of the solar wind. The possible linkage between solar magnetic field and elemental fractionation suggests the important role of the lower hybrid waves.

5. Acknowledgements

We thank the ACE SWICS-SWIMS instrument team and ACE Science Center for providing the ACE data. We thank SOHO and in particular the SOI. team at the Stanford University for making SOHO MDI data available. One of the author (Xuyu Wang) thanks Dr. Xuepu Zhao at Stanford University for providing the CSSS model and Dr. Yuzong Zhang in NOAC for help understanding the model. This work is supported by the National Natural Science Foundations of China (41074123), the National Basic Research Program of China (2006CB806303) and Y07024A900.

6. References

Aellig, M. R. et al. 1999, Solar Wind Nine, American Institute of Physics, 255
Aellig, M. R., Lazarus, A.J., & Steinberg, J. T. 2001, Geophys. Res. Lett., 28, 2767
Anders, E., & Grevesse, N. 1989, Geochim. Cosmochim. Acta 53, 197
Arge, C. N., & Mullan, D. J. 1998, Sol. Phys., 182, 293
Arnaud, M., & Raymond, J. 1992, ApJ, 398, 394
Bochsler,P., et al. 1998, ESA SP, 404, 37
Bochsler, P. 2007, Astron. Astrophys. Rev., 14, 1
Bogdan, Y. J., & Low, B. C. 1986, ApJ, 306, 271
Bryans, P., Badnell, N. R., Gorczyca, T. W., Laming, J. M., Mitthumsiri, W., & Savin, D. W. 2006, ApJS, 167, 343
Chen, J. 1996, JGR, 101, 27499
Chen, Y., Esser, R., & Hu Y. 2003, ApJ, 582, 467
Doschek, G. A. 1983, Sol. Phys., 86, 9
Doschek, G. A. & Laming, J. M. 2000, ApJ, 539, L71
Esser, R., & Edgar, R. J. 2000, ApJ, 532, L71
Esser, R., & Edgar, R. J. 2001, ApJ, 563, 1055
Feldman, U. 1992, Phys. Scr., 46, 202
Fisk, L.A., Schwadron, N. A., & Zurbuchen, T. H. 1999, J. Geophys. Res., 104, 19765
Fisk, L. A. 2003, J. Geophys. Res., 108(A4), 1157, doi:10.1029/2002JA009284
Fisk, L. A., & Zurbuchen, T. H. 2006, J. Geophys. Res., 111, A09115, doi: 10.1029/2005JA011575

Geiss, J., et al. 1994, Eos Trans. AGU, 75, #16, 278
Gopalswamy, N. et al. 2003, ApJ, 586, 562
H´enoux, J. C. 1998, Space Sci. Rev., 85, 215
Hildner,E. et al. 1976, Solar Physics, 48, 127
Kahler, S. W. 1992, Ann. Rev. in Astronomy and Astrophysics, 30, 113
Kartavykh, Y.Y., et al. 2007, ApJ, 671, 947
Kasper, J. C., Stevens, M. L., Lazarus, A. J., et al. 2007, ApJ, 660, 901
Klecker, B., et al. 2006, Adv. Space Res. 38, 493
Ko, Y.-K., Fisk, L.A., Geiss, J., Gloeckler, G., & Guhathakurta, M. 1997, Sol. Phys., 171, 345
Ko, Y.-K., Gloeckler, G., Cohen, C. M. S., & Galvin, A. B. 1999, J. Geophys. Res., 104, 17005, doi: 10.1029/1999A900112
Koutchmy, S. 1977, Solar corona, in Illustrated Glossary for Solar and Solar-Terrestrial Physics, edited by
A. Bruzek and C. J. Durrant, p. 39, D. Reidel, Nowell, Mass.
Laming, J. M. 2004, ApJ, 614, 1063
Laming, J. M., & Lepri, S. T. 2007, ApJ, 660, 1642
Levine, R. H. 1982, Sol. Phys. 79, 203
Liewer, P. C., Neugebauer, M., & Zurbuchen, T. H. 2004, Sol. Phys., 223, 209
Liu, Y. & Hayashi, K. 2006, ApJ, 640, 1135
Marsch, E., von Steiger, R., & Bochsler, P. 1995, A&A, 301, 261
McKenzie, J. F. & Feldman, U. 1992, ApJ, 389, 764
Meyer, J.-P., 1985, ApJS, 57, 151
Möbius, E., et al. 2003, Proc.28th Int. Cosmic Ray Conf. (Tsukuba) 6, 3273
Neugebauer, M., Liewer, P. C., Smith, E. J.,et al. 2002, J. Geophys. Res., 107, 1488,
doi:10.1029/2001JA000306
Owens, M. J., Schwadron, N. A., Crooker, N. U., Hughes, W. J., & Spence, H. E. 2007, Geophys. Res. Lett., 34, L06104, doi: 10.1029/2006GL028795
Parker, E. N. 1958, ApJ, 128, 664
Poduval, B., & Zhao, X.-P. 2004, J. Geophys. Res., 109(A8), A08102, doi: 10.1029/2004JA010384
Reames, D. V. 1999, Space Sci. Rev., 90, 413
Reisenfeld, D. B., Gosling, J. T., Forsyth, R. J., et al., 2003, Geophys. Res. Lett, 30(19), 8031
Richardson, I.G., & Cane, H. V. 2004 J. Geophys. Res., 109(A9), A09104, doi: 10.1029/2004JA010598
Schatten, K. H. 1971, Cosmic Electrodyn., 2, 232
Schwadron, N. A., Fisk, L. A., & Zurbuchen, T. H. 1999, ApJ, 521, 859
Scudder, J. D., & Olbert, S. 1983, In JPL Solar Wind Five, 163
Sheeley, N. R., 1996, ApJ, 469, 423
Smith, E. J., & Balogh, A. 1995, Geophys. Res. Lett., 22, 3317
Suess, S. T., & Smith, E. J., 1996, Geophys. Res. Lett., 23, 3267
Uzzo, M.,Ko, Y.-K., Raymond, J. C., et al., 2003, ApJ, 585, 1062
von Steiger, R., Wimmer Schweingruber, R. F., Geiss, J., et al., 1995 Adv. Space Res., 15, 3
Wang, X., Klecker, B., & Wurz, P., 2008, ApJ, 678, L145
Wang, Y. M. et al. 1998, ApJ, 508, 899
Widing, K. G. & Feldman, U. 1992, ApJ, 392, 715
Wurz, P., et al. 1998, Geophys. Res. Lett., 25, 2557

Wurz, P., et al. 1999, Phys. Chem. Earth(C), 24(4), 421
Wurz, P., Bochsler, P., Paquette, J. A., & Ipavich, F. M., 2003a, ApJ, 583, 489
Wurz, P., et al. 2003b, Solar Wind X, American Institute Physics, 679, 685
Yashiro, S., et al., 2004, J. Geophys. Res., 109, Issue A7, CiteID A07105
Zhao, X.-P., & Hoeksema, J. T., 1995, J. Geophys. Res., 100, 19
Zhao, X.-P., Hoeksema, J. T., & Rich, N. B. 2002, Adv. Space Res., 29, 411
Zhou, G.-P., Wang, J. X., & Cao, Z. L. 2003, A&A, 397, 1057

4

Solar Wind and Solar System Matter After Mission Genesis

Kurt Marti[1,*] and Peter Bochsler[2,3]
[1]*Department of Chemistry and Biochemistry, University of California, San Diego, La Jolla, California*
[2]*Physikalisches Institut, University of Bern*
[3]*Space Science Center and Department of Physics, University of New Hampshire, Durham, New Hampshire*
[1,3]*USA*
[2]*Switzerland*

1. Introduction

Elemental abundances in the Sun have been found to be generally consistent with those observed in carbonaceous meteorites (except volatiles), but the available solar system (SS) isotopic abundances are not uniform and studied objects in the inner and outer SS show variations in the isotopic abundances of some elements. Since the presolar cloud environment was dynamically evolving due to stellar additions of nucleosynthetic products and chemical reactions in a partially ionized medium, it is possible that matter was not uniform and not equilibrated at the time of SS formation. The isotopic abundances in the solar atmosphere and in the Sun cannot be accessed accurately by optical methods and it is necessary to rely on solar wind (SW) abundances. Some SW isotopic abundances (N, O and noble gases) have been determined in the form of implanted ions in catcher-foils returned by NASA's Apollo and Genesis missions, and also in some SW-exposed meteoritic matter, as well as in soils returned by lunar missions (see 4.).

The SS today is 4567 Ma old (Amelin et al., 2010) and is considered to have formed in a cluster of stars, embedded within a molecular cloud (Adams,2010). The clouds have lifetimes of several tens of millions of years or less (Hartmann et al., 2001). Within the clouds, the clusters themselves live for tens of millions of years and protoplanetary disks of one solar mass typically have lifetimes of 3 Ma (Williams and Cieza, 2011). Even the long-lived open clusters dynamically evaporate over hundreds of millions of years. As a result, the birth environment of the Sun has long since been dissipated. Nonetheless, the statistics of protoplanetary disk evolutions (Williams and Cieza, 2011) and various properties of our SS (Adams, 2010), coupled with an emerging understanding of star and planet formation processes, allow some visions of the birthplace. A mechanism for the production of crystalline silicates was suggested (Vorobyov, 2011), associated with the formation and destruction of massive fragments in young protostellar disks at radial distances of 50–100

* Corresponding Author

AU, and the annealing of small amorphous grains when the gas temperature exceeds the crystallization threshold of ~800 K.

Isotopic abundance studies in solar matter and in a variety of SS objects not only can provide information on where the elements were synthesized, but the same data are also important for the evaluation of current paradigms of star and planet formation. These issues have evolved over the years and today there exist rather detailed models for the origin of stars and planets. As discussed in Adams (2010) review, the properties of our SS are known in greater detail than those in other systems, and properties in our SS can be used to investigate initial birthplace conditions and to compare these to extrasolar environments that are being observed. The origin of the Oort cloud comets is being debated (Levison et al., 2010) and some models imply that accretion in the inner solar system was a protracted series of exchanges and probably not a single early event. However, even if some Oort cloud comets have been formed outside the solar system, they probably still formed from the same original cloud, and apart from some possible nucleosynthetic anomalies injected from rapidly evolving massive stars, their bulk composition will not be distinguishable from solar system matter. Nevertheless, the possible radial motion of Jupiter and Saturn (Walsh et al.,2011) implies a considerable redistribution of matter in the inner SS and exchanges with the outer parts at later times.

Isotopic composition studies of planetary bodies and meteorites have provided a powerful tool to unravel the history of the solar system. The chemical composition of the Sun with respect to refractory and moderately volatile elements seems well reflected in the so-called primitive meteorites. Whereas a very large body of data has been collected for the isotopic composition of meteorites, our knowledge of the isotopic composition of the Sun remains sketchy. Unfortunately, the isotopic composition of the solar atmosphere and the Sun cannot be accessed by optical methods. The only way to infer the solar isotopic composition is through analysis of solar particles, i.e., through investigations of the isotopic composition of the SW and of solar energetic particles. The Genesis mission of NASA has opened the window for extensive investigations of the SW nuclei and promises to solve some long-standing problems with respect to the isotopic composition of the Sun.

A quick look at the history of research on the elemental and isotopic abundances in SS matter shows that the agreement of solar spectroscopic data, and of abundances in other main sequence stars, with nonvolatile element abundances in carbonaceous chondritic meteorites has been interpreted to reflect a homogeneous SS environment (e.g. Arnold and Suess, 1969). These elemental and isotopic abundances (Suess and Urey, 1956) were used with much success (Burbidge et al., 1957; Cameron, 1957) in models of element synthesis. Although SS matter was taken to represent a well-mixed reservoir of nuclei that originated from several independent stellar sources, the research on isotopic abundances in meteorites from various sources and their precursor materials has shown that isotopic compositions of SS matter are non-uniform, and that this is especially true for the abundant light elements oxygen (O) and nitrogen (N). The recently determined isotopic abundances in the SW collected by Genesis (McKeegan et al., 2011; Marty et al., 2011) show that isotopic differences extend all the way to the center of our SS, the Sun itself. We will discuss the isotopic data for the Sun and some proposed models that may or may not account for the abundances observed in inner and outer SS matter and in planets. The isotopic make-ups of the heavy elements reflect largely differences in the mixing ratios of major nucleosynthetic components, but the presence of several minor anomalous isotopic signatures in meteorites raises questions of origin, degree of homogenization of SS matter, and the timing of accretion in the nebular environment which gave birth to the Sun and the SS. There are also suggestions that late events have modified the SS objects, such as during a late bombardment (e.g. Willbold et al., 2011). It has become increasingly clear that a solar system reference standard is required for measured isotopic abundances. The preferred selection of solar data for this purpose appears reasonable, when we consider that the Sun accounts for 99.8% of matter in the solar system, but at present we have SW isotopic data for only a few elements and the procedures to convert SW data to solar abundances need improvements. If these efforts are successful, abundances now determined in foils returned by NASA's Genesis mission can be expected to help resolve fundamental questions of how the solar system formed and evolved, and in what type of environment.

2. Isotopic fractionation processes in the Interstellar Medium (ISM) and the SS

Isotopic fractionation processes that are generating variabilities in the isotopic composition of solar system matter essentially originate from the differences in nuclear masses.

2.1 Isotope fractionation in chemical reactions

The existence of a non-vanishing zero-point energy of quantum-physical oscillators leads to energetically lower lying vibration states of the heavy isotope compared to the levels of a light isotope in an otherwise identical molecule. Consequently, the heavier isotope is more strongly bound and the equilibrium of a chemical reaction is somewhat shifted in the direction where the heavy isotope is more strongly bound. Similarly, isotopic shifts are observed in reaction rates. For astrophysical applications it is important to note that, even when a medium is extremely cold and the chemical equilibria among neutral species cannot be achieved due to the sluggish reaction kinetics, strong isotopic fractionation effects are expected to occur in ion-molecule reactions, which can proceed even at very low temperatures.

Considering the formation of solids during the birth of the solar system, physical-chemical processes such as adsorption of gases on grain surfaces, or condensation, sublimation and evaporation, can produce isotope effects, in the sense of heavier isotopes being trapped more readily than light isotopes. Again, generally these effects are considered to be "mass-dependent", i.e., an isotope whose mass differs by two mass units from the principal isotope is approximately twice as much affected as an isotope which differs by only one mass unit.

2.2 Mass-independent isotope fractionation

Strictly speaking, all isotopic fractionation effects discussed in this section are dependent on mass. However, in the literature the term "mass-independent" fractionation has been established for mechanisms, which do not discriminate between isotopes of a given element strictly according to their mass number.

Chemically identical molecules containing different isotopes are susceptible to different electromagnetic radiations, and consequently undergo chemical reactions with different time scales, depending on the flux of the corresponding frequency in the sensitive range.

Photo-dissociation of molecules is often a decisive first step in triggering chemical reactions in an astrophysical medium. In this context, molecules containing oxygen are of particular interest for mass-independent fractionation, because of the large variety of isotopic abundances. Whereas ^{16}O makes 99.76 percent, only 0.20 percent of natural oxygen is ^{18}O, and ^{17}O constitutes only 0.04 percent. While an interstellar molecular cloud or a stellar accretion disk might be opaque to the radiation contributing to the photo-dissociation of a molecule containing ^{16}O, it might be transparent to the equivalent radiation causing the dissociation of molecules containing the minor isotopes. Self-evidently, this can lead to a systematic isotopic discrimination of ^{16}O, but not of ^{17}O and ^{18}O for subsequent chemical reactions. Ca-Al rich minerals in meteorites, the first ones to condense in a hot medium in an intensive radiation field, appear to be depleted in ^{17}O and simultaneously in ^{18}O, compared to ordinary meteoritic material; or putting it inversely: These refractory minerals contain excesses of ^{16}O, indicating "mass-independent" fractionation of oxygen.

2.3 Isotope fractionations in the solar atmosphere, the solar corona, and the solar wind

Bochsler (2000) has investigated possible isotope effects related to secular gravitational settling of elements in the outer convective zone of the Sun. Despite the efficient vertical mixing of the outer convective zone, there is a tendency of heavier species to be slightly enriched at the bottom of the convective zone. This fractionation mechanism was extensively studied by Michaud and Vauclair (1991), and by Turcotte et al. (1998). Whereas the calculations of Bochsler (2000) are based on the modeled elemental depletions reported by Turcotte (1998), and were intended to be order-of-magnitude estimates, Turcotte and Wimmer (2002) used full-fledged solar models to compute the same effect. Although the effects turned out not to be very large, both studies clearly yielded isotopic fractionation factors, which deviate from the mass-dependent rule, in the sense that, ^{17}O was depleted in the outer convective zone relative to ^{16}O not by half the amount of ^{18}O. Similar results were found for all other chemical species containing more than two isotopes. This is due to the fact that, in addition to gravitation, which acts on the mass of a species, also radiation pressure plays a role, which is independent of the mass of a particle, and depending on the relative importance of the two agents, a deviation from strictly mass-dependent fractionation occurs. Fig.1 shows in a three-isotope plot the expected modification, due to gravitational settling, of the O isotopic composition of bulk solar oxygen (star symbol) to O present in the convective outer zone of the Sun. As indicated before, the effects are small, nevertheless they are important and above the detection limit for Genesis isotopic abundance measurements.

In the solar atmosphere, in the acceleration region of the solar wind, apart from the gravitational attraction, other forces come into play, among them wave pressure, Coulomb drag, the electric field due to the local separation of electrons and positively charged particles. Many sophisticated models have been developed to investigate the acceleration of heavy particles in this region. Bodmer and Bochsler (2000) have carried out a detailed study to investigate isotope effects in steady state models of the solar wind. For instance, they provide the $^{18}O/^{16}O$ fractionation effect in a typical coronal hole, and in a typical coronal streamer. These authors succeed in reproducing the typical elemental depletion of He/H as found in both streaming regimes. In the coronal streamer they find a depletion of ^{18}O relative to ^{16}O compared to the source, of 47 permil, and of 17 permil in a coronal hole associated solar wind.

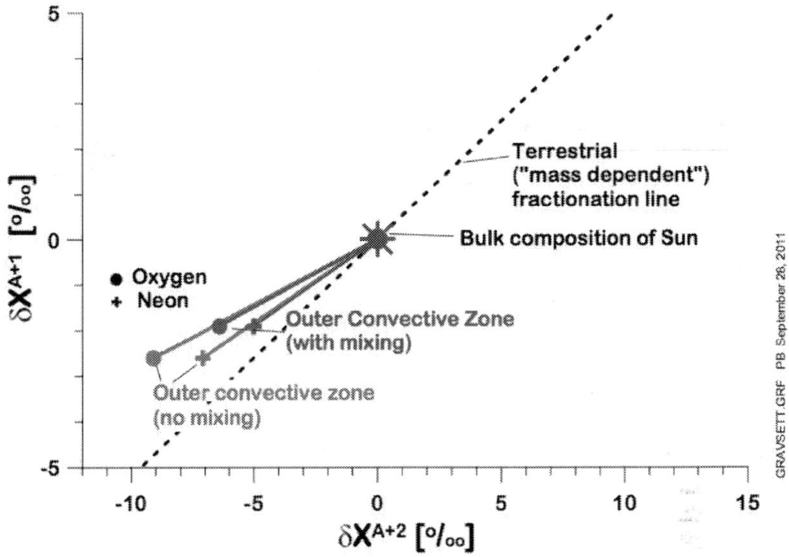

Fig. 1. Illustrates in a three-isotope plot the expected modification of the O isotopic composition, starting with bulk solar oxygen (star symbol) and comparing it to O in the convective outer zone (COZ) of the Sun, due to gravitational settling as modeled by Turcotte and Wimmer-Schweingruber (2002).

^{16}O, the lightest isotope, may be enriched in the SW relative to the heavier isotopes in this non-linear mass-dependent process. In order to predict the concomitant depletion of heavier isotopes, one would have to produce a full-fledged model of the acceleration of the solar wind, which is beyond the goal of the present paper. However, it is possible to make some estimates with the guidance of the paper of Bodmer and Bochsler (2000).

The relevant quantity is the Coulomb drag factor

$$H_x = \frac{2A_x - Z_x - 1}{Z_x^2}\sqrt{\frac{A_x + 1}{A_x}}$$

of a species, where A_x is the atomic number and Z_x is its charge at the relevant location near the coronal temperature maximum. In the case of oxygen $Z \approx 6$.

Following Bodmer and Bochsler (2000), and based on flux conservation after the coronal base, one finds with some algebra the following approximation:

$$\frac{f_{17}}{f_{16}} = \frac{\frac{f_{18}}{f_{16}}(H_{18} - H_{16})}{\frac{f_{18}}{f_{16}}(H_{18} - H_{17}) + (H_{17} - H_{16})}$$

Although, with the quantities given above, the relation between the relative enrichment factors f_{17}/f_{16} and f_{18}/f_{16} is close to $f_{17}/f_{16}=0.5\ f_{18}/f_{16}$, it is clear from the above equation that this is not exactly true. As in the case of gravitational settling in the outer convective zone, several forces are involved in a rather complicated manner. Again, as a consequence of this non-linearity, the isotopic fractionation pattern in an element with three stable isotopes, exhibits some deviation from the simple 'mass-dependent' fractionation (e.g., parallel to a terrestrial fractionation line). An extrapolation of Genesis results to the isotopic composition of the outer convective zone, or even the solar bulk composition needs refined modeling. On the experimental side, since the solar isotopic Mg composition is presumably very close to the average SS composition, a determination of the isotopic composition of Mg in the solar wind in Genesis foils would provide important clues regarding the importance of isotopic fractionation. Unfortunately, in situ measurements of the Mg isotopic composition in the SW have not been possible with the necessary accuracy (e.g. Bochsler et al. 1995, Kallenbach et al. 1998).

3. Review of isotopic signatures in meteoritic matter

3.1 Oxygen isotopes

The relative abundances of the three oxygen isotopes in SS matter vary, in particular the relative abundance of ^{16}O from being like that of the Earth to being depleted or enriched compared to the other two isotopes. One popular explanation for this variation is that dust and gas in the SS began with the same ^{16}O-rich composition, but that the solids evolved towards the terrestrial value. A study of the protosolar silicate dust from refractory CAI (Ca-Al-rich inclusions) concluded that primordial dust and gas differed in O isotopic compositions (Krot et al., 2010). In this interpretation dust had a different history than did gas before being incorporated into the SS. In another study the variable O isotopic compositions in SS matter was evaluated based on a random sampling hypothesis (Ozima et al., 2007). These authors inferred a common mean value of $\Delta^{17}O = 0$ (terrestrial fractionation line, see Fig. 1) for all planetary objects and conclude, that solar O and planetary O should be the identical, but distinct from CAI-oxygen. Further, O in chondrules, the high temperature components of chondrites, which presumably accreted early in SS history, was studied in detail (Libourel and Chaussidon, 2010; Connolly and Huss, 2010). Libourel and Chaussidon (2010) identified metal-bearing Mg-rich olivine aggregates to be among the precursors of Mg-rich chondrules of type I, the origin of which they consider to be condensation–evaporation processes in the nebular gas or in a planetary settings. Their Mg-rich olivines of type I chondrules, as well as isolated olivines from unequilibrated ordinary chondrites and carbonaceous chondrites, revealed the existence of several modes in the distribution of the $\Delta^{17}O$ values and the presence of a large range of mass fractionations. These authors conclude that oxygen isotopic compositions in Mg-rich olivines are unlikely of nebular origin (SS condensates) and suggest that such grains or aggregates might represent mm-sized fragments from disrupted first-generation differentiated objects, but Rudraswami et al. (2011) found no evidence in relict olivine grains to support a planetary origin. Connolly and Huss (2010) found significant variation in the O isotopic compositions in CR2 chondrite type II chondrules with no apparent relation to petrographic or geochemical data. These authors do not rule out that chondrules could have formed in different regions, but prefer an explanation that CR2 type II chondrules document changes in solid and gas compositions during formation. Liu et al. (2009) investigated the O records in the same individual refractory hibonites that were studied for ^{26}Al- and ^{10}Be-records (see 5.2). They found that oxygen isotopic compositions are highly ^{16}O-enriched, but are not derived from a homogeneous reservoir, as $\Delta^{17}O$ values span a range of −28‰ to −15‰.

Isotopic abundances implying extensive ^{16}O isotope fractionation were found in fine-grained matrix of a carbonaceous chondrite (Sakamoto et al. (2007) and these authors suggest that fractionation mechanism of this magnitude are important in tracing the origin of O isotopic variations. They conclude that this matrix was formed by oxidation of Fe-Ni metal and sulfides by heavy water that must have existed in the early SS. Their O data expand the range of observed $^{17,18}O/^{16}O$ variations to from -8% to +18%, relative to the terrestrial standard.

3.2 Nitrogen isotopes

Spectroscopic methods do not permit determinations of the N isotopic composition in the Sun with required accuracy. Most attempts have been indirect, based either on the analysis of the atmospheres of giant planets or of the SW nitrogen trapped in lunar soils. The recent determination of the N isotope ratio in SW, collected in concentrator foils during the Genesis mission (Marty et al. 2011), provides a SW reference. Busemann et al. (2006) found ^{15}N-rich compounds in meteoritic insoluble organic matter (IOM) that exceed enrichments already known to exist in interplanetary dust. Organic matter shows N isotopic variations that survived in meteorites, despite extensive alterations in the SS and on the parent bodies. Known SS objects exhibit values of $^{15}N/^{14}N$ ranging from 1.9 to 5.9 x 10^{-3}, inclusion of meteoritic material increases that range to 22 x 10^{-3}, and N implanted in the lunar surface reveals enigmatic variability in $^{15}N/^{14}N$ between 2.8 and 4.3 x 10^{-3} (Marti and Kerridge, 2010). N in the terrestrial planets is found to be variable, but relatively uniform when compared to asteroidal, cometary sources and Jupiter's atmosphere. The possible radial dependence in the SS was investigated, as well as a possible relationship with the D/H isotopic signature (Marty et al., 2010), but N in comet Wild-2 particles is heterogeneous (McKeegan et al., 2006) and cometary water with terrestrial D/H ratios (Hartogh et al., 2011) does not fit such a trend. Since the SS was formed in the collapse of a presolar cloud, isotope data in N-containing molecules should be useful tracers. Gerin et al. (2009) measured $^{15}N/^{14}N$ ratios in several dense cores of the interstellar medium (ISM) and found ratios in the range of 1.3 to 2.8 x 10^{-3}, which are low but overlap the range observed in meteorites.

3.3 Noble gas isotope abundances in SS reservoirs

For trapped heavy noble gases (Ar, Kr, Xe) uniform isotopic abundances are found in chondritic meteorites, while noble gas reservoirs in the inner planets atmospheres evolved and do not represent early trapped components. Heavy noble gas isotopic abundances in different meteorite classes are similar, although they were given different names: OC (ordinary chondritic; Lavielle and Marti, 1992), Q-type gases (Lewis et al., 1975) and Abee-type (Lee et al., 2009) in enstatite chondrites. Q-gases are located in carbonaceous carrier phases (Q), together with some interstellar grains (SiC, diamond) and are concentrated in acid-resistant residues (HF/HCl residues) (Lewis et al., 1975). Q-gases have a characteristic elemental pattern (Busemann et al.,2000) and show strong relative depletions of light gases when compared to solar abundances, for He by about seven orders of magnitude. The uniform isotopic abundances either indicate a common SS reservoir that differed from SW abundances, or they reveal a uniform presolar carrier phase. The Ne isotopic abundances are less uniform, and in phase Q $^{20}Ne/^{22}Ne$ ratios are in a range 10.1 to 10.7, clearly lower than in solar Ne. The evidence that chondrites contain IOM compounds with large ^{15}N excesses (see 3.2) may indicate a carrier and a presolar source. A presolar environment for Q-gas incorporation was also suggested by Huss and Alexander (1987) to account for the elemental fractionation pattern and the simultaneous presence of nucleosynthetic gas components. A similar elemental fractionation pattern without isotopic fractionations has been produced in the laboratory with gas trapping by carbon condensates (Niemeyer and Marti, 1981).

Enstatite chondrites (EC) show more variable elemental abundance patterns, with relatively high ratios Ar/Xe in some enstatite chondrites (Lee et al., 2009), as well as very low Ar/Xe ratios in an E3 chondrite (Nakashima et al., 2010). The Xe isotopic abundances in separated phases of EC's reveal the presence of small but variable components of solar Xe and of a nucleosynthetic component (HL-Xe, carried in diamonds). Lavielle and Marti (1992) suggested that also OC-Xe (or Q-Xe) data are consistent with mixtures of solar and HL-Xe components. Noble gas carriers in stardust were identified (e.g. Anders and Zinner, 1993) and include diamonds, SiC and graphite, containing different products of stellar synthesis.

3.4 Isotopic differences in Mg and Cr: Clues for extinct nuclides?

Magnesium and chromium isotopic data document decay products of ^{26}Al and ^{54}Mn (312 d half-life, an e-capture decay) and show that products from stellar synthesis were not well mixed in the presolar environment. Large ^{54}Cr excesses are residues of stellar synthesis, but a relation to $\Delta^{17}O$ (Trinquier et al., 2007) suggests that also ISM processes are documented, as well as a coupling to the s-process component in Mo (Dauphas et al., 2002). Extremely large ^{54}Cr anomalies were found in the acid-resistant residue of the CI chondrite Orgueil (Qin et al., 2011). These workers found that ^{54}Cr-rich regions are associated with sub-micron Cr oxide grains, likely spinels, and they suggest a Type II supernova origin and a heterogeneous distribution of the ^{54}Cr carrier.

Schiller et al. (2010) reported Mg isotope data for most classes of basaltic meteorites, and with the exception of four angrites and one diogenite, which have young ages or have low Al/Mg ratios, all bulk basaltic meteorites have ^{26}Mg excesses. The authors conclude that excesses record asteroidal formation of basaltic magmas with super-chondritic Al/Mg and confirm that radioactive decay of short-lived ^{26}Al was the primary heat source of melted planetesimals.

Villeneuve et al. (2011) found variable ^{26}Mg radiogenic enrichments and deficits, relative to terrestrial Mg, in separated refractory olivines from matrix samples and individual CAIs. These authors show that olivines formed in reservoirs enriched in ^{26}Mg from the decay of extinct ^{26}Al, while olivines in a pallasite show a deficit, and use the inferred crystallization ages to calculate the time of metal-silicate differentiation, which occurred only ~0.15 Ma after CAI formation.

3.5 Stellar synthesis products in heavy elements

Isotopic heterogeneities were observed in primitive chondrites and in chondrite components for Ba, Sm, Nd, Mo, Ru, Hf, Ti and Os, sometimes with conflicting results and interpretations.

Van Acken et al. (2011) suggest that Os is an ideal synthesis tracer because its abundances are affected by p-, r-, and s-processes; and since Os is a refractory element, it documents records from the earliest stages of condensation. They found that Os in less evolved enstatite and Rumuruti chondrites, representing end-members in oxidation state, shows similar deficits of the s-process component as some primitive carbonaceous and unequilibrated ordinary chondrites, while enstatite chondrites of higher metamorphic grades have terrestrial isotopic compositions. These authors report that laboratory-digestion-resistant presolar grains, most likely SiC, are carriers of anomalous Os and that presolar grains disintegrated during parent body processing. The magnitude of the anomalies requires a few ppm of presolar SiC with an unusual isotopic composition, possibly produced in a different stellar environment and injected into the region of formation. In other work on Os isotopes, Yokoyama et al. (2011) found excesses of the s-process component in acid residues that were enriched in insoluble organic matter (IOM), while they found terrestrial Os isotope compositions in bulk chondrites. Nearly all IOM-rich residues were enriched in s-process Os, prompting these authors to conclude that s-process-rich presolar grains (presolar SiC) are found in presolar silicate hosts from either red giant branch (RGB) or asymptotic giant branch (AGB) stars. Since they also found that Os that dissolves by weak acid leaching of bulk chondrites is enriched in r-process nuclides, they suggest that a fine-grained presolar silicate carrier phase formed from supernovae ejecta. Nucleosynthetic isotope variations in Mo as observed in a wide range of meteorites, but not between planets Earth and Mars, were recently reported by Burkhardt et al. (2011). There is a clear message in these Os and Mo isotope data, since s- and r- process nuclides were produced in different locations: either the respective carriers were not equilibrated, or the carriers formed in heterogeneous environments.

The elements Ba, Nd and Sm, all with allotments of p-, s-, and r-process products, are equally good tracers of heterogeneous distributions of nucleosynthetic products. Ordinary chondrites were found to be uniform in Ba isotopic abundances, but variations were found in phases of carbonaceous chondrites (Carlton et al., 2007). These authors suggest that isotopic variability observed in Ba, Nd and Sm in carbonaceous chondrites reflect distinct stellar nucleosynthetic contributions to early SS matter. Further, by using the ratio ^{148}Nd/^{144}Nd to correct for observed s-process deficiency, they found that the ^{146}Sm-^{142}Nd isochron (from alpha-decay of ^{146}Sm) is in agreement with earlier data, but that the ^{142}Nd abundance is deficient in these chondrites, compared to terrestrial rocks. Qin et al. (2011) analyse in detail correlated nucleosynthetic isotopic variations in Sr, Ba, Nd and Hf in a carbonaceous and in an ordinary chondrite and conclude that these variabilities are best explained by variable additions of pure s-process nuclides to a nebular composition slightly enriched in r-process isotopes compared to average SS material. Andreasen and Sharma (2007) found excesses of ^{135}Ba and ^{137}Ba in carbonaceous chondrites, but no anomalies in ^{130}Ba,^{132}Ba,^{138}Ba and in Sr isotopes. They conclude that carbonaceous chondrites have r-process excesses in ^{135}Ba and ^{137}Ba with respect to Earth, eucrite parent bodies and ordinary chondrites and suggest that the SS was heterogeneous beyond 2.7 AU, a region where carbonaceous chondrite parent bodies formed.

4. The SW collection mission: Genesis

NASA's Genesis mission collected SW atoms for more than two years and returned these for laboratory analyses. Positioned at the Sun-Earth L1 Lagrange point about 1.5 million km from Earth, the spacecraft was well beyond the Earth's atmosphere and magnetic field. Highest priorities of the mission were a determination of the abundances of the isotopes of oxygen and nitrogen in the SW (Burnett et al., 2003). On returning to Earth in 2004 with its payload, the capsule suffered an unplanned hard landing in Utah, shattering most of the collector materials and thereby greatly complicating the sample analysis. Isotopic abundances for noble gases, oxygen and nitrogen have now been published and studies on other elements are in progress.

The SW is the most relevant source of information on the isotopic composition of the Sun. In the pre-Genesis era a substantial amount of data has been obtained on the isotopic composition of the SW. Unfortunately, most of this data did not bring the required precision, i.e., a precision which enables distinction between the SW composition and the known SS isotopic signatures. The notable exception is the contribution of the Apollo-foil experiments, which sampled light noble gases during a similar phase of declining solar activity, and then analyzed these with laboratory mass spectrometers. Slightly, but significantly different data than Genesis were found, which still lack an explanation (Fig. 2). In situ measurements, in contrast to most foil experiments, are very valuable in providing high time resolution, allowing a good correlation with relevant solar wind parameters and, hence, a careful assessment of possible isotopic and elemental fractionation effects in the solar wind. Understanding these effects is crucial for the interpretation of most of the data, and difficult, if not impossible with any type of foil collection experiments.

4.1 SW noble gases collected by Genesis and in exposed SS surfaces

Heber et al. (2009) reported elemental and isotopic abundances of noble gases in the SW at L1 as collected by foils in the NASA Genesis mission. He, Ne and Ar were analyzed in diamond-like carbon on a silicon substrate (DOS) and were quantitatively retained in DOS and, with exception of He, also in Si. SW data presented by Heber et al. (2009) have the following isotopic composition:

$$^{3}He/^{4}He = 4.64 \times 10^{-4},\ ^{20}Ne/^{22}Ne = 13.78,\ ^{21}Ne/^{22}Ne = 0.0329,\ ^{36}Ar/^{38}Ar = 5.47.$$

Measured elemental ratios are $^{4}He/^{20}Ne = 656$, and $^{20}Ne/^{36}Ar = 42.1$.

The ratio $^{3}He/^{4}He$ reported by these authors agrees within uncertainties with long-term averages (Coplan et al., 1984; Bodmer and Bochsler, 1998) and does not indicate variations.

Fig.2 compares the SW Genesis data for $^{4}He/^{3}He$ and $^{4}He/^{20}Ne$ (Grimberg et al.,2008; Heber et al.,2009) with those of the Apollo-SWC experiment (Geiss et al. 2004). The Apollo data have been gathered in the declining phase of solar cycle 20, those of Genesis were collected during the declining phase of cycle 23. The Genesis data represent an average over 2.3 years exposure and are representative for in-ecliptic SW, while despite of considerably shorter exposures the Apollo-SWC experiments have sampled a period of comparable length (1969 through 1972) and suggest an intrinsic variability of the SW. Whereas the value by Grimberg et al. (2008) is compatible with the average and the scatter of the Apollo measurements, the determination of Heber et al. (2009) is incompatible with the Apollo values. It is not clear, whether the difference between these results can be explained with a real (secular) variability of the solar wind composition.

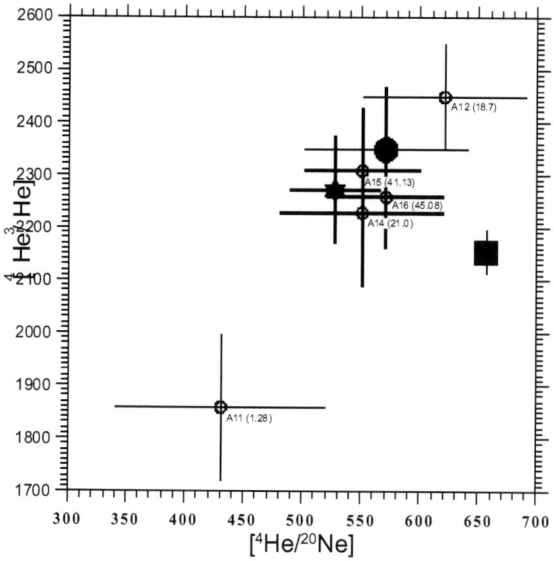

Fig. 2. Comparison of $^{4}He/^{3}He$ and $^{4}He/^{20}Ne$ ratios from the Apollo SW composition (circles) experiments and the Genesis mission. The exposed Genesis samples were bulk metallic glas (star symbol; Grimberg et al., 2008) and diamond-like carbon on silicate substrate (full square; Heber et al., 2009). The results from the Apollo missions are shown as circles with letters indicating the number of the mission and exposure times for the foils (in hours). The full circle shows the flux weighted average as given by Geiss et al. (2004).

The average $^{36}Ar/^{84}Kr$ ratio obtained by Vogel et al. (2011) from 14 individual analyses is 2390 ± 120 agrees with a preliminary ratio of 2030 measured by Meshik et al. (2009), while the preliminary upper limit (9.71) by these authors for the $^{84}Kr/^{132}Xe$ ratio compares well with 9.9 ± 0.3 given by Vogel et al. (2011). Vogel et al. (2011) reported an average ratio $^{36}Ar/^{38}Ar$ = 5.50 ± 0.01 which also agrees well with a ratio 5.501 ± 0.005 of Meshik et al. (2007). Reported ratios $^{86}Kr/^{84}Kr$ = 0.303 ± 0.001 and $^{129}Xe/^{132}Xe$ = 1.06 ± 0.01, at this time the only values available from Genesis, are consistent with solar data as observed in early etch steps in lunar ilmenite samples (Wieler and Baur, 1994). Martian interior samples (SNC meteorites) also contain a solar type Xe-S component (Ott, 1988; Mathew and Marti, 2001), while SW-implanted Xe components were observed in gas-rich meteorites like Pesyanoe (Marti, 1969; Mathew and Marti, 2003). Minor solar components were observed in 'sub-solar' gases in E-chondrites (Crabb and Anders, 1981; Busemann et al., 2003).

The solar photospheric abundances of the heavy noble gases that were not well known (Anders and Grevesse, 1989; Asplund et al., 2009; Lodders et al., 2009), can be inferred from SW data in Genesis foils. These data (Vogel et al., 2011) are in good agreement with ratios measured in the gas-rich meteorite Pesyanoe (Marti et al, 1972) that also provided the first isotopic data on solar-type Xe. Apparently, ion implantation processes and losses in asteroidal surfaces did affect light noble gases (e.g. the ratio Ne/Ar), while elemental abundance ratios of Ar, Kr and Xe were not much altered.

4.2 Oxygen in the SW and in the Sun

Oxygen isotopic abundances in the SW collected at L1 were measured in returned Genesis concentrator foils and corrected for fractionation according to radial position within the concentrator (McKeegan et al., 2011). The use of measured Ne isotopic data from adjacent samples for this correction is a good approximation, although the charge states for O and N are not identical (Bochsler, 2000), and the adopted mass-linear fractionation represents an approximation.

For $\Delta^{17}O$ in the SW, the displacement of measured ^{17}O data from the terrestrial fractionation line (see Fig. 1; $\Delta^{17}O = 0$), McKeegan et al.(2011) determined a mean value $\Delta^{17}O = -2.84\%$ for SW at L1, and after fractionation corrections they inferred an identical $\Delta^{17}O = -2.84\%$ datum for the Sun, implying that the average required fractionation correction is mass-dependent. However, as discussed earlier (in 2.3), a correction for gravitational settling in the outer convective zone (OCZ) is not linear in mass. The approach of forcing the solar value onto the correlation line for oxygen in CAIs appears quite arbitrary. Since the origin of CAIs is not known, this raises the issue of whether the implied assumption of having only two primitive O isotopic reservoirs in the SS, is appropriate. In the O three-isotope system it is always possible to interpret observed ratios by invoking a mass-dependent fractionation acting on mixtures of two distinct reservoirs, at least as long as such fractionations are within acceptable limits.

Note that a significant $\Delta^{17}O$ difference between the Sun on one hand, Earth, Moon and Mars on the other, contradicts the model of Ozima et al. (2007). It was discussed earlier that matter in the forming SS was recycled and mixed, at least in the inner parts, as is seen also from the close relationships in $\Delta^{17}O$ values in inner planetary reservoirs. O is the third most abundant element in the SS and it is not easy to keep solar and planetary oxygen apart. Reworking the total inventory of planetary oxygen in a non-mass-dependent process (e.g., by photo-dissociation of O-bearing molecules) is not easy. One may wonder what happened in such a procedure to volatile isotopic anomalies, such as Ne-E (^{22}Ne from short-lived ^{22}Na) in meteoritic grains.

4.3 Nitrogen in Genesis foils

The N isotope ratio in the SW at L1, determined by ion probe in foils of the Genesis concentrator (Marty et al., 2011), $^{15}N/^{14}N = 0.002178$ is much lower than that in the terrestrial atmosphere (0.00376). It differs from ratios observed in martian meteorites ALH84001 and Chassigny which, like the solar-type Xe isotopic abundances, were interpreted to represent primitive martian N (Mathew and Marti, 2001). The reported isotope ratio (0.002268) for the Sun (Marty et al., 2011) is obtained after corrections for an adopted fractionation for Coulomb drag (Bodmer and Bochsler, 1998), foil position in the concentrator, and gravitational settling. This value is close to that observed in osbornite of a meteoritic inclusion (Meibom et al., 2007), that the authors assume to represent an early SS condensate. Marty et al. (2011) suggest that the measured SW ratio characterizes nitrogen in the primitive SS, as a ^{15}N-depleted component is required to account for N in osbornite and for N in the atmosphere of Jupiter (Owen et al, 2001), and more generally to explain isotope variations in SS objects. The sources of heavy N, as observed in the inner SS, are currently not known, although presolar phases, indicating a range of $^{14}N/^{15}N$ ratios from ~50 to ~20,000 (Zinner et al., 2007), did make some contribution to N budgets in the SS.

5. Review of models of formation, origin and chronology

5.1 Time of formation of the Solar System (SS)

Determinations of the time of formation require the identification of solid objects that crystallized from homogeneous SS matter, and the recovery of undisturbed mineral isochron ages based on long-lived radionuclides. The most accurate determinations of absolute ages of the oldest SS objects currently are based on Pb-Pb ages, because of the precision in the half-lives of the progenitors ^{235}U and ^{238}U and the coupled age data of isotope pairs. The adopted most pristine solids for this purpose, chondrules and CAIs, were considered to have invariant SS abundance ratios of these two progenitors, which turned out to be an incorrect assumption (Brennecka et al., 2010). The first combined high-precision U and Pb isotopic data for a CAI, and U isotopic data for chondrules and whole rock fractions of the Allende meteorite (Amelin et al., 2010) show that the Allende meteorite bulk rock and chondrules data have distinctly lower ratios ($^{238}U/^{235}U$=137.747) than the CAI. The difference in the $^{238}U/^{235}U$ ratio of 0.129 between the CAI and chondrules and bulk meteorite document not only a difference in age, but shows that uncertainties in absolute ages of SS matter remain. The likely precursor, ^{247}Cm is formed in r-process nucleosynthesis in supernovae, and decays with a half-life of 15.6 million years (Ma) to ^{235}U. If it was present in the local ISM at the time of SS formation, and in a not completely homogenized environment, it can account for ^{235}U abundance variations. More data are required on the abundances and distribution of ^{235}U and ^{247}Cm to provide constraints on the interval between the last *r*-process nucleosynthetic event and the formation of the SS.

5.2 Presence of now- extinct nuclides in ISM and in SS matter

We consider specific isotopic excesses observed in SS matter due to extinct radionuclides with <10 Ma half-lives (given in parentheses, units Ma):

^{7}Be (53.44 days); ^{10}Be (1.39); ^{26}Al (0.717); ^{36}Cl (0.30); ^{41}Ca (0.10);
^{53}Mn (3.74); ^{60}Fe (2.62); ^{107}Pd (6.5); ^{182}Hf (8.90).

These short-lived, now-extinct, radioactive species indicate that a rather short time (< 1 Ma) must have elapsed between their production and their subsequent incorporation into early SS matter. ^{7}Be and ^{10}Be nuclides are not produced in stars and the half-life of fully ionized ^{7}Be may be much longer, because of delayed e-capture decay. ^{60}Fe is synthesized through successive neutron captures on Fe isotopes in neutron-rich environments inside massive stars, before or during their final evolution to core collapse supernovae. It can be detected after supernova ejections into the interstellar medium, from β-decays and gamma emission, like other radioactive isotopes ^{44}Ti, ^{56}Co, ^{26}Al. These nuclides provide evidence that nucleosynthesis is ongoing in the galaxy (The et al., 2006; Wang et al., 2007). Because of their short half-lives ^{44}Ti and ^{56}Co are detected as pointlike sources, e.g. in young type II supernova remnants (SNRs) Cas A and 1987A, respectively (Iyudin et al. 1994;). With its longer half-life ^{26}Al may propagate over significant distances and accumulate in the interstellar medium from many supernovae, until production and β-decay are in balance in the ISM, giving rise to a diffuse and galaxywide glow (Mahoney et al. 1982; Wang et al. 2007). ^{60}Fe from the same massive-star sources can be expected to follow ^{26}Al and new measurements (Wang et al. 2007) confirmed this. Although the detections of these isotopes are in agreement with the broad outlines of nucleosynthesis theory, there are also discrepancies in details. For example, ^{44}Ti

lives long enough that it should have been detected from several recent galactic supernovae, if these occur at a rate of ~2 per century. Also concentrations of ^{26}Al were seen in very young OB associations where even the most massive stars are not expected to have exploded, which led to the suggestion that most ^{26}Al production is associated with ejections in a pre-explosion phase of stellar evolution (Knödlseder et al. 1999).

Liu et al. (2009) studied ^{26}Al and ^{10}Be records in individual refractory chondritic hibonites. Spinel-hibonite spherules bear evidence of in situ ^{26}Al decay, whereas what the authors call PLAty-Crystals (PLACs) and Blue-aggregates either lack resolvable ^{26}Mg-excesses or exhibit ^{26}Mg-deficits by up to 4‰. They also found that eight out of 11 ^{26}Al-free PLAC grains record $^{10}B/^{11}B$ excesses that correlate with Be/B and the inferred initial $^{10}Be/^{9}Be$ ratio was $(5.1 \pm 1.4) \times 10^{-4}$. These data demonstrate that ^{10}Be cannot be used as a chronometer for these objects and that most of the ^{10}Be observed in CAIs must be produced differently, possibly by irradiation of precursor solid. The lack of ^{26}Al in PLAC hibonites indicates that ^{26}Al was not formed in the same process as ^{10}Be in PLAC, and Liu et al. (2009) conclude that these data indicate a very early formation of PLAC hibonites, prior to the incorporation of ^{26}Al.

Possible sources of short-lived radionuclides were discussed (Wasserburg et al., 2006; Tachibana et al., 2006; Wadhwa et al., 2007) and include production by energetic particle irradiation, stellar nucleosynthesis and mass-loss winds from Wolf-Rayet stars. In explosive synthesis generally several options exist and may not require multiple sources to provide a satisfactory match with observed abundances. However, most supernova models imply that if a supernova provided ^{26}Al and ^{41}Ca into the solar system, it would also have supplied 10-100 times the estimated ^{53}Mn abundance in the SS (e.g., Goswami and Vanhala 2000). The observed variability of abundances in meteorites raises questions regarding a widely accepted paradigm of SS formation, the gradual cooling of a collapsed molecular cloud and the sequential condensation of matter, growing from mineral grains to planets. Rather, the early solar nebula may have represented a dynamic assembly of domains, differentiating planetesimals and dust, which coexisted for a still poorly defined time. The time markers for these processes in principle are provided by abundances of short-lived isotopes as observed now as decay-products in SS matter.

5.3 Environment of formation

Adams (2010) has discussed the data for outer "edges" of the SS that provide further constraints on its dynamical past. The first edge is marked by the planet Neptune, which orbits with a semimajor axis a ≈ 30 AU, the range of planet formation. Beyond the last giant planet, the SS contains a large collection of smaller rocky bodies in the Kuiper Belt with orbits indicating much larger eccentricities and inclinations than the planetary orbits (Luu & Jewitt 2002). The Kuiper belt is dynamically excited and this property must be consistent with scenarios for SS birth, but these bodies contain relatively little of the total mass, which has been estimated to be 10–100 times less than Earth's (Bernstein et al. 2004). The outer boundary of the SS (inner boundary of the Kuiper belt) at 30 AU is thus significant and a drop-off is observed around 50 AU (Allen et al. 2000), the radial distance that roughly corresponds to the 2:1 mean motion resonance with Neptune. Because it is likely that additional bodies could have formed and survived beyond this radius, the existence of this edge at ~50 AU is important. At still greater distances, the SS contains a large, nearly spherical collection of comets known as the Oort cloud, a structure that extends to about

60,000 AU. Because the comets in the Oort cloud are loosely bound, gravitational perturbations from passing stars can easily disturb the cloud.

The presence of synthesized radionuclides such as ^{60}Fe (see 5.2) constrains the formation environment, since a moderately sized cluster of stars is required in order to contain a massive progenitor star for a supernova. Most models start with specific assumptions, like that Jupiter formed approximately at the current radial distance. There is information on one other Sun-like star with transient planets, Kepler-11, which has a closely packed six-member planetary system (Lissauer et al., 2011) and can provide useful information on dynamics and composition. In this case transient planets permit the determination of radii, and the inferred mass to size relations indicate substantial envelopes of light gases, but the largest planet (<300 earth masses) has a semi-major axis of only 0.46 AU. Lissauer et al. conclude that an in situ formation would require a massive protoplanetary disk of solids near the star and/or trapping of small solid bodies whose orbits were decaying towards the star as a result of gas drag. It would also require the accretion of significant amounts of gas by hot small rocky cores. Hydrodynamic calculations (Masset and Snellgrove, 2001; Walsh et al., 2011) have shown that giant planets can undergo a two-stage inward-then-outward migration. Further, it is not known whether matter from neighboring planetary systems in the star-forming cluster has been added to the SS, such as the Oort cloud of comets, as suggested by Levison et al. (2010).

5.4 Effects from an active early Sun

When a SS disk is exposed to external or internal UV radiation, the gas can be heated to sufficiently high temperatures to trigger and to drive flows; strong radiation fields also produce chemical changes. For example, oxygen isotopic changes are produced if the SS formed in an environment of intense FUV radiation fluxes. In one scenario, UV radiation may have produced selective photo-dissociation of CO within the collapsing protostellar envelope of the forming SS (Lee et al., 2008), or in an alternative scenario, the isotope selective photo-dissociation may have occurred at the surface of the nebula (Lyons and Young, 2005). Effects due to intense UV-radiation and/or plasma during an active phase of the Sun may also have affected matter in the inner SS. Shu et al. (2001) and Shang et al. (2000) proposed detailed models for effects by the active Sun, specifically the transport of CAI and chondrules. Lee et al (2009) proposed a scenario for loading enstatite grains at elevated temperatures with solar plasma in the inner SS, in a way similar to the proposed incorporation of Ar now observed in the atmosphere of Venus (Wetherill, 1981).

It seems plausible that an intense X-type solar wind, as proposed by Shu et al. (2001) led to a strong recycling and redistribution of condensates in the inner SS. Furthermore, it seems likely that this wind has established chemical and isotopic gradients with radial and latitudinal dependences. Also, it seems likely that this wind had a strong influence in shaping the atmospheric composition of the inner planets.

5.5 A look at the Interstellar Medium

It is not a purpose of this paper to discuss the literature of observations of molecules and of processes in the ISM. We refer the interested reader to a recent review of gas-phase processes and implications (Smith, 2011). However, since organic matter in the SS (primitive meteorites, comets) show large isotopic variations, it is important to determine the origin of this organic matter. Hydrogen in the ISM is observed to have huge enrichments of deuterium, and efficient processes at very low temperature capable to achieve enrichments are ion-molecule reactions. Regarding the elements O and N, we note that CO is the second most abundant molecule in interstellar clouds after H_2, with readily observable electronic transitions in the vacuum ultraviolet, vibrational bands in the infrared, and pure rotational lines in the millimeter-wave regime. The isotopic varieties of CO are used to constrain models of star formation, chemical networks, and stellar evolution. Isotopic O abundance measurements by high-resolution spectroscopy of X Persei with the Space Telescope Imaging Spectrograph (Sheffer et al., 2002) provided the first ultraviolet interstellar $^{12}O^{17}O$ column density data. The measured isotopomeric ratio is $^{12}C^{16}O/^{12}C^{17}O$ = 8700, while the detection of interstellar $^{12}C^{18}O$ establishes its isotopomeric ratio $^{12}C^{16}O/^{12}C^{18}O$ = 3000. These ratios are about five times higher than local ambient oxygen isotope ratios in the local ISM. The severe fractionation of rare species can be interpreted to show that both $^{12}C^{17}O$ and $^{12}C^{18}O$ are destroyed by photo-dissociation, whereas $^{12}C^{16}O$ avoids destruction through self-shielding depending on column densities. These authors also explain the small effects in the ratio $^{12}C/^{13}C$ by a balance between the photo-dissociation of $^{13}C^{16}O$ and its preferential formation via the isotope exchange reaction between CO and C^+.

Although large differences are observed between N isotopic abundances in different classes of meteorites, N is heavy when compared with the ratio in the SW, and the spectroscopic data on N in the ISM appear to be relevant. Also, N isotopic records in lunar surface samples which were exposed to the space environment at different times, including the recent exposure of South Ray samples (2 Ma ago), show that an influx of N into the inner SS has persisted to the present (Geiss and Bochsler, 1982; Marti and Kerridge, 2010). The observed abundances of C and N in implants of non-solar origin indicate low ratios C/N (Marti and Kerridge, 2010), distinct from ratios in carbonaceous chondrite matter. When looking for potential N-rich sources, we note that in Very Large Array observations of NH_3, Ragan et al. (2011) found in six infrared Dark-clouds at distances from 2 to 5 kpc that ammonia can serve as an excellent tracer of dense gas, as it shows no evidence of depletion, and they report an average abundance in these clouds of 8.1 x 10^{-7} relative to H_2. These authors found that although volume densities are on par with those in local star-formation regions of lower mass, they consist of much more mass, which induces very high internal pressures. Charnley & Rodgers (2002) showed that higher $^{15}N/^{14}N$ ratios are generated in high-density cores, where CO is depleted onto dust grains and N_2 remains in the gas phase. For prestellar cold cloud environments, with NH_3 freeze-out onto dust grain surfaces, models predict an enhancement in the gas-phase abundance of ^{15}N-bearing molecules. Bizzocchi et al. (2010) confirmed that this is the case for the N_2H^+ ion in a Taurus starless cloud, which is one of the best candidate sources for detection owing to its low central core temperature and high CO depletion. These authors infer a ratio $^{15}N/^{14}N$ that is about twice as large as the ratio inferred from NH_3 measurements and show that significant fractionations are observed under these conditions. They suggest that observed N isotopic variations observed in SS matter may be considered as a remnant of interstellar chemistry in the protosolar nebula, but also a transport mechanism is required for matter in the inner SS. Kwok and Zhang (2011) suggest that unidentified infrared emission features in circumstellar environments and in the ISM may arise from amorphous organic solids, of a type similar to IOM observed in chondrites (see 3.2). If correct, there may be a direct connection between meteoritic matter and the ISM.

6. Discussion

6.1 Noble gases as tracers of origin and evolution

Noble gases, like organic molecules discussed earlier, are useful tracers in studies of origin and evolution, due to the volatility (He and Ne) and the chemical inertness, but it is not possible to treat them as gas-phase-only elements, because trapped noble gases (Q-gases) are located in carbonaceous carriers that represent solid phases of an environment in which matter formed. On the other hand, solar-type gases as observed in planetary regoliths (lunar surface and gas-rich meteorites) represent implanted ions. Further, solar-type gases were observed in martian meteorites and in the atmosphere of Jupiter. The observed atmospheric abundances on the inner planets demonstrate complexities first because of the very low noble gas abundances, but mainly because of differences in isotopic abundances.

Reynolds (1963) introduced the term "xenology" for studies involving the abundances of 9 stable isotopes of Xe in solar system matter, which he expected to provide important information. One example is found in an apparently close relationship between trapped gas Q-Xe (or OC-Xe, Abee-Xe; see 3.3) and the solar Xe-S signature. Some papers (Lavielle and Marti, 1992; Marti and Mathew, 1998) considered mixing models of Xe-S and HL-Xe (component observed in presolar diamond) that were successful in accounting for observed components in meteorites, except for Xe abundances in Earth's atmosphere and Xe in chondritic metal (FVM-Xe; Marti et al., 1989). A model of an evolving environment such as one providing destruction of presolar diamonds (Huss, 1990) followed by processes of isotope homogenization is considered here. The lack of total homogenization of Xe in meteorites was documented in distinct isotopic abundances found in a ureilite (Wilkening and Marti, 1976), which indicates slightly variable mixing ratios of solar Xe-S and HL-Xe. The ^{182}Hf-^{182}W chronometric data for ureilites by Lee et al. (2009) provide a time-frame and show early differentiation of the ureilite parent body, shortly after the SS formed. Likewise, isotopic data for an environment where the Xe homogenization process had progressed even less is found in CB-carbonaceous chondrites (Nakashima et al., 2008; Nakashima and Nagao, 2009) with abundances somewhat closer to the solar Xe-S composition, specifically in a metal concentrate. Another longstanding xenology puzzle is found in monoisotopic excesses of ^{134}Xe in chondritic metal (Marti et al., 1989). This so-called FVM-Xe appears to relate to the just discussed variable mixing ratios of Xe-S and HL-Xe. In this case the FVM-Xe data is consistent with a mixture of Xe-S and fission-Xe, the latter with the specific signature of neutron-induced fission of ^{235}U (Marti et al., 1989). FVM-Xe was enriched in small grains, characteristics for surface locations, and may represent ions implanted together with fission Xe into metal grains. The neutron fluence required to produce the inferred amount of fission-Xe is high, however, and suggestive of an ISM environment (Hua et al., 2000). There is evidence for very early formation of chondritic metal (Lee and Halliday, 1996; Zolensky and Thomas, 1995).

Turning to planetary atmospheres, there are atmospheric data for the largest planet (Mahaffy et al., 2000), and the isotopic abundances of the larger isotopes are consistent with solar-type Xe, although the data have large uncertainties and their abundance relative to H is about 2.5 times larger. For the inner planets, there are atmospheric data for Mars and these are best compared to those in Earth's atmosphere. Remarkably, the Xe isotopic abundances in the atmospheres of the two planets are similar (except fission-affected heavy Xe isotopes), but differ strongly from those in other SS reservoirs. There are two distinct Xe components in the case of Mars (Mathew and Marti, 2001), solar-type Xe-S in its interior (in martian meteorites) and atmospheric Xe that is mass-fractionated (favoring the heavy isotopes) by 37.7 permil per amu, relative to Xe-S. Pepin (2006) favors a fractionation pattern based on hydrodynamic escape that fits the observed abundances well, without invoking a fission component in the martian atmosphere. For planet Earth Caffee et al. (1999) reported a very minor distinct Xe-S component in some well gases, but in this case the same hydrodynamic fractionation (Pepin, 2006) of solar Xe-S does not work. When considering isotopic fractionations in the gravity fields of these different size bodies, the option of a common origin can not be discarded. Lunar rocks that were never exposed to SW have shown that the composition of lunar Xe reveals an excellent match to the terrestrial atmosphere (Lightner and Marti, 1974). However, there is an unresolved question of whether Xe represents an indigenous component, or a contaminating phase with a terrestrial source (Niedermann and Eugster, 1990). The excellent agreement of lunar and terrestrial O isotopic abundances prompted Pahlevan and Stevenson (2007) to investigate the issue of isotopic exchanges between lunar and terrestrial matter by the giant impact and during the time required for the orbiting disk of magma and gas to condense into the moon. These calculations address oxygen isotopic exchanges, but also apply to exchanges of Xe. The time of formation of the moon was inferred from ^{182}Hf – ^{182}W data, but depends on the initial W isotopic composition, but likely was 60 Ma after formation of the SS (Touboul et al., 2007), consistent with data from coupled $^{146, 147}Sm$ – $^{142, 143}Nd$ systematics in lunar samples (Boyet and Carlson, 2007). Late collisions and accretion of matter by the moon remain options (Jutzi and Asphaug, 2011; Willbold et al., 2011). Also, if terrestrial H_2O was supplied by comets, as suggested by Hartogh et al. (2011), a supply of noble gases to the earth-moon system as well as to other inner planets can not be excluded.

6.2 Gas and condensed phase elements: Oxygen and nitrogen

In a self-shielding environment photons capable of dissociating $^{12}C^{16}O$ are attenuated and as a result ^{17}O and ^{18}O are preferentially dissociated from CO molecules (see 2.3 and 5.4). These atoms can recombine into H_2O molecules and produce heavy water (Clayton, 2002; Yurimoto and Kuramoto, 2004). If a mechanism exists, capable of separating newly formed heavy water from the remaining oxygen, this heavy signature can be exported. If H_2O vapor freezes onto grains and forms ice mantles while ^{16}O preferentially remains as CO in the gas, then a separation of solids from gas produces distinct isotope reservoirs. Yurimoto and Kuramoto (2004) discuss the observation of water ices in molecular clouds and suggest that heavy water in the SS probably was imported from its presolar molecular cloud and redistributed into the inner SS. Water ice and vapor was also observed in the disk around a young star (Hogerheijde et al., 2011). In the model of heavy water additions to SS matter, observed O isotopic signatures should plot on mixing-lines of the end-member components, the original presolar gas phase and the local water ices of the presolar molecular cloud. The O data reported for olivines of enstatite and carbonaceous chondrites (Weisberg et al., 2011; Ushikubo et al., 2011) determine a mixing line that is consistent with O data in Allende chondrules (Rudraswami et al., 2011) and also fits O in Earth's mantle, as well as O in the very heavy (presolar) water in a matrix sample (see 3.1).

Alternatively, Clayton (2002) suggests that self-shielding may have prevailed at the inner boundary of the solar nebula for intense UV radiation from the early active Sun, and that this radiation may have dissociated CO and also N_2. An effective separation and redistribution of residual gas and condensed phases must then account for the O and also N isotopic variations. These requirements apparently are met in the environments of the ISM or the presolar molecular cloud, but restrictions apply in the environment of the inner boundary of the SS. In simple models the observed isotopic composition depends on a mixing ratio, and mixing models of two components with different isotope ratios have been considered ever since the discovery of anomalous ^{17}O (Clayton et al., 1973). The large difference in the ratios $^{15}N/^{14}N$ of inner planets, meteorites and the SW (Marty et al., 2011) are even more difficult to explain in a SS environment. Further, the large variations in implanted N in lunar regolith samples require distinct sources of N. On the moon the smaller fraction of N has its origin in SW implants. In meteorites, and specifically in phases of the carbonaceous matrix where elements reveal isotopic signatures of stellar nuclear processes, another source is indicated, an origin in the presolar molecular cloud. Maret et al. (2006) suggest that N_2 abundances in the ISM are very small, because most N exists in atomic form (and in ices) and this would account for large fractionations, of a magnitude as observed in meteorites.

N concentrations observed in presolar grains of SiC, diamond and graphite range from permil to percent (Zinner et al., 2007), but these grains apparently are rare in chondrites. Busemann et al. (2006) working on phases of carbonaceous chondrites found very large isotopic anomalies in hydrogen and nitrogen, exceeding those found in interplanetary dust particles, and suggest an origin in the presolar molecular cloud or perhaps in the protoplanetary disk. The most extreme D/H values were found in IOM (Insoluble Organic Matter), and since these highly anomalous separates survived the chemical separations and of course all the parent body processing, IOM appears to represent robust matter that formed in special environments with properties as observed in the ISM. Remusat et al. (2007) found that the ratio D/H in IOM depends on the C-H bond dissociation energy, and they suggest that the observed correlation of D-enrichments does not indicate formations in the ISM, but rather equilibrium exchanges in a D-rich reservoir after IOM syntheses. However, since meteoritic D/H ratios are generally smaller than those observed in the ISM, it appears that a correlation with bond dissociation energies could also be due to hydrogen exchange reactions during reheating events. The inferred ^{15}N-rich source with low C/N ratios in the lunar regolith is not known. The lack of a relation of N isotope ratios in SS matter with either D/H or $\Delta\,^{17}O$ (Marty et al., 2011) do not indicate simple mixtures of a ^{15}N-rich end-member (like IOM) and a solar (^{15}N-depleted) component. In the scenario of a delivery of H_2O with a terrestrial H/D ratio by comets (Hartogh et al., 2011) a supply of N also has to be expected. Mumma and Charnley (2011) review the chemical composition and isotopic abundances in comets that indicate a general enrichment in ^{15}N. They conclude that an understanding of the nitrogen isotopic variations in volatile SS material demands more rigorous astronomical ground-truths, such as measurements of N isotope ratios in molecular clouds, and specifically in highly depleted cores that are forming low-mass protostars. In the previous discussion (6.1) an evolving environment (destruction of a pre-existing component) was considered, and a similar fate could affect N: a looked-after component does no longer exist, but the environment has changed by way of its destruction.

6.3 Origin of SS nucleosynthetic heterogeneities

Ti isotopic variations in matter of the inner SS permit some insight into the history of stellar products. Although the anomalies for isotopes ^{46}Ti and ^{50}Ti have different nucleosynthetic origins, the observed variations (Trinquier et al., 2009) permit to follow an evolutionary path for correlated mass-independent variations of ^{46}Ti and ^{50}Ti in bulk solid analyses. These authors conclude that the observed correlations imply that the presolar dust, inherited from the protosolar molecular cloud, must have been well mixed when SS solids formed, but that a subsequent process was necessary that imparted isotopic variability at the planetary scale. This process could have been thermal processing of molecular cloud material, like volatile element depletions in the inner SS, including selective destruction of thermally unstable, isotopically anomalous presolar components. As discussed earlier (see 3.5), the Os isotope homogeneity in bulk chondrites contrasts with isotopic heterogeneities observed in various other elements (Cr, Mo, Ba, Sm, and Nd), at the same level of resolution, while Os isotopic abundances are not uniform in separates of carbonaceous chondrites. It is not clear, whether different carriers were involved in the transport, or differences existed in chemical compositions or thermal processing during imports of matter into the inner SS.

7. Conclusions

Several of the observed isotopic abundance variations in SS matter are due to incomplete mixing of products from different stellar sources. These components were injected into and processed in the presolar cloud before the formation of the SS, and grains that escaped homogenization processes exhibit substantial variations in their isotopic make-ups. Some stellar products were radioactive nuclei with about Ma half-lives (^{10}Be, ^{26}Al, ^{53}Mn, ^{60}Fe) that are used in chronological applications that trace the sequence of events in the nebula. Multiple sources are indicated for these radionuclides, and the extent of homogenization is recorded in the decay products, which are often found to be variable, sometimes even missing. The early solar nebula may have represented a dynamic assembly of domains, differentiating planetesimals and dust, which coexisted for some time.

The paper first summarizes observed abundance variations for O, N, noble gases and some heavy elements (recorders of p-, s- and r-processes) and then assesses these abundances in the light of available SW abundance data as obtained by the Genesis and lunar missions. Some striking isotopic differences invite reevaluations of origin and paths of evolution, although some uncertainties remain between the measured SW data and real solar abundances that, although small, are nevertheless critical in some evaluations. The discussion focuses on large differences observed in isotopic data of O and N in different locations of the inner SS, in giant planets and in comets, and evaluates a model of changing mixing ratios of isotopically distinct forms of matter (gases and solids/ices). In this case the observed abundances in SS matter are determined by presolar nebular gas-phase data (presumably close to, but not identical to solar data) and the abundances in partially homogenized condensed phases (solids/ices). As discussed (see 6.2), water ices were also observed in the disk of a young star. The water ice signatures determine isotope ratios in mixtures, for O in a three-isotope diagram, and determine whether planetary and meteoritic O data plot on different mixing-lines. Reported O data in olivines of carbonaceous and enstatite chondrites define a mixing-line that is consistent with O data in chondrules and also fits oxygen ratios in the Earth's mantle and O in reported heavy (presolar nebula) water (see 3.1). Characteristic data for sources of today's observed, and also of some lost (by destruction) carriers of isotopically distinct components, characterize a presolar environment with incomplete equilibrations. Some authors suggest that carrier phases and

ISM reaction residues were initially located in the outer ranges of the solar nebula and then injected to the inner SS, possibly also at later times. An injection of ices would also have changed isotopic abundances of volatiles like N or heavy noble gases. A documentation of sources is not easy, as the local ISM has evolved over the 4567 Ma timeframe. If unidentified infrared emission features in circumstellar environments and in the ISM arise from amorphous organic solids (as observed in carbonaceous chondrites) as suggested, this data may provide a direct link between meteoritic matter and the ISM. Indications of coupled isotopic variations, like those for ^{54}Cr, s-process Mo and $\Delta^{17}O$, require detailed documentation of carrier phases. The available data for SS matter also may be incomplete (e.g. one comet with terrestrial water, and one meteorite with very heavy oxygen). In summary, the SW abundance determinations in foils from the Genesis mission provide essential reference data for investigations of abundance variations in SS matter.

8. Acknowledgment

PB acknowledges support by NASA SR&TGrant NNX09AW32G

9. References

Adams F.C. (2010), Annu. Rev. Astron. Astrophys. 48:47–85
Allen R.L, Bernstein G.M, Malhotra R. (2000), Astrophs. Journal 542:964
Amelin Y., Kaltenbach A., Iizuka T., et al. (2010), Earth and Planetary Science Letters 300 (2010) 343–350
Anders E. and Grevesse N. (1989). Geochim. Cosmochim. Acta 53, 197–214.
Andreasen R. and Sharma M. (2007), The Astrophysical Journal, 665,874-883,
Arnold J.R. and Suess H.E.(1969), Cosmochemistry, Ann. Rev. of Phys. Chemistry 20, 293-314.
Asplund M., Grevesse N., Sauval A. J. and Scott P. (2009). Annu. Rev. Astron. Astrophys. 47, 481–522.
Bernstein G.M, Trilling D.E, Allen R.L, Brown M.E, Holman M, Malhotra R.(2004), Astron. J. 128:1364
Bizzocchi L., Caselli P. and Dore L. (2010), Astronomy and Astrophysics 510, L5
Bochsler P. (2000), Rev. Geophys., 38, 247- 266
Bochsler P., Gonin M., Sheldon R.B., Zurbuchen T., Gloeckler G., Hamilton D.C., Collier M.R., and Hovestadt D. (1997), Proc. of the 8th International Solar Wind Conference, Dana Point, USA. 199-202.
Bodmer R. and Bochsler P. (1998), Astron. Astrophys. 337, 921.
Bodmer R., and Bochsler P. (2000), J. Geophys. Res. 105, 47-60.
Boyet M. and Carlson R. W. (2007), *Earth Planet. Sci. Lett.* 262, 505-516.
Brennecka, G.A.,Weyer, S.,Wadhwa,M., Janney, P.E., Zipfel, J., Anbar, A.D., (2010), Science 327,449–451.
Burbidge E.M., Burbidge G.R., Fowler W.A., Hoyle F. (1957), Rev.Mod.Phys.29,547.
Burkhardt C., Kleine T., Oberli F., Pack A., Bourdon B. and Wieler R. (2011), Earth and Planetary Science Letters 312, 390–400
Burnett D.S. *et al.* (2003), *Space Sci. Rev.* 105, 509.
Busemann H., Baur H. and Wieler R. (2000), Meteoritics and Planetary Science 35,949-913 (2000)
Busemann H., Baur H., and Wieler R. (2003) *Lunar & Planet. Sci.* XXXIV(LPI,#1665)
Caffee M. W., Hudson G. B., Velsko C., Huss G. R., Alexander E. C. J., and
Chivas A. R. (1999), *Science* 285, 2115-2118.

Carlson R.W., Boyet M. and Horan M. (2007), Science 316, 1175-1178

Cameron A. G W. (1957), Publ. Astron. Soc. Pacific 169, 201

Clayton R.N., Grossman L. and Mayeda T.K. (1973), Science 182, 485-488

Clayton R. N. (2002), Nature 415, 860–861.

Connolly H.C. and Huss G.R. (2010), Geochimica et Cosmochimica Acta 74, 2473–2483

Coplan M. A., Ogilvie K. W., Bochsler P. andGeiss J. (1984), SolarPhys. 93, 415–434.

Crabb J. and Anders E. (1981), *Geochim. Cosmochim. Acta* 45, 2443-2464.

Dauphas, N., Marty, B., & Reisberg, L. (2002) ApJ, 569, L139

Geiss J., Eberhardt P., Bühler F., Meister J., and Signer P. (1970) J. Geophys. Res. **75**, 5972

Geiss J. and Bochsler P. (1982), *Geochim. Cosmochim. Acta 46, 529-548.*

Geiss J., Bühler F., Cerutti H., Eberhardt P., Filleux Ch., Meister J., and Signer P. (2004) Space Sci. Rev. 110, 307

Gerin M., Marcelino N., Biver N., Roueff E., Coudert L. H., Elkeurti M., Lis D. C. and Bockelee-Morvan D. (2009), Astron. Astrophys. 498, L9 -L12.

Goswami, J. N., & Vanhala, H. A. T.(2000), in Protoatars and Planets IV, ed. V. Mannings, A. P. Boss, & S. S. Russell (Tucson: Univ. Arizona Press), 963

Grimberg A., Baur H., Bühler F., Bochsler P., and Wieler R. (2008) Geochim. Cosmochim. Acta **72**, 626-645

Hartmann L, Ballesteros-Paredes J, Bergin EA. (2001), *Ap. J.* 562,852

Hartogh P., Lis D.C., Bockelee-Morvan D. et al. (2011), Nature 478, 218-220.

Heber V.S., Wieler R., Baur H., Olinger C., Friedmann T.A. and Burnett D.S. (2009), GeochimicaetCosmochimicaActa73, 7414–7432

Hogerheijde M.R., Bergin E.A., Brinch C. et al. (2011), Science 334, 338-340.

Huss G.R. and Alexander, Jr., E. C. (1987), J. Geophys. Res. 92, E710–E716

Huss, G.R., and Scott, E. R. D. (2010), *The Astrophysical Journal* 713,1159-1166.

Hua X.-M., Lingenfelter R. E., Marti K., and Zytkow A. N. (2000), Astrophys. Journal 531,1081-1087.

Iyudin, A.F., Diehl R., Bloemen, H. et al. (1994), Astronomy&Astrophysics, 284, L1

Jutzi M. and Asphaug E. (2011), Nature 476, 69-72

Kallenbach R., Ipavich F.M., Kucharek H., Bochsler P., Galvin A.B., Geiss J., Gliem F., Gloeckler G., Grünwaldt H., Hefti S., Hilchenbach M., and Hovestadt D., (1998) Space Sci. Rev. 85, 357-370.

Krot, A. N., Nagashima, K., Ciesla, F. J., Meyer, B.S., Hutcheon, I.D., Davis A.M., Huss G.R. and Scott R.D. (2010), The Astrophysical Journal, 713:1159–1166.

Knödlseder, J., Bennett, K., Bloemen, H., et al. (1999), Astronomy&Astrophysics, 344, 68

Kwok S. and Zhang Y. (2011), Nature 479, 80-83

Lavielle, B., and Marti, K. (1992), J. Geophys. Res., 97, 20,875–20,881.

Lee D-C., Halliday A. N., Singletary S. J., and Grove T. L. (2009), *Earth Planet. Sci. Lett.* 288, 611-618.

Lee J.E., Bergin E.A., Lyons J.R. (2008), *Meteorit. Planet. Sci.* 43, 1351

Lee J-Y., Marti K. and Wacker J.F. (2009), JOURNAL OF GEOPHYS. RESEARCH, VOL. 114, E04003.

Levison H.F. *et al.* (2010), *Science* 329, 187-190

Lewis R. S., Srinivasan B. and Anders E. (1975), Science 190, 1251–1262.

Libourel Guy and Chaussidon Marc (2010), Earth and Planetary Science Letters 301 (2011) 9–21

Lissauer Jack J.,et al.(2011), Nature 470, 53-58

Liu M-C.,McKeegan K.D., Goswami J.N., Marhas K.K., Sahijpal S., Ireland T.R. and Davis A.M. (2009), Geochimica et Cosmochimica Acta 73, 5051-5079

Luu J.Xand Jewitt D.C.(2002), Annu. Rev. Astron. Astrophys. 40:63

Lyons J.R., Young E.D. (2005), Nature 435, 317
Mahaffy P.R., Niemann H.B., Alpert A. et al.(2000), J. Geophys. Research 105,E6, 15061-15071
Mahoney, W. A., Ling, J. C., Jacobson, A. S., & Lingenfelter, R. E. (1982), ApJ, 262, 742
Marti K., Wilkening L.L. and Suess H.E. (1972), Astrophys. Journal 173, 445-450
Marti K., Kim J. S., Lavielle B., Pellas P., and Perrin C. (1989), Z. Naturforsch. 44a, 963-967
Marti K. and Mathew K. J. (1998), Proc. Indian Acad. Sci. (Earth Planet. Sci.) 107, 425-431.
Marti K. and Kerridge J. (2010), Science 328, 1112
Marty B. et al. (2010), Geochim. Cosmochim. Acta 74, 340
Marty B., Chaussidon M., Wiens R.C., Jurewicz A.J.G. and Burnett D.S. (2011), Science. 332, 1533-1536
Mathew K.J. and Marti K.(2001), J. Geophys. Res. (Planets) 106, 1401-1422.
Mathew K.J. and Marti K. (2003), Meteoritics and Planetary Science 38, 627-643.
Maret S., Bergin E.A. and Lada C.J. (2006), Nature 442, 425-427.
Masset, F. and Snellgrove (2001), Mon. Not. R. Astron. Soc. 320, L55–L59.
McKeegan K.D. et al. (2006), Science 314, 1724-1728
McKeegan K.D. et al. (2011), Science 332, 1528-1532
Meibom A. et al. (2007), Astrophys. J. 656, L33.
Meshik A. P., Mabry J. C., Hohenberg C. M., Marocci Y., Pravdivtseva O. V., Burnett D. S., Olinger C. T., Wiens R. C., Reisenfeld D. B., Allton J. H., Stansbery K. M. and Jurewicz A. J. G. (2007), Science 318, 433–435
Meshik A. P., Hohenberg C. M., Pravdivtseva O. V., Mabry J. C., Allton J. H. and Burnett D. S. (2009), Lunar Planet. Sci. 40, Lunar Planet. Inst., Houston. #2037
Michaud G. and Vauclair S.(1991) *Solar Interior and Atmosphere,* edited by A. N. Cox, W. C. Livingston, and M. S. Matthews, pp. 304-325, Univ. of Ariz. Press.
Mumma M.J. and Charnley S.B. (2011), Annu. Rev. Astron. Astrophys. 49, 471-524
Nakashima D., Schwenzer S.P., Ott U.,and Ivanova M.A. (2008), Lunar & Planet. Science XXXIX, #1478
Nakashima D. and Nagao K. (2009) Antarctic Meteorite Symp.XXXII, NIPR, 54-55
Nakashima D., Ott U., El Goresy A.and Nakamura T. (2010), Geochimica et Cosmochimica Acta 74, 5134–5149
Niedermann S. and Eugster O. (1992), Geochim et Cosmochim Acta 56, 493-509
Niemeyer S. and Leich D. A. (1976), *Proc. Lunar Sci. Conf. 7th,* 587-597.
Niemeyer S. and Marti K. (1981), Proc. Lunar Planet. $ci. 12B, 1177-1188.
Owen T., Mahaffy P.R., Niemann H.B., Atreya S. and Wong M. (2001), Astrophys. J. 553, L77.
Ozima M., Podosek F.A., Hiuchi T., Yin Q.-Z., Yamada A. (2007), Icarus 186, 562-570
Pepin R. O. (2006), *Earth Planet. Sci. Lett.* 252, 1-14.
Qin L., Nittler L.R., Alexander C. M. O'D., Wang J., Stadermann F.J., Carlson R. W. (2011), Geochim. Cosmochim. Acta 75, 629–644
Qin L., Carlson R.W. and Alexander C.M.O'D. (2011), Geochimica et Cosmochimica Acta 75, 7806–7828
Ragan S., Bergin E. and Wilner D. (2011), Astrophys. Journal 736, 163
Remusat L., Palhol F., Robert F., Derenne S. and France-Lanord C. (2006), Earth and Planetary Science Letters 243, 15–25
Reynolds J.H. (1963), J. Geophys. Res. 68, 2939-2956
Rudraswami N.G., Ushikubo T., Nakashima D. and Kita N.T. (2011), Geochimica et Cosmochimica Acta 75, 7596–7611
Sakamoto N., Seto Y., Itoh S. et al. (2007), Science 317, 231-233
Schiller M., Baker J.A. and Bizzarro M. (2010), Geochimica et Cosmochimica Acta 74, 4844–4864
Sheffer Y., Lambert D.L. and Federman S.R. (2002), Astrophys. Journal, 574:L171–L174

Shang S, Shu F.H, Lee T, Glassgold AE. (2000), *Space Sci. Rev.* 92:153
Shu F.H, Shang S, Gounelle M, Glassgold A.E, Lee T. (2001), *Ap. J.* 548:1029
Smith I.W.M. (2011), Annu. Rev. Astron. Astrophys. 49, 29-66
Suess H.E. and Urey H.C. (1956), Rev. Mod. Phys. 28, 53
Tachibana, S., Huss, G. R., Kita, N. T., Shimoda, G. & Morishita, Y. 2006, ApJ, 639, L87
The, L. S., Clayton, D. D., Diehl, R., et al. (2006), Astronomy&Astrophysics, 450, 1037
Touboul M., Kleine T. , Bourdon B., Palme H. & R. Wieler R. (2007), Nature 450, 1206-1209.
Trinquier A., Birck J.-L. and Allegre C. J. (2007) Widespread ^{54}Cr heterogeneity in the inner solar system. Astrophys. J. 655, 1179–1185
Trinquier A., Elliott T., Ulfbeck D., Coath C, Krot A.N. and Bizzarro M . (2009), Science 324, 374-376
Turcotte S. et al. (1998), Astrophys. J. 504, 539-558
Turcotte S., Wimmer-Schweingruber R.F. (2002) J. Geophys. Res. 107, doi:10.1029/2002JA009418, 2002
Ushikubo T., Kimura M., Kita, N. T. and Valley J. W. (2011), Lunar Planet. Sci. XLII., #1183
Van Acken D., Brandon A.D. and Humayun M. (2011), Geochimica et Cosmochimica Acta 75 (2011) 4020–4036
Villeneuve J., Chaussidon M. and Libourel G. (2011), Earth and Planetary Science Letters 301, 107–116
Vogel N., Heber V.S., Baur H., Burnett D.S. and Wieler R. (2011), Geochimica et Cosmochimica Acta 75, 3057–3071
Vorobyov E.I. (2011), Astrophysical Journal Letters, 728, L45.
Wadhwa M., Amelin Y., Davis A.M. et al. (2007). In Reipurth, Jewitt & Keil, Eds.,pp. 835–48
Walsh K.J. et al. (2011), Nature 475, 206-209
Wang W., Harris M.J., Diehl R. et al. (2007), Astronomy&Astrophysics 469, 1005–1012
Wasserburg G.J, Busso M, Gallino R, Nollett KM. (2006), *Nucl. Phys. A* 777:5
Weisberg M.K., Ebel D.S., Connolly Jr. H.C. et al. (2011), Geochimica et Cosmochimica Acta 75, 6556–6569
Wetherill, G. W. (1981), Icarus, 46, 70- 80.
Wieler R. and Baur H. (1994), Meteoritics 29, 570–580.
Williams J.P. and Cieza L.A. (2011), Annu. Rev. Astron. Astrophys. 49, 67-117.
Wilkening L. L. and Marti K. (1976), *Geochim. Cosmochim. Acta* 40, 1465-1473.
Willbold M., Elliott T. and Moorbath S. (2011), Nature 477, 195-198
Yokoyama T., Alexander C.M.O'D. and Walker R.J. (2010), Earth and Planetary Science Letters 291 (2010) 48–59
Yurimoto H. and Kuramoto K. (2004), Science 305, 1763–1766
Zinner E., Amari S., Guinness R. et al. (2007), Geochim. Cosmochim. Acta 71, 4786-4813
Zolensky M. E. and Thomas K. L. (1995), Geochim. Cosmochim. Acta 59, 4707- 4712.

5

Measuring the Isotopic Composition of Solar Wind Noble Gases

Alex Meshik, Charles Hohenberg,
Olga Pravdivtseva and Donald Burnett
Washington University, Saint Louis, MO
California Institute of Technology, Pasadena, CA
USA

1. Introduction

It is generally accepted that the primitive Sun, which contains the vast majority of the mass of the solar system, has the same composition as the primitive solar nebula, and that the contemporary Sun has a similar composition except perhaps for light elements modified in main sequence hydrogen burning. The diversity of isotopic and elemental compositions now observed in various solar system reservoirs is most likely the result of subsequent modification and noble gases can provide us with valuable tools to understand the evolutionary paths leading to these different compositions. However, to do this we need to know the composition of the Sun with sufficient precision to delineate the different paths and processes leading to the variations observed and how the present solar wind noble gases may differ from that composition.

Solar optical spectroscopy, the main source of early knowledge about composition of the Sun, does not reveal isotopic information and noble gases do not have useful lines in the solar spectra except for He which was, interestingly, first found in the Sun by this method. Early estimations of solar abundances were based on the combination of photospheric spectral data and laboratory analysis of primitive meteorites which carry a clear signature of their original noble gases. This approach is justified by the fact that the CI chondrites, a rare class of primitive meteorites, and photospheric spectroscopy yield almost identical abundances of most nonvolatile elements. Since meteorites, which were formed by the preferential accretion of solids, clearly differ from the unfractionated solar nebula, the composition of primitive meteorites does not provide a suitable measure of solar system volatiles. Noble gases in meteorites are depleted by many orders of magnitude compared with the solar nebula and, although lunar soils and breccias, implanted with solar wind noble gases, did provide a needed ground truth, neither by themselves could provide a good values for solar volatiles. The first "best estimate" of solar abundances was found by interpolating between adjacent non-volatile elements (Anders & Grevesse, 1989), supplemented with the lunar data, and the later updates (Palme and Beer 1993, Grevesse & Sauval 1998, Lodders, 2010) provided presumably more reliable estimates, but all failed to supply precise isotopic, or even elemental, compositions of the solar noble gases.

Light solar wind noble gases were directly measured by mass spectrometers on various spacecrafts. The most recent of those missions were WIND, Ulysses, SOHO (Solar and Heliospheric Observatory and ACE (Advanced Composition Explorer), see NASA website and review papers (i.e. Wimmer-Schweingruber, 1999, 2001). But the flux is low for the heavier noble gases, and the compositions of the light gases are known to vary with energy, so none of these provided solar isotopic and elemental abundances with sufficient precision.

The Apollo and Luna missions delivered samples of solar wind (SW) accumulated over million years in the lunar regolith. Besides the shallowly implanted SW noble gases, these samples also contained deeper noble gases mainly produced by the spallation reactions of cosmic ray protons and secondary particles, and compositions may be modified by diffusion. In order to delineate the various components, these gases were extracted using stepped pyrolysis and analyzed in sensitive mass spectrometers operated in the static mode. At low temperatures the released gases were dominated by "surface-correlated", mostly SW, while at high temperatures they were mainly "volume-correlated", mostly spallation-produced, noble gases and other in-situ contributions such as radioactive decay. To determine SW compositions, isotope correlation analyses were used. In three-isotope correlation plots two component mixtures are distributed in linear arrays, while for three component mixtures the data fill two-dimensional figures whose apexes define the pure end-member compositions. Several independent analyses using slightly different databases, slightly different techniques and somewhat different assumptions yielded several slightly different compositions for the heavy solar noble gases Xe and Kr. These compositions are referred to as BEOC 10084 (Eberhardt et al., 1970), SUCOR (Podosek et al., 1971) and BEOC 12001 (Eberhardt et al., 1972). Later, in attempts to better separate SW and gases that resided more deeply, stepped extractions from grain-size separates of the lunar fines were carried out (Drozd et al, 1972; Behrmann, et al, 1973; Basford et al., 1973; Bernatowicz et al., 1979) and a lunar soil 71501 was studied using in-vacuo stepped-etching technique (Wieler & Baur., 1995). This technique called CSSE (for Closed System Stepped Etching) allowed a better depth resolution of SW noble gases while reducing the potential mass-fractionation during stepped pyrolysis. Beside lunar soils (Wieler et al., 1986), SW-rich meteorites (e.g. Pesyanoe) were studied and, after significant spallation and other corrections, these studies yielded yet another composition for heavy noble gases in the solar wind (Pepin et al., 1995). All of these determinations of SW noble gases were in general agreement but there were slight differences in composition and no general consensus as to which was best.

One of the major complications was the presence of two seemingly distinct SW noble gas components, apparently residing at different depths within a given target: the "normal" SW and the more deeply implanted, presumably the more energetic component, subsequently labeled SEP, for the Solar Energetic Particle component (not to be confused with SEP, a label for solar flares by the solar physics community). The SEP "component" was first identified by Black & Pepin (1969) and Black (1972) with an apparent $^{20}Ne/^{22}Ne$ ratio <11, much smaller than "normal" SW value of $^{20}Ne/^{22}Ne = 13.7 \pm 0.3$, and they then called it Ne-C or Ne-SF assuming that it was produced by Solar Flares. The low $^{20}Ne/^{22}Ne$ ratio was supported by direct Ne analyses in solar flares (Dietrich & Simpson, 1979, Mewaldt et al., 1981, 1984), so this interpretation gained even more supporters (i.e. Nautiyal et al., 1981, 1986, Benkert at al., 1993) but it had to be much lower in energy than these solar flares. Wieler et al (1986) suggested replacing the term SF with SEP (for Solar Energetic Particles) since the SEP must have energies intermediate between SW and SF ions, and because the SF flux was insufficient to produce the quantity of SEP observed. The proposed SEP component was assumed to have more energy than the typical SW of ~ 1 keV/nucleon but far less energy than the solar flares (~1 MeV/nucleon), commonly referred to as SEP particles by the space physics community, so this terminology was confusing.

There were still more problems in interpreting noble gases released from lunar soils, lunar breccias, and gas-rich meteorites even though they were dominated by the solar wind. Radiation damage and disruption caused by solar wind hydrogen in surfaces exposed to the SW for as little as tens of years leads to lattice defects, enhanced diffusive losses and accelerated surface erosion (compared with laboratory simulations on undamaged samples). With enhanced diffusive losses comes an exaggerated isotopic fractionation effect. The Apollo Solar Wind Composition (SWC) experiment was designed to measure the light solar wind noble gases on pristine metallic surfaces without such effects, and it was successfully carried out during Apollo 11-16, the first five lunar landing missions. In this experiment the Apollo crews exposed Al- and Pt-foils to the SW for up to 45 hours and the foils were returned to Earth where the directly implanted SW-He, Ne and Ar were analyzed in noble gas mass spectrometers (Geiss et al., 1972, 2004). Since the exposure was short enough to avoid saturation effects, it was too short for analyses of the least abundant Xe and Kr and even the Ar analysis ($^{36}Ar/^{38}Ar = 5.4 \pm 0.3$) was not sufficiently precise to delineate solar from terrestrial argon. Moreover, the light noble gases implanted in the foils were easily contaminated by small amounts of dust from the lunar regolith which contained large concentrations of noble gases implanted in the dust by the solar wind but altered by the processes just discussed. Although the dust was largely removed, and these foils represented the best solar wind data for the light noble gases at that time, the problem was not completely resolved.

The Genesis Space Mission (http://genesismission.jpl.nasa.gov/; Burnett et al, 2003) provided a dramatic improvement over the SWC experiment. With more than 400-times longer exposure, much purer collector materials, and free from contamination by most other components, it collected pure contemporary solar wind from outside of the terrestrial magnetosphere. Not only could the compositions of noble gases be determined with new precision, but many other elements could be measured as well including N and O. Ultrapure materials, prepared exclusively for the purpose of SW collection with low-blank analyses (Jurewicz at al., 2003), were exposed to the SW for 27 months at the L1 Lagrangian point, a pseudo-stable location which orbits with the Earth between the Sun and the Earth.

On September 8, 2004 the Genesis returned capsule landed although, due to a parachute failure (Genesis Mishap Investigation Report, 2005), it was not as "soft" as was originally planned. The "hard" landing caused significant delay in SW analyses because of the need to identify and clean several thousand fragments of broken collectors and, in many incidences, develop new techniques for the analyses. In spite of this, Genesis turned out to be a very successful mission with most of the original objectives met, in fact it was the first successful sample return mission since the Apollo era. We present here the results of a comprehensive analysis of SW noble gases collected by the Genesis SW-collectors. All of the analyses reported here were performed at Washington University in St. Louis using mass-spectrometers especially developed for Genesis and laser extraction techniques that continued to evolve during the course of analyses for the mission, descriptions of which are presented in this study.

2. Solar wind collection

Highly ionized solar wind ions, ~ 1keV/nucleon energy, can be effectively collected by most solids, penetrating the lattice, losing energy by scattering, and coming to rest at a depth (range) that is characteristic of the ion, its energy and the target material. Once an energetic ion penetrates a solid it becomes quickly stripped of all residual electrons. After it slows down sufficiently to pick up electrons from the lattice, its charge state is determined by its instantaneous energy and the composition of the target material. Each scattering outcome depends upon specific impact parameters and the interactions between the incident ions and the lattice electrons result in quantum exclusions of some otherwise available states. The constantly changing energy makes these effects even more complex so, in spite of long efforts of many renowned physicists (e.g. Fermi, 1940; Bohr, 1940; Knipp and Teller, 1941 and others), no analytical solution for the ranges of ions has been found. Instead, a Monte Carlo approach is commonly used to simulate each scattering, statistically tracking the trajectory and energy of a population of energetic ions penetrating solid materials and arriving at a distribution of expected ranges as a function of the ion, the initial energy and the target material.

SRIM (the **S**topping and **R**ange of **I**ons in **M**atter) is a suite of the computer codes which calculate the ranges of ions from 10 eV/nucleon to 2 GeV/nucleon in various materials based upon Monte Carlo simulations of successive scatterings, with intermediate trajectories and energies defining conditions for subsequent scatterings, leading to a final distribution of penetration depths (Ziegler et al., 2008, and available at www.srim.org). Figure 1 displays SRIM-2008 results for normal incident solar wind noble gases stopping in aluminum. As can be seen, while heavy noble gases penetrate deeper than lighter ones, all implanted SW-noble gases reside within the outermost 0.3 μm. Therefore, the active portion of the Genesis SW-collectors need not be thicker than a micron or so. Since such a thin foil cannot support itself

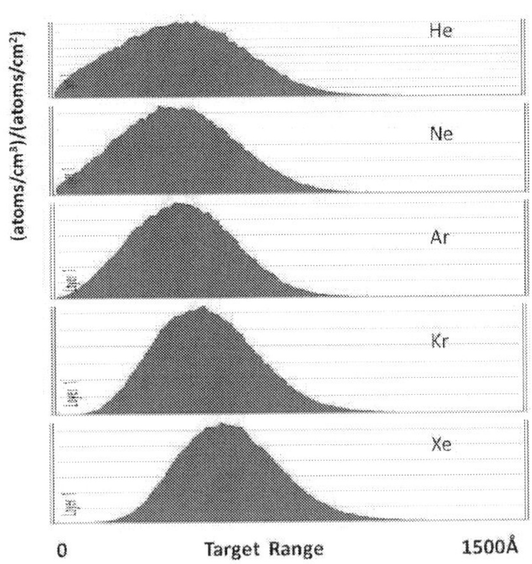

Fig. 1. Depth profiles of noble gases implanted into aluminum at 1kV/nucleon energy. Calculated by SRIM-2008 (Ziegler et al., 2008).

or withstand spacecraft mechanical stress, Genesis collectors are made using an active layer of collector material ~ 1 µm thick deposited on sapphire substrate material (Jurewicz at al., 2003). These "sandwiches" turned out to be very convenient for laser extraction techniques because the laser energy is confined to the coatings where all implanted SW reside. Thus, there is no thermal coupling to the substrate so the substrate does not contribute to background effects from either indigenous or trapped atmospheric noble gases, both of which are ubiquitously present at various levels in all materials including the substrate material used here. This was another significant improvement of the Genesis SW collectors over those of the Apollo SWC experiment in which the foils were self-supporting aluminum films ~ 15 µm thick with the noble gases extracted by pyrolysis (complete melting) of the foils, resulting in noble gas backgrounds at least 15 times higher than in the Genesis 1-µm ultrapure coatings on sapphire.

2.1 Aluminum solar wind collectors

Two types of SW-collectors were used in this study: AloS (~ 15 µm Aluminum deposited on Sapphire substrates) and PAC (Polished Aluminum Collectors). The latter consisted of 0.05" thick highly polished T6 6061 Al-alloy, a material not intended to function as a SW collector but as a thermal control surface. After the hard landing the PAC turned out to be the largest area available for the analyses of SW noble gases, especially important for the heavy noble gases.

Prior to Genesis mission, we had developed a unique method for collecting low-energy cometary volatiles by growing low-Z metal films on sapphire substrates (Hohenberg at al., 1997). This method utilized a technique we referred to as "active capture" and involved the "anomalous adsorption" of Xe and Kr at chemically active sites, permanently entrapping them in the growing metal films (Hohenberg et al., 2002). Anomalous adsorption is a term we use to distinguish the chemical bonding of heavy noble gases from conventional Van der Waals adsorption and requires the availability of unfilled bonds. In the course of refining the active capture technique, low background laser ablation methods were developed to extract noble gases from these films, during which backgrounds, trapping efficiencies and other properties of these films were extensively studied. A natural extension of this work led to the optimized Genesis SW collectors and recovery techniques of impinged SW gases.

The aluminum on Sapphire (AloS) collectors have many advantages over other thin films and over the polished aluminum collectors (PAC). First, Al has a relatively low melting point compared to other metallic films, requiring less laser power for ablation and therefore less energy deposited in the laser extraction cell which results in lower noble gas backgrounds (blanks), especially important for the low abundance heavy noble gases. Second, the low-Z of the target aluminum means that the backscatter of SW ions will be much smaller compared with other potential collectors such as Au, requiring a much smaller back scatter correction especially for the light noble gases where the projectile Z is also low. Third, aluminum is a good conductor, eliminating any charging effects. Finally, the rapid diffusion of hydrogen in Al (compared with Si and other collector materials) reduces lattice damage and lattice distortion effects caused by the huge amounts of SW hydrogen which can adversely affect the quantitative retention of the light noble gases. Moreover, these SW hydrogen effects are difficult to properly model or simulate so reducing the problem is the best way to minimize the effect (Meshik et al., 2000).

The main disadvantage of AloS is that the Al coating is somewhat fragile and can be easily damaged. Several scratched fragments of AloS have demonstrated measurable SW-He losses, especially for the light ^3He isotope, making the remaining He isotopically heavier (Mabry, 2008).

The solid aluminum SW collector, PAC (Figure 2), is a harder material than the AloS film, thus it is somewhat more robust, but the excellent thermal conduction properties of solid aluminum requires a UV laser to adequately couple to the aluminum and a short-pulse to deliver the energy faster than the energy dissipation from the laser pit ("explosive" degassing with each laser pulse). This places some constraints on the laser system but it also provides means for better depth profiling of the released gases.

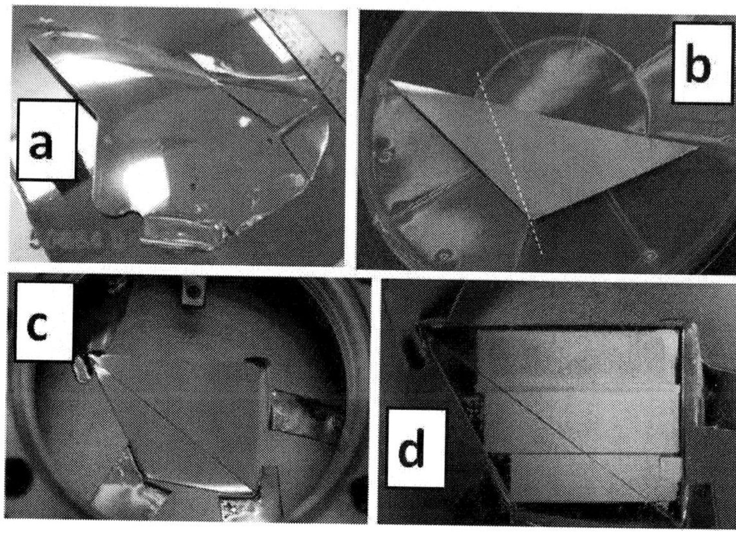

Fig. 2. Genesis Polished Aluminum Collector (PAC) cut in several fragments - (a). Fragment #4 was further split in two parts - (b), which were rearranged for UV-laser ablation and loaded into extraction cell - (c). Last panel (d) shows the PAC after analyses of noble gases released during the ablation.

3. Experimental

3.1 Extraction of noble gases from Genesis solar wind collectors

There are several ways of extracting noble gases from Genesis SW-collectors. The first is simple pyrolysis, melting of the material carrying the SW and extracting noble gases in a single step from the melt. The second is step-wise pyrolysis, increasing the temperature incrementally in steps, allowing extraction by enhanced diffusion, first from the weakly bound sites, usually the shallowly implanted noble gases (and most of the superficial contamination), then progressively from the more deeply implanted noble gases. This method was extensively used in the analyses of SW implanted into lunar soils and breccias, SW-rich (referred to as gas-rich) meteorites and the Apollo SWC-experiments. The third technique, mentioned above as CSSE (Wieler & Baur, 1994), which, in contrast with step-wise pyrolysis, is a step-wise etching method. This is carried out at constant temperature, allowing a better depth resolution of SW noble gases and eliminating some of the mass-fractionation caused by diffusion in step-wise pyrolysis. An elegant version of CSSE was developed for Genesis gold SW-collectors which were step-wise amalgamated *in vacuo* by mercury vapor, thus incrementally releasing SW gases. This technique was also used in analysis of Genesis AuoS (gold film on sapphire) and the solid gold foils (Pepin et al., 2011).

During the evolution of Genesis noble gas analyses, laser extraction techniques were further developed (Meshik et al., 2006) reducing the background even more for the PAC collectors. This provided an alternative to step-wise pyrolysis and CSSE. Gradually increasing the applied UV- laser power with each raster was thought to be capable of separating surface contamination from the more deeply implanted SW noble gases. However, there were considerable complications especially for the low abundant heavy noble gases which required the ablation of several square centimeters of the PAC SW collector to extract enough SW gases for precision measurements. During the process, sputtered aluminum from the collector was deposited both on the walls of the extraction cell and on the internal surface of the vacuum viewport, thus attenuating the laser power delivered to the sample while heating the viewport and the whole extraction cell. This progressively decreased the extraction efficiency of the laser and increased the noble gas background. During the course and evolution of these analyses several improvements were made which reduced the sputtering of collector material on the viewport (Figure 3) and the associated viewport

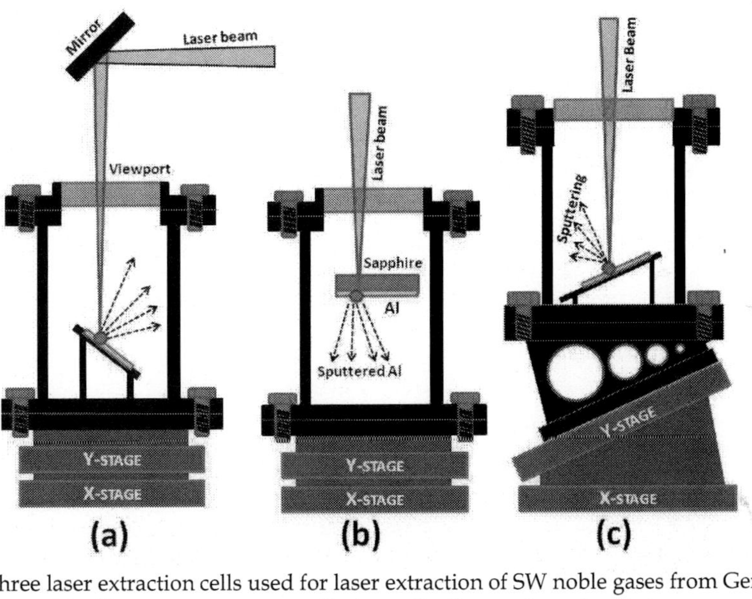

Fig. 3. Three laser extraction cells used for laser extraction of SW noble gases from Genesis Al-collectors. The X-stage moves perpendicular to the figure plane, the Y-stage moves from left to right. The angle between the laser beam and the normal to the ablated surface is $\pi/4$ in (a), π in (b, but the ablated Al does not reach the viewport) and $\pi/8$ (c). The cells (a) and (c) were used for ablation of both AloS and PAC, cell (b) is only suitable for AloS and other Genesis SW collectors with transparent sapphire substrates.

heating, reducing the background. All of these improvements have one thing in common: since the sputtered material emerges as $\sin^2$ of the incidence angle it is highly weighted in a direction perpendicular to the collector surface. By allowing the laser to pass through the viewport and hit the surface at an oblique angle, the sputtered aluminum largely ends up on other parts of the extraction cell, harmlessly away from the vacuum viewport, as shown in the drawing (Figure 3).

3.2 Purification of noble gases prior to isotope analysis

Although all of the Genesis collectors are made of ultrapure materials, some terrestrial heavy noble gases may be trapped at the interface between the sapphire substrate and the collector films. This can largely be avoided by refined extraction techniques which avoid that interface. Reduction of terrestrial contamination from material acquired during the hard landing was done by careful cleaning of the PAC to remove as much of this material as possible (Allton et al., 2006; Calaway et al., 2007). However, considering the fragility of the Al films, the AloS collectors were not extensively cleaned. The only surface treatment of the AloS was the mechanical removal of suspicious dust particles and water spots by repeated rinsing with acetone. Contamination was not a problem for the analyses of the abundant light noble gases, but for the less abundant Kr and Xe variable contamination resulted in backgrounds well above the blank levels observed in the same material not flown on the mission. One explanation for the elevated and non-reproducible behavior of the heavy noble gas background was a "brown stain", a thin Si-based polymerized coating, often observed on the flown collectors and other surfaces of the spacecraft (perhaps formed by UV-polymerization of surface contaminants). Ozone plasma treatments reduced Xe and Kr blank to some extent, but the "flown" and "not-flown" AloS collectors still had very different Kr and Xe backgrounds. It was first thought that some of the elevated blanks may have originated from the interface between the Al-film and sapphire substrate, mentioned previously but then why were the flight and non-flight materials so different?

It was finally realized that neither contamination by the "Utah mud" nor the "brown stain", could be responsible for the differences between the noble gas backgrounds of flight and the non-flight AloS collectors but it was the SW-hydrogen, the dominant SW component, that makes the major difference. The huge amounts of SW hydrogen, released during the Al-ablation of the AloS collector, interact with internal surfaces of the vacuum system and the getter material. Surface oxide removal and reduction of reacted getter alloy liberates significant quantities of noble gases which would otherwise be dormantly trapped there. Interestingly, ultrahigh vacuum systems are known to be efficiently cleaned with hydrogen at elevated temperatures. Therefore, any SW hydrogen implanted and now released from the AloS must be selectively removed from the vacuum system as quickly as possible to prevent excessive noble gas contamination by such surface "cleaning". To remove the hydrogen, we used a Pd finger, a 5 mm diameter Pd tube with 0.3 mm walls, with the interior exposed to the extraction system and the exterior exposed to the atmosphere, and this solved the problem. When the Pd tube was heated to 500°C it removed 99% of hydrogen from the system to the atmosphere in less than 1 minute. It is interesting that oxygen at atmospheric pressure is needed on the exterior of the tube to remove the hydrogen. The Pd finger was the main modification of the noble gas purification line which otherwise is similar to that used in conventional noble gas mass spectrometry.

3.3 Measurements of light noble gases

Precise isotopic analyses of light noble gases require a mass spectrometer with high resolving power (to separate $^{20}Ne^+$ from $^{40}Ar^{++}$ & HF^+ and to separate $^3He^+$ from HD^+ & H^{3+}) and a large dynamic range with minimal pressure effects. No commercially available mass spectrometer satisfied our requirements. Therefore, in this study we used a modified 90° magnetic sector mass-spectrometer, the "SuperGnome" (Hohenberg, 1980). It has highly sensitive GS-61 ion source (Baur, 1980), without electron focusing magnet, which results in an extremely small mass discrimination. Because the extraction fields are small and the ions originate on a cone leading to the same trajectory, all of ions have the same energy and follow nearly the same path. This leads to a tight cluster of trajectories, with no removal by source defining slits, so the instrument has a nearly 100% ion transmission. Since few ions are removed by the slits, it also has low memory effects and long useful counting times. In contrast, a widely used Nier-type ion source has significant and often non-reproducible isotope mass discrimination, lower sensitivity and typically 10-50% ion transmission, which implies that 2 to 9 of every 10 ions are wasted.

The disadvantage of GS-61 ion source is caused by the same things as its advantage. Since the ions originate in a region of low electric field gradient, they are slow at being extracted and then follow the same trajectory. Thus, when the pressure is higher the ion density is fairly large, causing space-charge effects. This effect is non-linear with sensitivity, transmission propensity for double charging caused by variable space charge effects in the ionization region. Space charge effects are not present for the heavy noble gases where ion density is low. However, for SW He and Ne, exacerbated by copious quantities of SW hydrogen, space charge effects can be severe and the extended time in the ionization region leads to significant and variable formation of $^{20}NeH^+$ which interferes with the low abundance $^{21}Ne^+$. The only way to correct for pressure effects in Ne and He measurements is to match the composition of sample to that of an independently known reference standard. In the course of this study we used artificial mixture of helium isotopes with $^3He/^4He = 6.5 \times 10^{-4}$ (manufactured and certified by ChemGas, France), which is much closer to actual SW ratio than to atmospheric. This helium standard was mixed with atmospheric Ne-He to match SW He/Ne ratio, and with hydrogen to simulate the H abundance in these collectors.

Mass resolution was set to ~ 200 so isobaric interferences from doubly ionized CO_2^{++} and $^{40}Ar^{++}$ were present on $^{22}Ne^+$ and $^{20}Ne^+$, respectively, the latter interference was significantly reduced by running the ion source at 48 eV electron energy. The doubly-charged correction factors were typically $^{40}Ar^{++}/Ar^+ = 0.006$ and $CO_2^{++}/CO_2^+ = 0.02$. The correction for interferences at ^3He was more complex since it came from both HD^+ and H^{3+} which are not present at constant proportion and, in fact, also pressure dependent. Luckily, after hydrogen removal, helium becomes the most abundant SW noble gas so the corrections for interference at m/e=3 never exceeded ~10% and the hydride corrections greatly reduced. $^3He/^4He$ and $^{21}Ne/^{20}Ne$ ratios were corrected for small effects due to high counting rates and deadtime, typically from 10 to 12 ns, corresponding to ~ 1% correction at a 1 MHz count rate.

Argon analyses did not present any of these problems because Ar was cryogenically separated from He and Ne (and H) eliminating any of the pressure effects mentioned above. The SW $^{36}Ar/^{38}Ar$ ratio is close to the atmospheric value, and the terrestrial contamination can be accurately subtracted using ^{40}Ar which is absent in SW in measurable amounts.

Isobaric interferences from HCℓ on m/e= 36 and 38 were very small and, since all of our ultrahigh vacuum pumping lines used oil free scroll pumps, magnetically levitated (lubricant-free) turbomolecular pumps and ion pumps, eliminating most hydrocarbon interferences, the ubiquitous hydrocarbon interference at m/e=38 was not present. All of these factors contributed to the fact that our first measured SW $^{36}Ar/^{38}Ar$ ratio was 5.501 ± 0.005 (1σ) (Meshik et al., 2007) which remains the most accurate value for SW argon measured to date.

3.4 Measurements of heavy noble gases

To measure the isotopic compositions of the heavy noble gases, which are present in such low abundances in the solar wind, a special multi-collector version of the Noblesse mass spectrometer was specially constructed for us by NU Instruments. It utilized a "bright", Nier-type, ion source with ~70% ion transmission and unique fast electrostatic zoom lens allowing us to change the effective spacing between isotopes. Since different noble gases have different spacing between the isotopes on the focal plane, variable isotope spacing allows us to use a multiple-dynode collector system to simultaneously measure isotopes of different noble gases. Eight continuous dynode electron multipliers from Burle™, and one Faraday cup collector on the high mass side provided for the simultaneous counting of 9 different ion beams. The high sensitivity of this instrument, 1.8×10^{-16} cm³ STP ^{132}Xe/cps is ~ 3-times higher than that of the SuperGnome, and the 8 multipliers, made this instrument ideal for the low count rate measurements of Genesis SW Kr and Xe. Moreover, the zoom lens allowed Kr and Xe to be measured simultaneously. However, the miniature Burle electron multipliers are mounted just few mm apart, allowing no room for electrostatic shielding so they do suffer from some crosstalk with > 50,000 count/s ion beams. This configuration is, therefore, not as suitable for He and Ne when a high dynamic range is more important but, for the heavy noble gases, when the counting statistic represents the major source of errors, Multi-Noblesse excels.

The Noblesse mass spectrometer has a counting half-life for Xe of ~ 17 minutes, almost 3 times shorter than SuperGnome instrument, reflecting its higher sensitivity, and its Nier-type source makes memory effects more pronounced in the Noblesse. To minimize these effects, and to correct for them, only small spikes of atmospheric Xe and Kr were ever admitted into this mass spectrometer for calibration and all vacuum lines, extraction, purification and pumping systems were assembled from new parts which were never exposed to any isotopically anomalous noble gases. Whenever possible, these parts were internally electropolished to minimize isobaric contaminations and pumping lines were made as short as possible with no pipes being thinner than ¾" in diameter for maximum conductance. Additionally, the high voltage power supply for the ion source was modified to be switched on simultaneously with the beginning of measurement, providing a more precise "time zero" when the gas inside the mass spectrometer has not been yet altered by counting and memory growth. The configuration of ion collector for heavy noble gas measurements is shown in Table 1.

There is a potential problem associated with hydrocarbon interference at m/e = 78 due to the omnipresent C_6H_6 (benzene) which is not completely resolved from ^{78}Kr. Attempts to correct for benzene using hydrocarbons measured at m/e=79 and 77, which were measured anyway, (step 4 in Table 1), were not successful. Luckily ^{78}Kr, the lightest stable Kr isotope,

Magnet, Zoom Lens	Ion collectors (EM - electron multipliers, FC - Faraday Cup):								
	FC	EM1	EM2	EM3	EM4	EM5	EM6	EM7	EM8
B1, Z1		^{136}Xe	^{134}Xe	^{132}Xe	^{130}Xe	^{128}Xe	^{126}Xe	^{124}Xe	
B2, Z2				^{131}Xe	^{129}Xe				
B3, Z3		^{86}Kr		^{84}Kr	^{83}Kr	^{82}Kr		^{80}Kr	
B4, Z4			^{84}Kr	^{83}Kr	^{82}Kr		^{80}Kr		^{78}Kr
B5, Z5	^{40}Ar				^{38}Ar		^{37}Cl		^{36}Ar

Table 1. Assignment of Noblesse ion collectors for Measurements of Genesis heavy noble gases: All isotopes of Xe, Kr and Ar can be measured in five steps of the magnetic field (B) and associated zoom lens (Z) settings. At least one more step (not shown) is needed for the baseline measurement. All Kr isotopes (except ^{86}Kr and ^{78}Kr) are measured twice by different electron multipliers, providing an internal check for the multiplier performance. Switching from one step to another takes less than 2 seconds.

passes through the outermost edge of the zoom lens and, when the electrostatic fringe field of the zoom lens is intentionally distorted, we can measure ^{78}Kr slightly off-center, where the contribution of benzene is more negligible (Figure 4). This distortion does not affect the other Kr isotopes which pass through the middle part of the lens, but it does provide means for partially resolving the benzene interference and obtaining a valid ^{78}Kr measurement.

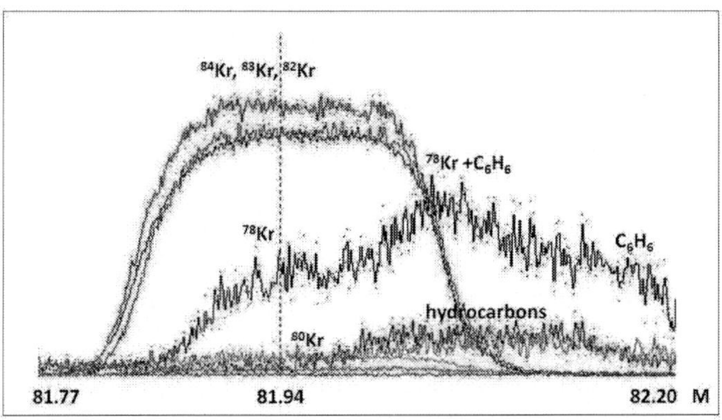

Fig. 4. Simultaneous detection of Kr isotopes 84, 83, 82, 80 and 78. Intentional fringe field distortion of the electrostatic zoom lens allows measurement of ^{78}Kr without significant benzene contribution. Vertical scales (count rates) are different for different isotopes. Horizontal axis is mass for central (axial) isotope, ^{82}Kr in this case. The assignment of ion collectors corresponds to step 4 of Table 1.

However, even after solution of the benzene interference problem at ^{78}Kr, measuring all heavy noble gas isotopes in one run without separation remained difficult because of the limited dynamic range of the miniature electron multipliers and because of the pressure effects in the ion source (although these were much less for the Nier-type ion source in Noblesse than for the GS-61 ion source in the SuperGnome). Additionally, there is the "change-of-charge" effect that interferes with the measurement of ^{80}Kr. As mentioned in 3.3, doubly charged $^{40}Ar^{++}$ interferes with singly charged $^{20}Ne^{+}$ but another effect of doubly charged $^{40}Ar^{++}$ interferes with ^{80}Kr. A small fraction of $^{40}Ar^{++}$ ions can pick up an electron from the source defining slits, becoming $^{40}Ar^{+}$ but with the double energy, thus following the same trajectory as $^{80}Kr^{+}$. This effect is clearly detectible whenever Kr is measured in the presence of ^{40}Ar. Therefore, Ar must be cryogenically separated from Kr, although complete Ar removal cannot be achieved without losing a small fraction of Kr and fractionating the rest. At a temperature of -125°C for activated charcoal trap ~2% of the original Ar is still present so an additional measure is required to further reduce the "change-of-charge" effect on ^{80}Kr. This was done by a reduction of the electron energy from 100 eV to 75 eV at the cost of ~ 10% sensitivity loss. Luckily, the solar wind contains very little, if any, ^{40}Ar so most of the "change-of-charge" problems occur during the calibration of the mass spectrometer.

4. Results and discussion

4.1 Depth profiles of light noble gases

Measuring the composition of noble gases as a function of implantation depth required a uniform laser ablation of the same area of SW collector with each step incrementally increased in the power density delivered to the target. Our frequency quadrupled NdYAG laser (Powerlite-6030 from Continuum™) delivered ~ 10 mJ of 266 nm in 7ns pulses at 30Hz. The best power stability (shot-to-shot) of 12% (barely sufficient for depth profiling) was achieved only at maximum power and only after about a ½-hour "warm-up" period. Several methods were used to control the power: From a pair of rotatable polarizers to attenuating the output power by series of parallel fused quartz plates, each reflecting a few percent of incident beam. However, best results were achieved by selecting delay times of from 125 ms to 300 ms between the flash lamp and the Pockels cell varying the oscillator cavity gain curve of the NdYAG rod.

During the UV-laser step profiling, the laser remained stationary while the extraction cell, mounted on a X-Y-stage moved back and forth (Fig. 3). The stage was programmed to keep velocity constant (typically, 3 mm/s). A fast shutter (computer controlled) blocks the laser during the U-turn of the stage to prevent the power density delivered at the edges of rastered area from increasing beyond that delivered elsewhere. All of the computer codes to control the shutter and the Newport stage via GPIB interface were written in Labview 7.1.

To avoid any contribution of noble gases from the walls of the ablation pit due to stage instability or from beam bleed, with the potential for heating of the un-degassed aluminum adjacent to the rastered area as power increases, each subsequent raster area was made progressively smaller. Therefore, the gas amounts were normalized to the area specific for each step. An example of a completed stepped-power UV laser extraction is shown at the top of the Figure 2d. Depth profiles for He, Ne (preliminarily reported by Meshik et al., 2006; Mabry et al., 2007) and Ar (this work) are assembled in Figure 5.

Solar wind ions are bound to the solar magnetic field and, thus, all ions are implanted with equal velocity so that all SW noble gases (Figure 5) show the same general pattern: The lighter isotopes of each gas (^{3}He, ^{20}Ne, ^{21}Ne and ^{36}Ar) are implanted at shallower depths than the heavier isotopes (^{4}He, ^{22}Ne, and ^{38}Ar), in general agreement with SRIM-2008 simulations for ions implanted at the same velocity, therefore at slightly different energies.

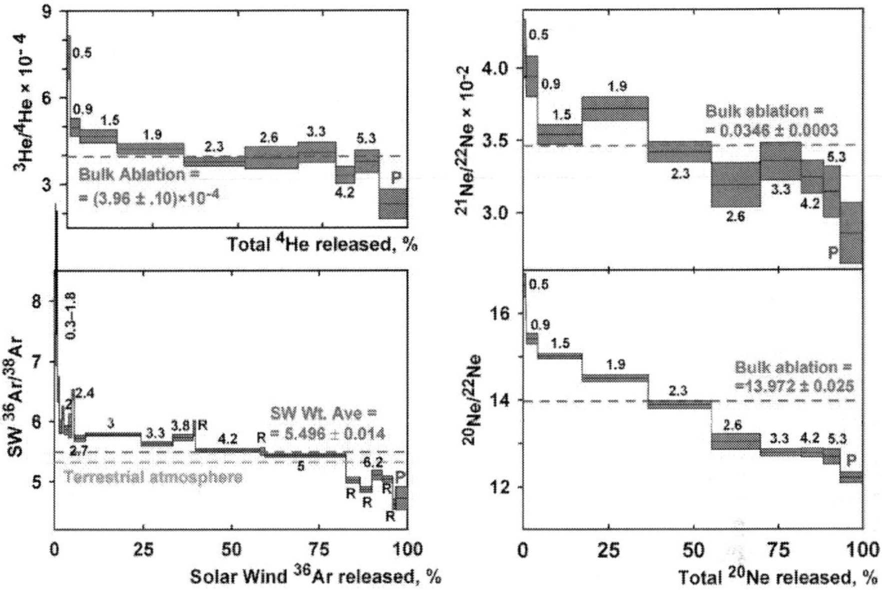

Fig. 5. Isotopic composition of light noble gases extracted from PAC using stepped power UV-laser ablation. Argon in each step has been corrected for the atmospheric contamination assuming that ^{40}Ar is absent in the SW. He and Ne are not corrected, since there is no way to determine the atmospheric contribution to each individual step. Numbers indicate laser output in mJ, R is re-raster with the same power, P stands for pyrolysis made after the completion of laser ablation. Dashed lines show sum of all steps in Ar plot and bulk IR-ablation values of AloS. Error bars are 1σ.

Similar patterns were reported for ^{20}Ne/^{22}Ne ratios measured by the CSSE extractions from the BMG (Bulk Metallic Glass, $Zr_{58.5}Nb_{2.8}Cu_{15.6}Ni_{12.8}Al_{10.3}$) Genesis SW collector (Grimberg et al., 2006). Analogous isotopic effect has been observed in SW He, Ne and Ar in the lunar regolith samples (e.g. Benkert et al., 1993) and, as we discussed in the Introduction, was interpreted at that time as indicating the presence of two distinct solar wind components: (1) Conventional solar wind, SW, and (2) A more energetic, thus more deeply implanted, high energy tail of the solar wind, referred to as Solar Energetic Particles, SEP. Until recently this interpretation was widely accepted, and even became incorporated into noble gas text books (Ozima & Podosek., 2002; Noble Gases in Geochemistry and Cosmochemistry, 2002) as a distinct component. Very few papers (e.g. Becker, 1995) recognized that solar wind isotope ratios will naturally get heavier with implantation depth. Genesis results clearly support this realization and a distinct SEP component is no longer necessary.

Argon was extracted from PAC using 23 steps of UV-laser ablation with some on them being repeat extractions made at the same output laser power. These were the first analyses made using Noblesse multi-collector mass spectrometer. A record low value for the ^{40}Ar/^{36}Ar ratio of 1.12 was found in step #16. This is the most pure SW-Ar (lowest ^{40}Ar/^{36}Ar ratio) ever observed for a natural sample, demonstrating the ability of the laser stepped-power technique to separate SW-Ar from terrestrial contamination, mainly present at the surface of the SW-collector. The total SW $^{36}Ar/^{38}Ar$ = 5.496 ± 0.011 (calculated as weighted sum of all steps) is indistinguishable from $^{36}Ar/^{38}Ar$ = 5.501 ± 0.005 measured in AloS using one step IR-laser extraction (Meshik et al., 2007). Considering that these two measurements were made two years apart using two different mass spectrometers and two different laser extraction techniques, this agreement gives us strong confidence that this is a true SW-Ar composition. Both of these SW $^{36}Ar/^{38}Ar$ analyses agree well with SW-Ar measured independently in different Genesis collectors: AuoS (Gold on Sapphire), DOS (Diamond-like-carbon on Sapphire) and CZ-Si (Czochralski-grown Si). The timeline of SW Ar measurements (Figure 6) demonstrates the high precision of the Genesis results compared to all of the pre-Genesis measurements. Only after Genesis we can confidently conclude that the SW $^{36}Ar/^{38}Ar$ ratio is significantly higher than that in the terrestrial atmosphere, suggesting atmospheric losses in the early evolution of the Earth's atmosphere.

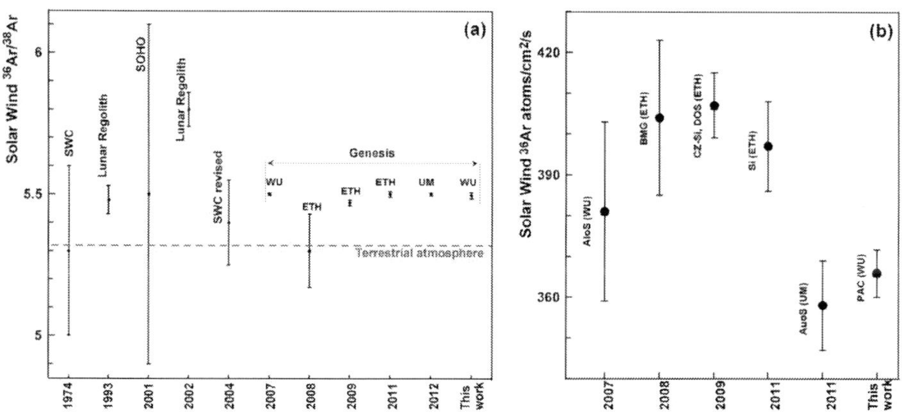

Fig. 6. Comparison of pre-Genesis analyses of SW $^{36}Ar/^{38}Ar$ ratios: (a) (Cerutti, 1974; Benkert et al., 1993; Weygand et al., 2001; Palma et al., 2002; Geiss et al., 2004) with Genesis measurements (Meshik et al., 2007; Grimberg et al., 2008; Heber et al., 2009; Vogel et al., 2001; Pepin et al., 2012; this work). Abbreviations next to Genesis data points stand for laboratories were the analyses were performed: WU - Washington University, ETH - Eidgenössischen Technischen Hochschule Zürich, UM - University of Minnesota. All Genesis results, except for the early, ETH analysis, agree with each other and demonstrate significantly higher precision than those based upon pre-Genesis data. SW ^{36}Ar fluxes at L1 station (b) are measured in different Genesis targets by different laboratories. All error bars are 1σ.

Solar wind ^{36}Ar fluxes are in reasonable, although not perfect, agreement (all are within 3σ). Interestingly, the lower values of SW ^{36}Ar fluxes are found in metal films (Al and Au) while the 8% higher fluxes are observed in nonmetallic materials (Figure 6). Future experiments will show if this difference is real or an experimental artifact.

The stepped-power laser extraction techniques were developed and refined during the evolution of these Genesis analyses and some of the properties of these techniques were realized only after the experiment was completed. One interesting observation was made from the pyrolysis of the PAC which had already been degased by stepped-power laser extractions. Total melting was initially carried out to ensure that all of the SW noble gases had been completely removed by the laser extraction but, as it turned out, this was not the case. Several percent of the SW gases remained present in the PAC even after laser extraction to a depth much greater than the solar wind implantation. UV-laser has sufficient power to extract almost 100% of noble gases from PAC in one extraction step, but when the power increased step-by-step the extraction is no longer complete. About 3.4% of ^{36}Ar, 6.8% of ^{20}Ne and 8.3% of ^{4}He are still present in PAC after 23-steps of laser extraction (step P in Figure 5) and released only by total pyrolysis of the remaining piece. Interestingly, this is more than it could be expected from SRIM simulations (Figure 1) of the solar wind implantation: so this is an extraction effect, not only an implantation effect. Microscopic observations of laser rastered areas of PAC show that the laser raster did not really make an excavation with a flat bottom, but melted, and re-melted the Al several times, evaporating only a part of it. This heating causes enhanced diffusion of gases in the melt and, since light gases move faster than the heavy ones, more He goes into the remaining Al than Ne (and Ar). In other words, heating from the stepped-power laser technique modifies the original distribution of SW noble gases, making the profile wider and deeper with each step so, in this sense, the technique has some properties similar to traditional step-wise pyrolysis. Therefore, the interpretation of stepped-power laser extractions is not as straightforward as we would like it to be because of the modification of the distribution by the extraction itself and perhaps some fractionation effects since the implanted light noble gases are more easily mobilized. One way to reduce the problem is to use laser pulses much shorter than the 7 ns used in this work to more explosively degas the material without as much heating.

The degree of diffusion losses of noble gases in Genesis SW collectors depends on the material and the thermal history. A step-wise pyrolysis is indicative of such losses. Figure 7 shows the cumulative release of ^{20}Ne implanted at 20 keV into the different Genesis materials: PAC, AloS and BMG. The first two materials, Al alloy and pure Al, are significantly less retentive compared to amorphous (below~1000°C) BMG. These Ne release profiles can be used to estimate SW-Ne losses in real Genesis materials.

$^{20}Ne/^{22}Ne$ ratios and fluxes of SW ^{20}Ne, measured in the St. Louis, Minnesota and Zürich laboratories, are shown on Figure 8. Although all measured Ne isotope ratios agree to within 3σ, there is a trend suggesting that the higher $^{20}Ne/^{22}Ne$ ratios seem to correspond to the higher ^{20}Ne fluxes, and the PAC seems to suggest a lower ^{20}Ne flux than either the AloS, BMG or CZ-Si. Given the different thermal diffusion properties of the Genesis collectors (Figure. 7), this seems to make sense. Since exposure times were identical, the lower apparent SW-Ne fluxes indicate some loss of SW Ne. If such losses do occur, the lighter isotope, ^{20}Ne in this case, will escape preferentially for two reasons: (1) it is implanted at shallower depth and, (2) since it is lighter, it is slightly more mobile than ^{22}Ne, thus more susceptible to diffusive loss. Moreover, broadening of the original depth distribution will be more significant for ^{20}Ne than for ^{22}Ne. This has been confirmed by comparison of two fragments of PAC Genesis sample analyzed at different conditions. One was unbaked prior to analysis, another was kept in vacuum for 10 days at 220°C resulting in a lower ^{20}Ne content and a lower $^{20}Ne/^{22}Ne$ ratio. A long-term He diffusion experiment in which a sample of PAC was baked at 240°C for 322 days (38% of the duration of the Genesis collector exposure time) showed large losses of He, confirming significant diffusive losses of light

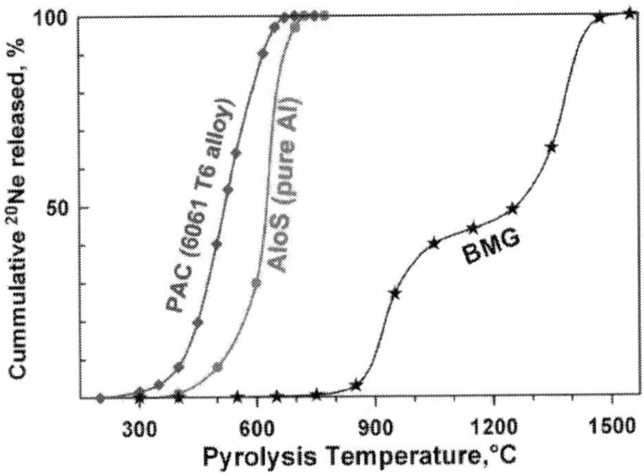

Fig. 7. Release profiles of Ne implanted into different SW collectors at 1 keV/nucleon. Each temperature step was maintained for 30 min. The difference in release curves is the basis for estimation of thermal gas losses and the average temperatures experienced by the Genesis collectors.

noble gases from the PAC although, in that experiment, Ne was not measured (Mabry, 2009). These observations, and the verifying experiments, all point out that some Ne losses, and consequent isotope fractionation, must have occurred with the PAC collector. Although the "low" $^{20}Ne/^{22}Ne$ ratios observed in the PAC agree more closely with the previous "lunar" ratios (cf. Benkert et al., 1993), we believe the higher $^{20}Ne/^{22}Ne$ ratios observed in the AloS collectors, being less modified; provide a better measurement of the modern solar wind. Given a solar wind flux of about 10^7 protons/cm^2/s, lunar surface material is quickly saturated with solar wind hydrogen to the point that, without extensive diffusive redistribution, the implanted solar wind hydrogen atoms will outnumber the host lattice atoms in a broad region near the end of its range in only a few tens of years. This means, among other things, extensive lattice damage and enhanced surface erosion with associated effects on the diffusion and retention of the implanted light noble gases. We, therefore, expect large and variable diffusive losses from lunar soils and regolith samples. In addition, even though the foils were carefully cleaned, the Apollo Solar Wind Composition Experiment is still susceptible to contamination by fine lunar dust that contains both diffusively modified solar wind Ne and spallation-produced Ne. Thus, we conclude that AloS, CZ-Si and DOS measurements should provide the definitive composition of the modern solar wind Ne. However, the Ne measured in the Zurich laboratory in the Si and DOS collectors, which are expected to be equally retentive, appear to be slightly heavier than those measured in the AloS collectors (St. Louis) and in the AuoS collectors (Minneapolis), as shown in Figure 8. At present time we do not have a reasonable explanation. However, a higher resolution (~1500) mass-spectrometer is expected to be installed at Washington University in the future, it will be capable of resolving $^{40}Ar^{++}$ from $^{20}Ne^{+}$, removing one of the uncertainties in Ne analysis. Re-analysis of Genesis SW neon using this instrument will provide an opportunity for better precision and exploration of any apparent discrepancy in the SW $^{20}Ne/^{22}Ne$ ratios obtained by the different laboratories.

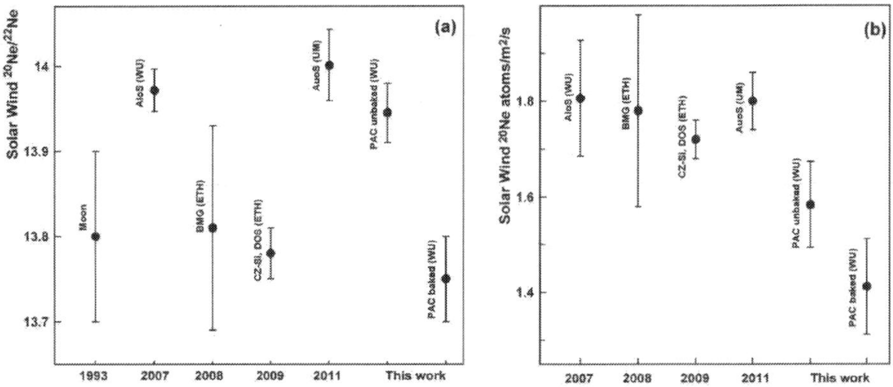

Fig. 8. Solar wind $^{20}Ne/^{22}Ne$ ratios (a) and ^{20}Ne fluxes at L1 station (b) measured by different laboratories in different Genesis SW-collector materials. All abbreviations are the same as in Fig. 6. "Lunar" SW-Ne is from Benkert et al., 1993. A diffusion experiment demonstrates that PAC kept in vacuum for 10 days at 220°C may lose some Ne, preferentially ^{20}Ne.

Helium is the most abundant noble gas in the SW. It is also the lightest, the most susceptible to diffusive loss and, because it has the largest relative difference in masses of its two isotopes, it is the most indicative of isotopic mass fractionation. All Genesis He analyses and some "pre"- Genesis results are shown in Figure 9. Both isotope ratios and apparent SW He fluxes are scattered much more than would be justified by the statistical uncertainties.

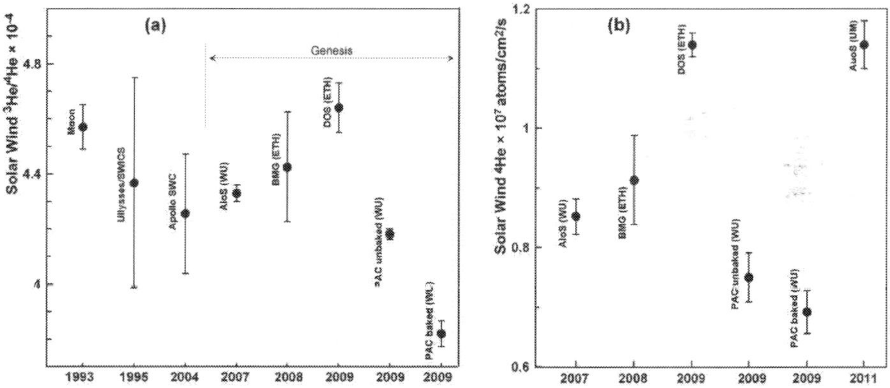

Fig. 9. Isotopic compositions (a) and fluxes (b) at the L1 point measured in different Genesis SW collectors in St. Louis, Minnesota and Zürich labs. Lunar and Apollo SWC SW data are from Benkert et al., 1993 and Geiss et al., 2004. The 500 day average He composition from Ulysses/SWICS is from Bodmer et al., 1995. Aluminum collectors (both AloS and PAC) are from Mabry et al., 2007 and Mabry., 2009. PAC (baked) was kept at 240°C for 322 days, ~38% of total Genesis collection time. Helium losses from all Al-collectors are evident and apparently are accompanied by preferential losses of ^{3}He.

Systematical errors, not reflected in the data, evidently exist in this figure. Both types of Al collectors have significantly lower concentrations of SW-He (Figure 9b), demonstrating the diffusive losses expected from the thermal release profiles shown in Figure 7. The AloS and the PAC have the lowest $^{3}He/^{4}He$ ratios observed among the SW collectors and pre-Genesis SW-He determinations, suggesting that aluminum diffusively loses He at the temperature of the exposed collector surfaces and, as expected from the diffusive properties shown in Figure 7, the PAC loses more He than the AloS collector.

The real-time diffusion experiment conducted by Mabry (2009) confirms the poor SW-He retention properties of the PAC at the elevated collection temperature. Both the $^{3}He/^{4}He$ ratios and the He concentrations (reflecting apparent He fluxes) were significantly lower following a 322 day vacuum bake (38% of the Genesis mission) at 240°C. Even the unbaked reference sample of the PAC demonstrates the lowest measured apparent He fluxes and $^{3}He/^{4}He$ ratios indicating significant He losses during the Genesis collection period, not surprising since the temperatures of Genesis PAC and AloS collectors were estimated to be around 165°C (Mabry, 2009). Therefore, none of the Genesis aluminum collectors completely retain solar wind He or preserve the original $^{3}He/^{4}He$ ratios, both can only be considered as lower limits. Among the other SW-He collectors, DOS (Diamond-like Carbon on Sapphire) CZ-Si (Czochralski-grown silicon) and gold (both AuoS and foil), DOS is probably the best choice since it does not require as high backscattering corrections (up to 35% for Au). AloS however do not demonstrate significant Ne losses and completely retain SW Ar. Therefore, AloS was the choice material for analyses of SW heavy noble gases.

4.2 Heavy noble gases

The large concentration of the light SW noble gases He and Ne in the Genesis collectors meant that corrections for atmospheric or other contaminations were usually negligible. Argon from the collectors contained significant terrestrial contributions but since the solar wind has negligible ^{40}Ar, and the terrestrial isotopic ratios are well known, this can easily be removed to leave pure solar wind argon. For krypton and xenon, which are far less abundant in the solar wind, the terrestrial contamination becomes a serious problem and there is no "terrestrial only" isotope to identify the trapped component. In fact, the compositions of SW and terrestrial noble gases are not significantly different so partitioning by isotopic composition is not possible. Our original intention was to use stepped-power laser extraction to separate any superficial surface-correlated contamination from the more deeply implanted SW noble gases. A complicating factor is the low abundances of the heavy noble gases in the SW which requires analyzing very large areas of the collectors for precise measurements in stepped-power laser extractions. The conventional way to document terrestrial Xe and Kr contributions is to analyze reference (non-flight) SW collectors, manufactured in the same way as the flight collectors, utilizing the same procedures and raster areas. The Xe and Kr signals measured in these non-flight coupons would then be a proxy for blanks in the actual collectors. Unfortunately, the AloS collectors were manufactured in several batches and after the "hard" landing of the Genesis return capsule it became challenging to pair flight and non-flight AloS material. That said, a more severe problem was found. In the laser extraction experiments it was observed that the Xe and Kr blanks were neither proportional to the raster areas nor were they very reproducible. It was soon realized that the large quantities of implanted SW hydrogen released from SW collectors reacted with the getter material. Since the SAES getters were produced by sintering in an inert atmosphere, this liberated dormant Xe and Kr from the getters. The quantity of hydrogen was so large that it could not be separated cryogenically in the sample system. Removing hydrogen from the flight tube using palladium (described in 3.2) significantly reduced this problem, but did not eliminate it completely, so new techniques for minimizing terrestrial noble gases had to be developed.

The alternative approach for blank correction is based on the significant difference in $^{84}Kr/^{132}Xe$ ratios between the terrestrial atmosphere (27.78; Ozima & Podosek., 2002) and the solar wind (9.55, Meshik et al., 2009). In the case of binary mixtures of SW and terrestrial components, the $^{84}Kr/^{132}Xe$ can be used as a measure of terrestrial contribution. Since the Washington University multi-collector mass spectrometer, the laser extraction cells and the purification system have never seen any isotopically anomalous gases, we are limited to these two compositions (with negligible mass fractionation). A capability to simultaneously measure both heavy noble gases in a single run (3.4) was needed to use this approach, but this was the plan all along. Laser extraction was done in a single step using maximal power to ensure the complete extraction of SW noble gases, to provide maximum signal and to minimize the analysis time. Xe and Kr were cryogenically separated from at least 98% of the Ar using activated "Berkeley" charcoal finger kept at -125°C, which reduced the change-of-charge effect at ^{80}Kr. Both PAC and AloS collectors were analyzed in different laser cells (shown in Figure 3) using pulsed laser extraction at two wavelengths: 266 nm for PAC and 1064 nm for AloS. It was realized that Kr and Xe may not be trapped in atmospheric proportion, with Xe usually more "sticky" than Kr, but it was assumed that they would probably not be isotopically fractionated to any significant degree (an assumption that could be checked later). To determine the actual trapped $^{84}Kr/^{132}Xe$ ratio we assumed that, for all 24 samples analyzed, this ratio was constant. Equations (1) and (2) describe binary mixtures between SW and terrestrial trapped gases for each measurement:

$$^{84}Kr_{SW} = {}^{84}Kr_M \times \left[\left(\frac{^{132}Xe}{^{84}Kr}\right)_M - \left(\frac{^{132}Xe}{^{84}Kr}\right)_T\right] \Big/ \left[\left(\frac{^{132}Xe}{^{84}Kr}\right)_T - \left(\frac{^{132}Xe}{^{84}Kr}\right)_{SW}\right] \quad (1)$$

$$^{132}Xe_{SW} = {}^{132}Xe_M \times \left[\left(\frac{^{84}Kr}{^{132}Xe}\right)_M - \left(\frac{^{84}Kr}{^{132}Xe}\right)_T\right] \Big/ \left[\left(\frac{^{84}Kr}{^{132}Kr}\right)_T - \left(\frac{^{84}Kr}{^{132}Xe}\right)_{SW}\right] \quad (2)$$

Here SW refers to Solar Wind, M to Measured and T to Trapped (or Terrestrial) and the two unknowns are $(^{132}Xe/^{84}Kr)_{SW}$ and $(^{132}Xe/^{84}Kr)_T$. With two equations only two measurements are needed to determine the values for these ratios but, for the 24 measurements available, the system is over-determined. A multi-variance solution is obtained from minimization of the standard deviations of the SW fluencies and the most probable values for $(^{132}Xe/^{84}Kr)_{SW}$, and correspondingly, $^{132}Xe_{SW}$ and $^{84}Kr_{SW}$ were obtained. The best convergence, shown in Figure 10a, was achieved at $(^{84}Kr/^{132}Xe)_{trapped}$ = 24.4 (Figure 10b), only 12% lower than the terrestrial atmosphere, a value confirming our assumption of no significant isotopic mass fractionation in this component.

All Kr and Xe isotopic analyses are shown in Figures 11 and 12, respectively, and Table 2 presents final results. All of the data in these figures show consistent results even though

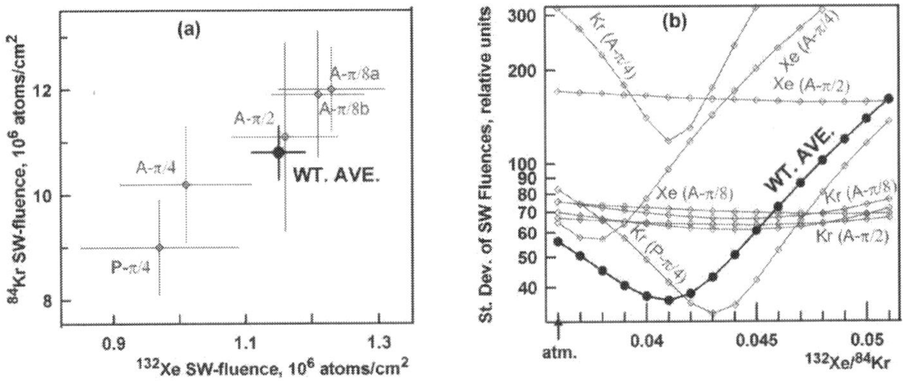

Fig. 10. The best convergence of SW fluencies (a) has been achieved at $^{132}Xe/^{84}Kr = 0.041$ (b), providing our current fluence estimate: $(1.15 \pm 0.04) \times 10^6$ ^{132}Xe atoms/cm^2 and $(1.08 \pm 0.05) \times 10^7$ ^{84}Kr atoms/cm^2.

^{86}Kr	^{84}Kr	^{83}Kr	^{82}Kr	^{80}Kr	^{78}Kr
.3012 (4)	≡ 1	.2034 (2)	.2054 (2)	.0412 (2)	.00642 (5)

^{136}Xe	^{134}Xe	^{132}Xe	^{131}Xe	^{130}Xe	^{129}Xe	^{128}Xe	^{126}Xe	^{124}Xe
.3003 (6)	.3692 (7)	≡ 1	.8263 (13)	.1649 (4)	1.0401(10)	.0842 (3)	.00417(9)	.00492 (7)

Table 2. Isotopic composition of heavy noble gases in solar wind measured in aluminum Genesis collectors. Errors are 1σ.

they represent both types of aluminum SW collectors, were analyzed in different extraction cells under different conditions, using two different pulsed laser wavelengths, and were performed several months apart.

The isotopic composition of solar wind heavy noble gases from the Genesis collectors (this work) can be compared with solar wind Xe and Kr previously inferred from lunar surface material (c.f. Pepin et al., 1995, Figure 13).

A few first-order observations can be made: The isotopic ratios of heavy SW noble gases implanted by lunar regolith over millions of years are indistinguishable from the contemporary SW observed in the Genesis collectors to within < 1%. This sets an upper limit to possible temporal variations of SW Kr and Xe. The small isotope differences we do observe suggest that SW-Kr inferred from the lunar regolith is slightly heavier than that we measure in Genesis while no such trend is observed for Xe. SW compositions inferred from lunar regolith may be more subjective to systematic error, and they are less precise than those measured by Genesis, at least at the major isotopes. For instance, the trend we see in this comparison is suggestive of some diffusive loss of Kr from the lunar regolith, not the case for the more retentive Xe. However, in order to test whether this effect is real, Kr compositions inferred from the lunar regolith should be revisited.

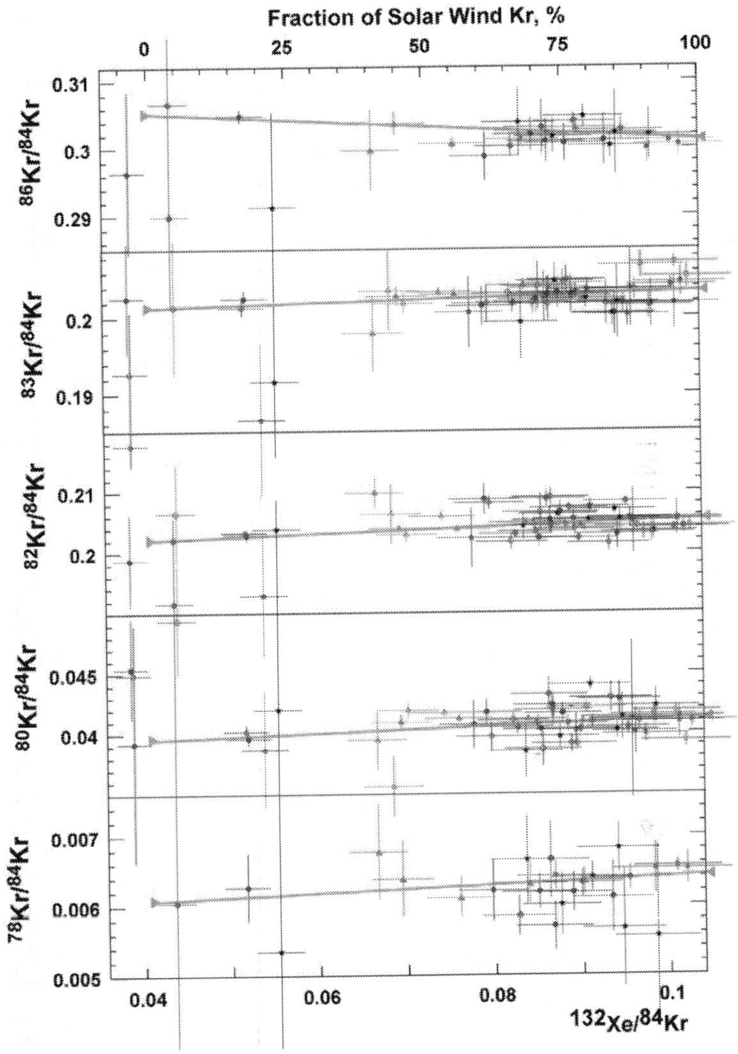

Fig. 11. Kr isotopic composition measured in Genesis Al collectors. Fitting line forced trough the estimated trapped component, the ordinate intercept gives isotopic composition of the solar wind. Different colors correspond to different experimental conditions, which within statistical errors result in the same SW composition.

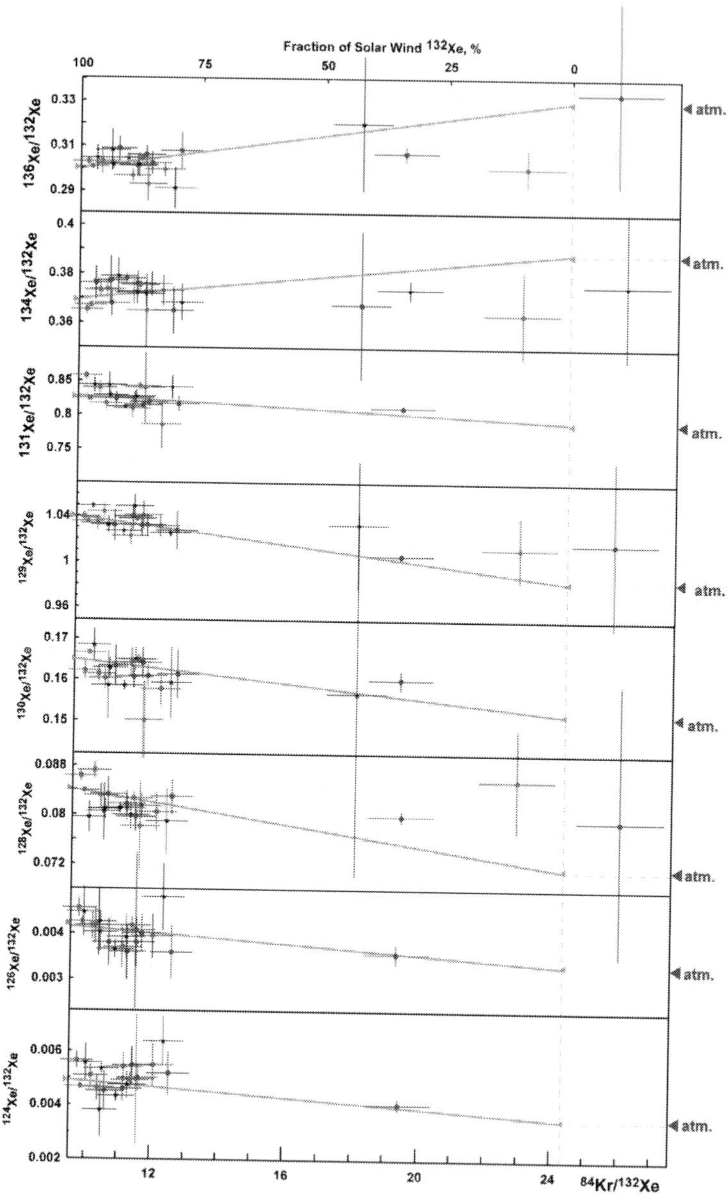

Fig. 12. Xe isotopic composition in Genesis Aluminum collectors. Fitting line forced trough the estimated trapped component, ordinate intercept gives isotopic composition of the solar wind. Different colors correspond to different experimental conditions.

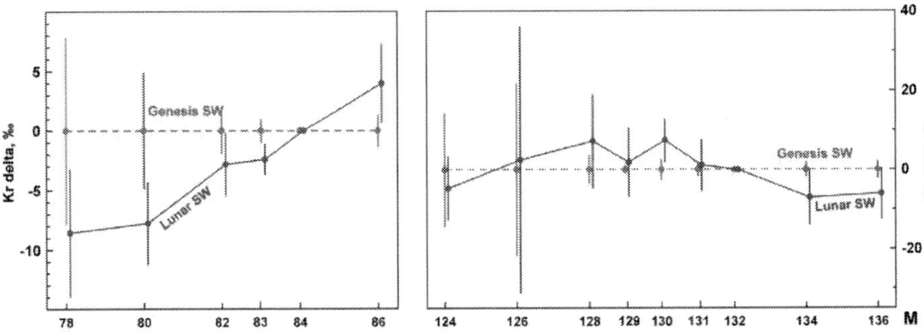

Fig. 13. "Lunar" SW (Pepin at al, 1995) vs. Genesis SW (this work). Although normalized to the Genesis composition as a delta plot, the errors shown (1σ) are not propagated, so overlapping errors infer consistency.

5. Conclusion

Depth profile of solar wind argon retrieved from Genesis Al collector confirms the isotopic fractionation which occurs during implantation at constant velocity, an effect previously observed for helium and neon. This eliminates the need for a unique heavy SEP noble gas component thought to be required from analyses of lunar surface material. New, more precise Ar analyses, and recent results from two independent laboratories, confirm our earlier published value for the $^{36}Ar/^{38}Ar$ ratio in the solar wind. The unique analytical capability developed at Washington University to simultaneously analyze all Xe and Kr isotopes allows us to determine the composition of heavy noble gases in the solar wind to a precision exceeding previous values inferred from lunar surface material.

6. Acknowledgments

We are in debt to Judith Allton, Patti Burkett, Phil Freedman, Amy Jurewicz, Karen McNamara, Melissa Rodriguez, John Saxton, Eileen Stansbery, Rainer Wieler, Roger Wiens Dorothy Woolum, and all members of Genesis Science Team for their invaluable help and support. This work was supported by NASA grant NNX07AM76G.

7. References

Allton, JH., Calaway, MJ., Hittle, JD., Rodriguez, MC., Stansbery, EK., & McNamara, KM. (2006) Cleaning surface particle contamination with ultrapure water (UPW) megasonic flow on Genesis array collectors. Lunar Planet. Sci. XXXVII, 2324

Anders, E., Grevesse, N. (1989) Abundances of elements: meteoritic and solar. *Geochimica et Cosmochimica Acta*, Vol. 53, pp. 197-214

Basford, JR., Dragon, JC., Pepin, RO., Coscio, MR., & Murthy, VR. (1973) Krypton and xenon in lunar fines. *Proceedings of 4th Lunar and Planetary Science Conference, Houston, Texas*, pp. 1915-1955

Baur, H. (1980) Numerische simulation und praktische erprobung einer rotations symmetrischen lonenquelle für gasmassenspektrometer. Abhandlung zur Erlangung des Titles eines Doktors der Naturwissenschaften der Eidgenössischen Technischen Hochschule Zürich., Diss. ETH No. 6596

Becker, RH. (1998) Effects of grain-surface erosion on solar wind isotopic ratios and SEP mixing ratio in extraterrestrial samples. *Proceedings of 29th Lunar and Planetary Science Conference, Houston, Texas*, abstract #1329

Behrmann, C., Drozd, R., Hohenberg, CM., & Ralston. C. (1973) Extinct lunar radioactivities: xenon from ^{244}Pu and ^{129}I in Apollo 14 breccias. *Earth & Planetary Science Letters*, Vol. 17, p. 446

Benkert, J-P., Baur, H., Signer, P., & Wieler. R. (1993) He, Ne, and Ar from the solar wind and solar energetic particles in lunar ilmenites and pyroxenes. *Journal of Geophysical Research*, Vol. 98, pp. 13147-13162

Bernatowicz, TJ., Hohenberg, CM., & Podosek. FA. (1979) Xenon component organization in 14301. *Proceedings of 10th Lunar and Planetary Science Conference, Houston, Texas*, pp. 1587-1616

Black, DC. (1972) On the origin of trapped helium, neon and argon isotopic variations in meteorites - I. Gas-rich meteorites, lunar soil and breccias. *Geochimica et Cosmochimica Acta*, Vol. 36, pp. 347-375

Black, DC., & Pepin, RO. (1969) Trapped Ne in meteorites - II. *Earth & Planetary Science Letters*, Vol. 6, pp. 395-405

Bohr, N. (1940) Scattering and stopping of fission fragments. *Physical Reviews*, Vol. 58, 654-655

Bodmer, R., Bochsler, P., Geiss, J., von Stieger, R., & Gloecker, G. (1995) Solar-wind helium isotopic composition from SWICS Ulyses. *Space Science Reviews*, Vol. 72, pp. 61-64

Burnett, DS., Barraclough, BL., Bennett, R., Neugebauer, M., Oldham, LP., Sasaki, CN., Sevilla, D., Smith, N., Stansberry, E., Sweetnam, D., & Wiens, RC. (2003) The Genesis Discovery Mission: Return of Solar matter to Earth, *Space Science Reviews*, Vol. 105, pp. 509-534

Calaway, MJ., Burnett, DS., Rodriguez, MC., Sestak, S., Allton, J.H., & Stansbery, K. (2007) Decontamination of Genesis Array materials by ozone cleaning. Lunar and Planetary Science XXXVIII 1627.pdf

Cerutti, H. (1974) Die Bestimmung des Argons im Sonnenwind aus Messungen an den Apollo-SWC-Folien, PhD Thesis, University of Bern, Switzerland

Dietrich, WF., & Simpson, JA. (1979) The isotopic and elemental abundances of neon nuclei accelerated in solar flares. Astrophys. J 231 :L91-L95

Drozd, R., Hohenberg, CM., & Ragan, D. (1972) Fission xenon from extinct 244Pu in 14301. *Earth & Planetary Science Letters*, Vol. 15, p. 338

Eberhardt, P., Geiss, J., Graf, H., Grögler, N., Krähenbühl, U., Schwaller, H., Schwarzmüller, J., & Stettler, A. (1970) Trapped solar wind noble gases, exposure age and K/Ar-age in Apollo II lunar fine material. Proc. Apollo II Lunar Sci. Conf., Geochimica & Cosmochimica Acta, Suppl. 1, Vol. 2, pp.1037-1070. Pergamon.

Eberhardt, P., Geiss, J., Graf, H., Grögler, N., Mendia, MD., Mörgeli, M., Schwaller, H., & Stettler, A. (1972) Trapped solar wind noble gases in Apollo 12 lunar fines 12001 and Apollo 11 breccia 10046. *Proceedings of 3rd Lunar and Planetary Science Conference, Houston, Texas, Geochimica et Cosmochimica Acta*, Suppl. 3, Vol. 2, pp. 1821-1856

Fermi, E. (1940) The ionization loss of energy in gases and condensed materials. *Physical Review,* Vol. 57, No. 6, pp. 485-493

Geiss, J., Bühler, F., Cerutti, H., Eberhardt, P., & Filleux, CH. (1972) Solar Wind Composition Experiment, Apollo 16 Prelimenary Science Report, NASA SP-315, 14-1–14-10

Geiss, J., Bühler, F., Cerutti, H., Eberhardt, P., Filleux, CH., Meister, J., & Signer, P. (2004) The Apollo SWC experiment: Results, conclusions, consequences. *Space Science Reviews,* Vol. 110, pp. 307-335.

Grevesse, N., & Sauval, AJ. (1998) Standard solar composition. *Space Science Reviews,* Vol. 85, pp. 161-174

Grimberg, A., Baur, H., Bochsler, P., Bühler, F., Burnett, DS., Hays, CC., Heber VS., Jurewicz, AJG., & Wieler R. (2006) Solar wind neon from Genesis: implication for the lunar noble gas record. *Science,* Vol. 314, pp. 1133-1135

Grimberg, A., Baur, H., Bühler, F., Bochsler, P., & Wieler R. (2008) Solar wind helium, neon, and argon isotopic and elemental composition: Data from the metallic glass flown on NASA's Genesis mission. *Geochimica et Cosmochimica Acta,* Vol. 72, pp. 626-645

Heber, VS., Wieler, R., Baur, H., Olinger, C., Friedmann, TA., & Burnett, DS. (2009) Noble gas composition of the solar wind as collected by the Genesis mission. *Geochimica et Cosmochimica Acta,* Vol. 73, pp. 7414-7432

Hohenberg, CM. (1980) High sensitivity pulse-counting mass-spectrometer system for noble gas analysis. *Revue of Scientific Instruments,* Vol. 51, No. 8, pp. 1075-1082

Hohenberg, CM., Thonnard, N., Kehm, K., Meshik, AP., Berryhill, & A., Glenn, A. (1997) Active capture of low-energy volatiles: Bringing back gases from cometary encounter. *Proceedings of 27th Lunar and Planetary Science Conference, Houston, Texas,* pp. 581-582

Hohenberg, CM., Thonnard, N., & Meshik, A. (2002) Active capture and anomalous adsorption: New mechanisms for the incorporation of heavy noble gases. *Meteoritic and Planetary Science,* Vol. 37, pp. 257-267

Jurewicz, AJG., Burnett, DS., Wiens, RC., Friedmann, TA., Hays, CC., Hohlfelder, RJ., Nishizumi, K., Stone, JA., Woolum, DS., Becker, R., Butterworth, AL., Campbell, AJ., Ebihara, M., Franchi, IA., Heber, V., Hohenberg, CM., Humayun, M., McKeegan, KD., Mcnamara, K., Meshik, A., Pepin, RO., Schlutter, D., & Wieler, R. (2003) The Genesis solar-wind collector materials. *Space Science Reviews,* Vol. 105, pp. 535-560

Knipp, J., & Teller, E. (1941) On energy loss of heavy ions. *Physical Review,* Vol. 59, No. 8, pp. 659-669

Lodders, K. (2010) Solar system abundances of the elements. In: Principles and Perspectives in Cosmochemistry, Goswami and Reddy (eds), Astrophysics and Space Science Proceedings, Springer-Verlag Berlin Heidelberg, pp. 379-417 (ISBN 978-3-642-10351-3)

Mabry, JC. (2009) Solar wind helium, neon, and argon in Genesis Aluminum collectors. PhD Thesis, Washington University in Saint Louis, USA

Mabry, JC., Meshik, AP., Hohenberg, CM., Marrocchi, Y., Pravdivtseva, OV., Wiens, RC., Olinger, C., Reisenfeld, DB., Alton, J., Basten, R., McNamara, K., Stansbery, E., & Bernett, DS. (2007) Refinement and implications of noble gas measurements from Genesis. *Proceedings of 38th Lunar and Planetary Science Conference, Houston, Texas,* Abstract #2412

Mewaldt, RA., Spalding, JD., Stone, EC., & Vogt, RE. (1981) High resolution measurements of solar flare isotopes. Proc 17th Int. Cosmic Ray Conf. 3:131-135

Mewaldt, RA., Spalding, JD., & Stone, EC. (1984) A high resolution study of the isotopes of solar flare nuclei. *The Astrophysical Journal,* Vol. 280, pp. 892-901

McKeegan, KD., Kallio, APA., Heber, VS., Jarzebinski, G., Mao, PH., Coath, CD., Kunihiro, CD., Wiens, RC., Nordholt, JE., Moses, RW., Reisenfeld, DB. Jr., Jurewicz, AJG., & Burnett, DS. (2011) The oxygen isotopic composition of the Sun inferred from captured solar wind. *Science,* Vol. 332, pp. 1528-1532

Marty, B., Chaussidon, M., Wiens, RC., Jurewicz, AJG., & Burnett, DS. (2011) A ^{15}N-poor isotopic composition for the solar system as shown by Genesis solar wind samples. *Science,* Vol. 332, pp. 1533-1536

Meshik, AP., Hohenberg, CM., Burnett, DS., Woolum, DS., & Jurewicz, AJG. (2000) Release profile as an indicator of solar wind neon loss from Genesis collectors. *Proceedings of 63rd Annual Meeting of the Meteoritical Society,* Abstract #5142

Meshik, AP., Marrocchi, Y, Hohenberg, CM., Pravdivtseva, OV., Mabry, JC., Allton, JH., Bastien, R., McNamara, K, Stansbery E, & Burnett. DS. (2006) Solar neon released from Genesis aluminum collector during stepped UV-laser extraction and step-wise pyrolysis. 69th Annual Meeting of the Meteoritical Society, Abstract #5083

Meshik, A., Mabry, J, Hohenberg, C., Marrocchi, Y, Pravdivtseva, O., Burnett, D, Ollinger, C., Wiens, R., Reisenfeld, D., Allton, J., McNamara, K, Stansbery E, & Jurewicz A. (2007) Constraints on Neon and Argon isotopic fractionation in solar wind. *Science* 318, 433-435

Meshik, AP., Hohenberg, CM., Pravdivtseva, OP., Mabry, JC, Allton, JH., & Burnett, DS. (2009) Relative abundances of heavy noble gases from the polished aluminum solar wind collector on Genesis. *40th Lunar and Planetary Science Conference, Houston, Texas,* Abstract #2037

National Aeronautics and Space Administration (2005) Genesis Mishap Investigation Board Report. Available from http://www.nasa.gov/pdf/149414main_Genesis_MIB.pdf

Nautiyal, CM., Padia, JT., Rao, MN., & Venkatesan, TR. (1981) Solar flare neon: Clues from implanted noble gases in lunar soils and rocks. *Proceedings of 12ᵛ Lunar and Planetary Science Conference, Houston, Texas*, pp. 627-637

Ozima, M., & Podosek, FA. (2002) *Noble gas geochemistry* (2nd edition). Cambridge University Press, ISBN 0-521-80366-7, USA

Palme, H., & Beer, H. (1993) Abundances of the elements in the solar system. In: Landorf-Börnstein, Group VI: Astronomy and Astrophysics, Voigt HH (ed), 3(a) Springer, Berlin, p 196-221

Palma, RL., Becker, RH., Pepin, RO., & Schlutter, DJ. (2002) Irradiation records in regolith materials, II: Solar wind and solar energetic particle components in helium, neon, and argon extracted from single lunar mineral grains and from Kapoeta howardite by stepwise pulse heating. *Geochimica et Cosmochimica Acta*, Vol. 66, pp. 2929-2958

Pepin, RO., Becker, RH., & Rider, PE. (1995) Xenon and krypton in extraterrestrial regolith soils and in the solar wind. *Geochimica et Cosmochimica Acta*, Vol. 59, pp. 4997-5022

Pepin, RO., Schlutter, DJ., & Becker, RH. (2012) Helium, neon and argon composition of the solar wind as recorded in gold and other Genesis collector materials. *Geochimica et Cosmochimica Acta (in press).*

Podosek, FA., Huneke, JC., Burnett, DS., & Wasserburg, GJ. (1971) Isotopic composition of xenon and krypton in the lunar soil and in the solar wind. *Earth Planetary Science Letters*, Vol. 10, pp. 199-216

Porcelli, D., Ballentine, CJ., & Wieler, R. (Editors). (2002) *Noble Gases in Geochemistry and Cosmochemistry*. The Mineralogical Society of America, ISBN 0-939950-59-6, USA

Vogel, N., Heber, VS., Baur, H., Burnett, DS., & Wieler, R. (2011) Argon, krypton and xenon in the bulk solar wind as collected by the Genesis mission. *Geochimica et Cosmochimica Acta*, Vol. 75, pp. 3057-3071

Weygand, JM., Ipavich, FM., Wurz, P., Paquette, JA., & Bochsler, P. (2001) Determination of the 36Ar/38Ar isotopic abundance ratio of the solar wind using SOHO/CELLAS/MTOF. *Geochimica et Cosmochimica Acta*, Vol. 65, pp. 4589-4596

Wieler, R., & Baur, H. (1995) Fractionation of Xe, Kr, and Ar in the solar corpuscular radiation deduced by closed system etching of lunar soils. *Astrophysical Journal*, Vol. 453, pp. 987-997

Wieler, R., Baur, H., & Signer, P. (1986) Noble gases from solar energetic particles revealed by closed system strpwise etching of lunar minerals. *Geochimica et Cosmochimica Acta*, Vol. 50, pp. 1997-2017

Wimmer-Schweingruber, RF., Bochsler, P., & Wurz, P. (1999) Isotopes in the solar wind : New results from ACE, SOHO, and WIND Solar Wind 9:147-152

Wimmer-Schweingruber, RF., & Bochsler, P. (2001) Lunar soils: a long term archive for the galactic environment of the heliosphere. In: Solar and galactic composition. Wimmer-Schweingruber RF (ed) AIP Conference Proceedings, Vol. 598, Am Inst Phys, Melville, New York, pp. 399-404

Ziegler, JF., Biersack, JP., & Ziegler, MD. (2008) SRIM - The Stopping and Range of Ions in Matter. 398p. ISBN-13 978-0-9654207-1-6

6

Solar Wind Noble Gases in Micrometeorites

Takahito Osawa
Quantum Beam Science Directorate,
Japan Atomic Energy Agency (JAEA)
Japan

1. Introduction

Most extraterrestrial materials discovered on the Earth have no solar wind noble gases. In fact, only four types of extraterrestrial materials contain noble gases attributed to the solar wind or its fractionated component: gas-rich meteorites, lunar materials collected by the Apollo missions, asteroid samples returned from Itokawa by the Hayabusa mission, and micrometeorites. Except for micrometeorites, all of these have a specific history of solar wind irradiation on the surface of their parent bodies. On the other hand, solar wind noble gases in micrometeorites are implanted during orbital evolution in interplanetary space. Micrometeorites have a different origin and irradiation history from the other three materials and from typical meteorites, meaning that these tiny particles that fell on the Earth can provide us valuable information about the activity of the solar system. Of all the analytical methods in planetary science, noble gas analysis of extraterrestrial materials is one of the most useful, because the analysis can reveal not only their origin and age but also their history of irradiation by galactic and solar cosmic rays and solar wind. In particular, the most reliable positive proof of an extraterrestrial origin for micrometeorites is the solar wind noble gases. In this chapter, solar wind noble gases trapped in micrometeorites are reviewed.

1.1 Nomenclature of extraterrestrial dust

First, the nomenclature of extraterrestrial dust must be explained because the peculiar technical terms in the field of planetary science are perplexing for researchers belonging to different scientific fields. The main terms for extraterrestrial dust are *micrometeorite*, *interplanetary dust particle* (*IDP*), *cosmic spherule*, and *cosmic dust*. *Micrometeorite* can indicate all types of extraterrestrial dust collected on the Earth, but is mainly used to indicate extraterrestrial dust corrected in polar regions. *IDPs* are very small dust particles (<30 μm in diameter) collected in the stratosphere by airplane and are often called *stratospheric dust particles* or *Brownlee particles*. *Cosmic spherules* are small spherical particles recovered from deep-sea sediment, polar regions, and sedimentary rocks. Their spherical shape is due to severe heating during atmospheric entry. Tiny spherical particles found in sedimentary rocks are generally called *microspherules*, *microkrystite*, or *microtektites*. *Unmelted micrometeorites* indicates micrometeorites other than the cosmic spherules, whose shape is irregular. *Cosmic dust* indicates all types of extraterrestrial dust, including intergalactic dust, interstellar dust, interplanetary dust, and circumplanetary dust. *Extraterrestrial dust* is another versatile term synonymous with *cosmic dust*, but it is not as widely used as *cosmic dust*.

Micrometeorite is thought to be the best term representing extraterrestrial dust in this chapter for a few reasons. First, the cosmic dusts with solar wind noble gases reviewed here are not intergalactic dusts or interstellar dusts. Second, the Antarctic micrometeorites that are the main target of this paper are not IDPs. Therefore, the word *micrometeorite* adequately represents all types of cosmic dust that contain solar wind noble gases.

1.2 Collection of micrometeorites

It was already suspected in the Middle Ages that a large number of dusty objects existed in interplanetary space. Zodiacal light is a faint glow that extends away from the sun in the ecliptic plane of the sky, visible to the naked eye in the western sky shortly after sunset or in the eastern sky shortly before sunrise. Already in 1683, Giovanni Domenico Cassini presented the correct explanation of this prominent light phenomenon visible to the human eye. Its spectrum indicates it to be sunlight scattered by interplanetary dust orbiting the sun. It is called "counter-glow" or "Gegenschein" in German (Yamakoshi, 1994). The zodiacal light contributes about a third of the total light in the sky on a moonless night. The sky is, however, seldom dark enough for the entire band of zodiacal light to be seen. Micrometeorites in interplanetary space, contributors to the zodiacal light, are constantly produced by asteroid collisions and liberated from the sublimating icy surfaces of comets. Since the radiation pressure of the sun is sufficient to blow submicron grains (beta meteoroids) out of the solar system, only larger grains (20–200 μm) contribute to the zodiacal light. Poynting-Robertson drag causes larger grains to depart from Keplerian orbits and to spiral slowly toward the sun.

Micrometeorites are the main contributors of extraterrestrial material accreted on the Earth. The accretion rate of cosmic dust particles has been estimated by various means so far, and the values calculated in those reports are different. There is, however, no difference in the conclusion that micrometeorites are the primary extraterrestrial deposit on Earth. Published reports estimating the accretion rate of extraterrestrial matter are well summarized in an appendix table of Peucker-Ehrenbrink (1996). For example, Love and Brownlee (1993) determined the mass flux and size distribution of micrometeoroids in the critical submillimeter size range by measuring hypervelocity impact craters found on the space-facing end of the gravity-gradient-stabilized Long Duration Exposure Facility (LDEF) satellite. A small-particle mass accretion rate of 40,000 ± 20,000 tons/yr was obtained. In another estimate, a Japanese micrometeorite research group carefully picked up Antarctic micrometeorites and accurately counted their numbers, yielding accretion rates of 5,600–10,400 tons/yr (Yada et al., 2001).

Although such a large amount of micrometeorites is continuously supplied to the Earth, micrometeorites have been collected in places where extraterrestrial particles are concentrated and/or terrestrial dust is rare, such as the deep sea, the stratosphere, and polar regions. It is very difficult to discover micrometeorites in inhabitable areas that are contaminated by artificial and terrestrial dusts. Since E. Nishibori collected micrometeorites in Antarctica in 1957–1958 (Nishibori and Ishizaki, 1959), a large number of micrometeorites have been recovered from the Antarctic and Greenland ice sheets and northern Canada

(Theil and Schmidt, 1961; Shima and Yabuki, 1968; Maurette et al., 1986, 1987, 1991; Koeberl and Hagen, 1989; Cresswell and Herd, 1992; Taylor et al., 1997, 1998; Nakamura et al., 1999; Yada and Kojima, 2000; Iwata and Imae, 2002; Rochette et al., 2008; Carole et al., 2011). Antarctic micrometeorites (AMMs) have larger sizes (50–300 μm) than the IDPs captured in the stratosphere (<30 μm). Since most of the mass accreted by the Earth is contained in larger particles (50–400 μm) (Kortenkamp and Dermott, 1998), AMMs represent the interplanetary dust population well.

2. Solar wind noble gases in deep-sea sediment

Isotopic noble gas study on micrometeorites was difficult for a long time because of terrestrial contamination and the small sizes of micrometeorites. Measurements on single cosmic particles had to wait for great improvement of analytical devices. Therefore, the first noble gas study on micrometeorites was a measurement on deep-sea sediments in which micrometeorites were concentrated. The first noble gas isotopic study on deep-sea sediments was performed by Merrihue (1964). Magnetic and nonmagnetic separates of modern red clays from the Pacific Ocean were analyzed using a glass extraction and purification system, and excess ^{3}He and ^{21}Ne were discovered. The reported $^{3}He/^{4}He$ ratios (shown as $^{4}He/^{3}He$ in Merrihue's paper) are clearly higher than that of the terrestrial atmosphere, and a relatively high $^{20}Ne/^{22}Ne$ ratio (11.0 ± 1.0) is reported in the 1000°C step of the magnetic separate. $^{40}Ar/^{36}Ar$ ratios lower than that of the atmosphere in the 1000°C and 1400°C steps of the magnetic separate (268 ± 7 and 172 ± 8) were clearly detected, indicating the presence of extraterrestrial materials. This excellent research for the first time presented overwhelming evidence that extraterrestrial materials with extraterrestrial noble gases had accumulated in the deep-sea sediments. Nine years later, Krylov et al. (1973) reported He isotopic compositions of fifteen oceanic oozes recovered from various regions of the Pacific and Atlantic oceans and the iceberg-melting region of Greenland, which were analyzed by researchers in the Soviet Union. The isotopic ratios for Pacific red clays are tens or a hundred times that found in the various crustal rocks. On the other hand, Atlantic red clays have low $^{3}He/^{4}He$ ratios of 2–3 × 10^{-6} and no ^{3}He anomaly was found in the Greenland samples. They believed that the likely source for the elevated ^{3}He content in the Pacific Ocean sediments is cosmic rather than the hypothetical ^{3}He from the mantle in the clays. The idea was confirmed by studying nitric-acid–treated ooze, which had the same order of $^{3}He/^{4}He$ ratios as untreated ooze. Indeed, the high $^{3}He/^{4}He$ ratios found in the red clays should be attributed to micrometeorites.

After these two reports, research in the field stagnated for a long time, and these important researches were forgotten completely. Japanese researchers, however, renewed study in the field in the 1980s. Ozima et al. (1984) measured thirty-nine sediments from twelve different sites, ten sites from the western to central Pacific and two sites from the Atlantic Ocean. They found $^{3}He/^{4}He$ ratios higher than 5 × 10^{-5} for six sites and concluded that the very high $^{3}He/^{4}He$ ratios in the sediments reflected the input of extraterrestrial materials. Amari and Ozima (1985) subsequently reported a He anomaly in deep-sea sediments, and they rediscovered that the carrier of exotic He was concentrated in magnetic fractions, which was consistent with the result of Merrihue's analysis. Since most terrestrial particles are nonmagnetic, magnetic cosmic dusts are concentrated in magnetic separates. They concluded that the ferromagnetic separates are essentially magnetite using thermomagnetic analyses. They also performed a stepwise degassing experiment, which suggested that He is trapped fairly tightly. Amari and Ozima (1988) analyzed magnetic fractions separated from four deep-sea sediments from the Pacific Ocean. Notably, the study presented Ne and Ar isotopic compositions of the sediments. In all the samples, the $^{20}Ne/^{22}Ne$ ratios were constant (11.6 ± 0.6) in most temperature steps. This result should now be interpreted as being caused by a mixing of solar wind (SW) and implantation-fractionated solar wind (IFSW) components, although they concluded that the Ne was from a unique component. $^{40}Ar/^{36}Ar$ ratios lower than that of the atmosphere, 296, were evidently detected in high-temperature fractions of all samples, indicating the existence of extraterrestrial Ar. They concluded from the $^{20}Ne/^{22}Ne$ ratios and thermal release patterns of He that the extraterrestrial noble gases are implanted solar flare particles.

Fukumoto et al. (1986) determined elemental abundances and isotopic compositions of noble gases in separates and acid-leached residues of deep-sea sediments collected on a cruise of R/V Hakureimaru, Geological Survey of Japan. A $^{3}He/^{4}He$ ratio of $(2.73 \pm 0.06) \times 10^{-4}$ was detected for the magnetic separate B2M. Nitric acid treatment did not affect the isotopic ratio, and the $^{3}He/^{4}He$ ratio of the leached sample B2M-1 is $(2.74 \pm 0.08) \times 10^{-4}$, suggesting that the acid did not attack the carrier of the high $^{3}He/^{4}He$ ratio. Ne isotopic compositions show that the extraterrestrial materials in the sediments were affected by SW component rather than cosmic-ray spallation. Extraterrestrial Ar was detected in the acid-leached residue B1-3, whose $^{40}Ar/^{36}Ar$ was 194.3 ± 52.2. Matsuda et al. (1990) carried out stepwise extraction analyses for the magnetic separate and 3M-HCl–leached residues of the same sample used by Fukumoto et al. (1986). Extraterrestrial He and Ne were observed in most temperature steps of all samples. The magnetic separate lost about 75% of its ^{3}He without a drastic change in its isotopic ratios when it was dissolved in 3M HCl at room temperature for two days, and a sample more severely etched for six days had similar elemental and isotopic compositions of He and Ne to those of the two-day-etched sample, indicating that the extraterrestrial He and Ne should be concentrated in fine particles and/or on the surface of the magnetic grains. These studies performed by Japanese institutes clarified that extraterrestrial materials with solar-derived He and Ne are concentrated in deep-sea sediments and that the most plausible candidate for the carrier of the extraterrestrial noble gas is micrometeorites accreted on the Earth.

Reported $^{3}He/^{4}He$ ratios are summarized in Fig. 1. There are some differences in the isotopic ratios among the reports, and the ratio gradually increased with the year of the study, with the exception of the data from Merrihue (1964), reflecting the improvement in sample separation. Very high $^{3}He/^{4}He$ ratios were consistently detected in the study by Matsuda et al. (1990) because of their use of magnetic separation (they analyzed only 0.53% by weight of the dry sediment) and acid leaching. Such physical and chemical separations concentrated the extraterrestrial materials that exist in deep-sea sediment. The $^{3}He/^{4}He$ ratios reported by Merrihue (1964) are clearly too high, and the true values should be lower than the reported ratios. Ne isotopic compositions are also summarized in Fig. 2. The plots are distributed between the values of the SW and IFSW, indicating that the extraterrestrial materials in deep-sea sediments do not have chondritic noble gas compositions and that cosmogenic Ne is not dominant. The remarkably low $^{21}Ne/^{22}Ne$ ratios detected in the magnetic separates from deep-sea sediments are clearly consistent with the isotopic compositions of individual micrometeorites.

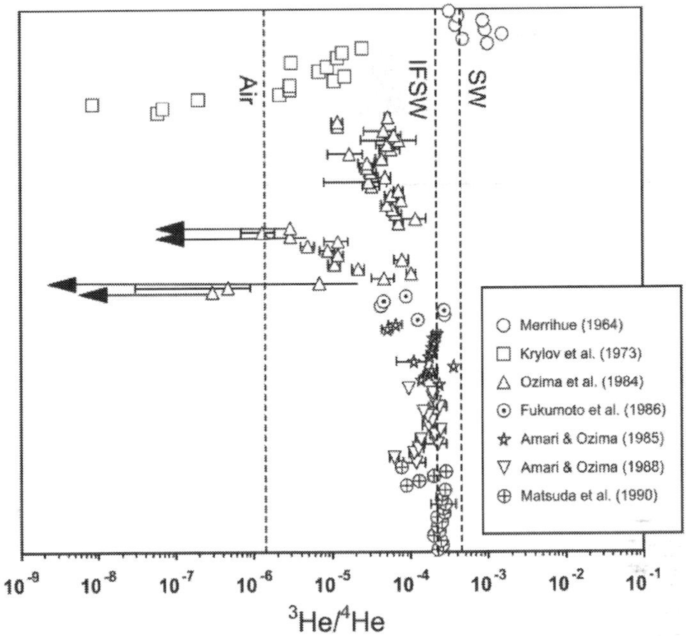

Fig. 1. Reported ^{3}He/^{4}He ratios of deep-sea sediments. Dotted lines show the isotopic ratios of the terrestrial atmosphere at 1.4×10^{-6}, implantation-fractionated solar wind (IFSW) at 2.17×10^{-4} (Benkert et al., 1993), and solar wind (SW) at 4.53×10^{-4} (Heber et al., 2008).

3. Solar wind noble gases detected in individual unmelted micrometeorites

Since noble gas isotope analysis for a single micrometeorite is very difficult because of the extremely small amount of noble gases in a particle, a mass spectrometer with high sensitivity and low background is required to determine accurate isotopic ratios of noble gases released from individual micrometeorites. The first attempt to measure single micrometeorites from deep Pacific Ocean sediments was made by Nier et al. (1987, 1990). They measured He and Ne in deep Pacific particles collected directly from the ocean floor with a 300 kg towed magnetic sled. The samples used were bulk magnetic fines that passed through a 100 μm sieve (they called them "deep Pacific magnetic fines") and individual particles larger than 100 μm in diameter. The individual particles were irregular, and their elemental composition, mineralogy, and texture were consistent with those of meteoritic materials. They measured thirty-five magnetic fines and six individual particles and suggested the possibility that there could be several types of extraterrestrial particles present in the magnetic fines. The most significant result in the paper was the extremely high He isotopic ratios observed in the 1600°C steps of the magnetic fines and individual particles. They attributed the exotic noble gas compositions to solar flare particles.

IDPs collected from the stratosphere have provided valuable information on extraterrestrial noble gases trapped in cosmic dust particles. The first report concerning noble gas

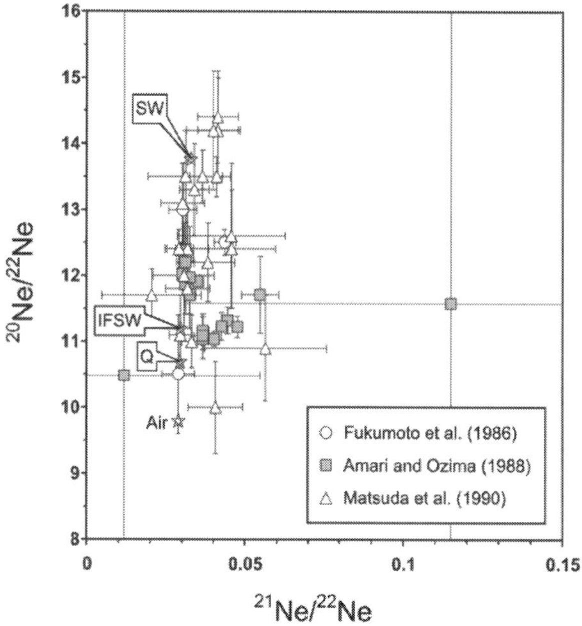

Fig. 2. Three-isotope plot of Ne for deep-sea sediments. SW and IFSW data are from Heber et al. (2008) and Benkert et al. (1993), respectively.

compositions of IDPs is that by Rajan et al. (1977). They detected very high concentrations of ^{4}He ranging from 0.002 to 0.25 cm^{3} STP/g in ten stratospheric particles collected by NASA U-2 aircraft and asserted that the particles were extraterrestrial and that some or all of them were exposed to solar wind for at least 10–100 years. Hudson et al. (1981) selected thirteen chondritic stratospheric particles and measured Ne, Ar, Kr, and Xe by stepwise heating at 1400°C, 1500°C, and 1600°C. The ^{20}Ne/^{36}Ar ratio in the particles is 9 ± 3, indicating the presence of solar-type light noble gas. On the other hand, the ^{132}Xe concentration of ~10^{-7} cm^{3} STP/g and the heavy noble gas elemental pattern suggested a substantial contribution from planetary sources. This is the only report on Kr and Xe in extraterrestrial dusts before Osawa et al. (2000).

The first noble gas measurement for individual IDPs was performed by Nier and Schlutter (1989). They measured He and Ne isotopic compositions for sixteen individual stratospheric particles. The samples were wrapped in a small piece of previously degassed Ta foil, and noble gases were extracted by heating, which was accomplished by passing an electric current directly through the foil. Except for one sample, the IDPs had ^{3}He/^{4}He ratios of 1.5–4.3 × 10^{-4}. The average of the ^{20}Ne/^{22}Ne ratio was 12.0 ± 0.5. In the next stage, they performed stepwise heating for fragments from twenty individual particles to clarify the origin of the particles using the release pattern of ^{4}He (Nier and Schlutter, 1992). Twelve of the IDP fragments contained an appreciable amount of ^{4}He, 50% of which was released by the time the particles were heated to approximately 630°C. Four IDP fragments contained appreciably less ^{4}He, and this was released at a higher temperature. The remaining four fragments had too little ^4He to permit a determination. This result suggested that the parent IDPs of the twelve particles that contained an appreciable amount of ^4He suffered very little heating in their descent and are likely of asteroidal origin, although one cannot rule out the possibility that at least some of them had a cometary origin and entered the Earth's atmosphere at a grazing angle. Nier and Schlutter later performed pulse-heating sequences for twenty-four individual IDPs to learn about the thermal history of the particles and distinguish between IDPs of asteroidal and cometary origin. In this investigation, fifteen of twenty-four particles had $^3He/^4He$ ratios above 10^{-3}, and the highest value, 2×10^{-2}, was found in L2011 D7. They had no explanation for this anomaly.

Kehm et al. (1998a) performed combined trace element and light noble gas measurements on fourteen IDPs from the L2036 stratospheric collector using a laser gas-extraction system and a synchrotron X-ray microprobe. The Ne isotopic compositions in these IDPs were dominated by implanted solar components including SW and IFSW Ne. The Ar isotopic compositions of six large IDPs (>25 μm in their longest dimension) demonstrated enrichment in solar components. Low ^4He contents were observed in five particles that exhibited Zn depletion, indicating severe heating and volatile loss during atmospheric entry. Kehm et al. (1998b) later performed trace element and noble gas measurements on ten large IDPs (~20 μm). They suggested preferential He loss during atmospheric entry heating in this study. Kehm et al. (1999) performed noble gas measurements on JJ-91 IDPs and presented major differences between the result of their measurements and the data of Nier and Schlutter (1993). Kehm et al. (1999) did not detect an anomalously high $^3He/^4He$ ratio in a fragment of 2011 cluster 11, in which a very high $^3He/^4He$ ratio was detected by Nier and Schlutter (1993). However, the reasons for the differences were not clear. Kehm et al. (2002) measured noble gases in 32 individual IDPs, and the ^4He, ^{20}Ne, and ^{36}Ar contents were determined for 31 IDPs. The noble gas elemental compositions were consistent with the presence of fractionated solar wind, but the isotopic compositions were unknown.

Ne isotopic compositions of individual unmelted micrometeorites collected from seasonal lakes on the Greenland ice sheet were reported by Olinger et al. (1990). The extraterrestrial origin of the particles was confirmed by the isotopic data. Maurette et al. (1991) reported Ne isotopic compositions of unmelted and partially melted micrometeorites recovered from Antarctic blue ice. Stuart et al. (1999) measured He isotopes in forty-five putative micrometeorites in the size range of 50–400 μm recovered from Antarctic ice. They determined the He isotopic compositions of twenty-six particles. Pepin et al. (2000, 2001) reported He, Ne, and Ar isotopic ratios for many IDPs and discussed the extremely high ^3He concentration found in some large cluster particles by Nier and Schlutter (1993). They proposed several possibilities to explain the overabundance of ^3He. The noble gas research group at the University of Tokyo reported isotopic compositions of noble gases including Ar, Kr, and Xe for individual unmelted AMMs using a highly established mass spectrometer with a laser gas extraction system (Osawa and Nagao, 2002a, 2002b; Osawa et al., 2000, 2001, 2003). These studies clarified that many micrometeorites contain not only extraterrestrial He and Ne but also extraterrestrial Ar. It is, however, very difficult to detect extraterrestrial Kr and Xe because the concentrations of heavy noble gases are extremely low and the effect of adsorbed terrestrial atmosphere cannot be ignored. Osawa and Nagao (2003) and Osawa et al. (2010) reported noble gas compositions of individual cosmic spherules recovered from Antarctica, and about 40% of the cosmic spherules preserved extraterrestrial noble gases, although their noble gas concentrations were very low due to severe heating.

3.1 He isotopic ratios of micrometeorites

Compiled He isotope data for unmelted AMMs and IDPs are depicted in Fig. 3. The data on IDPs with strikingly high $^3He/^4He$ ratios reported by Nier and Schlutter (1993) are excluded here. The $^3He/^4He$ ratios in the AMMs and IDPs are plotted against the concentrations of ^4He in this figure. The range of ^4He concentrations extends from 10^{-6} to 10 cm^3 STP/g, which may reflect the degree of entry heating for each AMM and IDP. The $^3He/^4He$ ratios of most AMMs are distributed between those of SW and IFSW value, showing the presence of SW He, but there is no significant correlation between the isotopic ratios and ^4He concentration. Since the SW noble gas is thought to become saturated in the surface layer of a small particle in interplanetary space within about a few decades (e.g., Hudson et al., 1981), solar-wind-derived He is implanted in the surface of AMMs and IDPs. It is, however, notable that the isotopic ratios are not clustered around the SW value, and more than half of the particles have $^3He/^4He$ ratios lower than that of SW. This is due to isotopic fractionation during solar wind ion implantation and the loss of the surface layer of the particles during atmospheric entry. The surface layers of the micrometeorites were preferentially heated and ablated by flash heating (e.g., Love and Brownlee, 1991). However, the SW He in the micrometeorites had not been completely extracted by the heating, and the remaining solar-wind-derived He proves the extraterrestrial origin of the AMMs and IDPs.

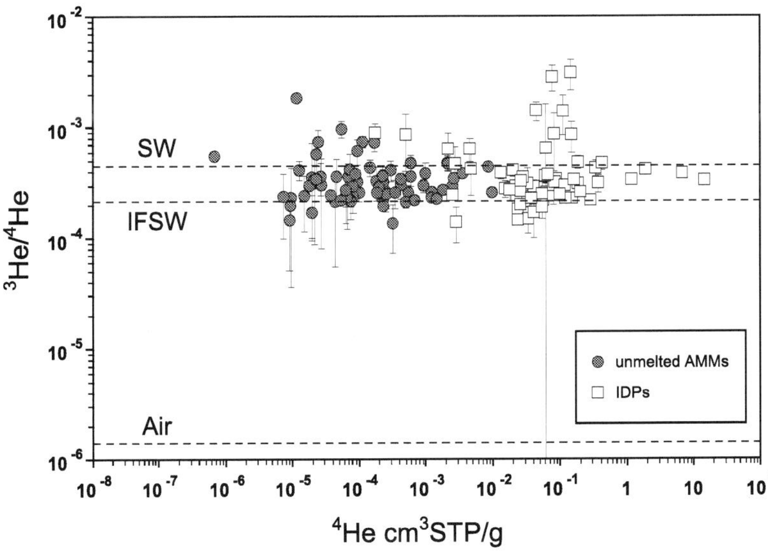

Fig. 3. ^4He concentration and $^3He/^4He$ ratio of unmelted AMMs and IDPs. IDP data are from Nier and Schlutter (1990, 1992) and Pepin et al. (2000, 2011). Unmelted AMM data are from Stuart et al. (1999), Osawa and Nagao (2002b), and Osawa et al. (2003).

The very large difference in ^4He concentration between AMMs and IDPs is remarkable; IDPs have a much higher concentration of ^4He than do AMMs, but the $^3He/^4He$ ratio of most IDPs falls in a similar range to that of AMMs. The large difference in ^4He concentration is mainly caused by the size range; ^4He concentrations in cosmic dust particles correlate with their grain sizes (Stuart et al., 1999). IDPs are smaller than AMMs and have a higher surface area/volume ratio than do AMMs. Since the mechanism of accumulation of SW noble gases in micrometeorites is ion implantation, the concentration of SW noble gases depends on surface area. A high surface area/volume ratio thus causes a high noble gas concentration. A secondary reason for the high He concentration of IDPs is the lower heating temperature; IDPs can escape severe heating because of their low weight and density. He loss in AMMs occurs in response to the thermal decomposition of phyllosilicates and diffusive loss and bubble rupture during atmospheric entry, rather than melting (Stuart et al., 1999). Aqueous alteration in the Antarctic snow can be another possible cause of He loss in AMMs. For example, jarosite [$KFe_3(SO_4)_2(OH)_6$], a by-product mineral resulting from aqueous alteration of sulfide minerals, is observed in ~43% of the AMMs collected from 30,000 year old glacial ice (Terada et al., 2001), and these AMMs have lower He concentrations than AMMs collected from fresh snow, indicating He loss due to aqueous alteration (Osawa and Nagao, 2002). Osawa et al. (2003) reported that jarosite-bearing AMMs have relatively low concentrations of ^{4}He, suggesting loss of He during long-term storage in ice. However, since jarosite is not often found in AMMs, aqueous alteration in ice is not the main cause of the low He concentration of AMMs.

Although the He isotopic ratios of most AMMs and IDPs simply reflect solar-derived He, it is not possible to completely deny the contributions of other components such as planetary He and cosmogenic ^{3}He, an additional component found in some IDPs. In addition, isotopic fractionation during entry deceleration heating should be taken into consideration. Some AMMs and IDPs have higher ^{3}He/^{4}He ratios than that of SW. These probably reflect cosmogenic ^{3}He because the ^{3}He/^{4}He ratio of cosmogenic He is very high, about 0.2. Since cosmogenic ^{3}He is more strongly retained in a micrometeorite than SW He, which exists mostly in the surface layer because of the low energy of solar wind, the ^{3}He/^{4}He ratio is elevated by the preferential loss of solar-wind–derived He. If cosmogenic ^{3}He does not exist in the AMMs, the ^{3}He/^{4}He ratio will approach the ratio of IFSW after the loss of the surface layer of the micrometeorites (Grimberg et al., 2008). The cosmogenic ^{3}He concentrations of some unmelted AMMs with relatively high ^{3}He/^{4}He ratios are much lower than those of IDPs with high concentrations of cosmogenic ^{3}He of over 5×10^{-6} cm^3 STP/g (Pepin et al., 2001). Strikingly high ^{3}He/^{4}He ratios, possibly due to some unknown reservoir, were reported for some IDPs (Nier and Schlutter, 1993; Pepin et al., 2000). For example, the IDP L2011D7 has a low ^{4}He content (3.4×10^{-12} cm^{3}STP) and an unusually high ^{3}He/^{4}He ratio ($(2.0 \pm 0.3) \times 10^{-2}$; Nier and Schlutter, 1993). Kehm et al. (1999), however, did not detect such anomalously high ^{3}He/^{4}He ratios in individual IDP grains separated from the same cluster IDP L2011. In their measurement, nine of eleven IDPs had high He content ($0.7–7 \times 10^{-10}$ cm^{3}STP) and low ^{3}He/^{4}He ratios; the He compositions correspond to those of typical IDPs shown in Fig. 3. The high ^{3}He/^{4}He ratios found in the enigmatic IDPs are thus very problematic. If the large overabundance of ^{3}He is to be attributed to cosmogenic ^{3}He, extremely long periods of cosmic-ray irradiation time are required (Pepin et al., 2001). It is noted that the lack of Ne isotopic data obstructs the interpretation of the problem of excess ^{3}He in IDPs. Even if the enigmatic IDPs are excluded in this discussion, the excess ^{3}He concentrations of AMMs are clearly low compared to those of IDPs. Since the low concentration of cosmogenic ^{3}He presumably indicates preferential loss of He due to severe entry heating, ^{3}He exposure ages of AMMs are not reliable, in contrast to those of IDPs.

The geometric average of $^3He/^4He$ ratios of AMMs, 3.10×10^{-4}, is slightly lower than that of IDPs, 3.55×10^{-4}, which may also reflect the difference in the degree of surface loss or heating during atmospheric entry. This result is consistent with the large difference in He concentration between the two micrometeorite series. Note that a geometric average is more suitable for evaluating the representative He isotopic ratio of micrometeorite samples than an arithmetic mean because the distributions of $^3He/^4He$ ratios of AMMs and IDPs are evidently not normal distributions. In conclusion, unmelted AMMs and IDPs preserve extraterrestrial He derived from energetic implantation of solar wind, but the effects of gas loss and fractionation cannot be ignored. SW He trapped in micrometeorites found on the Earth does not, therefore, represent pure solar wind.

It is extremely difficult to detect solar wind noble gases in the totally melted cosmic spherules because most volatiles have been depleted by harsh heating during atmospheric entry. It is, however, surprising that extraterrestrial He, Ne, and Ar still remain in some cosmic spherules (Osawa and Nagao, 2003; Osawa et al., 2010). Fig. 4 shows the ^4He contents and the $^3He/^4He$ ratios of unmelted AMMs and cosmic spherules. Since only 29 of 130 spherules preserved detectable amounts of ^3He, the ^4He contents of the spherules presented in Fig. 4 do not reflect the distribution of the noble gas contents of all spherules. Even the ^4He contents of the gas-rich cosmic spherules shown in the figure are much lower than those of unmelted AMMs. All of the gas-rich cosmic spherules have $^3He/^4He$ ratios higher than that of terrestrial air within one sigma error, proving their extraterrestrial origin. Furthermore, many spherules have He isotopic ratios close to that of SW, as do the unmelted micrometeorites, indicating that the spherules have preserved solar-derived He in

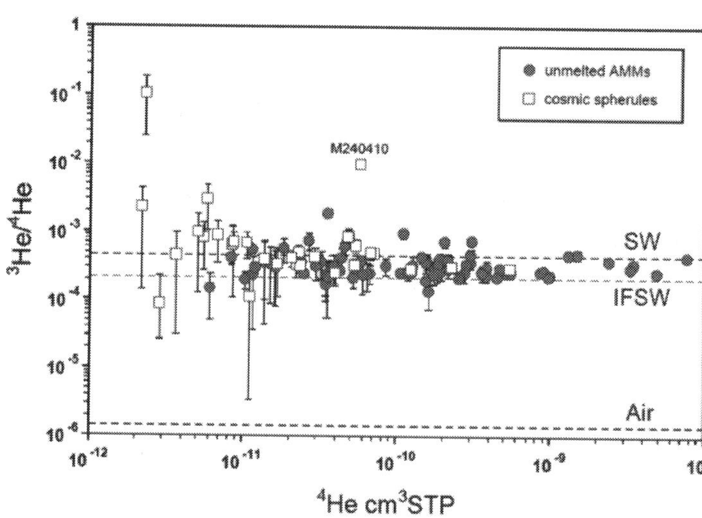

Fig. 4. Relationship between ^4He content and $^3He/^4He$ ratio of unmelted AMMs and cosmic spherules. Unmelted AMM data are from Stuart et al. (1999), Osawa and Nagao (2002b), and Osawa et al. (2003). Cosmic spherule data are from Osawa and Nagao (2003) and Osawa et al. (2010).

spite of their severe heating. This result implies that the spherules are small particles in interplanetary space and not fragments of meteorites fallen to the Earth, as solar-gas-rich meteorites are quite rare. Osawa et al. (2010) discovered an exotic cosmic spherule, M240410, which has an extraordinarily high $^3He/^4He$ ratio ($(9.7 \pm 1.1) \times 10^{-3}$) and high ^3He content (5.53×10^{-13} cm^3STP) that resulted from cosmogenic production of ^3He. Such high isotopic ratios have not been found in unmelted micrometeorites, indicating that this specific spherule may have an exceptional history. The highest $^3He/^4He$ ratio reported to date in an unmelted micrometeorite is $(1.843 \pm 0.050) \times 10^{-3}$ (Stuart et al., 1999), which is much lower than that of M240410.

3.2 Ne isotopic ratios of micrometeorites

Ne isotope data on micrometeorites can provide information on solar wind, fractionated solar wind, and cosmogenic nuclides. These three components can be separated using a diagram because Ne has three stable isotopes in contrast with He, which has only two. The Ne isotopic composition is thus useful for separating SW components from cosmogenic nuclides, but the Ne concentration of micrometeorites is much lower than the He concentration.

Fig. 5 displays Ne isotopic compositions of unmelted micrometeorites, IDPs, and cosmic spherules. It is remarkable that most micrometeorite data are clustered around the IFSW value and show no cosmogenic ^{21}Ne within the error limit, indicating short exposure ages. Several micrometeorites have $^{21}Ne/^{22}Ne$ ratios higher than that of SW; for example, two exceptional Dome Fuji AMMs have long cosmic-ray exposure (CRE) ages (>100 Myr). However, most micrometeorites have exposure ages shorter than 1 Myr (Osawa and Nagao, 2002a). An enigmatic cosmic spherule, M240410, has an extremely high concentration of cosmogenic ^{21}Ne and was calculated to have a very long CRE age of 393 Myr when 4π exposure to galactic and solar cosmic rays was taken into consideration, indicating that the source of the particle may have been an Edgeworth-Kuiper belt object (Osawa et al., 2010). The Ne isotopic compositions of several unmelted micrometeorites are close to, or above, the SW $^{20}Ne/^{22}Ne$ ratio of 13.77 (Heber et al., 2008). These are Greenland micrometeorite compositions reported by Olinger et al. (1990), and the high $^{20}Ne/^{22}Ne$ ratios are due to the overestimation of CO_2^{++} interference. Hence, the SW-like Ne compositions detected in some micrometeorites do not indicate the presence of unfractionated solar wind, and the solar-derived Ne in all types of micrometeorites is partially depleted and fractionated.

The effect of partial loss of Ne can be observed in a trend in the $^{20}Ne/^{22}Ne$ ratio. The average $^{20}Ne/^{22}Ne$ ratios of IDPs, unmelted micrometeorites, and cosmic spherules are 11.92, 11.39, and 10.57, respectively; the difference in the isotopic ratios among the three micrometeorite groups may reflect the degree of atmospheric entry heating. The smaller IDPs (~20 μm) experienced lower entry temperatures compared to the larger micrometeorites (~100 μm) because the maximum temperature during the trajectory depends on particle radius (e.g., Rizk et al., 1991). The average $^{20}Ne/^{22}Ne$ ratio of cosmic spherules is lower than the IFSW ratio, 11.3, reflecting contamination by the terrestrial atmosphere. Although noble gases in cosmic spherules are considerably depleted by severe flash heating, some spherules preserved solar-wind-derived He and Ne, suggesting that the cosmic spherules have been exposed to solar wind and/or solar flares before atmospheric entry and that they are not simple atmospheric entry ablation fragments of meteorites.

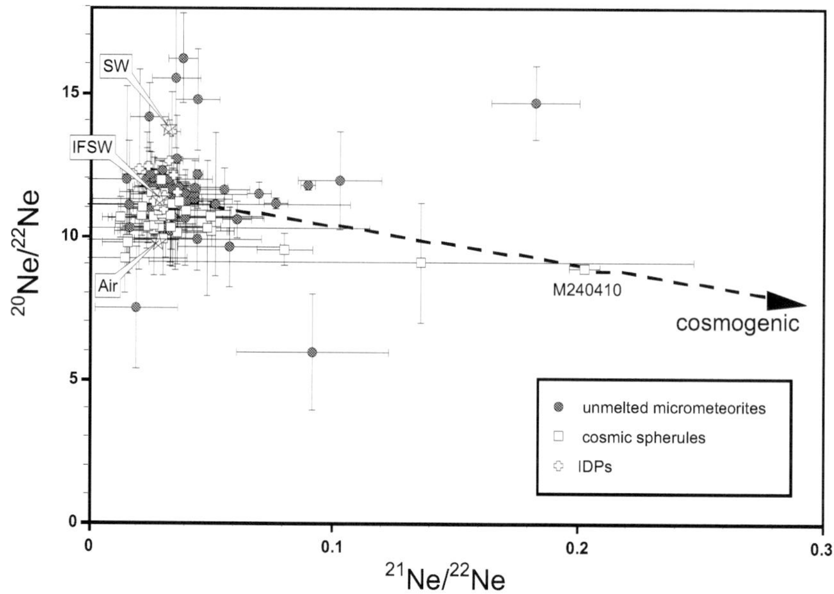

Fig. 5. Three-isotope plot of Ne for unmelted AMMs, cosmic spherules, and IDPs. Unmelted AMM data are from Olinger et al. (1990), Osawa and Nagao (2002b), and Osawa et al. (2003). Cosmic spherule data are from Osawa and Nagao (2003) and Osawa et al. (2010). IDP data are from Pepin et al. (2000). An arrow shows the direction of cosmogenic Ne.

3.3 Ar isotopic ratios of micrometeorites

Ar isotopic compositions of individual micrometeorites were reported only by two groups, at Washington University and the University of Tokyo (Kehm et al., 1998a; Osawa and Nagao, 2002a, 2002b, 2003; Osawa et al., 2000, 2001, 2003, 2010). Merrihue (1964) reported a low $^{40}Ar/^{36}Ar$ ratio (172 ± 5 in the 1400°C fraction) in a magnetic separate of Pacific red clay and suggested that it contains meteoritic material, but that the data do not correspond to those of a single micrometeorite. Since Ar has three stable isotopes, as does Ne, the Ar isotopic compositions of micrometeorites can clarify the contributions of more than two components. A three-isotope plot of Ar for unmelted AMMs and cosmic spherules is presented in Fig. 6. IDP data from Kehm et al. (1998a) are not plotted in this diagram because of the lack of raw data. All unmelted micrometeorites with detectable amounts of Ar have $^{40}Ar/^{36}Ar$ ratios lower than that of the terrestrial atmosphere, 296, confirming their classification as extraterrestrial because terrestrial materials with $^{40}Ar/^{36}Ar$ ratios lower than that of terrestrial air are very few. Although the Ar isotopic compositions of cosmic spherules have large uncertainties due to the very low Ar concentrations, the $^{40}Ar/^{36}Ar$ ratios of many spherules are lower than the atmospheric value. This indicates that extraterrestrial Ar is detectable for these samples because significant gas loss and terrestrial contamination do not overwhelm the extraterrestrial Ar completely.

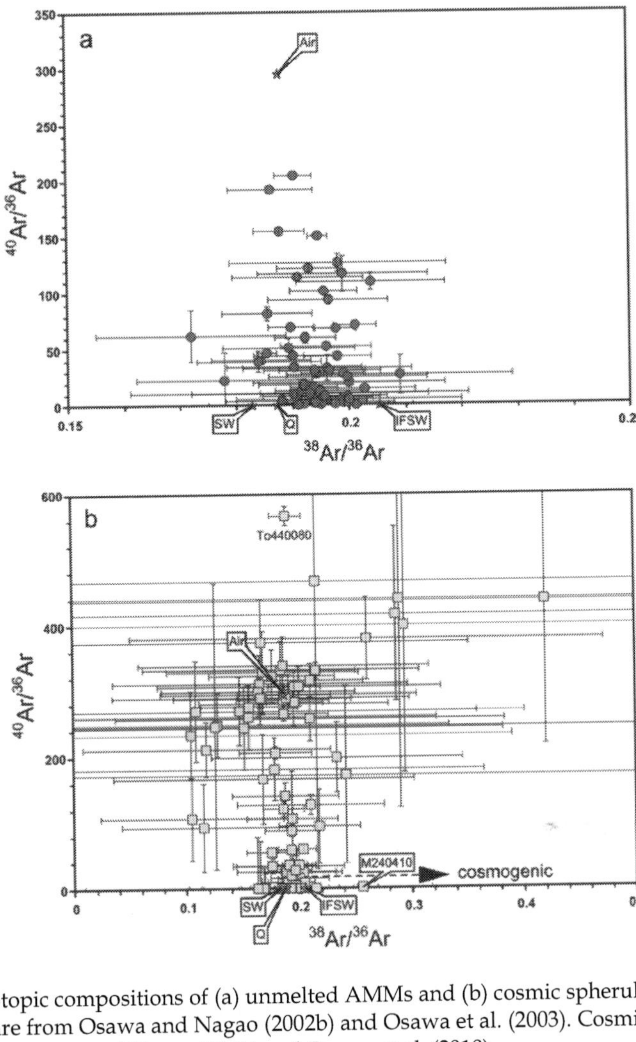

Fig. 6. Ar isotopic compositions of (a) unmelted AMMs and (b) cosmic spherules. Unmelted AMM data are from Osawa and Nagao (2002b) and Osawa et al. (2003). Cosmic spherule data are from Osawa and Nagao (2003) and Osawa et al. (2010).

Cosmic-ray–produced spallogenic ^{38}Ar was detected only in spherule M240410, which has a detectable amount of cosmogenic ^{3}He and ^{21}Ne. The concentration of cosmogenic ^{38}Ar of the spherule is 4.8×10^{-8} cm^3 STP/g, and the CRE age calculated with 2π irradiation is 382.1 Myr (Osawa et al., 2010). All micrometeorites other than this exceptional spherule have no cosmogenic ^{38}Ar, even the unmelted AMMs with relatively high $^{21}Ne/^{22}Ne$ ratios, presumably due to the lower rate of cosmic-ray production of ^{38}Ar than that of ^{21}Ne (Eugster, 1988).

The Ar isotopic composition of unmelted AMMs is composed of three components: terrestrial atmosphere, IFSW, and a component of primordial trapped Ar, such as the Q component (Osawa and Nagao, 2002b). Q-noble gas is the main component of heavy noble gases in primitive chondrites hosted by the phase Q, which is an oxidizable phase of a residue of treatment with hydrochloric acid and hydrofluoric acid (e.g., Lewis et al. 1975; Ott et al., 1981; Huss et al., 1996). $^{38}Ar/^{36}Ar$ ratios that are relatively high compared to the SW value are observed in unmelted AMMs, and the average $^{38}Ar/^{36}Ar$ ratio of 0.193 is higher than the Q-Ar value of 0.187 (Busemann et al., 2000). This indicates the presence of IFSW Ar, in agreement with the IFSW-like Ne composition shown in Fig. 5. The contribution of unfractionated SW component is small, and fractionated absorbed air need not be considered. In contrast with the cases for He and Ne, the contribution of the primordial trapped Ar component is detectable.

About 40% of cosmic spherules and most unmelted AMMs preserved detectable amounts of extraterrestrial Ar but were affected by atmospheric contamination; most ^{40}Ar in the micrometeorites was dominantly derived from the terrestrial atmosphere. It is not obvious that there exists radiogenic ^{40}Ar produced in situ because $^{40}Ar/^{36}Ar$ ratios higher than those of the Q or solar components can be explained by atmospheric contamination (Osawa and Nagao, 2002b), and the concentrations of potassium in AMMs are low (Nakamura et al., 1999; Kurat et al., 1994). The enigmatic spherule To440080, however, has an exceptionality high $^{40}Ar/^{36}Ar$ ratio (566.3 ± 14.8), in spite of the presence of IFSW-like Ne. The high isotopic ratio is clearly due to radiogenic ^{40}Ar. This spherule has a high ^{36}Ar concentration (6.5×10^{-7} cm^3 STP/g) in spite of its high $^{40}Ar/^{36}Ar$ ratio, although meteorites with such high ^{36}Ar concentrations generally have lower $^{40}Ar/^{36}Ar$ ratios than this spherule. An IFSW ^{36}Ar contribution of approximately 50% is calculated from the concentration of ^{20}Ne, if a $^{20}Ne/^{36}Ar$ ratio of 47 is adopted as the IFSW ratio (Murer et al., 1997). If this estimation is correct, the original $^{40}Ar/^{36}Ar$ ratio of this spherule was over 1000, and this spherule undoubtedly originated in a different type of the parent body than did the other micrometeorites.

The contributions of the three Ar components (air, Q, and IFSW) in unmelted AMMs can be estimated using a simple mixing model. In this estimation, all of the ^{40}Ar is assumed to be atmospheric because the $^{40}Ar/^{36}Ar$ ratios of the IFSW and Q components are inaccurate but assumed to be very low. Atmospheric ^{36}Ar and ^{38}Ar are thus probably overestimated, but they contribute only 5% and 4% of the total Ar, respectively. The contribution of the Q component is comparable to that of IFSW component, and the average contributions of ^{36}Ar and ^{38}Ar of the Q component are found to be 45% and 47% of the total Ar, respectively (Osawa et al., 2002). Since $^{38}Ar/^{36}Ar$ ratios of cosmic spherules have large uncertainties, as shown in Fig. 6(b), it is difficult to differentiate the contribution of IFSW Ar from that of primordial trapped Ar in individual spherules. The contribution from the Q component may be comparable with that from IFSW component, as it is in the case of unmelted AMMs, since there is no sign that the original noble gas compositions of the cosmic spherules (other than To440080 and M240410) are different from those of the unmelted AMMs. In conclusion, the low $^{40}Ar/^{36}Ar$ ratios of micrometeorites are not only due to solar wind irradiation.

3.4 Kr and Xe in micrometeorites

Since Kr and Xe concentrations of single micrometeorites are extremely low, their isotopic compositions cannot be determined accurately, and Kr and Xe isotopic ratios of micrometeorites typically have uncertainties larger than 20% (Osawa and Nagao, 2002b).

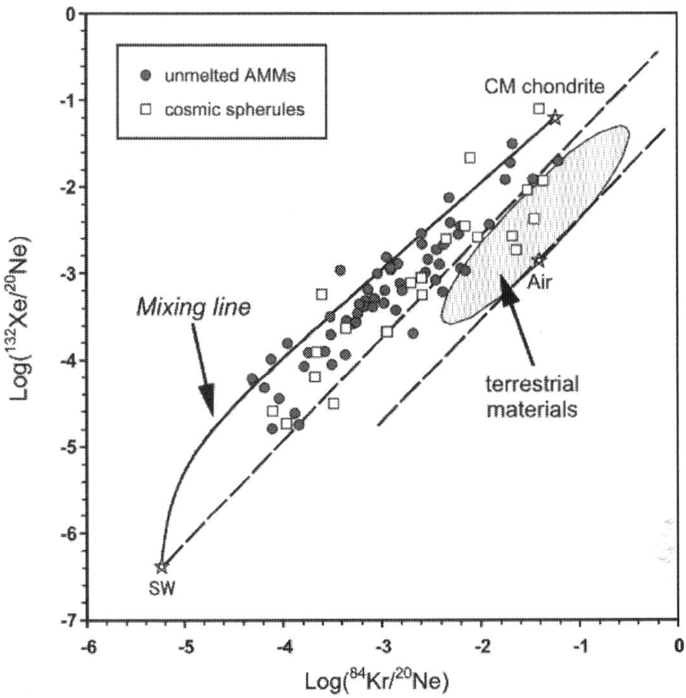

Fig. 7. $^{84}Kr/^{20}Ne$ ratios versus $^{132}Xe/^{20}Ne$ ratios on logarithmic scale. Dotted lines show theoretical fractionation lines of terrestrial air and SW component established by mass-dependent Rayleigh distillation. A solid line shows a mixing of SW and CM chondrite compositions. Air is from Ozima and Podosek (2002). SW data is represented by the 71501 low-temperature regime in Becker et al. (1989). CM2 chondrite is represented by Belgica-7904 (Nagao et al., 1984). Unmelted AMM data are from Osawa and Nagao (2002b) and Osawa et al. (2003). Cosmic spherule data are from Osawa and Nagao (2003) and Osawa et al. (2010).

In addition, Kr and Xe have no large isotopic anomalies, in contrast with the cases for light noble gases. Indeed the mean values of the Kr and Xe isotopic ratios of micrometeorites are identical within the error to the atmospheric values (Osawa et al., 2000; Osawa and Nagao, 2002a, 2002b). Although micrometeorites may preserve solar-derived Kr and Xe, the isotopic compositions of Kr and Xe are useless to identify the solar component. Even the rocky grains of the asteroid Itokawa recovered by the Hayabusa spacecraft have no Kr and Xe attributable to solar wind, although terrestrial contamination of the samples is very low (Nagao et al., 2011).

The noble gas elemental composition including ^{84}Kr and ^{132}Xe is, however, useful for identifying the sources of heavy noble gases. The relative abundances of ^{20}Ne, ^{84}Kr, and ^{132}Xe are depicted in Fig. 7 on a logarithmic scale. All terrestrial materials are distributed below the theoretical mass fractionation line of SW noble gases because the abundance of terrestrial Xe is low, having been selectively depleted by unknown causes (the so-called "missing Xe"). Extraterrestrial materials can thus be distinguished using the diagram. Most of the unmelted AMM data points do not overlap the area representing terrestrial materials, indicating an extraterrestrial origin of the unmelted AMMs. Most of the unmelted AMMs are distributed above the mass fractionation line of SW noble gases. On the other hand, a few cosmic spherules are plotted in the area representing terrestrial materials, indicating contamination by terrestrial atmosphere.

The solid line shows mixing between SW and the primordial trapped component represented by the noble gas composition of a CM2 chondrite, Belgica-7904 (Nagao et al., 1984)). The noble gas composition of Belgica-7904 mainly reflects the Q component for Kr and Xe and the HL component for Ne. HL gas is a primitive component trapped in presolar diamonds. SW data is substituted for IFSW data in the diagram, under the assumption that there is no difference between IFSW and SW value since the noble gas elemental abundance of IFSW component is unclear. Most unmelted AMMs are distributed between the SW-CM2 chondrite mixing line and the mass fractionation line of SW noble gases. The figure clearly shows that both the primordial trapped component and the SW component are preserved in the micrometeorites. The noble gas compositions of the micrometeorites are thus explained by mixing of three components: a primordial trapped component, SW, and terrestrial contamination. The contribution of each component can be roughly estimated using the simple mixing model. If unfractionated air is assumed in the calculation, the average contributions of atmospheric ^{84}Kr and ^{132}Xe are 1.5% and 2% of the total Kr and Xe, respectively. These values are, however, not accurate because air adsorbed on the surface of micrometeorites should be fractionated and its noble gas elemental ratios cannot be determined accurately (Osawa and Nagao, 2002b). If the elemental compositions of adsorption-fractionated air are arbitrarily set to be $^{84}Kr/^{20}Ne = 0.1$ and $^{132}Xe/^{20}Ne = 0.0043$, the mean contribution of the fractionated air is only 0.6% of the total ^{84}Kr and ^{132}Xe. 99% of ^{132}Xe and 95% of ^{84}Kr in micrometeorites is due to the primordial trapped component, and the contribution of SW component for Kr and Xe is very low (Osawa et al., 2003). This estimation implies that it is almost impossible to identify the SW Kr and Xe from the isotopic compositions of Kr and Xe.

4. Conclusion

Development of noble gas mass spectrometers has enabled the analysis of single micrometeorites, and noble gas isotopic research has revealed that most micrometeorites collected on the Earth preserved detectable amounts of SW-derived He, Ne, and Ar. However, Kr and Xe are dominated by the primordial component, and solar-derived Xe is almost negligible. The anomalously high $^{3}He/^{4}He$ ratio and solar-wind-like Ne isotopic composition observed in deep-sea sediments are caused by abundant micrometeorites accumulated on the bottom of the ocean. SW noble gases in micrometeorites were energetically implanted into the surface of micrometeorites in interplanetary space during orbital evolution, but they were partially depleted and fractionated by atmospheric entry heating. Noble gases in cosmic spherules were considerably depleted by harsh heating. The short CRE ages of most micrometeorites inferred from the lack of cosmogenic ^{21}Ne and ^{38}Ar show that the duration of solar wind exposure is less than 1 Myr. Since the terrestrial ages of IDPs and AMMs recovered from fresh Antarctic snow are very low, the trapped SW noble gases in these micrometeorites reflect the composition of recent solar wind.

5. References

Amari S. & Ozima M. (1985) Search for the origin of exotic helium in deep-sea sediments. *Nature*, Vol.317, pp. 520 - 522

Amari S. & Ozima M. (1988) Extra-terrestrial noble gases in deep sea sediments. *Geochim. Cosmochim. Acta*, Vol.52, pp. 1087-1095

Becker R. H. & Pepin R. O. (1989) Long-term changes in solar wind elemental and isotopic ratios: A comparison of two lunar ilmenites of different antiquities. *Geochim. Cosmochim. Acta*, Vol. 53, pp. 1135-1146

Benkert J. -P.; Baur H.; Signer P. & Wieler R. (1993) He, Ne, and Ar from the solar wind and solar energetic particles in lunar ilmenites and pyroxenes. *J. Geophys. Res.*, Vol.98, E7, pp. 13147-13162.

Busemann H.; Baur H. & Wieler R. (2000) Primordial noble gases in "phase Q" in carbonaceous and ordinary chondrites studied by closed-system stepping etching. *Meteorit. Planet. Sci.* Vol.35, pp. 949-973.

Carole C.; Luigi F. & Taylor S. (2011) Vestoid cosmic spherules from the South Pole Water Well and Transantarctic Mountains (Antarctica): A major and trace element study. *Geochim. Cosmochim. Acta*, Vol.75, pp. 1199-1215

Cresswell R. G. & Herd R. K. (1992) Canadian Arctic Meteorite Project (CAMP): 1990. *Meteoritics*, Vol.27, pp. 81-85

Eugster O. (1988) Cosmic-ray rates for 3He, 21Ne, 38Ar, 83Kr and 126Xe in chondrites based on 81Kr-Kr exposure ages. *Geochim. Cosmochim. Acta*, Vol. 52, pp. 1649-1662

Fukumoto H.; Nagao K. & Matsuda J. (1986) Noble gas studies on the host phase of high $^{3}He/^{4}He$ ratios in deep-sea sediments. *Geochim. Cosmochim. Acta*, Vol.50, pp. 2245-2253

Grimberg A.; Baur H.; Bühler F.; Bochsler P. & Wieler R. (2008) Solar wind helium, neon, and argon isotopic and elemental composition: Data from the metallic glass flown on NASA's Genesis mission. *Geochim. Cosmochi. Acta*, Vol.72, pp. 626–645

Heber V. S.; Baur H.; Bochsler P.; Burnett D. S.: Reisenfeld D. B.: Wieler R., & Wiens R. C. (2008) Helium, neon, and argon isotopic and elemental composition of solar wind regimes collected by GENESIS: Implications on fractionation processes upon solar wind formation. *Lunar and Planetary Science*, XXXIX, pp. 1779

Hudson B.; Flynn G. J.; Fraundorf P.; Hohenberg C. M. & Shirck J. (1981) Noble gases in stratospheric dust particles: Confirming of extraterrestrial origin. *Science*, Vol.211, pp. 383-386

Huss G. R.; Lewis R. S. & Hemkin S. (1996) The "normal planetary" noble gas component in primitive chondrites: Compositions, carrier and metamorphic history. Geochim. Cosmochim. Acta, Vol.60, pp. 3311–3340

Iwata N. & Imae N. (2002) Antarctic micrometeorite collection at a bare ice region near Syowa Station by JARE-41 in 2000. *Antarct. Meteorite Res.*, Vol.15, pp. 25-37.

Kehm K.; Flynn G. J.; Sutton S. R. & Hohenberg C. M. (1998a) Combined noble gas and trace element measurements in single IDPs from the L2036 collector. *Lunar Planet. Sci.*, XXIX, 1970

Kehm K.; Flynn G. J.; Sutton S. R. & Hohenberg C. M. (1998b) Helium, neon, and argon measured in large stratospheric dust particles. *Meteorit. Planet. Sci.*, Vol.33, A82

Kehm K.; Flynn G. J.; Hohenberg C. M.; Palma R. L.; Pepin R.; Schlutter G. J.; Sutton S. R. & Walker R. M. (1999) A consortium investigation of possible cometary IDPs. *Lunar Planet. Sci.*, XXX, 1398

Kehm K.; Flynn G. J.; Sutton S. R. & Hohenberg C. M. (2002) Combined noble gas and trace elements on individual stratospheric interplanetary dust particles. *Meteorit. Planet. Sci.*, Vol.37, pp. 1323-1335

Koeberl C. & Hagen E. H. (1989) Extraterrestrial spherules in glacial sediment from the Transantarctic Mountains, Antarctica: Structure, Mineralogy, and chemical composition. *Geochim. Cosmochim. Acta*, Vol.53, pp. 937-944

Kortenkamp S. J. & Dermott S. F. (1998) Accretion of interplanetary dust particles by the Earth. *Icarus*, Vol.135, pp. 469-495

Krylov A. Y.; Mamyrin B. A.; Khabarin L. A.; Mazina T. I. & Silin Y. I. (1974) Helium isotopes in oceanfloor bedrock, *Geochemistry Int.* Vol.11, pp. 839–843 (Translated from Geokhimiya 2, 284-288)

Kurat G.; Koeberl C.; Presper T.; Brandstätter F. & Maurette M. (1994) Petrology and geochemistry of Antarctic micrometeorites. *Geochim. Cosmochim. Acta*, Vol.58, pp. 3879-3904.

Lewis R. S.; Srinivasan B. & Anders E. (1975) Host phase of a strange xenon component in Allende. *Science*, Vol.190, pp. 1251–1262

Love S. G. & Brownlee D. E. (1991) Heating and thermal transformation of micrometeoroids entering the earth's atmosphere. *Icarus*, Vol. 89, pp. 26-43

Matsuda J.; Murota M. & Nagao K. (1990) He and Ne isotopic studies on the extraterrestrial material in deep-sea sediments. *J. Geophys. Res.*, Vol.95, pp. 7111-7117

Maurette M.; Hammer C.; Brownlee D. E.; Reeh N. & Thomsen H. H. (1986) Placers of cosmic dust in the blue ice lakes of Greenland. *Science*, Vol.233, pp. 869-872

Maurette M.; Jehanno C.; Robin E. & Hammer C. (1987) Characteristics and mass distribution of extraterrestrial dust from Greenland ice cap. *Nature*, Vol.328, pp. 699-702

Maurette M.; Olinger C.; Michel-Levy M. C.; Kurat G.; Pourchet M.; Brandstätter F. & Bourot-Denise M. (1991) A collection of diverse micrometeorites recovered from 100 tonnes of Antarctic blue ice. *Nature*, Vol.351, pp. 44-47

Merrihue C. (1964) Rare gas evidence for cosmic dust in modern Pacific red clay. *Ann. N. Y. Acad. Sci.*, Vol.119, pp. 351-367

Nagao K.; Inoue K. & Ogata K. (1984) Primordial rare gases in Belgica-7904 (C2) carbonaceous chondrite. *Proc. Ninth Symp. Antarct. Meteorites*, Vol.35, pp. 257-266

Nagao K.; Okazaki R.; Nakamura T.; Miura Y. N.; Osawa T.; Bajo K.; Matsuda S.; Ebihara M.; Ireland T. R.; Kitajima F.; Naraoka H.; Noguchi T.; Tsuchiyama A.; Uesugi M.; Yurimoto H.; Zolensky M.; Shirai K.; Abe M.; Yada T.; Ishibashi Y.; Fujimura A.; Mukai T.; Ueno M.; Okada T.; Yoshikawa M. & Kawaguchi J. (2011) Irradiation History of Itokawa Regolith Material Deduced from Noble Gases in the Hayabusa Samples. Science, Vol.333, pp. 1128-1131

Nakamura T.; Imae N.; Nakai I.; Noguchi T.; Yano H.; Terada K.; Murakami T.; Fukuoka T.; Nogami K.; Ohashi H.; Nozaki W.; Hashimoto M.; Kondo N.; Matsuzaki H.; Ichikawa O. & Ohmori R. (1999) Antarctic micrometeorites collected at the Dome Fuji Station. *Antarct. Meteorite Res.*, Vol.12, pp. 183-198

Murer, C. A.; Baur, H.; Signer, P. & Wieler, R. (1997) Helium, Neon, and Argon abundances in the solar wind: In vacuo etching of meteoritic iron-nickel. *Geochim. Cosmochim. Acta*, Vol.61, pp. 1303-1314.

Nier A. O.; Schlutter D. J. & Brownlee D. E. (1987) Helium and neon isotopes in extraterrestrial particles. *Lunar Planet. Sci.*, XVIII, 720-721

Nier A. O. & Schlutter D. J. (1989) Helium and Neon isotopes in stratospheric particles. *Lunar Planet. Sci.*, XX, 790-791

Nier A. O. & Schlutter D. J. (1990) Helium and neon isotopes in stratospheric particles. *Meteoritics*, Vol.25, pp. 263-267

Nier A. O. & Schlutter D. J. (1992) Extraction of helium from individual interplanetary dust particles by step-heating. Meteoritics, Vol.27, pp. 166-173

Nier A. O. & Schlutter D. J. (1993) The thermal history of interplanetary dust particles collected in the Earth's stratosphere. *Meteoritics*, Vol.28, pp. 675-681

Nishibori E. & Ishizaki M. (1959) Meteoritic dust collected at Syowa Base, Ongul island, east coast of Lützow-Holm bay, Antarctica. *Antarctic Record*, Vol.7, pp. 35-38

Olinger C. T.; Maurette M.; Walker R. M. & Hohenberg C. M. (1990) Neon measurements of individual Greenland sediment particles: proof of an extraterrestrial origin and comparison with EDX and morphological analyses. *Earth Planet. Sci. Lett.*, Vol.100, pp. 77-93

Osawa T. & Nagao K. (2002a) On low noble gas concentrations in Antarctic micrometeorites collected from Kuwagata Nunatak in the Yamato meteorite ice field. *Antarct. Meteorite Res.*, Vol.15, pp. 165-177.

Osawa T. & Nagao K. (2002b) Noble gas compositions of Antarctic micrometeorites collected at the Dome Fuji Station in 1996 and 1997. *Meteorit. Planet. Sci.*, Vol.37, pp. 911-936.

Osawa T. & Nagao K. (2003) Remnant Extraterrestrial Noble Gases in Antarctic Cosmic Spherules. *Antarct. Meteorite Res.*, Vol.16, pp. 196-219

Osawa T.; Nagao K.; Nakamura T. & Takaoka N. (2000) Noble gas measurement in individual micrometeorites using laser gas-extraction system. *Antarct. Meteorites Res.*, Vol.13, pp. 322-341

Osawa T.; Kagi H. & Nagao K. (2001) Mid-Infrared transmission spectra of individual Antarctic micrometeorites and carbonaceous chondrites. *Antarct. Meteorite Res.*, Vol.14, pp.71-88

Osawa T.; Nakamura T. & Nagao K. (2003) Noble gas isotopes and mineral assemblages of Antarctic micrometeorites collected at the meteorite ice field around the Yamato Mountains. *Meteorit. Planet. Sci.*, Vol.38, pp. 1627–1640

Osawa T.; Yamamoto Y.; Noguchi T.; Iose A. & Nagao K. (2010) Interior textures, chemical compositions, and noble gas signatures of Antarctic cosmic spherules: Possible sources of spherules with long exposure ages. *Meteorit. Planet. Sci.*, Vol.45, pp. 1320-1339

Ott U.; Mack R. & Chang S. (1981) Noble-gas-rich separates from the Allende meteorite. *Geochim. Cosmochim. Acta*, Vol.45, pp. 1751–1788

Ozima M. & Podosek F. A. (2002) *Noble gas geochemistry*. pp. 12-13, Cambridge University Press, Cambridge, UK.

Ozima M.; Takayanagi M.; Zashu S. & Amari S. (1984) High $^{3}He/^{4}He$ ratio in ocean sediments. *Nature*, Vol.311, pp. 448-450.

Pepin R. O.; Palma R. L. & Schlutter D. J. (2000) Noble gases in interplanetary dust particles, I: The excess herium-3 problem and estimates of the relative fluxes of solar wind and solar energetic particles in interplanetary space. *Meteorit. Planet. Sci.* Vol.35, pp. 495-504.

Pepin R. O.; Palma R. L. & Schlutter, D. J. (2001) Noble gases in interplanetary dust particles, II: Excess helium-3 in cluster particles and modeling constraints on interplanetary dust particles exposures to cosmic-ray irradiation. Meteorit. Planet. Sci. 36, 1515-1534.Peucker-Ehrenbrink B. (1996) Accretion of extraterrestrial matter during the last 80 million years and its effect on the marine osmium isotope record. *Geochim. Cosmochim. Acta,* Vol.60, pp. 3187-3196

Rajan R. S.; Brownlee D. E.; Tomandl D.; Hodge P. W.; Harry Farrar IV & Britten R. A. (1977) Detection of 4He in stratospheric particles gives evidence of extraterrestrial origin. *Nature,* Vol.267, pp. 133-134.

Rizk B.; Hunten D. M. & Engel S. (1991) Effects of size-dependent emissivity on maximum temperature during micrometeorite entry. *J. Geophys. Res.* Vol.96, pp. 1303-1314.

Rochette P.; Folco L.; Suaveta C.; van Ginneken M.; Gattacceca J.; Perchiazzi N.; Braucher R.; & Harveyd R. P. (2008) Micrometeorites from the transantarctic mountains. *Proc. Natl. Acad. Sci. USA,* Vol.105, pp. 18206-18211

Shima M. & Yabuki H. (1968) Study on the extraterrestrial material at Antarctica (I). *Antarctic Record* Vol.33, pp. 53-64

Stuart F. M.; Harrop P. J.; Knot S. & Turner G. (1999) Laser extraction of helium isotopes from Antarctic micrometeorites: Source of He and implications for the flux of extraterrestrial 3He to earth. *Geochim. Cosmochim. Acta,* Vol.63, pp. 2653-2665.

Taylor S.; Lever J. H.; Harvey R. P. & Govoni J. (1997) Collecting Micrometeorites from the South Pole Water Well. CRREL Report 97-1, U. S. Army Cold Regions Research and Engineering Laboratory, Hanover, New Hampshire, USA.

Taylor S.; Lever J. H. & Harvey R. P. (1998) Accretion rate of cosmic spherules measured at the South Pole. *Nature,* Vol.392, pp. 899-903.

Terada K.; Yada T.; Kojima H.; Noguchi T.; Nakamura T.; Murakami T.; Yano H.; Nozaki W.; Nakamura Y.; Matsumoto N.; Kamata J.; Mori T.; Nakai I.; Sasaki M.; Itabashi M.; Setoyanagi T.; Nagao K.; Osawa T.; Hiyagon H.; Mizutani S.; Fukuoka T.; Nogami K.; Ohmori R. & Ohashi H. (2001) General characterization of Antarctic micrometeorites collected by the 39th Japanese Antarctic Research Expedition: Consortium studies of JARE AMMs (III). *Antarct. Meteorite Res.,* Vol.14, pp.89-107.

Theil E. & Schmidt R. A. (1961) Spherules from the Antarctic icecap. *J. Geophys. Res.,* Vol.66, 307-310

Yada T. & Kojima H. (2000) The collection of micrometeorites in the Yamato Meteorite Ice Field of Antarctica in 1998. *Antarct. Meteorite Res.,* Vol.13, pp.9-18.

Yada T.; Nakamura T.; Takaoka N.; Noguchi T. & Terada K. (2001a) Terrestrial accretion rates of micrometeorites in the last glacial period. *Antarctic Meteorites,* XXVI, pp. 159-161

Yamakoshi K. (1994) *Extraterrestrial dust,* Kluwer Academic Publishers, Terra Scientific Publishing Company, ISBN: 079-2322-94-0, Tokyo, Japan

Part 3

The Solar Wind Dynamics – From Large to Small Scales

7

Multifractal Turbulence in the Heliosphere

Wiesław M. Macek
Faculty of Mathematics and Natural Sciences, Cardinal Stefan Wyszyński University and Space Research Centre, Polish Academy of Sciences
Poland

1. Introduction

The aim of the chapter is to give an introduction to the new developments in turbulence using nonlinear dynamics and multifractals. To quantify scaling of turbulence we use a generalized two-scale weighted Cantor set (Macek & Szczepaniak, 2008). We apply this model to intermittent turbulence in the solar wind plasma in the inner and the outer heliosphere at the ecliptic and at high heliospheric latitudes, and even in the heliosheath, beyond the termination shock. We hope that the generalized multifractal model will be a useful tool for analysis of intermittent turbulence in the heliospheric plasma. We thus believe that multifractal analysis of various complex environments can shed light on the nature of turbulence.

1.1 Chaos and fractals basics

Nonlinear dynamical systems are often highly sensitive to initial conditions resulting in chaotic motion. In practice, therefore, the behavior of such systems cannot be predicted in the long term, even though the laws of dynamics unambiguously determine its evolution. Chaos is thus a non-periodic long-term behavior in a deterministic system that exhibits sensitivity to initial conditions. Yet we are not entirely without hope here in terms of predictability, because in a dissipative system (with friction) the trajectories describing its evolution in the space of system states asymptotically converge towards a certain invariant set, which is called an attractor; strange attractors are fractal sets (generally with a fractal dimension) which exhibit a hidden order within the chaos (Macek, 2006b).

We remind that a fractal is a rough or fragmented geometrical object that can be subdivided in parts, each of which is (at least approximately) a reduced-size copy of the whole. Strange attractors are often fractal sets, which exhibits a hidden order within chaos. Fractals are generally *self-similar* and independent of scale (generally with a particular fractal dimension). A multifractal is an object that demonstrate various self-similarities, described by a multifractal spectrum of dimensions and a singularity spectrum. One can say that self-similarity of multifractals is point dependent resulting in the singularity spectrum. A multifractal is therefore in a certain sens like a set of intertwined fractals (Macek & Wawrzaszek, 2009).

1.2 Importance of multifractality

Starting from seminal works of Kolmogorov (1941) and Kraichnan (1965) many authors have attempted to recover the observed scaling laws, by using multifractal phenomenological

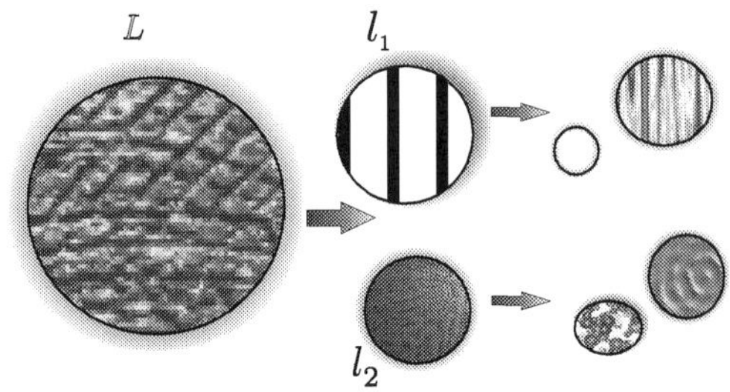

Fig. 1. Schematics of binomial multiplicative processes of cascading eddies.

models of turbulence describing distribution of the energy flux between cascading eddies at various scales (Carbone, 1993; Frisch, 1995; Meneveau & Sreenivasan, 1987). In particular, multifractal scaling of this flux in solar wind turbulence using Helios (plasma) data in the inner heliosphere has been analyzed by Marsch et al. (1996). It is known that fluctuations of the solar magnetic fields may also exhibit multifractal scaling laws. The multifractal spectrum has been investigated using magnetic field data measured *in situ* by Advanced Composition Explorer (ACE) in the inner heliosphere (Macek & Wawrzaszek, 2011a), by Voyager in the outer heliosphere up to large distances from the Sun (Burlaga, 1991; 1995; 2001; 2004; Macek & Wawrzaszek, 2009), and even in the heliosheath (Burlaga & Ness, 2010; Burlaga et al., 2006; 2005; Macek et al., 2011).

To quantify scaling of solar wind turbulence we have developed a generalized two-scale weighted Cantor set model using the partition technique (Macek, 2007; Macek & Szczepaniak, 2008), which leads to complementary information about the multifractal nature of the fluctuations as the rank-ordered multifractal analysis (cf. Lamy et al., 2010). We have investigated the spectrum of generalized dimensions and the corresponding multifractal singularity spectrum depending on one probability measure parameter and two rescaling parameters. In this way we have looked at the inhomogeneous rate of the transfer of the energy flux indicating multifractal and intermittent behavior of solar wind turbulence. In particular, we have studied in detail fluctuations of the velocity of the flow of the solar wind, as measured in the inner heliosphere by Helios (Macek & Szczepaniak, 2008) and ACE (Szczepaniak & Macek, 2008), and Voyager in the outer heliosphere (Macek & Wawrzaszek, 2009; Macek & Wawrzaszek, 2011b), including Ulysses observations at high heliospheric latitudes (Wawrzaszek & Macek, 2010).

2. Methods for phenomenological turbulence model

2.1 Turbulence cascade scenario

In this chapter we consider a standard scenario of cascading eddies, as schematically shown in Figure 1 (cf. Meneveau & Sreenivasan, 1991). We see that a large eddy of size L is divided into two smaller *not necessarily equal* pieces of size l_1 and l_2. Both pieces may have different probability measures, p_1 and p_2, as indicated by the different shading. At the n-th stage we have 2^n various eddies. The processes continue until the Kolmogorov scale is reached (cf. Macek, 2007; Macek et al., 2009; Meneveau & Sreenivasan, 1991). In particular, space filling turbulence could be recovered for $l_1 + l_2 = 1$ (Burlaga et al., 1993). Ideally, in the inertial region of the system of size L, $\eta \ll l \ll L = 1$ (normalized), the energy is not allowed to be dissipated directly, assuming $p_1 + p_2 = 1$, until the Kolmogorov scale η is reached. However, in this range at each n-th step of the binomial multiplicative process, the flux of kinetic energy density ε transferred to smaller eddies (energy transfer rate) could be divided into nonequal fractions p and $1-p$.

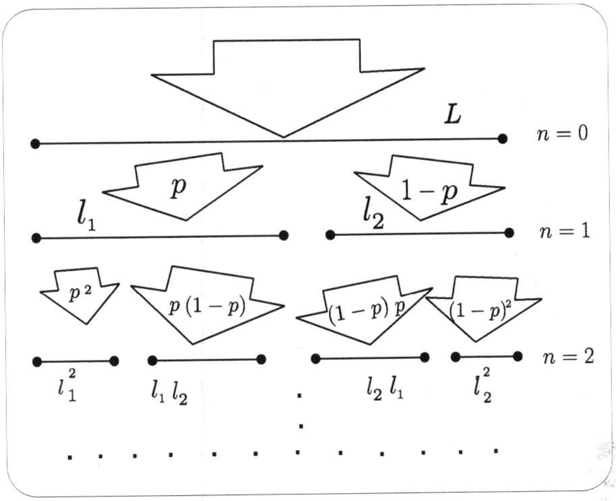

Fig. 2. Two-scale weighted Cantor set model for asymmetric solar wind turbulence.

Naturally, this process can be described by the generalized weighted Cantor set as illustrated in Figure 2, taken from (Macek, 2007). In the first step of the two-scale model construction we have two eddies of sizes $l_1 = 1/r$ and $l_2 = 1/s$, satisfying $p/l_1 + (1-p)/l_2 = 1$, or equivalently $rp + s(1-p) = 1$. Therefore, the initial energy flux ε_0 is transferred to these eddies with the different proportions: $rp\varepsilon_0$ and $s(1-p)\varepsilon_0$. In the next step the energy is divided between four eddies as follows: $(rp)^2\varepsilon_0$, $rsp(1-p)\varepsilon_0$, $sr(1-p)p\varepsilon_0$, and $s^2(1-p)^2\varepsilon_0$. At nth step we have $N = 2^n$ eddies and partition of energy ε can be described by the relation (Burlaga et al., 1993):

$$\varepsilon = \sum_{i=1}^{N} \varepsilon_i = \varepsilon_0 (rp + s(1-p))^n = \varepsilon_0 \sum_{k=0}^{n} \binom{n}{k} (rp)^{(n-k)} (s(1-p))^k. \tag{1}$$

2.2 Comparison with the p-model

The multifractal measure (Mandelbrot, 1989) $\mu = \varepsilon / \langle \varepsilon_L \rangle$ (normalized) on the unit interval for (a) the usual one-scale p-model (Meneveau & Sreenivasan, 1987) and (b) the generalized two-scale cascade model is shown in Figure 3 ($n = 7$), taken from (Macek & Szczepaniak, 2008). It is worth noting that intermittent pulses are much stronger for the model with two different scaling parameters. In particular, for non space-filling turbulence, $l_1 + l_2 < 1$, one still could have a multifractal cascade, even for unweighted (equal) energy transfer, $p = 0.5$. Only for $l_1 = l_2 = 0.5$ and $p = 0.5$ there is no multifractality.

2.3 Energy transfer rate and probability measure

In the first step of our analysis we construct multifractal measure (Mandelbrot, 1989) defining by using some approximation the transfer rate of the energy flux ε in energy cascade

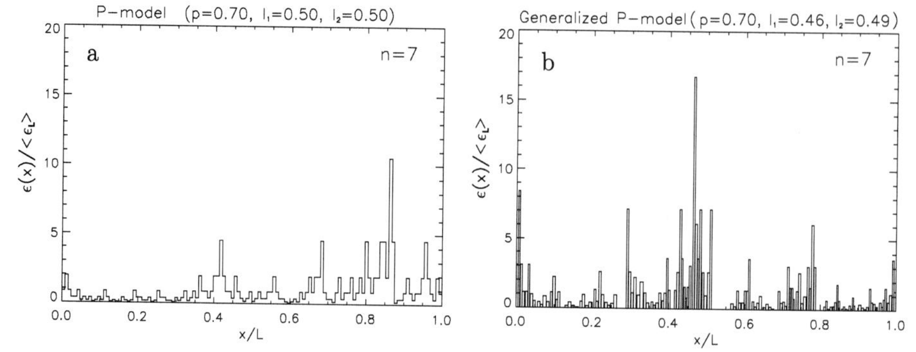

Fig. 3. The multifractal measure $\mu = \varepsilon / \langle \varepsilon_L \rangle$ on the unit interval for (a) the usual one-scale p-model and (b) the generalized two-scale cascade model. Intermittent pulses are stronger for the model with two different scaling parameters.

(Macek & Wawrzaszek, 2009; Wawrzaszek & Macek, 2010). Namely, given a turbulent eddy of size l with a velocity amplitude $u(x)$ at a point x the transfer rate of this quantity $\varepsilon(x, l)$ is widely estimated by the third moment of increments of velocity fluctuations (e.g., Frisch, 1995; Frisch et al., 1978)

$$\varepsilon(x, l) \sim \frac{|u(x+l) - u(x)|^3}{l}, \tag{2}$$

where $u(x)$ and $u(x + l)$ are velocity components parallel to the longitudinal direction separated from a position x by a distance l. Recently, limitations of this approximation are discussed and its hydromagnetic generalization for the Alfvénic fluctuations are considered (Marino et al., 2008; Sorriso-Valvo et al., 2007).

Now, we decompose the signal in segments of size l and then each segment is associated to an eddy. Therefore to each ith eddy of size l in the turbulence cascade we associate a probability measure defined by

$$p(x_i, l) \equiv \frac{\varepsilon(x_i, l)}{\sum_{i=1}^{N} \varepsilon(x_i, l)} = p_i(l). \tag{3}$$

This quantity can be interpreted as a probability that the energy is transferred to an eddy of size l. As is usual, at a given position $x = v_{\text{sw}} t$, where v_{sw} is the average solar wind speed, the temporal scales measured in units of sampling time can be interpreted as the spatial scales $l = v_{\text{sw}} \Delta t$ (Taylor's hypothesis).

2.4 Structure of interplanetary magnetic fields

Let us take a stationary magnetic field $B(t)$ in the equatorial plane. We can again decompose this signal into time intervals of size Δt corresponding to the spatial scales $l = v_{\text{sw}} \Delta t$. Then to each time interval one can associate a magnetic flux past the cross-section perpendicular to the plane during that time. In every considered year we use a discrete time series of daily averages, which is normalized so that we have $\langle B(t) \rangle = \frac{1}{N} \sum_{i=1}^{N} B(t_i) = 1$, where $i = 1, \ldots, N = 2^n$ (we take n = 8). Next, given this (normalized) time series $B(t_i)$, to each interval of temporal scale Δt (using $\Delta t = 2^k$, with $k = 0, 1, \ldots, n$) we associate some probability measure

$$p(x_j, l) \equiv \frac{1}{N} \sum_{i=1+(j-1)\Delta t}^{j\Delta t} B(t_i) = p_j(l), \tag{4}$$

where $j = 2^{n-k}$, i.e., calculated by using the successive average values $\langle B(t_i, \Delta t) \rangle$ of $B(t_i)$ between t_i and $t_i + \Delta t$ (Burlaga et al., 2006).

2.5 Scaling of probability measure and generalized dimensions

Now, for a continuous index $-\infty < q < \infty$ using a q-order total probability measure (Macek & Wawrzaszek, 2009)

$$I(q, l) \equiv \sum_{i=1}^{N} p_i^q(l) \tag{5}$$

and a q-order generalized information $H(q, l)$ (corresponding to Renyi's entropy) defined by Grassberger (1983)

$$H(q, l) \equiv -\log I(q, l) = -\log \sum_{i=1}^{N} p_i^q(l) \tag{6}$$

one obtains the usual q-order generalized dimensions (Hentschel & Procaccia, 1983) $D_q \equiv \tau(q) \,/\, (q-1)$, where

$$\tau(q) = \lim_{l \to 0} \frac{H(q, l)}{\log(1/l)}. \tag{7}$$

2.6 Generalized measures and multifractality

Using Equation (5), we also define a one-parameter q family of (normalized) generalized pseudoprobability measures (Chhabra et al., 1989; Chhabra & Jensen, 1989)

$$\mu_i(q, l) \equiv \frac{p_i^q(l)}{I(q, l)}. \tag{8}$$

Now, with an associated fractal dimension index $f_i(q, l) \equiv \log \mu_i(q, l) / \log l$ for a given q the multifractal singularity spectrum of dimensions is defined directly as the averages taken with respect to the measure $\mu(q, l)$ in Equation (8) denoted from here on by $\langle \ldots \rangle$ (skipping a subscript av)

$$f(q) \equiv \lim_{l \to 0} \sum_{i=1}^{N} \mu_i(q, l)\, f_i(q, l) = \langle f(q) \rangle \tag{9}$$

and the corresponding average value of the singularity strength is given by Chhabra & Jensen (1989)

$$\alpha(q) \equiv \lim_{l \to 0} \sum_{i=1}^{N} \mu_i(q, l)\, \alpha_i(l) = \langle \alpha(q) \rangle. \tag{10}$$

Hence by using a q-order mixed Shannon information entropy

$$S(q, l) = -\sum_{i=1}^{N} \mu_i(q, l) \log p_i(l) \tag{11}$$

we obtain the singularity strength as a function of q

$$\alpha(q) = \lim_{l \to 0} \frac{S(q,l)}{\log(1/l)} = \lim_{l \to 0} \frac{\langle \log p_i(l) \rangle}{\log(l)}, \tag{12}$$

Similarly, by using the q-order generalized Shannon entropy

$$K(q,l) = -\sum_{i=1}^{N} \mu_i(q,l) \log \mu_i(q,l) \tag{13}$$

we obtain directly the singularity spectrum as a function of q

$$f(q) = \lim_{l \to 0} \frac{K(q,l)}{\log(1/l)} = \lim_{l \to 0} \frac{\langle \log \mu_i(q,l) \rangle}{\log(l)}. \tag{14}$$

One can easily verify that the multifractal singularity spectrum $f(\alpha)$ as a function of α satisfies the following Legendre transformation (Halsey et al., 1986; Jensen et al., 1987):

$$\alpha(q) = \frac{d\,\tau(q)}{dq}, \qquad f(\alpha) = q\alpha(q) - \tau(q). \tag{15}$$

2.7 Two-scale weighted Cantor set

Let us now consider the generalized weighted Cantor set, as shown in Figure 2, where the probability of providing energy for one eddy of size l_1 is p (say, $p \leq 1/2$), and for the other eddy of size l_2 is $1 - p$. At each stage of construction of this generalized Cantor set we basically have two rescaling parameters l_1 and l_2, where $l_1 + l_2 \leq L = 1$ (normalized) and two different probability measure $p_1 = p$ and $p_2 = 1 - p$. To obtain the generalized dimensions $D_q \equiv \tau(q)/(q-1)$ for this multifractal set we use the following partition function (a generator) at the n-th level of construction (Halsey et al., 1986; Hentschel & Procaccia, 1983)

$$\Gamma_n^q(l_1, l_2, p) = \left(\frac{p^q}{l_1^{\tau(q)}} + \frac{(1-p)^q}{l_2^{\tau(q)}} \right)^n = 1. \tag{16}$$

We see that after n iterations, $\tau(q)$ does not depend on n, we have $\binom{n}{k}$ intervals of width $l = l_1^k l_2^{n-k}$, where $k = 1, \ldots, n$, visited with various probabilities. The resulting set of 2^n closed intervals (more and more narrow segments of various widths and probabilities) for $n \to \infty$ becomes the weighted two-scale Cantor set.

For any q in Equation (16) one obtains $D_q = \tau(q)/(q-1)$ by solving numerically the following transcendental equation (e.g., Ott, 1993)

$$\frac{p^q}{l_1^{\tau(q)}} + \frac{(1-p)^q}{l_2^{\tau(q)}} = 1. \tag{17}$$

When both scales are equal $l_1 = l_2 = \lambda$, Equation (17) can be solved explicitly to give the formula for the generalized dimensions (Macek, 2006a; 2007)

$$\tau(q) \equiv (q-1)D_q = \frac{\ln[p^q + (1-p)^q]}{\ln \lambda}. \tag{18}$$

For space filling turbulence ($\lambda = 1/2$) one recovers the formula for the multifractal cascade of the standard $p-$model for fully developed turbulence (Meneveau & Sreenivasan, 1987), which obviously corresponds to the weighted one-scale Cantor set (Hentschel & Procaccia, 1983), (cf. Macek, 2002, Figure 3) and (Macek et al., 2006, Figure 4).

2.8 Multifractal formalism

Theory of multifractals allows us an intuitive understanding of multiplicative processes and of the intermittent distributions of various characteristics of turbulence, see (Wawrzaszek & Macek, 2010). As an extension of fractals, multifractals could be seen as objects that demonstrate various self-similarities at various scales. Consequently, the multifractals are described by an infinite number of the generalized dimensions, D_q, as depicted in Figure 4 (a) and by the multifractal spectrum $f(\alpha)$ sketched in Figure 4 (b) (Halsey et al., 1986). The generalized dimensions D_q are calculated as a function of a continuous index q (Grassberger, 1983; Grassberger & Procaccia, 1983; Halsey et al., 1986; Hentschel & Procaccia, 1983). This parameter q, where $-\infty < q < \infty$, can be compared to a microscope for exploring different regions of the singular measurements. In the case of turbulence cascade the generalized dimensions are related to inhomogeneity with which the energy is distributed between different eddies (Meneveau & Sreenivasan, 1991). In this way they provide information about dynamics of multiplicative process of cascading eddies. Here high positive values of q emphasize regions of intense energy transfer rate, while negative values of q accentuate low-transfer rate regions. Similarly, high positive values of q emphasize regions of intense magnetic fluctuations larger than the average, while negative values of q accentuate fluctuations lower than the average (Burlaga, 1995).

An alternative description can be formulated by using the singularity spectrum $f(\alpha)$ as a function of a singularity strength α, which quantify multifractality of a given system (e.g., Ott, 1993). This function describes singularities occurring in considered probability measure attributed to different regions of the phase space of a given dynamical system. Admittedly, both functions $f(\alpha)$ and D_q have the same information about multifractality. However, the singularity multifractal spectrum is easier to interpret theoretically by comparing the experimental results with the models under study.

2.9 Degree of multifractality and asymmetry

The difference of the maximum and minimum dimension, associated with the least dense and most dense points in the phase space, is given by

$$\Delta \equiv \alpha_{\text{max}} - \alpha_{\text{min}} = D_{-\infty} - D_{\infty} = \left| \frac{\log(1-p)}{\log l_2} - \frac{\log(p)}{\log l_1} \right|. \tag{19}$$

In the limit $p \to 0$ this difference rises to infinity. Hence, it can be regarded as a degree of multifractality, see (e.g., Macek, 2006a; 2007). The degree of multifractality Δ is naturally related to the deviation from a strict self-similarity. That is why Δ is also a measure of intermittency, which is in contrast to self-similarity (Frisch, 1995, ch. 8). In the case of the symmetric spectrum using Equation (18) the degree of multifractality is

$$\Delta = D_{-\infty} - D_{+\infty} = \ln(1/p - 1)/\ln(1/\lambda). \tag{20}$$

In particular, the usual middle one-third Cantor set without any multifractality is recovered with $p = 1/2$ and $\lambda = 1/3$.

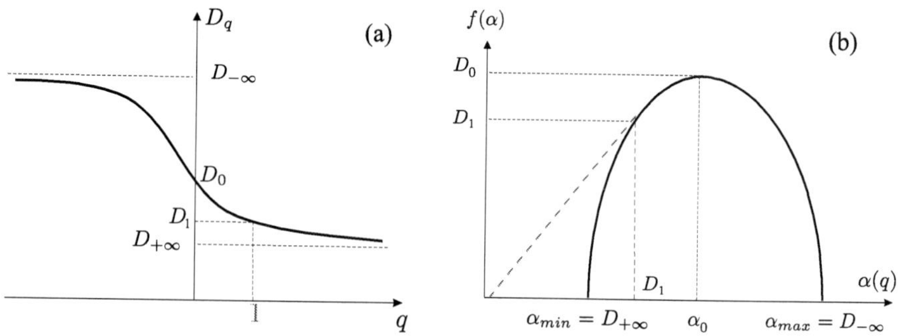

Fig. 4. (a) The generalized dimensions D_q as a function of any real q, $-\infty < q < +\infty$, and (b) the singularity multifractal spectrum $f(\alpha)$ versus the singularity strength α with some general properties: (1) the maximum value of $f(\alpha)$ is D_0; (2) $f(D_1) = D_1$; and (3) the line joining the origin to the point on the $f(\alpha)$ curve, where $\alpha = D_1$ is tangent to the curve (Ott, 1993).

Moreover, using the value of the strength of singularity α_0 at which the singularity spectrum has its maximum $f(\alpha_0) = 1$ we define a measure of asymmetry by

$$A \equiv \frac{\alpha_0 - \alpha_{\min}}{\alpha_{\max} - \alpha_0}. \tag{21}$$

3. Solar wind data

We have analyzed time series of plasma velocity and interplanetary magnetic field strength measured during space missions onboard various spacecraft, such as Helios, ACE, Ulysses, and Voyager, exploring different regions of the heliosphere during solar minimum and solar maximum.

3.1 Solar wind velocity fluctuations

3.1.1 Inner heliosphere

The Helios 2 data using plasma parameters measured *in situ* in the inner heliosphere at distances 0.3 – 0.9 AU from the Sun have been been reported by Schwenn (1990). Using these data, with sampling time of 40.5 s, the X-velocity (mainly radial) component of the plasma flow, $u = v_x$, with 4514 points, (two-day) sample, has been investigated by Macek (1998; 2002; 2003) and Macek & Redaelli (2000) for testing of the solar wind attractor (Macek, 1998). The Alfvénic fluctuations with longer (several-day) samples have been studied by Macek (2006a; 2007) and Macek et al. (2005; 2006). Further, to study turbulence cascade, Macek & Szczepaniak (2008) have selected four-day time intervals of v_x samples (each of 8531 data points) in 1976 (solar minimum) for both slow and fast solar wind streams.

The results for data obtained by ACE in the ecliptic plane near the libration point $L1$, i.e., approximately at a distance of $R = 1$ AU from the Sun and dependence on solar cycle have been discussed by Szczepaniak & Macek (2008). They have studied the multifractal measure obtained using $N = 2^n$, with $n = 18$, data points interpolated with sampling time of 64 s, for (a) solar minimum (2006) and (b) solar maximum (2001), correspondingly. For example, the

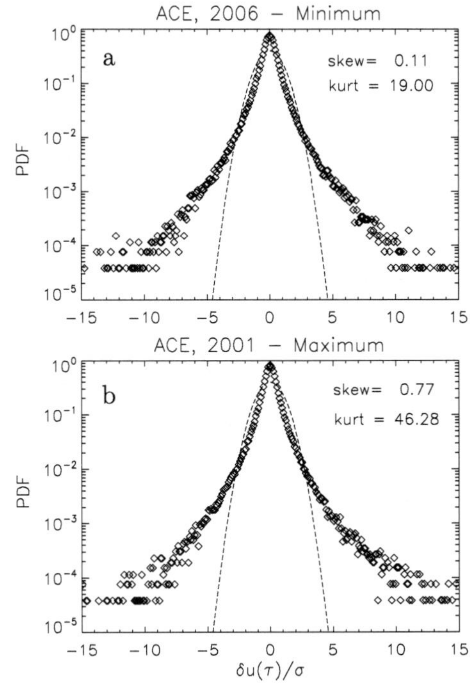

Fig. 5. Probability density functions of fluctuations of the solar wind radial velocity for (a) solar minimum (2006) and (b) solar maximum (2001), correspondingly, as compared with the normal distribution (dashed lines).

obtained probability density functions of fluctuations of the solar wind velocity is shown in Figure 5, taken from (Szczepaniak & Macek, 2008).

In addition, Macek et al. (2009) have analyzed ACE data separately for both slow and fast solar wind streams. They have selected five-day time intervals of v_x samples, each of 6750 data points, for both slow and fast solar wind streams during solar minimum (2006) and maximum (2001). In Figure 6, taken from (Macek et al., 2009), we show the time trace of the multifractal measure $p(t_i, \Delta t) = \varepsilon(t_i, \Delta t) \ / \ \sum \varepsilon(t_i, \Delta t)$ given by Equations (2) and (3) and obtained using v_x samples (in time domain) as measured by ACE for the slow (a) and (c) and fast (b) and (d) solar wind during solar minimum (2006) and maximum (2001), correspondingly. One can notice that intermittent pulses are somewhat stronger for data at solar maximum. This results in fatter tails of the probability distribution functions as shown in Figure 5, for solar maximum and minimum with large deviations from the normal distribution (dashed lines).

3.1.2 Outer heliosphere

Macek & Wawrzaszek (2009) have tested asymmetry of the multifractal scaling for the wealth of solar wind data provided by another space mission. Namely, they have analyzed time series of velocities measured by Voyager 2 at various distances from the Sun, 2.5, 25, and 50 AU, selecting long (13-day) time intervals of v_x samples, each of 2^{11} data points, interpolated

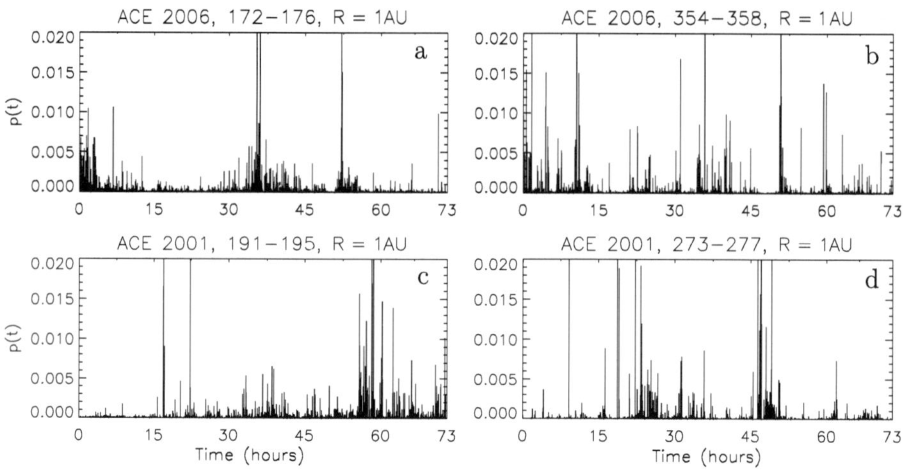

Fig. 6. The time trace of the normalized transfer rate of the energy flux $p(t_i, \Delta t) = \varepsilon(t_i, \Delta t) \ / \ \sum \varepsilon(t_i, \Delta t)$ obtained using data of the v_x velocity components measured by ACE at 1 AU for the slow (a) and (c) and fast (b) and (d) solar wind during solar minimum (2006) and maximum (2001), correspondingly.

with sampling time of 192 s for both slow and fast solar wind streams during the following solar minima: 1978, 1987–1988, and 1996–1997. The same analysis has been repeated for the Voyager 2 data for the slow solar wind during the solar maxima: 1981, 1989, 2001, at 10, 30, and 65 AU, correspondingly. This has allowed us to investigate the dependence of the multifractal spectra on the phase of the solar cycle (Macek & Wawrzaszek, 2011b).

3.1.3 Out of ecliptic

It is worth noting that Ulysses' periodic (6.2 years) orbit with perihelion at 1.3 AU and aphelion at 5.4 AU and latitudinal excursion of $\pm 82^\circ$ gives us a new possibility to study both latitudinal and radial dependence of the solar wind (Horbury et al., 1996; Smith et al., 1995).

Wawrzaszek & Macek (2010) have determined multifractal characteristics of turbulence scaling such as the degree of multifractality and asymmetry of the multifractal singularity spectrum for the data provided by Ulysses space mission. They have used plasma flow measurements as obtained from the SWOOPS instrument (Solar Wind Observations Over the Poles of the Sun). Namely, they have analyzed the data measured by Ulysses out of the ecliptic plane at different heliographic latitudes ($+32^\circ \div +40^\circ$, $+47^\circ \div +48^\circ$, $+74^\circ \div +78^\circ$, $-40^\circ \div -47^\circ$, $-50^\circ \div -56^\circ$, $-69^\circ \div -71^\circ$) and heliocentric distances of $R = 1.4 - 5.0$ AU from the Sun, selecting twelve-day v_x samples, each of 4096 data points, with sampling time of 242 s $\approx$4 min, for solar wind streams during solar minimum (1994 - 1997, 2006 - 2007).

3.2 Magnetic field strength fluctuations

Macek & Wawrzaszek (2011a) have tested for the multifractal scaling of the interplanetary magnetic field strengths, B, for ACE data at 1 AU from the Sun. In case of ACE the sampling time resolution of 16 s for the magnetic field is much better than that for the Voyager data, which allow us to investigate the scaling on small scales of the order of minutes.

The calculated energy spectral density as a function of frequency for the data set of the magnetic field strengths $|B|$ consisting of about 2×10^6 measurements for (a) the whole year 2006 during solar minimum and (b) the whole year 2001 during solar maximum is illustrated in Figure 1 of the paper by Macek & Wawrzaszek (2011a). It has been shown that the spectrum density is roughly consistent with this well-known power-law dependence $E(f) \propto f^{-5/3}$ at wide range of frequency, f, suggesting a self-similar fractal turbulence model often used for looking at scaling properties of plasma fluctuations (e.g., Burlaga & Klein, 1986). However, it is clear that the spectrum alone, which is based on a second moment (or a variance), cannot fully describe fluctuations in the solar wind turbulence (cf. Alexandrova et al., 2007). Admittedly, intermittency, which is deviation from self-similarity (e.g., Frisch, 1995), usually results in non-Gaussian probability distribution functions. However, the multifractal powerful method generalizes these scaling properties by considering not only various moments of the magnetic field, but the whole spectrum of scaling indices (Halsey et al., 1986).

Therefore, Macek & Wawrzaszek (2011a) have analyzed time series of the magnetic field of the solar wind on both small and large scales using multifractal methods. To investigate scaling properties in fuller detail, using basic 64-s sampling time for small scales, they have selected long time intervals of $|B|$ of interpolated samples, each of 2^{18} data points, from day 1 to 194. Similarly, for large scales they have used daily averages of samples from day 1 to 256 of 2^8 data points. The data for both small and large scale fluctuations during solar minimum (2006) and maximum (2001) are shown in Figure 2 of the paper by Macek & Wawrzaszek (2011a).

4. Results and discussion

4.1 Multifractal model for plasma turbulence

For a given q, we calculate the generalized q-order total probability measure $I(q,l)$ of Equation (5) as a function of various scales l that cover turbulence cascade (cf. Macek & Szczepaniak, 2008, Equation (2)). On a small scale l in the scaling region one should have, according to Equations (5) to (7), $I(q,l) \propto l^{\tau(q)}$, where $\tau(q)$ is an approximation of the ideal limit $l \to 0$ solution of Equation (7) (e.g., Macek et al., 2005, Equation (1)). Equivalently, writing $I(q,l) = \sum p_i (p_i)^{q-1}$ as a usual weighted average of $\langle (p_i)^{q-1} \rangle_{av}$, one can associate bulk with the generalized average probability per cascading eddies

$$\bar{\mu}(q,l) \equiv \sqrt[q-1]{\langle (p_i)^{q-1} \rangle_{av}}, \tag{22}$$

and identify D_q as a scaling of bulk with size l,

$$\bar{\mu}(q,l) \propto l^{D_q}. \tag{23}$$

Hence, the slopes of the logarithm of $\bar{\mu}(q,l)$ of Equation (23) versus $\log l$ (normalized) provides

$$D_q(l) = \frac{\log \bar{\mu}(q,l)}{\log l}. \tag{24}$$

4.1.1 Inner heliosphere

In Figure 7 we have depicted some of the values of the generalized average probability, $\log \bar{\mu}(q,l)$, for the following values of q: 6, 4, 2, 1, 0, -1, -2. These results are obtained using

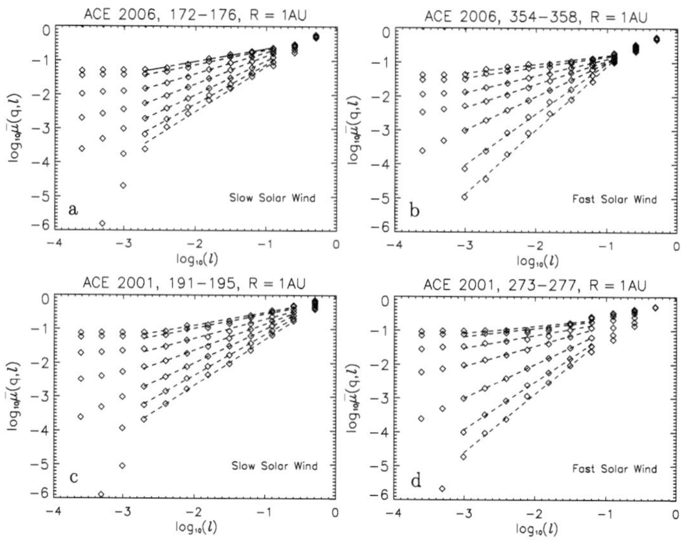

Fig. 7. Plots of the generalized average probability of cascading eddies $\log_{10} \bar{\mu}(q,l)$ versus $\log_{10} l$ for the following values of q: 6, 4, 2, 1, 0, -1, -2, measured by ACE at 1 AU (diamonds) for the slow (a) and (c) and fast (b) and (d) solar wind during solar minimum (2006) and maximum (2001), correspondingly.

data of the v_x velocity components measured by ACE at 1 AU (diamonds) for the slow (a) and (c) and fast (b) and (d) solar wind during solar minimum (2006) and maximum (2001), correspondingly. The generalized dimensions D_q as a function of q are shown in Figure 8. The values of D_q given in Equation (24) are calculated using the radial velocity components $u = v_x$ (cf. Macek et al., 2005, Figure 3).

In addition, in Figures 9 and 10 we see the generalized average logarithmic probability and pseudoprobability measures of cascading eddies $\langle \log_{10} p_i(l) \rangle$ and $\langle \log_{10} \mu_i(q,l) \rangle$ versus $\log_{10} l$, as given in Equations (3) and (8). The obtained results for the singularity spectra $f(\alpha)$ as a function of α are shown in Figure 11 for the slow (a) and (c), and fast (b) and (d) solar wind streams at solar minimum and maximum, correspondingly. Both values of D_q and $f(\alpha)$ for one-dimensional turbulence have been computed directly from the data, by using the experimental velocity components.

For $q \geq 0$ these results agree with the usual one-scale p-model fitted to the dimension spectra as obtained analytically using $l_1 = l_2 = 0.5$ in Equation (18) and the corresponding value of the parameter $p \simeq 0.21$ and 0.20, 0.15 and 0.12 for the slow (a) and (c), and fast (b) and (d) solar wind streams at solar minimum and maximum, correspondingly, as shown by dashed lines. On the contrary, for $q < 0$ the p-model cannot describe the observational results (Marsch et al., 1996). Macek et al. (2009) have shown that the experimental values are consistent also with the generalized dimensions D_q obtained numerically from Equations (22-24) and the singularity spectra $f(\alpha)$ from Equations (10) and (14) for the weighted two-scale Cantor set using an asymmetric scaling, i.e., using unequal scales $l_1 \neq l_2$, as is shown in Figure 8 and 11 (a), (b), (c), and (d) by continuous lines. In this way we confirm the universal shape of the multifractal spectrum, as illustrated in Figure 4.

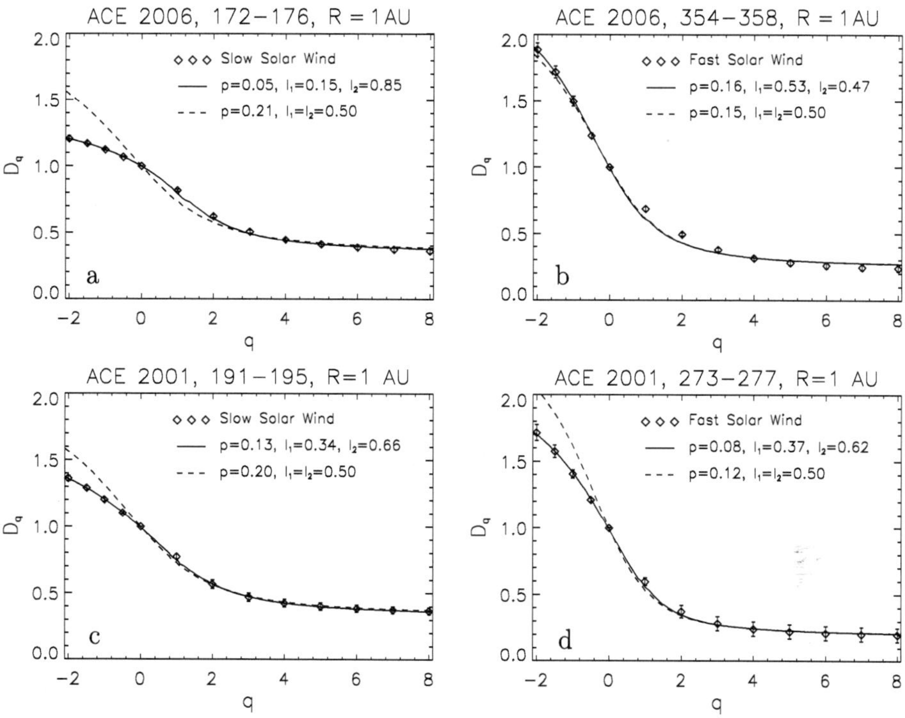

Fig. 8. The generalized dimensions D_q as a function of q. The values obtained for one-dimensional turbulence are calculated for the usual one-scale (dashed lines) p-model and the generalized two-scale (continuous lines) model with parameters fitted to the multifractal measure $\mu(q,l)$ obtained using data measured by ACE at 1 AU (diamonds) for the slow (a) and (c) and fast (b) and (d) solar wind during solar minimum (2006) and maximum (2001), correspondingly.

	Slow Solar Wind	Fast Solar Wind
Solar Min.	$\Delta = 1.22, A = 2.21$	$\Delta = 2.56, A = 0.95$
Solar Max.	$\Delta = 1.60, A = 1.33$	$\Delta = 2.31, A = 1.25$

Table 1. Degree of Multifractality Δ and Asymmetry A for Solar Wind Data in the Inner Heliosphere

We see from Table 1 that the degree of multifractality Δ and asymmetry A of the solar wind in the inner heliosphere are different for slow (Δ = 1.2 – 1.6) and fast (Δ = 2.3 – 2.6) streams; the velocity fluctuations in the fast streams seem to be more multifractal than those for the slow solar wind (the generalized dimensions vary more with the index q) (Macek et al., 2009). On the other hand, it seems that in the slow streams the scaling is more asymmetric than that for the fast wind. In our view this could possibly reflect the large-scale scale velocity structure. Further, the degree of asymmetry of the dimension spectra for the slow wind is rather anticorrelated with the phase of the solar magnetic activity: A decreases from 2.2 to

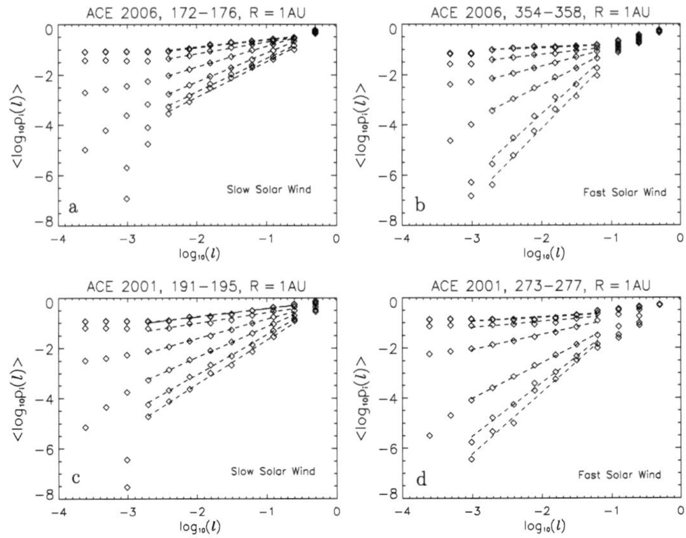

Fig. 9. Plots of the generalized average logarithmic probability of cascading eddies $\langle \log_{10} p_i(l) \rangle$ versus $\log_{10} l$. for the following values of q: 2, 1, 0, -1, -2, measured by ACE at 1 AU (diamonds) for the slow (a) and (c) and fast (b) and (d) solar wind during solar minimum (2006) and maximum (2001), correspondingly.

1.3, and only weakly correlated for the fast wind; only the fast wind during solar minimum exhibits roughly symmetric scaling, $A \sim 1$, i.e., one-scale Cantor set model applies.

4.1.2 Outer heliosphere

These results are obtained using data of the v_x velocity components measured by Voyager 2 during solar minimum (1978, 1987–1988, 1996–1997) at various distance from the Sun: 2.5, 25, and 50 AU. In this way, the singularity spectra $f(\alpha)$ are obtained directly from the data as a function of α as given by Equations (12) and (14) and the results are presented in Figure 12, taken from (Macek & Wawrzaszek, 2009), for the slow (a), (c), and (e) and fast (b), (d), and (f) solar wind, correspondingly.

Heliospheric Distance (Year)	Slow Solar Wind	Fast Solar Wind
2.5 AU (1978)	$\Delta = 1.95, A = 0.91$	$\Delta = 2.12, A = 1.54$
25 AU (1987–1988)	$\Delta = 2.02, A = 0.98$	$\Delta = 2.93, A = 0.66$
50 AU (1996–1997)	$\Delta = 2.10, A = 1.14$	$\Delta = 1.94, A = 0.95$

Table 2. Degree of Multifractality Δ and Asymmetry A for Solar Wind Data in the Outer Heliosphere During Solar Minimum.

We see from Table 2 that the obtained values of Δ obtained from Equation (19) for the solar wind in the outer heliosphere are somewhat different for slow and fast streams. For the fast wind (not very far away from the Sun) D_q fells more steeply with q than for the slow wind, and therefore one can say that the degree of multifractality is larger for the fast wind.

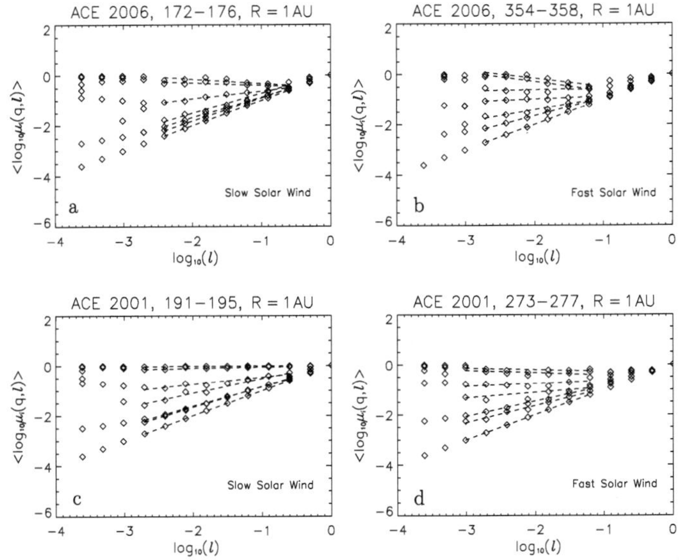

Fig. 10. Plots of the generalized average logarithmic pseudoprobability measure of cascading eddies $\langle \log_{10} \mu_i(q,l) \rangle$ versus $\log_{10} l$ for the following values of q: 2, 1, 0, -1, -2, measured by ACE at 1 AU (diamonds) for the slow (a) and (c) and fast (b) and (d) solar wind during solar minimum (2006) and maximum (2001), correspondingly.

In general, we see that the multifractal spectrum of the solar wind is only roughly consistent with that for the multifractal measure of the self-similar weighted symmetric one-scale weighted Cantor set only for $q \geq 0$. On the other hand, this spectrum is in a very good agreement with two-scale asymmetric weighted Cantor set schematically shown in Figure 2 for both positive and negative q. Obviously, taking two different scales for eddies in the cascade, one obtains a more general situation than in the usual p-model for fully developed turbulence (Meneveau & Sreenivasan, 1987), especially for an asymmetric scaling, $l_1 \neq l_2$.

4.1.3 Out of ecliptic

Wawrzaszek & Macek (2010) have observed a latitudinal dependence of the multifractal characteristics of turbulence. The calculated degree of multifractality and asymmetry as a function on heliographic latitude for the fast solar wind are summarized in Figure 13 and 14 (a) with some specific values listed in Table 3 (the case of the slow wind is denoted by an asterisk). We see that the degree of multifractality Δ and asymmetry A of the dimension spectra of the fast solar wind out of the ecliptic plane are similar for positive and the corresponding negative latitudes. Therefore, it seems that the values of these multifractal characteristics exhibits some symmetry with respect to the ecliptic plane. In particular, in a region from 50° to 70° we observe a minimum of the degree of multifractality (intermittency). This could be related to interactions between fast and slow streams, which usually can still take place at latitudes from 30° to 50°. Another possibility is appearance of first new solar spots for a subsequent solar cycle at some intermediate latitudes. At polar regions, where the pure fast streams are present, the degree of multifractality rises again. It is interesting that a similar behavior of

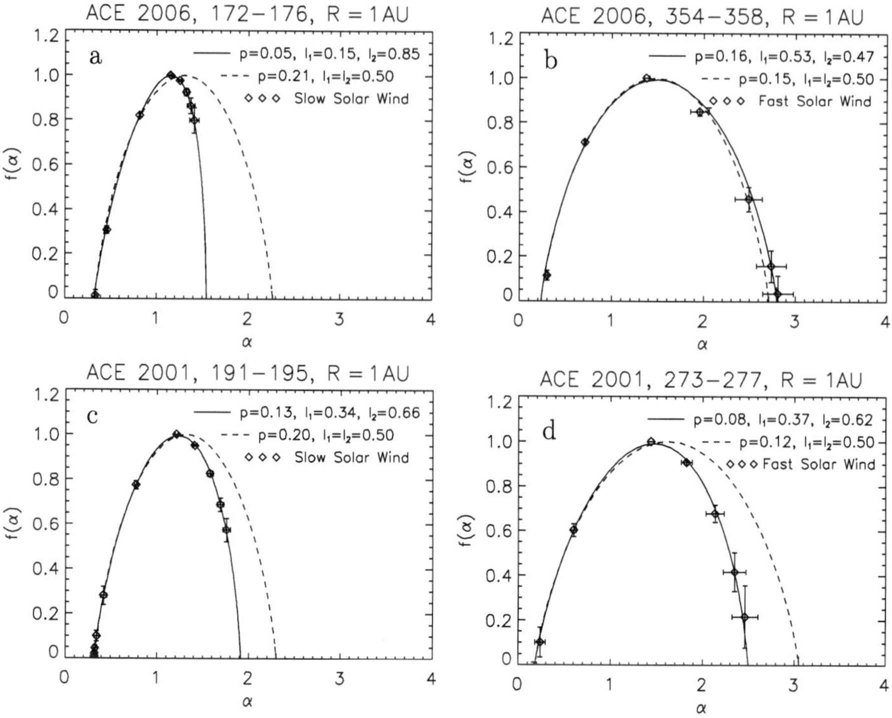

Fig. 11. The corresponding singularity spectrum $f(\alpha)$ as a function of α.

flatness, which is another measure of intermittency, has been observed at high latitudes by using the magnetic data (Yordanova et al., 2009).

Further, the degree of multifractality and asymmetry seem to be somewhat correlated. We see that when latitudes change from $+32^\circ \div +40^\circ$ to $-50^\circ \div -56^\circ$ then Δ decreases from 1.50 to 1.27, and the value of A changes only slightly from 1.10 to 1.07. Only at very high polar regions larger than 70° this correlation ceases. Moreover, the scaling of the fast streams from the polar region of the Sun exhibit more multifractal and asymmetric character, $\Delta = 1.80$, $A = 0.80$, than that for the slow wind from the equatorial region, $\Delta = 1.52$, $A = 1.14$ (in both cases we have relatively large errors of these parameters). In Figure 14 (b) we show how the parameters of the two-scale Cantor set model p and l_1 (during solar minimum) depend on the heliographic latitudes, rising at $\sim 50^\circ$ and again above $\sim 70^\circ$. It is clear that both parameters seem to be correlated.

Let us now compare the results obtained out of ecliptic with those obtained at the ecliptic plane using the generalized two-scale cascade model. First, as seen from Table 1, our analysis of the data obtained onboard ACE spacecraft at the Earth's orbit, especially in the fast solar wind, indicates multifractal structure with the degree of multifractality of $\Delta = 2.56 \pm 0.16$ and the degree of asymmetry $A = 0.95 \pm 0.11$ during solar minimum (Macek et al., 2009). Similar values are obtained by Voyager spacecraft, e.g., Table 2, at distance of 2.5 AU we have $\Delta = 2.12 \pm 0.14$ and $A = 1.54 \pm 0.24$, and in the outer heliosphere at 25 AU we

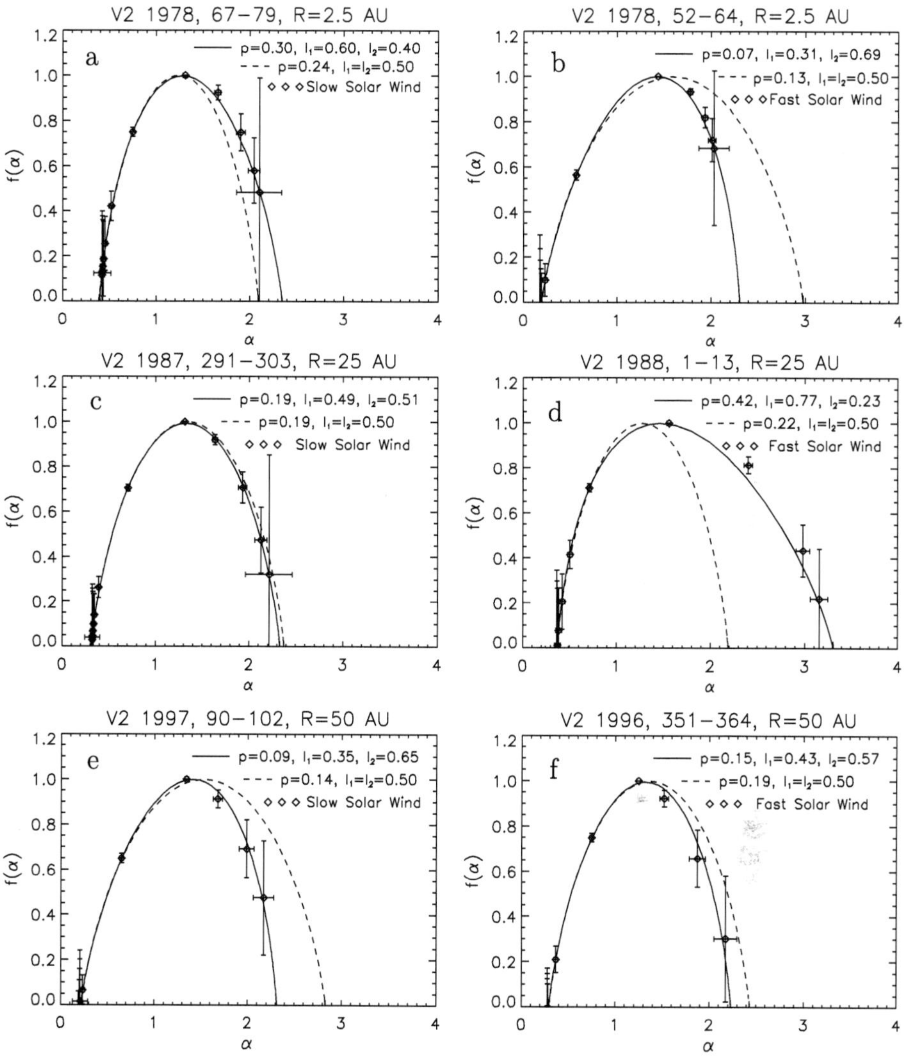

Fig. 12. The singularity spectrum $f(\alpha)$ calculated for the one-scale p-model (dashed lines) and the generalized two-scale (continuous lines) models with parameters fitted to the multifractal measure $\mu(q,l)$ using data measured by Voyager 2 during solar minimum (1978, 1987–1988, 1996–1997) at 2.5, 25, and 50 AU (diamonds) for the slow (a, c, e) and fast (b, d, f) solar wind, correspondingly.

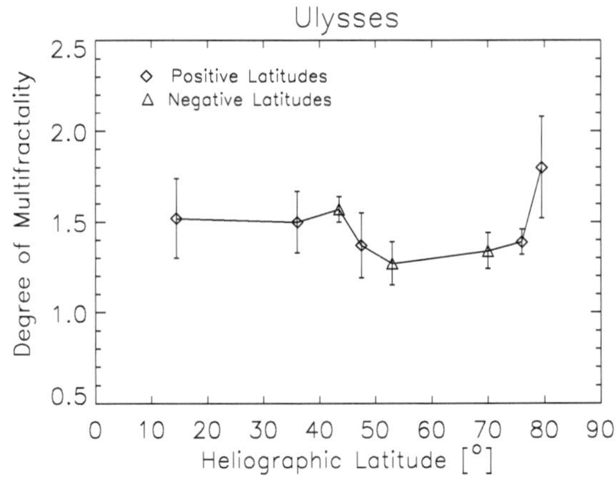

Fig. 13. Degree of multifractality Δ (continuous line) for the slow (at 15°) and fast (above 15°) solar wind during solar minimum (1994 - 1996, 2006 - 2007) in dependence on heliographic latitude below (triangles) and above (diamonds) the ecliptic.

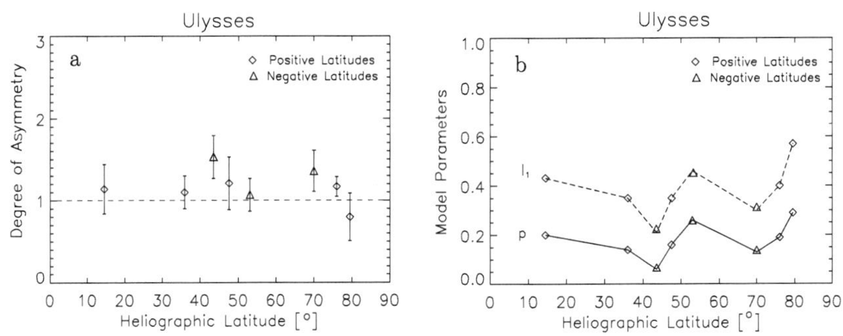

Fig. 14. (a) Degree of asymmetry A and (b) change of two-scale model parameters p (continuous line) and l_1 (dashed line) in dependence on heliographic latitude during solar minimum (1994 - 1996, 2006 - 2007).

have also large values $\Delta = 2.93 \pm 0.10$ and rather asymmetric spectrum $A = 0.66 \pm 0.11$ (Macek & Wawrzaszek, 2009). We see that at high latitudes during solar minimum in the fast solar wind, Table 3, we observe somewhat smaller degree of multifractality and intermittency as compared with those at the ecliptic, Table 2.

These results are consistent with previous results confirming that slow wind intermittency is higher than that for the fast wind (e.g., Sorriso-Valvo et al., 1999), and that intermittency in the fast wind increases with the heliocentric distance, including high latitudes (e.g., Bruno et al., 2003; 2001). In addition, symmetric multifractal singularity spectra are observed at high latitudes, in contrast to often significant asymmetry of the multifractal singularity spectrum at the ecliptic wind. This demonstrate that solar wind turbulence may exhibit somewhat different scaling at various latitudes resulting from different dynamics of the ecliptic and polar winds. Notwithstanding of the complexity of solar wind fluctuations it appears that

Heliographic Latitude	Heliocentric Distance	Multifractality Δ	Asymmetry A
$+14° \div +15°$ (1997)*	4.9 AU	1.52 ± 0.22	1.14 ± 0.30
$+32° \div +40°$ (1995)	1.4 AU	1.50 ± 0.17	1.10 ± 0.20
$+47° \div +48°$ (1996)	3.3 AU	1.37 ± 0.18	1.21 ± 0.32
$+74° \div +78°$ (1995)	1.8 – 1.9 AU	1.39 ± 0.07	1.17 ± 0.12
$+79° \div +80°$ (1995)	1.9 – 2.0 AU	1.80 ± 0.28	0.80 ± 0.29
$-40° \div -47°$ (2007)	1.6 AU	1.57 ± 0.07	1.53 ± 0.20
$-50° \div -56°$ (1994)	1.6 – 1.7 AU	1.27 ± 0.12	1.07 ± 0.20
$-69° \div -71°$ (2006)	2.8 – 2.9 AU	1.34 ± 0.10	1.36 ± 0.25

Table 3. Degree of Multifractality Δ and Asymmetry A for the Energy Transfer Rate in the Out of Ecliptic Plane for the Fast Solar Wind (the case of the slow wind is denoted by an asterisk).

the standard one-scale p model can roughly describe these nonlinear fluctuations out of the ecliptic, hopefully also in the polar regions. However, the generalized two-scale Cantor set model is necessary for describing scaling of solar wind intermittent turbulence near the ecliptic.

4.2 Multifractal model for magnetic turbulence

In the inertial region the q-order total probability measure, the partition function in Equation (5), should scale as

$$\sum p_j^q(l) \sim l^{\tau(q)}, \tag{25}$$

with $\tau(q)$ given in Equation (7). In this case Burlaga (1995) has shown that the average value of the qth moment of the magnetic field strength B at various scales $l = v_{SW}\Delta t$ scales as

$$\langle B^q(l)\rangle \sim l^{\gamma(q)}, \tag{26}$$

with the similar exponent $\gamma(q) = (q-1)(D_q - 1)$.

For a given q, using the slopes $\gamma(q)$ of $\log_{10}\langle B^q\rangle$ versus $\log_{10} l$ in the inertial range one can obtain the values of D_q as a function of q according to Equation (26). Equivalently, as discussed in Subsection 2.8, the multifractal spectrum $f(\alpha)$ as a function of scaling indices α indicates universal multifractal scaling behavior.

4.2.1 Inner heliosphere

Macek & Wawrzaszek (2011a) have shown that the degree of multifractality for magnetic field fluctuations of the solar wind at ~1 AU for large scales from 2 to 16 days is greater than that for the small scales from 2 min. to 18 h. In particular, they have demonstrated that on small scales the multifractal scaling is strongly asymmetric in contrast to a rather symmetric spectrum on the large scales, where the evolution of the multifractality with the solar cycle is also observed.

4.2.2 Outer heliosphere and the heliosheath

Further, the results for the multifractal spectrum $f(\alpha)$ obtained using the Voyager 1 data of the solar wind magnetic fields in the distant heliosphere beyond the planets, at 50 AU (1992, diamonds) and 90 AU (2003, triangles), and after crossing the heliospheric shock, at

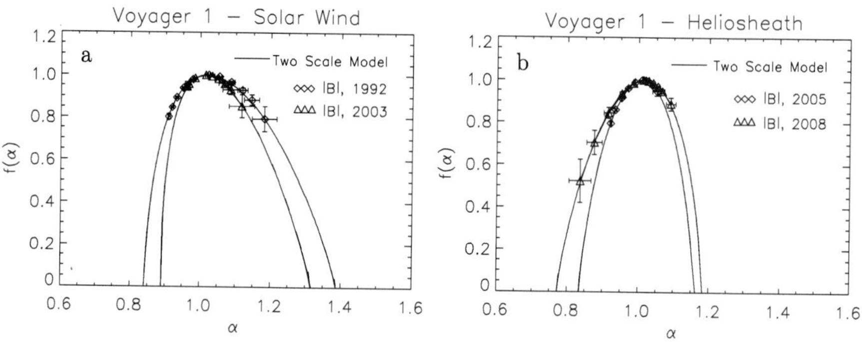

Fig. 15. The multifractal singularity spectrum of the magnetic fields observed by Voyager 1 (a) in the solar wind near 50 and 90 AU (1992, diamonds, and 2003, triangles) and (b) in the heliosheath near 95 and 105 AU (2005, diamonds, and 2008, triangles) together with a fit to the two-scale model (solid curve), suggesting change of the symmetry of the spectrum at the termination shock.

95 AU (2005, diamonds) and 105 AU (2008, triangles), are presented in Figures 15 (a) and (b), correspondingly, taken from (Macek et al., 2011). It is worth noting a change of the symmetry of the spectrum at the shock relative to its maximum at a critical singularity strength $\alpha = 1$. Because the density of the measure $\epsilon \propto l^{\alpha-1}$, this is related to changing properties of the magnetic field density ϵ at the termination shock. Consequently, a concentration of magnetic fields shrinks resulting in thinner flux tubes or stronger current concentration in the heliosheath.

Macek et al. (2011) were also looking for the degree of multifractality Δ in the heliosphere as a function of the heliospheric distances during solar minimum (MIN), solar maximum (MAX), declining (DEC) and rising (RIS) phases of solar cycles. The obtained values of Δ roughly follow the fitted periodically decreasing function of time (in years, dotted), $20.27 - 0.00992t + 0.06\sin((t-1980)/(2\pi(11)) + \pi/2)$, with the corresponding averages shown by continuous lines in Figure 16, taken from (Macek et al., 2011). The crossing of the termination shock (TS) by Voyager 1 is marked by a vertical dashed line. Below are shown the Sunspot Numbers (SSN) during years 1980–2008. We see that the degree of multifractality falls steadily with distance and is apparently modulated by the solar activity, as noted by Burlaga et al. (2003).

Macek & Wawrzaszek (2009) have already demonstrated that the multifractal scaling is asymmetric in the outer heliosphere. Now, the degree of asymmetry A of this multifractal spectrum in the heliosphere as a function of the phases of solar cycles is shown in Figure 17, taken from (Macek et al., 2011); the value $A = 1$ (dotted) corresponds to the one-scale symmetric model. One sees that in the heliosphere only one of three points above unity is at large distances from the Sun. In fact, inside the outer heliosphere prevalently $A < 1$ and only once (during the declining phase) the left-skewed spectrum ($A > 1$) was clearly observed. Anyway, it seems that the right-skewed spectrum ($A < 1$) before the crossing of the termination shock is preferred. As expected the multifractal scaling is asymmetric before shock crossing with the calculated degree of asymmetry at distances $70 - 90$ AU equal to $A = 0.47 - 0.96$. It also seems that the asymmetry is probably changing when crossing the termination shock ($A = 1.0 - 1.5$) as is also illustrated in Figure 15, but owing to large errors

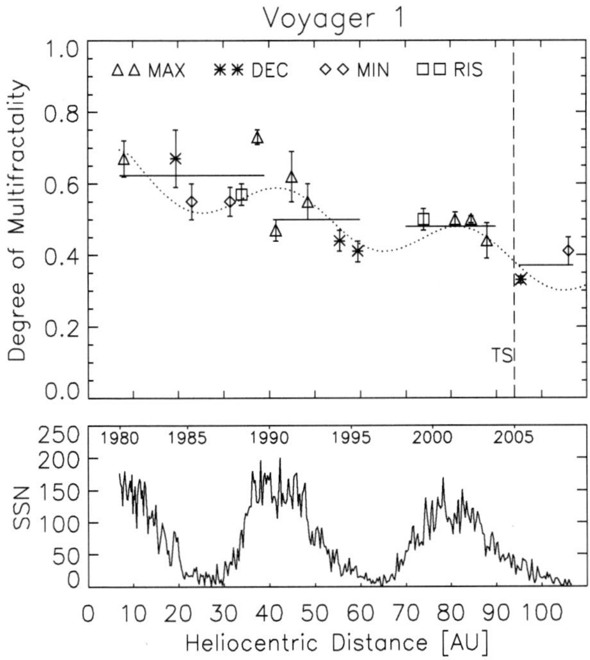

Fig. 16. The degree of multifractality Δ in the heliosphere versus the heliospheric distances compared to a periodically decreasing function (dotted) during solar minimum (MIN) and solar maximum (MAX), declining (DEC) and rising (RIS) phases of solar cycles, with the corresponding averages shown by continuous lines. The crossing of the termination shock (TS) by Voyager 1 is marked by a vertical dashed line. Below is shown the Sunspot Number (SSN) during years 1980–2008.

bars and a very limited sample, symmetric spectrum is still locally possible in the heliosheath (cf. Burlaga & Ness, 2010).

4.3 Degree of multifractality and asymmetry

For comparison, the values calculated from the papers by Burlaga et al. (2006), Burlaga & Ness (2010), and Macek et al. (2011) (Two-Scale Model) are also given in Figure 18. One sees that the degree of multifractality for fluctuations of the interplanetary magnetic field strength obtained from independent types of studies are in surprisingly good agreement; generally these values are smaller than that for the energy rate transfer in the turbulence cascade ($\Delta = 2 - 3$), Table 2, taken from (Macek & Wawrzaszek, 2009). Moreover, it is worth noting that our values obtained before the shock crossing, $\Delta = 0.4 - 0.7$, are somewhat greater than those for the heliosheath Δ = 0.3–0.4. This confirms the results presented by Burlaga et al. (2006) and Burlaga & Ness (2010). In this way we have provided a supporting evidence that the magnetic field behavior in the outer heliosphere, even in a very deep heliosphere, may exhibit a multifractal scaling, while in the heliosheath smaller values indicate possibility toward a monofractal behavior, implying roughly constant density of the probability measure.

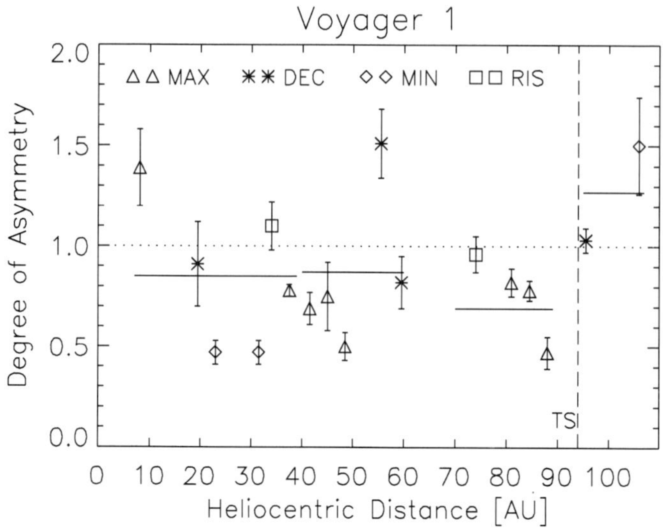

Fig. 17. The degree of asymmetry A of the multifractal spectrum in the heliosphere as a function of the heliospheric distance during solar minimum (MIN) and solar maximum (MAX), declining (DEC) and rising (RIS) phases of solar cycles, with the corresponding averages denoted by continuous lines; the value $A = 1$ (dotted) corresponds to the one-scale symmetric model. The crossing of the termination shock (TS) by Voyager 1 is marked by a vertical dashed line.

Termination shock ⇒ Heliosheath

	7 – 40 AU	40 – 60 AU	70 – 90 AU	95 – 107 AU
Burlaga	$\Delta = 0.64$ $A = 0.69$	$\Delta = 0.69$ $A = 0.63$	$\Delta = 0.69$ $A = 0.63$	$\Delta = 0.34$ $A = 0.89$
Two-Scale Model	$\Delta = 0.55 - 0.73$ $A = 0.47 - 1.39$	$\Delta = 0.41 - 0.62$ $A = 0.51 - 1.51$	$\Delta = 0.44 - 0.50$ $A = 0.47 - 0.96$	$\Delta = 0.33 - 0.41$ $A = 1.03 - 1.51$

Heliopause ⇒

Fig. 18. The degree of multifractality Δ and asymmetry A for the magnetic field strengths in the outer heliosphere and beyond the termination shock.

5. Conclusions

In this chapter we have studied the inhomogeneous rate of the transfer of the energy flux indicating multifractal and intermittent behavior of solar wind turbulence in the inner and outer heliosphere, also out of ecliptic, and even in the heliosheath. It has been shown that the generalized dimensions and multifractal spectra for solar wind are consistent with the generalized p-model for both positive and negative q, but rather with different scaling parameters for sizes of eddies, while the usual p-model can only reproduce the spectrum for $q \geq 0$. We have demonstrated that intermittent pulses are stronger for the asymmetric scaling. In fact, using the two-scale weighted Cantor set model, which is a convenient tool to investigate this scaling, one can confirm the characteristic shape of the universal multifractal singularity spectrum; as seen in Figure 4, $f(\alpha)$ is a downward concave function of scaling indices α. In our view, this obtained shape of the multifractal spectrum results not only from the nonuniform probability of the energy transfer rate but mainly from the multiscale nature of the cascade.

It is well known that the fast wind is associated with coronal holes, while the slow wind mainly originates from the equatorial regions of the Sun. Consequently, the structure of the flow differs significantly for the slow and fast streams. Hence the fast wind is considered to be relatively uniform and stable, while the slow wind is more turbulent and quite variable in velocities, possibly owing to a strong velocity shear (Goldstein et al., 1995). Macek et al. (2009) have shown that the velocity fluctuations in the fast streams seem to be more multifractal than those for the slow solar wind.

By investigating Ulysses data Wawrzaszek & Macek (2010) have shown that at high latitudes during solar minimum in the fast solar wind we observe somewhat smaller degree of multifractality and intermittency as compared with those at the ecliptic. Moreover, the degree of multifractality and asymmetry of the fast solar wind exhibit latitudinal dependence with some symmetry with respect to the ecliptic plane. Both quantities seem to be correlated during solar minimum for latitudes below 70°. The multifractal singularity spectra become roughly symmetric. The minimum intermittency is observed at mid-latitudes and is possibly related to the transition from the region where the interaction of the fast and slow streams takes place to a more homogeneous region of the pure fast solar wind.

It is worth noting that the multifractal scaling is often rather asymmetric (Helios, ACE, and Voyager). In particular, the fast wind during solar minimum exhibits strong asymmetric scaling. Moreover, both the degree of multifractality and degree of asymmetry are correlated with the heliospheric distance and we observe the evolution of multifractal scaling in the outer heliosphere (cf. Burlaga, 1991; 2004; Burlaga et al., 2003; Macek & Wawrzaszek, 2009).

Macek et al. (2011) have shown for the first time that the degree of multifractality for magnetic field fluctuations of the solar wind falls steadily with the distance from the Sun and seems to be modulated by the solar activity. Moreover, in contrast to the right-skewed asymmetric spectrum with singularity strength $\alpha > 1$ inside the heliosphere, the spectrum becomes more left-skewed, $\alpha < 1$, or approximately symmetric after the shock crossing in the heliosheath, where the plasma is expected to be roughly in equilibrium in the transition to the interstellar medium. In particular, before the shock crossing, especially during solar maximum, turbulence is more multifractal than that in the heliosheath.

Hence one can hope that the new more general asymmetric multifractal model could shed light on the nature of turbulence and we therefore propose this model as a useful tool for analysis of intermittent turbulence in various environments.

6. Acknowledgments

This work has been supported by the Polish National Science Centre (NCN) and the Ministry of Science and Higher Education (MNiSW) through Grant NN 307 0564 40. We would like to thank the plasma instruments teams of Helios, Advanced Composition Explorer, Ulysses, and Voyager, and also the magnetic field instruments teams of ACE and Voyager missions for providing experimental data.

7. References

Alexandrova, O., Carbone, V., Veltri, P. & Sorriso-Valvo, L. (2007). Solar wind Cluster observations: Turbulent spectrum and role of Hall effect, *Planet. Space Sci.* 55: 2224–2227.

Bruno, R., Carbone, V., Sorriso-Valvo, L. & Bavassano, B. (2003). Radial evolution of solar wind intermittency in the inner heliosphere, *J. Geophys. Res.* 108(A3): 1130.

Bruno, R., Carbone, V., Veltri, P., Pietropaolo, E. & Bavassano, B. (2001). Identifying intermittency events in the solar wind, *Planet. Space Sci.* 49: 1201–1210.

Burlaga, L. F. (1991). Multifractal structure of the interplanetary magnetic field: Voyager 2 observations near 25 AU, 1987-1988, *Geophys. Res. Lett.* 18: 69–72.

Burlaga, L. F. (1995). *Interplanetary magnetohydrodynamics*, New York: Oxford Univ. Press.

Burlaga, L. F. (2001). Lognormal and multifractal distributions of the heliospheric magnetic field, *J. Geophys. Res.* 106: 15917–15927.

Burlaga, L. F. (2004). Multifractal structure of the large-scale heliospheric magnetic field strength fluctuations near 85 AU, *Nonlinear Processes Geophys.* 11: 441–445.

Burlaga, L. F. & Klein, L. W. (1986). Fractal structure of the interplanetary magnetic field, *J. Geophys. Res.* 91: 347–350.

Burlaga, L. F. & Ness, N. F. (2010). Sectors and large-scale magnetic field strength fluctuations in the heliosheath near 110 AU: Voyager 1, 2009, *Astrophys. J.* 725: 1306–1316.

Burlaga, L. F., Ness, N. F. & Acuña, M. H. (2006). Multiscale structure of magnetic fields in the heliosheath, *J. Geophys. Res.* 111: A09112.

Burlaga, L. F., Ness, N. F., Acuña, M. H., Lepping, R. P., Connerney, J. E. P., Stone, E. C. & McDonald, F. B. (2005). Crossing the termination shock into the heliosheath: Magnetic fields, *Science* 309: 2027–2029.

Burlaga, L. F., Perko, J. & Pirraglia, J. (1993). Cosmic-ray modulation, merged interaction regions, and multifractals, *Astrophys. J.* 407: 347–358.

Burlaga, L. F., Wang, C. & Ness, N. F. (2003). A model and observations of the multifractal spectrum of the heliospheric magnetic field strength fluctuations near 40 AU, *Geophys. Res. Lett.* 30: 1543.

Carbone, V. (1993). Cascade model for intermittency in fully developed magnetohydrodynamic turbulence, *Phys. Rev. Lett.* 71: 1546–1548.

Chhabra, A. B., Meneveau, C., Jensen, R. V. & Sreenivasan, K. R. (1989). Direct determination of the f(α) singularity spectrum and its application to fully developed turbulence, *Phys. Rev. A* 40(9): 5284–5294.

Chhabra, A. & Jensen, R. V. (1989). Direct determination of the f(α) singularity spectrum, *Phys. Rev. Lett.* 62(12): 1327–1330.

Frisch, U. (1995). *Turbulence. The legacy of A.N. Kolmogorov*, Cambridge: Cambridge Univ. Press, U. K.

Frisch, U., Sulem, P.-L. & Nelkin, M. (1978). A simple dynamical model of intermittent fully developed turbulence, *J. Fluid Mech.* 87: 719–736.

Goldstein, B. E., Smith, E. J., Balogh, A., Horbury, T. S., Goldstein, M. L. & Roberts, D. A. (1995). Properties of magnetohydrodynamic turbulence in the solar wind as observed by Ulysses at high heliographic latitudes, *Geophys. Res. Lett.* 22: 3393–3396.

Grassberger, P. (1983). Generalized dimensions of strange attractors, *Phys. Lett. A* 97: 227–230.

Grassberger, P. & Procaccia, I. (1983). Measuring the strangeness of strange attractors, *Physica D* 9: 189–208.

Halsey, T. C., Jensen, M. H., Kadanoff, L. P., Procaccia, I. & Shraiman, B. I. (1986). Fractal measures and their singularities: The characterization of strange sets, *Phys. Rev. A* 33(2): 1141–1151.

Hentschel, H. G. E. & Procaccia, I. (1983). The infinite number of generalized dimensions of fractals and strange attractors, *Physica D* 8: 435–444.

Horbury, T. S., Balogh, A., Forsyth, R. J. & Smith, E. J. (1996). Magnetic field signatures of unevolved turbulence in solar polar flows, *J. Geophys. Res.* 101: 405–413.

Jensen, M. H., Kadanoff, L. P. & Procaccia, I. (1987). Scaling structure and thermodynamics of strange sets, *Phys. Rev. A* 36: 1409–1420.

Kolmogorov, A. (1941). The local structure of turbulence in incompressible viscous fluid for very large Reynolds' numbers, *Dokl. Akad. Nauk SSSR*, 30: 301–305.

Kraichnan, R. H. (1965). Inertial-range spectrum of hydromagnetic turbulence, *Phys. Fluids* 8: 1385–1387.

Lamy, H., Wawrzaszek, A., Macek, W. M. & Chang, T. (2010). New multifractal analyses of the solar wind turbulence: Rank-ordered multifractal analysis and generalized two-scale weighted Cantor set model, *in* M. Maksimovic, K. Issautier, N. Meyer-Vernet, M. Moncuquet, & F. Pantellini (eds), *Twelfth International Solar Wind Conference*, Vol. 1216 of *American Institute of Physics Conference Series*, pp. 124–127.

Macek, W. M. (1998). Testing for an attractor in the solar wind flow, *Physica D* 122: 254–264.

Macek, W. M. (2002). Multifractality and chaos in the solar wind, *in* S. Boccaletti, B. J. Gluckman, J. Kurths, L. M. Pecora & M. L. Spano (eds), *Experimental Chaos*, Vol. 622 of *American Institute of Physics Conference Series*, pp. 74–79.

Macek, W. M. (2003). The multifractal spectrum for the solar wind flow, *in* M. Velli, R. Bruno, F. Malara & B. Bucci (eds), *Solar Wind Ten*, Vol. 679 of *American Institute of Physics Conference Series*, pp. 530–533.

Macek, W. M. (2006a). Modeling multifractality of the solar wind, *Space Sci. Rev.* 122: 329–337.

Macek, W. M. (2006b). Multifractal solar wind, *Academia* 3: 16–18.

Macek, W. M. (2007). Multifractality and intermittency in the solar wind, *Nonlinear Processes Geophys.* 14: 695–700.

Macek, W. M., Bruno, R. & Consolini, G. (2005). Generalized dimensions for fluctuations in the solar wind, *Phys. Rev. E* 72(1): 017202.

Macek, W. M., Bruno, R. & Consolini, G. (2006). Testing for multifractality of the slow solar wind, *Adv. Space Res.* 37: 461–466.

Macek, W. M. & Redaelli, S. (2000). Estimation of the entropy of the solar wind flow, *Phys. Rev. E* 62: 6496–6504.

Macek, W. M. & Szczepaniak, A. (2008). Generalized two-scale weighted Cantor set model for solar wind turbulence, *Geophys. Res. Lett.* 35: L02108.

Macek, W. M. & Wawrzaszek, A. (2009). Evolution of asymmetric multifractal scaling of solar wind turbulence in the outer heliosphere, *J. Geophys. Res.* 114(13): A03108.

Macek, W. M. & Wawrzaszek, A. (2011a). Multifractal structure of small and large scales fluctuations of interplanetary magnetic fields, *Planet. Space Sci.* 59: 569–574.

Macek, W. M. & Wawrzaszek, A. (2011b). Multifractal two-scale Cantor set model for slow solar wind turbulence in the outer heliosphere during solar maximum, *Nonlinear*

Processes Geophys. 18(3): 287–294.
URL: *http://www.nonlin-processes-geophys.net/18/287/2011/*

Macek, W. M., Wawrzaszek, A. & Carbone, V. (2011). Observation of the multifractal spectrum at the termination shock by Voyager 1, *Geophys. Res. Lett.* 38: L19103.

Macek, W. M., Wawrzaszek, A. & Hada, T. (2009). Multiscale multifractal intermittent turbulence in space plasmas, *J. Plasma Fusion Res. SERIES* 8: 142–147.

Mandelbrot, B. B. (1989). Multifractal measures, especially for the geophysicist, *Pure Appl. Geophys.* 131: 5–42.

Marino, R., Sorriso-Valvo, L., Carbone, V., Noullez, A., Bruno, R. & Bavassano, B. (2008). Heating the solar wind by a magnetohydrodynamic turbulent energy cascade, *Astrophys. J.* 677: L71–L74.

Marsch, E., Tu, C.-Y. & Rosenbauer, H. (1996). Multifractal scaling of the kinetic energy flux in solar wind turbulence, *Ann. Geophys.* 14: 259–269.

Meneveau, C. & Sreenivasan, K. R. (1987). Simple multifractal cascade model for fully developed turbulence, *Phys. Rev. Lett.* 59: 1424–1427.

Meneveau, C. & Sreenivasan, K. R. (1991). The multifractal nature of turbulent energy dissipation, *J. Fluid Mech.* 224: 429–484.

Ott, E. (1993). *Chaos in dynamical systems*, Cambridge: Cambridge Univ. Press, U. K.

Schwenn, R. (1990). Large-scale structure of the interplanetary medium, *in* R. Schwenn, E. Marsch (eds), *Physics of the Inner Heliosphere I*, Vol. 20, Berlin: Springer-Verlag, pp. 99–181.

Smith, E. J., Balogh, A., Lepping, R. P., Neugebauer, M., Phillips, J. & Tsurutani, B. T. (1995). Ulysses observations of latitude gradients in the heliospheric magnetic field, *Adv. Space Res.* 16: 165–170.

Sorriso-Valvo, L., Carbone, V., Veltri, P., Consolini, G. & Bruno, R. (1999). Intermittency in the solar wind turbulence through probability distribution functions of fluctuations, *Geophys. Res. Lett.* 26: 1801–1804.

Sorriso-Valvo, L., Marino, R., Carbone, V., Noullez, A., Lepreti, F., Veltri, P., Bruno, R., Bavassano, B. & Pietropaolo, E. (2007). Observation of inertial energy cascade in interplanetary space plasma, *Phys. Rev. Lett.* 99(11): 115001.

Szczepaniak, A. & Macek, W. M. (2008). Asymmetric multifractal model for solar wind intermittent turbulence, *Nonlinear Processes Geophys.* 15: 615–620.

Wawrzaszek, A. & Macek, W. M. (2010). Observation of the multifractal spectrum in solar wind turbulence by Ulysses at high latitudes, *J. Geophys. Res.* 115(14): A07104.

Yordanova, E., Balogh, A., Noullez, A. & von Steiger, R. (2009). Turbulence and intermittency in the heliospheric magnetic field in fast and slow solar wind, *J. Geophys. Res.* 114(A13): A08101.

8

Field-Aligned Current Mechanisms of Prominence Destabilization

Petko Nenovski
National Institute of Geophysics, Geodesy and Geography, Sofia Bulgaria

1. Introduction

Solar wind is initiated from active solar flare regions. Solar flares start with an active-region prominence: cold dense plasma associated with an arcade of looped magnetic field lines. A prominence can be stable for hours or days but sometimes it takes on a violent evolution: as the magnetic field structure slowly evolves, the loop becomes more and more stretched and the prominence starts rising slowly until the configuration becomes highly unstable and reconnection initiates, causing a large, impulsive energy release which will make the prominence erupt more rapidly; as reconnection continues both the erupting prominence and the plasma on the newly formed loops are heated tremendously causing very bright X-ray emissions. Some flares are accompanied by coronal mass ejections where plasma piled up in the corona above the rising magnetic loops is ejected by the intense energy bursts associated with reconnecting field lines. The lower-lying prominence which consists of chromospheric material might be ejected as well. Coronal mass ejections (CME) give rise to large disturbances propagating outward in the solar wind. This picture points out that prominences are a basic source of solar wind in which the Earth's and planet magnetospheres and other non-magnetized planets are continuously immersed. When emitted in a direction which brings it on a collision course with the Earth, a CME will have a profound impact on the outskirts of the terrestrial atmosphere: the magnetosphere and the ionosphere (Bellan, 2004).

Prominences are arched structures protruding from the solar surface. They are known to consist of plasma-filled magnetic flux tubes and occur at many different scales. Also, they are imbedded in a hot corona via a so-called prominence-corona transition region. Quiescent prominences (QPs) are those prominences located in the solar corona and are denser and cooler objects than their surroundings and relatively quiet if they are viewed at large time scales. Prominences invariably are located parallel to and above a photospheric, magnetic polarity reversal line, although not every part of a reversal line has a prominence above it. Polarimetric observation has shown that a prominence is threaded by a largely horizontal magnetic field with a principal component along the prominence length (Leroy 1989; Tandberg-Hanssen 1995). Therefore, prominences are naturally classified by whether their magnetic fields thread in the same or opposite direction relative to the direction of the underlying photospheric bipolar field. These two types of prominences, described as normal and inverse, respectively, imply distinct topologies for the coronal magnetic fields around the prominence.

The precise knowledge of prominence plasma parameters (temperature, density, ionisation ratio and magnetic field) is very important for prominence theory and related problems (Tandberg-Hanssen 1995). Typical values of QP temperature lie in the range 5000-8000 K (Hirayama 1985a; Zhang et al. 1987; Mein & Mein 1991). Lower temperatures (T ~ 4300 K) have been reported by Hirayama (1985b). An interesting tendency of increasing prominence temperature toward the outer edges, where temperatures may reach values of 10^4 to $2x10^4$ K, has been pointed out by Hirayama (1971). The prominence plasma density is not as well known as the temperature because of observational difficulties. The electron density is in the range of 10^{10} –10^{11} cm^{-3} (Hirayama 1972), but these values have been found not to be precise (Vial 1986). The registered densities depend on both the prominence type and the method used (Hirayama 1985b). Electron densities ten times smaller (Bommier et al. 1986) and larger than 10^{11} cm^{-3} (N_e ~ $10^{11.3}$ cm^{-3}) (Landman 1984a) have been found. The hydrogen density of QP lies in the range of (3÷6)x10^{11} cm^{-3} (Landman 1984a; Vernazza 1981; Hirayama 1986) The ionisation ratio n_{HII}/n_{HI} usually varies in the interval of 1-3. According to Landman (1984b) it is in the range of 0.05 -1, although higher values (~3) have been obtained by Vial (1982).

The prominence magnetic field plays an important role in the QP formation, stability and dynamics. Its magnitude and orientation depend on the FAC flowing within the prominence body. The first measurements of the longitudinal component of the magnetic field (along the prominence axis) have been done by Zirin & Severny (1961) using the Zeeman effect. The results based on this effect were summarized by Tandberg-Hanssen (1995). With the use of the Hanle effect, a new technique has been developed to measure the longitudinal and transversal prominence magnetic field (Leroy 1985). The absolute magnetic field strength in the QPs is generally found to be in the range from a few Gauss to 10 G, occasionally reaching 20 or 30 G.

Another characteristic of the QPs is the presence of internal motions during their stable period, i.e. when they are not activated. The velocity field of a QP is due to three main types of motions - vertical, horizontal and oscillations. The vertical motions can be divided into two classes - downward flows and upward flows. Downward motions are registered in prominences at the limb (Engvold 1976; Cui Shu et al. 1985) and have values of about 5 kms^{-1} . Spectrographic studies (Kubot 1980) give an average value of 0.7 kms^{-1}. Upward flows of 0.5÷5 kms^{-1} are observed in disk filaments (Mein 1977; Schmieder et al. 1984; Schmieder et al. 1985). Mass flows along filament threads, with a flow velocity in the range 5-25 km s^{-1}, have also been observed (Zirker et al. 1998; Lin et al. 2003, 2005). Zirker et al. (1998) and Lin et al. (2003) have detected flows in opposite directions within adjacent threads, a phenomenon known as counterstreaming. Some observations show fast (50 kms^{-1}) horizontal motions (inclined to the axis of the prominence at about 20°) at the edges of the filaments (Malherbe et al. 1983). Usually, the velocity of horizontal motions in QP is 10 to 20 kms^{-1}.

The presence of oscillatory motions in QPs is a proven observational fact (Tsubaki 1989; Vrsnak 1993; Vrsnak & Ruzdjak 1994). A rough classification of the prominence oscillations, made on the basis of oscillation amplitude, divides them into two main classes; large amplitude (20 kms^{-1}) oscillations affecting the whole object, and small amplitude (20 kms^{-1})

oscillations affecting the fibril or a restricted area within the prominence (Oliver 1999). The oscillations may have a large range of periods - from a few minutes to hours (Bashkirtsev et al. 1983; Bashkirtsev & Mashnich 1984; Weihr et al. 1984). Some observers have detected oscillations and travelling waves in individual threads or groups of threads (Yi et al. 1991; Yi & Engvold 1991; Lin 2004; Lin et al. 2007), with periods typically between 3 and 20 minutes. The prominence fibril structure also exhibits periodic variations (Tsubaki et al. 1988); individual fibrils may oscillate with their own periods (Thomson & Schmieder 1991; Yi et al. 1991). There is also some evidence that velocity oscillations are more easily detected at the edges of prominences or where the material seems fainter than in the prominence body (Tsubaki & Takeuchi 1986; Tsubaki et al. 1988).

Recently, high-resolution images of solar filaments (e.g., Lin et al. 2003, 2005, 2007) clearly show the existence of horizontal fine-structures within the filament body. This observational evidence suggests that prominences are composed of many field-aligned *threads*. These threads are usually skewed with respect to the filament long axis by an angle of 20° on average, although their orientation can vary significantly within the same prominence (Lin 2004). The observed thickness, d, and length, l, of threads are typically in the ranges $0.2'' < d < 0.6''$ and $5'' < l < 20''$ (Lin et al. 2005). Since the observed thickness is close to the resolution of present-day telescopes, it is likely that even thinner threads could exist. According to some models, a thread is believed to be part of a larger magnetic coronal flux tube which is anchored in the photosphere (e.g., Ballester & Priest 1989), with denser and cooler material near its apex, i.e., the observed thread itself. The process that leads to the formation of such structures however, is still unknown (e.g. Lin etal, 2005;2007).

The prominence plasma is thus a complex system in terms of temperature, density, plasma motion and specific magnetic field configuration. However, if we take into account the prominence fine structure, they *must be treated as dynamic and nonhomogenous formations of field-aligned currents (FACs)*. The FAC characteristics are usually unknown and nowadays there are no methods to detect them. The FAC are intrinsically connected with the magnetic field configuration. The subject of this review is the stability problem of field-aligned currents (FAC) flowing along prominence body.

In conventional MHD theory field-aligned currents (FAC) are supported by Alfven waves. There is another source of FAC, which is primarily due to the Alfven wave gradient existing at the plasma boundaries. In addition to the Alfven modes, the MHD surface waves, supported by these boundaries, can also carry field-aligned currents. The surface wave FAC should be dislocated, either at the prominence boundaries or within the prominence body. A physical interpretation of surface wave FAC as a boundary phenomenon, i.e. as FAC flowing into the plasma boundary structures, has also been given (Nenovski 1996). The MHD surface waves have been formerly examined as a source of field-aligned electric currents and they are suggested to promote QP destabilization via a field-aligned current (FAC) intensification process (Nenovski et al., 2001). The SW can, however, carry FAC only under certain conditions. The MHD surface wave should be a non-axial mode, i.e. it has to propagate obliquely to the ambient magnetic field. The surface wave FAC propagation or the surface wave group velocity is however directed practically along the magnetic field.

These "wave induced" currents tend to be concentrated at the prominence periphery, i.e. they are surface field-aligned currents. Such currents can be intensified in the case when surface wave bouncing processes at the prominence feet are possible and conditions for negative reflection (under negative dielectric properties conditions) are present. This surface FAC yields an increase of the azimuthal component of the magnetic field, which in turn may destabilize the whole QP system (Nenovski et al., 2001). The preferable situations for such a "switch on" of QP destabilization process require a plasma density increase at the footpoints and/or a low density plasma condition in the prominence body. The magnetic field strength requirements (threshold) depend on the neutral density variations and the neutral-plasma density ratio (Nenovski et al., 2001).

2. Intensification of the FACs, filamentation and dissipation

On the other hand, the process of wave FAC intensification if it starts will be followed inevitably by FAC density increase and hence instability due to non-linearity effects may emerge. Consequently there are various non-linear effects. The most dangerous one is the structural instability known as Bennett's pinch instability due to the interaction of the current with its own magnetic field. This is explainable by imagining the FAC as being composed of smaller individual components of that current all travelling within the prominence body, therefore a net inward pressure on the surface of this body will be generated. A further intensification of field-aligned currents could be followed by compression and distortion of plasma and ambient magnetic field at some points of the prominence body. As a consequence, the prominence structure composed by field-aligned current density and magnetic field flux where the field-aligned current flows can finally be disrupted. We advice the reader to read about the pinch effect initially examined by Northrup (1907) and also major developments by Bennett published in 1934 where an analysis of the full radial pressure balance in a static Z-pinch has been executed.

Examining in detail the field-aligned current density distribution along the prominence body the following considerations should be taken into account. First, the field-aligned currents supported by either Alfven wave, or MHD surface wave, would not occupy the whole cross section of the prominence body - hence the Bennett's instability configuration could not be realized. Second, along the flux tube coinciding with the prominence body the magnetic field flux $\Phi=B.S$ is constant, but the magnetic field intensity B would be inversely proportional to changes in the prominence cross section S. The cross section S changes along the prominence body. Therefore, the field-aligned current propagation could be followed by field-aligned current density changes (increase or decrease). Depending on the field aligned current density distribution, a field-aligned current filament mechanism would starts and be operative at points where FAC density becomes maximized.

2.1 Distribution of elementary currents

Note that a FAC structure mechanism embedded in the ambient FAC structure coexisting with the prominence is considered. Thus we will examine possible FAC structure formation process being easily initiated in the regions where the FAC intensities are sufficiently high, e.g. at the chromosphere-corona transition regions just above the chromosphere. The influence of the chromosphere on the FAC structures of course is through its conductivity (Appendix A). It is expected that pre-existing chromospheric density inhomogeneities of various scales could also influence an initiation of the FAC structure formation process. The latter are usually of irregular characteristics and probably of local extent. Initially we neglect them.

Before going to examine FAC structure mechanism by itself, let us introduce some relevant physical quantities. In previous analytical treatments of the plasma vortex structures (e.g. Southwood and Kivelson, 1993) less theoretical attention has been paid to the microphysics of the filament/vortex formation mechanism. We define the vorticity Ω by

$$\Omega \equiv \frac{\vec{B}_0 \cdot (\nabla \times \vec{v})}{B_0} \tag{1}$$

where $\boldsymbol{B_0}$ is the undisturbed magnetic field, $\boldsymbol{v}$ is the plasma velocity. Under MHD approach, three sources of field-aligned currents are known (Sato, 1982):

$$B\partial / \partial z(j_{||} / B) = \rho d / dt(\vec{B} \cdot \nabla \times \vec{v} / B^2) +$$
$$(2 / B)\vec{B} \times (\nabla P / B^2 + \rho / B^2 d\vec{v} / dt) \cdot \nabla B - \vec{B} \times (\rho / B^2 d\vec{v} / dt) \cdot \nabla N / N$$

Where v, $\rho = MN$ and P are the velocity, density and kinetic pressure of the plasma treated as MHD medium; the coordinate z is oriented along the magnetic field $\boldsymbol{B_0}$. Once a large-scale FAC is generated it flows along the magnetic field fluxes satisfying some boundary conditions. We will study effects coming from the FAC intensity changes down the magnetic field lines. The first source to the FAC intensity change is connected to the vorticity $\boldsymbol{\Omega}$ (see the first term). The second and third ones are due to gradients ∇N, ∇P, ∇B connected with the magnetic flux configurations and the inertial term dv/dt (the time derivative d/dt is equal to $\partial/\partial t + v.d/dr$). The time derivative d/dt is reduced to the $\boldsymbol{v.d/dr}$ term, i.e. we should take into account velocity gradients only along the plasma flow v. Thus inertial term does not contribute essentially to the large-scale FAC dynamics unless the triple scalar product $\boldsymbol{B_0}\times(\boldsymbol{v.d/dr})\boldsymbol{v}.\nabla B_0(N)$ yields in some cases considerable FAC input. Hence inertial term will contribute to the FAC changes only in the presence steep gradients ∇B, or ∇N (∇P). Indeed, steep gradients $\nabla B_0(N)$ are encountered at boundary crossings. The boundary regions (where FAC structures are located) by itself are however characterized with internal gradients $\nabla B_0(N)$ less than the steep gradients existing at their edges. Therefore, excluding fast time variations the vorticity $\boldsymbol{\Omega}$ appears to be main source of steady-state large-scale FAC intensity changes along the field lines. Then, performing the divergence operator on

$$\vec{E} + \vec{v} \times \vec{B} = 0 \tag{2}$$

one obtains

$$\nabla \cdot \vec{E} = -\vec{B} \cdot \nabla \times \vec{v} = -B\Omega \tag{3}$$

So

$$\Omega = -\rho / \varepsilon_0 B_0 = -eN_0 / \varepsilon_0 B_0 \tag{4}$$

where ρ is the equivalent charge density (see Parker, 1979); N_0 is the corresponding number density. This relation states that the vorticity Ω is equivalent to a charge density ρ. Using the charge conservation condition we could connect the charge ρ with the field-aligned current $j_{||}$: $div j = -d\rho/dt$. The FACs actually discharge the charges (built in the FAC source regions) by carrying them along the magnetic field lines. If we neglect the magnetic field line curvature and other gradient effects the current $j_{\perp}$ that flows perpendicular to the magnetic field lines is equal to zero (Shkarofsky et al., 1966). Therefore

$$div\vec{j} = div\vec{j}_{||} = \partial j_{||} / \partial z = -d\rho / dt \tag{5}$$

In connection to plasma convective vortex motion *v*, the field-aligned currents could be thought of as current discharges of electric charges being accumulated in the source regions. We will here interrelate the current discharges concept with the charge formalism because the latter is well developed. Now we are looking for FAC dynamics along the magnetic field lines. There are FAC structures of zero frequency (e.g. Nenovski, 1996). The time derivative d/dt is then given by $v_{||}\ d/dz$, where $v_{||}$ is the field-aligned charge velocity along the magnetic field lines. Eq. (5) relates quantitatively the field-aligned current density $j_{||}$ and the vorticity $\boldsymbol{\Omega}$.

$$\vec{j}_{||} = \varepsilon_0 B_0 \vec{v}_{||} \Omega \tag{6}$$

It is assumed that $v_{||}$ in eq. (6) coincides exactly with the Alfven speed v_A. The current density $j_{||}(z)$ depends obviously on the vortex $\boldsymbol{\Omega}$ and the Alfven velocity v_A distribution. The FAC carriers are of course ions or electrons. These charges however interact to each other and therefore the FAC 'elements' ($dj_{||}$) are subjected to the same forces as the charges. Their motion along the magnetic field lines produces in fact FAC elements of different sign. An ensemble of FAC elements with positive and negative signs (two-component ensemble) could be intervened in that supposedly possesses all the features of the classical Coulomb system of electrons and ions. Under this concept the distribution of the FAC elements is treated in terms of associated charge distribution. The accompanying plasma vortices (1) are to be consequently described by an appropriate stream function.

After having defined the vorticity $\boldsymbol{\Omega}$ and its relevance to the charge concept we need to clarify the forces of an ensemble of field-aligned currents elements dI. In magnetohydrodynamics (MHD), the current elements have finite length *dl*. This is due to chaotic, thermal motion of the charges. Each charge does not flow continuously along the magnetic field lines thus forming an infinite current line. Instead, its trajectory along the field line is interrupted at some distance. Due to thermal motion inherent to all charges in the system, new charge appears in place of the given charge that continues the charge motion along the field lines. We assume that an ensemble of current elements, or charges, which are continuously flowing along the field lines, builds a field-aligned current of infinite length. In order to estimate quantitatively the FAC distribution, the FAC elements interaction should be examined. The field-aligned current structure of infinite length will then consist of 'bundles' of one-polarity current elements each of length *dl*. Indeed, it is well known that the force that these field-aligned current elements experience obeys to the Biot-Savart laws, i.e.

$$F = \frac{(dI \sin\theta) \cdot dl}{r^2} = \frac{j_{||} \sin\theta dV}{r^2} \tag{7}$$

where $dI = j_{||}.dS$ is the current strength of a single current element, dl ($dl << r$), dS and $dV = dl.dS$ are its length, square and volume, respectively. The angle θ denotes the angle between the axis of the current elements and the point of field action. In the case of ensemble of current elements of opposite polarity and equal density N_0, the interaction process at very small angles θ will be screened from other current elements and then it will be less effective than in vacuum. Thus we assume that in an ensemble of current elements the interaction process takes place at angles θ nearly equal to $\pi/2$. The size of the interaction region cannot be much less than $(dV)^{1/3}$, i.e. $r > (N_0)^{-1/3}$. The square dependence on r however suggests that the description of the current element ensemble is analogous to that of the classical Coulomb system. A correspondence exists when the conventional charge in the Coulomb law stands for the charge $(j_{\|} \sin\theta) dV \sqrt{\varepsilon_o \mu_0}$ where μ_0 is the magnetic permeability. The volume element dV is expressed by the plasma density: $dV = N_0^{-1}$. For comparison with the classical Coulomb system of electrons and ions, in our case each current element is considered to be aligned along the magnetic field lines (an axial symmetry). After having determined the equivalence between the FAC elements and the charges that they carry, the FAC filament distribution appears to be e.g. a Boltzman-like. For this purpose the Biot-Savart law (7) guarantees the equivalence with the Coulomb law.

In order to understand the physics of filamentation, the field-aligned currents are examined as an ensemble of elementary currents. This allows using an approach similar to that of the spatial distribution of particles in a classical Coulomb system. According to (3) we have

$$\nabla \cdot \vec{E} = -\vec{B}_0 \cdot (\nabla \times \vec{v}) = -B_0(\Omega(+) + \Omega(-)) \tag{8}$$

where $\Omega(+)$ and $\Omega(-)$ correspond to vortices densities of positive and negative signs (with respect to the undisturbed magnetic field $\boldsymbol{B}_0$). The vortex distribution could be determined by assuming a Boltzman-like distribution of the relevant single charge q $(j_{\|}\sin\theta) dV \sqrt{\varepsilon_o \mu_0} \rightarrow q$ Hence we have

$$\Omega(\pm) = \pm \left({}^{eN_0(\pm)} / {}_{\varepsilon_0 B_0} \right) \exp\left({}^{\vec{j}_{||}(\pm)\cdot \vec{A}} / {}_{N_0(\pm)\cdot\Theta} \right) \tag{9}$$

where $N_0(\pm)$ is the number density of the positive (negative) vortices; $\boldsymbol{A}_\infty$ is the potential of a zero 'charge' defined by $q(\boldsymbol{A}_\infty) = 0$; Θ is the Gibb's distribution module. The sum of $\Omega(+)$ and $\Omega(-)$ of course will represent the total vortex density at given point. The solar plasma system is inhomogeneous and often in a non-equilbrium state. Then, the $\Omega(+)$ and $\Omega(-)$ distributions depends on the conditions under which the system exists. The particle distributions and associated vortices $\Omega(\pm)$ are assumed to be stationary and take different form. In the frequent current/neutral sheets configurations where the magnetic field is nearly zero the particles obey to the Speiser's distribution (Speiser, 1965; 1967). Particle distributions adjacent to the current/neutral sheets could be modeled as isotropic ones and may be approximated by either a velocity exponential ($f \sim \exp[-(E/\varepsilon)^{1/2}]$, where E is the particle energy and ε is related to thermal energy E_T in a parametric manner ($\varepsilon = \varepsilon(E_T)$), or kappa function ($f \sim [1+E/kE_T]^{-k-1}$) instead of Maxwellian one (e.g. Christon et al., 1988). The forth-coming FAC structure formation process could be described either velocity exponential, or kappa distributions. In these cases the corresponding constants ε or E_T stands for the Gibb's module Θ, defined for equilibrium state. In a non-equilibrium state distributions $\Omega(+)$ and $\Omega(-)$ (9) will incorporate additional terms which will indicate possible (first-order) anisotropy in the non-equilibrium FAC distribution. Further we examine disturbances $\Omega(+)$ and $\Omega(-)$ to an initial equilibrium state particle distribution, the self-interaction of current elements, $j_{/\!/}(\pm)/N_0(\pm)$ and subsequent vortex formation processes. Their distribution will be then governed by a self-consistent vector potential ψ (Balescu, 1975), where ψ stands for $\boldsymbol{j}_{||}(\pm).\boldsymbol{A}/N_0(\pm)$. Because the quasi-neutrality condition is fulfilled for both charges and currents in plasmas, i.e. their densities $N_0(+)$, $N_0(-)$ coincide, we study only configurations of equal current strength $I(\pm)$ of both polarities, $I(\pm) = I_0$, We define $N_0(\pm) = N_0$. Our basic equation of the field-aligned current elements dI of density N_0 then reads

$$\begin{aligned} &\Delta\vec{A} + \mu_0\vec{j}_{||}\sinh(\vec{j}_{||}\cdot\vec{A}/N_0\Theta) = 0 \Rightarrow \\ &\Delta(\vec{j}_{||}\cdot\vec{A}/N_0\Theta) + (\mu_0 j_{||}{}^2/N_0\Theta)\sinh(\vec{j}_{||}\cdot\vec{A}/N_0\Theta) = 0 \end{aligned} \tag{10}$$

Here $\boldsymbol{j}_{||}.\boldsymbol{A}/N_0\Theta = q(\boldsymbol{A}.\boldsymbol{v}_{||})$ where $v_{||}$ coincides with the Alfven velocity v_A. The quantity $\boldsymbol{A}.\boldsymbol{v}_{||}$ has a meaning of potential 'φ' which governs the distribution of charges q being given by (1). The coefficient λ^2 ($\lambda^2 = \mu_0(j_{||})^2/N_0\Theta$) is proportional to the squared single charge q and charge density N_0 (Nenovski et al., 2003)

The current density $\boldsymbol{j}_{||}(z)$ depends obviously on the vortex Ω and the Alfven velocity v_A distribution (see (6)). Thus it varies along the magnetic field lines. Both quantities Ω and v_A are determined by the prominence magnetic field geometry and the plasma density distribution. The FAC flowing along a magnetic field usually occupies a flux tube of given cross section $\boldsymbol{S}$. Because the magnetic field lines are parallel to the borders, $\boldsymbol{B}_0.\boldsymbol{S}$ = constant, then the FAC intensity will increase proportionally to the magnetic field magnitude $\boldsymbol{B}_0$. For given plasma density distribution $N = N(z)$, where z is the distance, we could express the vortex Ω and the Alfven wave v_A distribution along the field lines by geometry factors $L_{1,2}(z)$. Thus $\boldsymbol{\Omega}(z) = \Omega_0 L_1(z)$ and $\mathbf{v_A}(z) = \mathbf{v_{A0}} L_2(z)$, where z_0 is a starting point, e.g. the FAC source point. Quantitatively, the basic relation that accounts for the FAC distribution along the magnetic field lines has the form

$$\vec{j}_{||}(z) = \vec{j}_{0,||}L(z/z_0) \tag{11}$$

where the geometry factor $L(z/z_0)$ can be determined observationally. Thus, due the prominence magnetic field geometry the nonlinear term in eq. (10) will steeply change its magnitude $\lambda^2(z)$. It is noteworthy to mention that this feature persists irrespective of the fact whether the FAC structure has wave, or static nature. Therefore, the FAC density increase is the principal cause of the filament/vortex formation process due to self-interaction of the FAC elements.

The existence of nonlinearity suggests a FAC structure formation process that under certain conditions generates new spatial distribution of FAC vortice (thread) structures. By solving it we obtain various solutions governing the FAC structure formation process. This enables us to determine the FAC densities needed for FAC filament formation. Let us now consider the magnitude of the coefficient $1/\lambda^2(z)$ that controls the nonlinearly of our FAC system and, if possible, to obtain its actual magnitude. Before doing so, it is noteworthy to mention that eq. (10) is analogous to the nonlinear Poisson-Boltzman (PB) equation (Debye and Hueckel, 1923; Balescu, 1975)

$$\Delta\Psi = r_{De}^{-2}\sinh(\Psi) \tag{12}$$

It describes the spatial trend of the self-consistent electrostatic potential and the plasma number density. $\Psi = q(\varphi-\varphi_{\infty})/kT$ is the normalized self-consistent 'electric' potential: $E = -\nabla\varphi$; r_{De} is the Debye radius. Of course, this potential Ψ corresponds to the self-consistent electric potential in the classical Coulomb system (Ecker, 1972). Eqs. (10,12) are also identical to the Euler's equation for stationary flows of incompressible fluids (Kaptsov, 1988). It is well known also that solutions of such equations yield vortex structures of different scales. It is well known that such system possesses collective behavior, i.e. long-ranged correlative effect caused by the electric field produced by the charged particles (electrons and ions) emerges over the entire plasma system. It has already been shown that in such a system, under certain conditions periodic plasma number density distributions of some scales appear. These structures are dependent on the nonlinear term sinh(Ψ) and its coefficient, the Debye length r_{De}(e.g. Martinov et al., 1984, Martinov et al., 1986). In the derived nonlinear equation of the FAC element self-interaction (10) our coefficient λ^2 is equivalent to the squared Debye length in (12). In order to understand the physical meaning of the coefficient λ^2 let us examine the expression for the Debye length r_{De} in the classical Poisson-Boltzman equation. The latter is proportional to

$$N_0 e^2 / \varepsilon_0 kT \tag{13}$$

where e is the single charge, T the prominence temperature. In eq. (13) the N_0e quantity stands for the 'charge' density ρ where the charge e stands for $j_{II} dV\sqrt{\varepsilon_0\mu_0} = \frac{j_{II}}{N_0}\sqrt{\varepsilon_0\mu_0}$.

Because of the relationship (6), the vorticity Ω stands for this eN_0 quantity. We replace eN_0 with its equivalent q and obtain

$$N_0 q^2 / \varepsilon_0 kT = q\rho / \varepsilon_0 kT = (\varepsilon_0 / kT)(\vec{B}_0 \cdot \nabla \times \vec{v}) \tag{14}$$

It follows that the coefficient that determines the nonlinearity of the field-aligned current filament formation is determined by

$$\lambda^2 \equiv (N_0 q)/(\mu_0 j_{||}{}^2) = (cv_T / v_A)^2 / (v_A\Omega)^2 = \beta(c / v_A)^2 / (c / \Omega)^2, \tag{15}$$

where c is the light velocity, v_T the plasma thermal velocity, and β the ratio of plasma to magnetic pressure. Thus we have a fully determined coefficient needed for studying the FAC filament formation process.

Examining the physical parameters entering in (15) one can see that under given field-aligned current and vortex geometry, the filament formation process would be initiated at the minimum β value point. The smallest values of β are for 'cold' plasma, $\beta < m/M$, m and M are the electron and ion mass, respectively. It follows that the filament formation process will most easily appear at chromospheric heights. On the other hand, when the field-aligned current penetrates within corona, an increase of the FAC density $j_{||}$ (11) (by a factor z/z_0) enhances additionally the magnitude of the coefficient (15) (depending on λ^{-2}) governing the nonlinearity effect. Thus, we could suppose that at these heights the nonlinearity effect appears first. Each physical object described by such a Coulomb system would admit different periodic spatial distributions, which yield different structures of the system.

Remind that magnetohydrodynamic equilibrium state is governed by the nonlinear equation in the form (Shafranov, 1963)

$$\Delta A = -k\frac{dP}{dA}$$

where pressure $P = p + B_{||}^2/2\mu_0$ and P is an arbitrary functions of A (**A** is of course the vector magnetic potential, $p(A)$ and $B_{||}^2/2\mu_0$ are the plasma and the magnetic field pressures) have been considered. In an equilibrium state both the plasma pressure $p(A)$ and the magnetic field $B_{||}(A)$ determine the needed function $P(A)$. The latter determines the kind of nonlinearity which governs the vector potential distribution (Parker, 1979). The coefficient k of course characterizes the nonlinearity of the system and is an unknown parameter. In literature, several forms of the nonlinear function $G(A) = dP/dA$ have been chosen (e.g. Kriegman and Reiss, 1978; Streltsov et al, 1990). Thus we find a concordance of eq. (10) with the basic equation in magnetohydrodynamics.

2.2 Field-aligned current (FAC) filamentation instability

The above theoretical considerations represent a good basis for a simple construction of FAC filamentation process that starts at certain threshold values of the FAC density being intensified in the prominence body. It is found that the filament formation process starts at (Nenovski et al, 2003):

$$a^2 / \lambda^2 = \pi^2(4 + a^2 / b^2)$$

(λ is determined by (15) and b coincides with the size of the longer size of the rectangle of the field-aligned current localization region (a,b)). The lowest magnitude for a filament formation process is achieved at aspect ratio $r \equiv a/b \to 1$, i.e. for tube-like FAC structures. For sheet-like FAC structures the geometry ratio is high and the filament formation process is less attainable. Using eq. (10) we are able to determine the minimum 'charge' density ρ, or the field-aligned current density magnitude $j_{||,crit}$ needed for an initiation of fillament formation process. Having assumed square FAC region, the filament formation begins at (Nenovski et al, 2003)

$$j_{||,crit}a \geq 0.63(v_T / v_{T0})\ [\mathrm{A/m}] \tag{16}$$

where v_{T0} is the plasma 'thermal' velocity that corresponds to ambient prominence temperature T_0 assumed to be equal to 1x10^4 °K. Under prominence conditions the actual temperature varies and can reach values up to 60000 °K, e.g. T_0 is between 1-6 eV. Let us evaluate qualitatively the magnetic field strength produced by such filament current. Assuming that the azimuthal magnetic field $B_\perp$ produced by FAC filament is proportional to the quantity $\mu_0 j_{||} a$ and that the filament current size is obviously of thousand km and less we obtain the following critical values for the $B_\perp$ magnitudes: 10^{-6}-10^{-5} T. The actual magnitudes of $B_\perp$ are, of course, higher.

The critical current intensity depends on the current region size a (the smaller one) and therefore, it increases when a decreases. The threshold value current magnitude is controlled by the plasma 'thermal' velocity v_T of the current carriers. The newly formed FAC structures (filaments) are of scale a/m, or b/n where (m,n) characterize the number of vortices in the x and y directions. Hence, the size a in (16) should be replaced by a factor F

$$F = \frac{a}{\sqrt{m^2 + n^2 a^2 / b^2}}$$

This means that the critical (threshold) field-aligned current density $j_{\|,crit}$ (m,n) will increase with decreasing FAC structure scales. Factor F is thus a measure of the FAC filaments scaling with respect to the whole size of the initial FAC structure. Note that under the Earth's magnetosphere conditions this factor can raise up to two orders, where the current density in FAC structures span the interval from $\mu A/m^2$ to 10^{-4} A/m^2 (Iijima and Potemra, 1976).

It is noteworthy to mention that the dependence of $j_{\|,crit}$ on the thermal velocity can be thought as a counterpart of the Bennett's formula that states that the pinching effect of the field-aligned current is balanced by the plasma thermal pressure. The latter is however proportional to the squared thermal velocity, i.e. the threshold FAC strength (I) for pinching is proportional to v_T. Only in the case of *one polarity* (ambient) current I (in our examination it corresponds to $m = 1$, $n = 1$) the threshold value field-aligned current density coincides quantitatively with the Bennett formula.

A comparison of the suggested FAC structure (filaments) formation model with some observation sets of FAC structures/thread events is now possible. Remind that the ambient (large-scale) FAC structures might be hidden and that the final FAC filament/vortex structures will dominate. It follows that upward and downward FACs are generated within the ambient FAC geometry.

It is noteworthy to mention that an analogue of the examined FAC filamentation process could be found under the Earth's magnetosphere conditions (Nenovski et al 2003). There is a variety of FAC structure data which yield evidences for such a coexistence of Birkeland's FACs known also as Regions 1 (R1) and Region 2 (R2), and periodical FACs of smaller scales (Arshinkov et al. 1985; Ohtani et al. 1994). The measured FAC intensities exceed the threshold value ones (16). According to experimental results, these FAC structures are superimposed onto a larger-scale FAC system and the background FAC density measured in the FAC region 1 (R1) is three times less than the FAC sheet density of smaller scales. Taking into account that the FAC sheet intensity and thickness are correspondingly 0.2 A/m and 200 km we derive 3.33×10^{-7} A/m^2. This value exceeds 2.5 times the threshold value one. Hence, there are experimental evidences that the FAC filament formation process might take place.

2.3 Consequences of FAC filamentation instability

In our approach a hypothesis for a statistical distribution of FAC elements with Gibbs distribution module Θ is introduced. FAC elements governed by equilibrium-like or stationary forms of distribution are also allowed for. The considered FAC elements interact to each other in the same way as in the classical Coulomb system. For this ensemble, long-ranged correlative effects are consequently expected. All the FAC filaments (upward and downward) yield totally zero field-aligned current that corresponds to the quasi-neutrality condition of the charges in plasmas. The negative charges (electrons) compensate the positive charges of the ions. By using a 'charge' concept of the FAC elements we derived *quantitatively* an estimate of the parameter λ^2 which stands for the non-specified coefficient k introduced in the Grad-Shafranov equation of magnetohydrodynamic (MHD) equilbria. In the 'charge' approach adopted by us this parameter (λ^{-2}) depends on the density N_0 of the FAC filaments and the thermal energy T of the dominant plasma/prominence constituent. In the plane perpendicular to the ambient magnetic field the FAC element distribution could be described as if there were no magnetic field. This assumption could be accepted as far as there are no other effects – magnetic field curvatures, inertial or drift motion, etc. The governing physical parameter is the temperature T of the plasma environment. Hence, the separation of FAC elements into FAC structures/filaments is controlled by this temperature. An increase of FAC intensity along the magnetic field lines is however the principal factor for FAC structure/thread formation that is taken into account. When the FAC travels through the chromosphere-solar corona regions the plasma temperature T and the corresponding plasma parameter β could change. Both the FAC intensity and β changes along the magnetic field fluxes will control FAC structure formation processes. Preferable conditions for the examined FAC structure formation processes are expected to be initiated at heights approaching the chromosphere-corona transition region. Therefore, the FAC filament formation due to self-interaction process is expected to be a dominant feature at heights just above the chromosphere.

2.4 FAC structures generation in inhomogeneous plasma flows

Differential movements of the photospheric footpoints of magnetic flux in the solar atmosphere are studied both experimentally and theoretically during last three decades (Wu et al, 1983). FAC structures or threads can thus be generated by such processes of differential motions embedded in the chromosphere. The shear motion in the solar atmosphere will generate non-potential magnetic field, i.e. the magnetic field is twisted (Wu et al, 1983). Experimental evidences exist which yield a power dependence between the velocity and the magnetic field localization of type $v = \alpha B^{\beta}$ ($\beta \neq 1$) (Stenflo, 1976). Observations in H_{α} lines of the fine structures of Quescent Prominences (QP) reveal that various twisted magnetic fluxes in the form of filaments or threads (along the QP body) exist. The arch-like QP and loops, for example, consist of one or more stable magnetic flux ropes which exist few weeks or several months. What dynamic processes are responsible for their filament formation and destabilzation? Do these filament (thread) structures untwist and disappear subsequently by changes in the external conditions? These and other similar problems are not fully modelled even in the frame of MHD. The dynamical responses to photospheric shear motions, i.e. movements of footpoints of flux tubes have numerically been examined by Wu et al. (1983). In their study a build-up of magnetic energy up to 4 times faster than the rate of other modes (kinetic, potential, etc.) is shown. The volume where the magnetic energy is growing after the introduction of the shear, is limited, the magnetic field gradient is correspondingly concentrated in a narrow slab in the vicinity of the neutral line of the shear motion (Wu et al, 1983). Wu et al (1983) have suggested these magnetic field intensification processes as a source of flare energy. The quiescent prominences (QP) is interestingly to study because of possible steady-state twisted magnetic flux structures coexisting with the external conditions of simultaneous inhomogeneous shear flows, magnetic fields of inverse polarity and so on.

Theoretical considerations of the connection between the internal structures and the external conditions are needed. It is well-known that QPs are situated over the dark filaments (on solar disk) where the magnetic field polarity inversion and velocity shear are presented. It is possible also that the QP footpoints are subjected to both movement and rotation. It is expected that the inhomogeneity in the steady flow has to influence the structure behavior of the QP more dramatically. For example, a velocity shift at the plasma boundary can lead to the Kelvin-Helmholtz instability. Under velocity shift conditions there are always gradients and a boundary of finite extent evolves. The correct understanding of the physical processes connected with the QP internal magnetic structures requires to taking into account more explicitly the plasma and the magnetic field inhomogeneities. From theoretical point of view the finite spatial scales of the boundary inhomogeneities will induce new structures with own velocity and magnetic field characteristics. To best of our knowledge the structure formation problem in the presence of velocity and magnetic field inhomogeneities is not thoroughly examined.

2.4.1 Theoretical modeling

Consider a magnetic flux loop surrounded by inhomogeneous flow. The background magnetic field $\boldsymbol{B}$ is oriented along the z-axis. The plasma flow $\boldsymbol{V}$ can be modelled in two ways. In the first case, the inhomogeneity (the gradients) is oriented along the x-axis only (Case I), in the second one, the velocity gradient is along the radius (a cylindrical geometry)(Case II). The second case is responsible for the plasma rotation structure effects, as pointed by the well-known laboratory experiments (Nezlin et al, 1987; Nezlin and Snezhkin, 1993; Nezlin, 1994). In our opinion, theoretical consideration of flux tube structure formation suitable for the QP destabilization can simply be illustrated by Case I. Let us imagine a flux tube and choose a cross-section of some radius r_0. Denote the two ends of the tube diameter (along axis y) of this cross-section by points S and D. We consider the following plasma flow-(magnetic) flux tube geometry: An inhomogeneous flow is coming at point S, streams around the flux tube, say to the right, and drops out at point D. Another flow is coming in the cross-section from point D and goes round the rest of the flux tube on the left side and reaching point S disappears (a shear flow model). Thus, the magnetic flux is streamed entirely and its boundary of radius r_0 can be represented as a streamline. For convenience, the considered tube radius r_0 is assumed to be equal to 1 (in dimensionless presentation). The following problems are raised. Do such magnetic tubes (ropes) exist in the presence of velocity shears? Do velocity itself and its gradient cause (generate) filament structures and/or twisted magnetic (threads) inside the flux ropes?

Thus, our task is to study the internal structure evolution and stability of such tubes. We note that because the QP length L exceeds considerably the QP cross thickness we do not account for the possible differentiation along this tube length. In order to answer to these questions we study the structure formation in the presence of velocity gradients across the main magnetic field $\boldsymbol{B}_0$. The model consists of plasma flow V_0 (along the y-axis) perpendicular to the $\boldsymbol{B}_0$. The unperturbed magnetic filed is along the z-axis but an additional component $\boldsymbol{B}_\perp$ is assumed. The $\boldsymbol{B}_\perp$ vector makes with the velocity vector V_0 an angle smaller than $\pi/2$. We assume an x-dependent velocity given as a series

$$V = V_0 + V_0^{'}x + \frac{1}{2}V_0^{''}x^2 + ... \tag{17}$$

It is convenient to examine separately possible plasma structure formation due to constants V_0, V_0' (velocity gradient) and so on. For a description of the filament (thread) formation under velocity shear conditions (17) the set of reduced magnetohydrodynamics (RMHD) is allowed for. It is known that in the RMHD case the velocity and magnetic vector potential components parallel to the background magnetic field suffice to describe the self-consitent evolution of the plasma. These potentials φ and A should comprise all inhomogeneity characteritics depending on (x,y) coordinates. The initial set of RMHD reads

$$\begin{aligned}
&\partial\rho / \partial t + [\varphi,\rho] = [\rho_0,\varphi] \\
&\partial / \partial t(\nabla_{\perp}^{2}\varphi) + [\varphi,\nabla_{\perp}^{2}\varphi] = \rho_0^{-2}[\rho,P_0] \\
&-[\rho^{-1}\nabla_{\perp}^{2}A, A - B_{\perp}x] + B_0\nabla_{\perp}(\rho^{-1}\nabla_{\perp}\partial A / \partial x) \\
&\partial A / \partial t + [\varphi, A - B_{\perp}x] = B_0\partial\varphi / \partial x
\end{aligned} \tag{18}$$

The following assumptions are inferred. The time changes are slow compared to the transit time $t_A \equiv r_{0,max}/v_A$ where v_A is the Alfven velocity: $((\partial/\partial t)^{-1} > O(t_A))$; the z-changes are comparable to time changes $(v_A\partial/\partial z)^{-1} \approx (\partial/\partial t)^{-1} < t_A^{-1})$; the plasma parameter β ($\beta \equiv 2\mu_0 P/B^2$) is assumed small. Following the Strauss analysis (Strauss, 1977) the flow velocity should be considered as incompressible and of course perpendicular to the background magnetic field. In seeking localized in (x,y) plane solutions of (18) a proper motion of the modes along the y-axis as assumed, i.e. $\partial/\partial t = -u\partial/\partial y$. The derived mode structure is inclined at angle ϑ to the undisturbed magnetic field $\mathbf{B}_0$, ($\tan\vartheta \equiv B_0/B_{\perp}$).

Let us now study separately three cases a) V_0 = const; b) $V_0 = 0$, but $V_0' \neq 0$ (a shear). Later on, $V_0 \neq 0$, and $V_0' \neq 0$ are considered. In the first case the set basic equations (18) reduces to

$$(1 - m^2\mu_0^{-1}\rho^{-1})\Delta(\varphi_0 - V_0x) = \pm k^2(\varphi_0 - V_0x) - px \tag{19}$$

where a linear dependence between φ_0 and A is allowed ($A_0 = m\varphi_0$) and a coefficient p which cumulates the terms due to the perpendicular magnetic field $\mathbf{B}_{\perp}$ and constant pressure gradient ∇P is introduced. The coefficients m is equal to $B_{\perp}/V_0$. In the case of uniform flow V_0 a dipole (modon) structure, is possible (Strauss, 1977). Note that under uniform velocity conditions (V_0 = const, $V_0' = 0$) interesting six-cell type (Nenovski, 2008) structures are possible. In the latter case a dipole structure of smaller size still exists in the centre and four-cell, sheet-like structures of various sizes which surround the central structure appear (Nenovski, 2008). Note that the dipole structure is only a vestige and the four-sheet features of the cells dominate. Note also that the field-aligned currents (FAC) supported by the such a structure formation are of inverse polarity of both side of the velocity flow and resemble the well-known FAC region 1 and 2 picture by Iijima and Potemra (1976) in the Earth's magnetophere that emerges under certain velocity parameters.

In the second case ($V_0 = 0$, and $V_0' \neq 0$) we again seek localized solutions. The following φ_1 - A_1 relation however holds

$$2(\varphi_1 + V_0'x^2 + u^2 / V_0') = V_0'(A1 - B_{\perp}x)^2 / B_{\perp}^2 \tag{20}$$

Structures of smaller spatial scales in comparison to the external velocity gradient scales are inferred. The greater the velocity gradients are, the greater the induced FAC intensity is. Figure 1 and 2 illustrate quadrupole structure and its complication (an octopole structure).

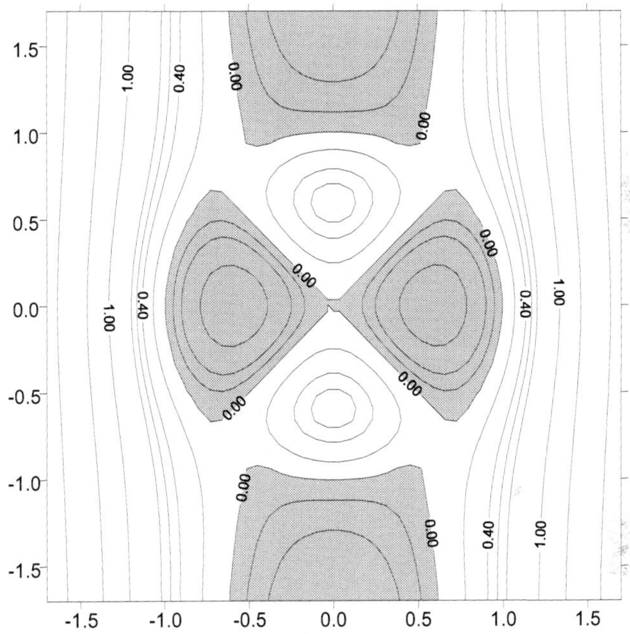

Fig. 1. Example of FAC qadrupole structure in a flux tube of dimensionless radius 1 immersed in shear velocity flow (V_0=0, $V_0^{'} \neq 0$). Hatched and non-hatched area denotes FAC and vortex structures of opposite sign.

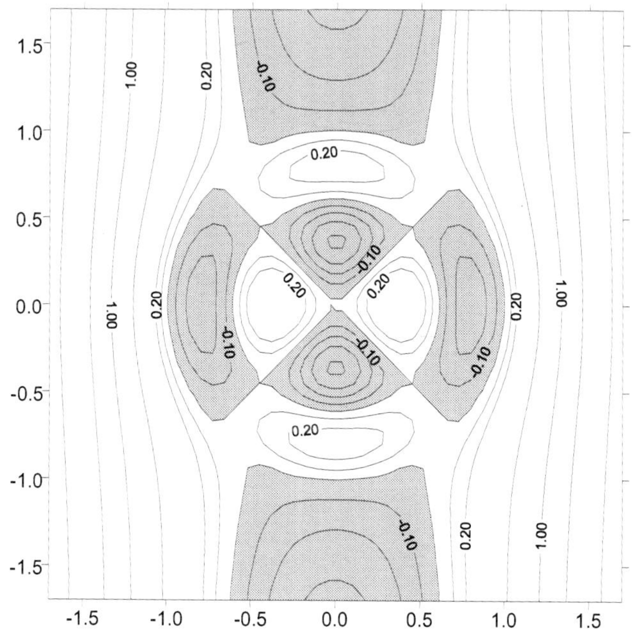

Fig. 2. Example of FAC octopole structure in a flux tube of radius 1 immersed in shear velocity flow (V_0=0, $V_0^{'}\neq 0$). Hatched and non-hatched area denotes FAC and vortex structures of opposite sign.

Further complications are expected under simultaneous considerations of V_0 and V_0'. Both factors suffice to yield more complex structures. The magnetic field variations in the radius r which govern the FAC localization supported by such filaments are illustrated in Figures 3 and 4. We note that transition filaments (e.g. a mixture of dipole and quadrupole structures) arise.

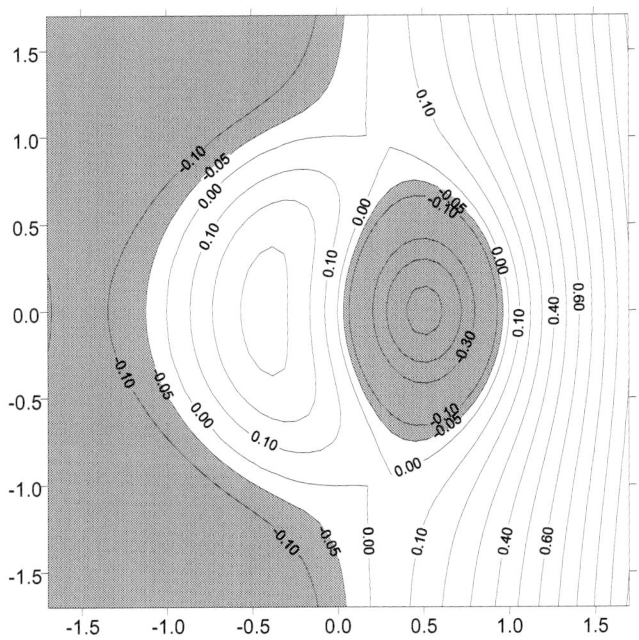

Fig. 3. FAC vortex/thread structures under inhomogeneous flow conditions for $V_0' r_0/V_0 = 0.08$. Hatched and non-hatched area denotes FAC and vortex structures of opposite sign.

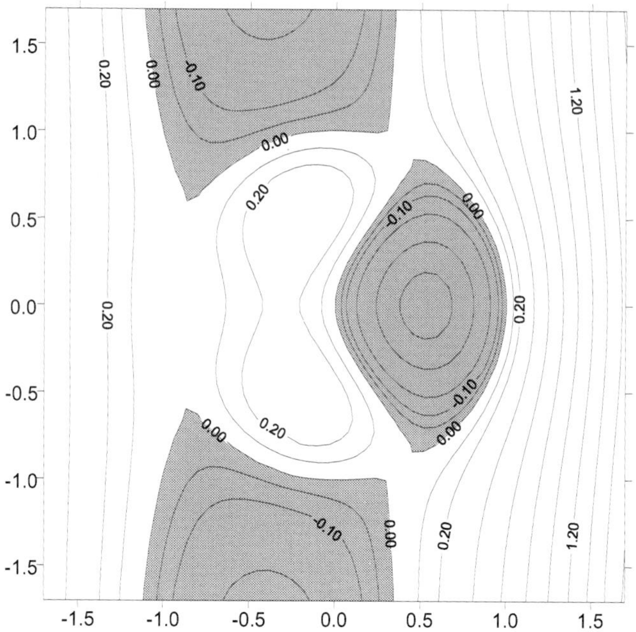

Fig. 4. FAC vortex/thread structures under inhomogeneous flow conditions. $V_0' r_0 / V_0 = 0.2$. Hatched and non-hatched area denotes FAC and vortex structures of opposite sign.

The transitional forms are: modified dipole structure and three-cell structure. The three-cell structure consists of a cell (filament) of intense plasma motion and intense FAC, and two weakened cells of equal polarity which are partly connected.

2.4.2 Filament (sub)structures or threads

The above results show that the structures complicate with increasing the inhomogeneity effects, i.e. by including next derivatives (V_0'', V_0''', etc.). There are considerations that the inclusion of odd-number derivatives only would give to one twisted thread of smaller size enveloping the flux tube z-axis. The structure modes examined here are presumably generated by velocity and velocity shear motion in the photosphere under or adjacent to the arch-shaped or loop QP footpoints. The velocity itself and its gradients at the chromosphere level could considerably influence the structure evolution.

Now we arrive closer to possible sources for the structures formation/destabilization processes in the solar atmosphere. The effect comes from the possibility the field-aligned currents to be intensified by filamentation processes. Another possible mechanism is the feeding of the field-aligned currents at the chromosphere) (Nenovski et al, 2001). This feeding enhances the filament velocity/magnetic field gradients and vice versa. The magnitude of the perpendicular component of the magnetic field increases and the pitch angle ϑ increases, as well. As a consequence, a QP (flux tube) destabilization due to filament FAC intensification processes may result.

2.5 Joule heating

The main factors that change the conditions of FAC structure generation and their dynamical evolution approaching the chromosphere heights are the indispensable intensification of the large-scale field-aligned current system and the transition from high to low β plasma conditions along their propagation direction. FAC filament formation processes might appear. The latter could facilitate drastically the FAC structures of different scales that will appear under appropriate FAC threshold values. The FAC filament formation expected at heights above the chromosphere would be the most favorable mechanism of FAC instability and subsequent enhanced (Joule) dissipation at the chromosphere heights.

The Joule heating is due to the Pedersen current and has been already described (e.g. Rees et. al., 1983). Its connection with the FAC distribution $j_{\|}$ is given by

$$j_{||} = -\nabla \cdot \vec{J}_{chromo} = -\nabla \cdot (\Sigma_P \vec{E} + \Sigma_H \vec{B} \times \vec{E}) \tag{21}$$

where

$$\vec{J}_{chromo} = \int dz j_{chromo}$$

Σ_P and Σ_H are the Pedersen and Hall conductances (integrated along z Pedersen and Hall conductivities), $\boldsymbol{E}$ is the electric field strength. The electromagnetic energy dissipation in the chromosphere (treated as a resistance) is thus

$$Q = \Sigma_P E^2 \tag{22}$$

The temperature increase, ΔT, connected with the dissipation rate Q is determined by

$$Q = C_p \Delta T \tag{23}$$

where Cp is the heat capacity of the medium. The latter is dependent on the neutral (hydrogen) density N_n distribution in height. At chromosphere heights the proton drift/electric field distribution might still be considered in frozen field condition, i.e. the electric field $\boldsymbol{E}$ is dependent on the plasma drift and the ambient magnetic field. This means that knowing the Pedersen conductivity σ_P and the ion drift $\boldsymbol{V}$ at the given height the temperature change ΔT (that depends on the Joule heating mechanism) might be determined.

At which heights are the greatest changes in the proton temperature ΔT expected? Let us model the profile of the temperature change ΔT using the chromosphere density, the Pedersen conductivity and the proton-neutral collision frequency distribution in height. The Pedersen conductivity σ_P is given by

$$\sigma_P = Ne^2 \left(\frac{\nu_{pn}}{\nu_{pn}^2 + \omega_{cp}^2} + \frac{\nu_e}{\nu_e^2 + \omega_{ce}^2} \right)$$

where N is the chromosphere density concentration, ν_{pn} and ν_e are the proton-neutral and electron collision frequencies, ω_{cp} and ω_{ce} are the proton and electron cyclotron frequencies.

At heights where the cyclotron frequencies are greater than corresponding collision frequencies, the Pedersen conductance is proportional to

$$\sigma_P \cong Ne^2(\frac{\nu_{pn}}{\nu_{pn}^2 + \omega_{cp}^2}) \cong Ne^2 \frac{\nu_{pn}}{\omega_{cp}^2}$$

Having in mind that proton collision frequency ν_{pn} is determined by the density N_n and the proton temperature T_p, and that the heat capacity Cp is proportional also to N_n, the temperature change ΔT at given height z is thus given by:

$$\Delta T(z) \equiv Q(z) / C_p(z) = \sigma_P(z)E^2 / C_p(z) \propto NT_p \,. \tag{24}$$

The latter result suggests that the temperature changes follow the plasma density changes at least at heights where the Pedersen conductivity approximation is applicable. This certainly includes the chromosphere-corona transition heights. The above considerations suggest that the Joule heating depending on the FAC geometry and intensity, should result both in horizontal and vertical distributions. Thus, both the observable FAC intensity distribution (plasma vortex) and the ion temperature changes in heights might be used as benchmarks of the QP destabilization mechanisms.

The overall dissipation rate is thus proportional to the squared field-aligned current density $j_{||}$ multiplied by squared factor F:

$$W_{diss} \geq (j_{||,crit}F)^2 / \sigma_P \tag{25}$$

at heights where Pedersen conductivity σ_P is different from zero. It follows that the FAC dissipation steepens with the decrease of the FAC scale. It is concluded that the dissipation and subsequent heating events are much more effective in the finest FAC structures expected in thread-like QP configurations. An increase of dissipation accompanied with temperature increase *however* might oppose the FAC filamentation process due to a temperature increase (the threshold values (eq. 16) subsequently increase). It is suggested that FAC filamentation and associated heating events are therefore counteracting processes and the both processes represent an essential component of QP destabilization mechanisms. If the QP destabilization event occurs, this leads to a subsequent generation of solar wind and/or coronal mass ejection (CME) events.

3. Conclusion

The MHD structure and wave generation is thought to be among the various processes initiated by the photospheric MHD disturbances, which propagate upward through the QP feet. We propose here a possible way of generating FAC (sub)structures in an arch-type QP by the MHD approach.

An emergence of FAC structures due to interactions of field-aligned current elements (subunits) that form the whole FAC system itself is demonstrated theoretically. We used the charge concept to build 'charge' counterparts responsible for the FAC filament formation mechanism. At some specific value of the control parameter, $\lambda^2 \equiv \mu_0(j_{||})^2/N_0\Theta$, these FAC elements bifurcate in new stationary states. Thus, a hierarchy of threshold value values for the emergence of FAC structures of different scales is derived. We demonstrated that the FAC structure formation appears as a counterpart of the pinching effect. Another difference is a co-existence of FAC structures of different scales that is possible above the threshold values. This FAC filament formation is considered in the term of successive stationary states that would evolve at the end to enhanced FAC dissipation. It is found that the threshold value for an initiation of the FAC structure formation depends on the plasma 'thermal' velocity and it is easily attainable for low β plasma conditions just above the chromosphere. The stationary FAC structures examined thus are an appropriate modelling of the dynamical transition of both the plasma and FAC vortex/thread structures from their initial states toward new ones during their journey through the chromosphere-corona transition regions.

The relevance of our problem of the FAC structure formation to the MHD point of view is pointed out, as well. In the Grad-Shafranov theory of the magnetohydrodynamic equilbria problem the stream function equation for the steady two-dimensional flow of non-viscous plasmas is exploited. It will therefore govern stationary (equilibrium) magnetohydrodynamic structures. A comparison shows the identity of the two equations and grounds our 'charge concept' approach with the magnetohydrodynamic equilibria problem. Our 'charge concept' approach determines quantitatively the nonlinearity coefficient, i.e. $\lambda^2 \equiv \mu_0(j_{\|})^2/N_0\Theta$. Note that in previously developed inertial/kinetic Alfven wave models (e.g. Chmyrev et al., 1988, Knudsen, 1996) this nonlinearity coefficient (denoted there by k^{-1}) corresponds to the squared plasma electron inertia c/ω_{pe} being dependent on plasma density only. In contrast to the above-mentioned inertial/kinetic Alfven wave model, our general examination states that FAC/vortex/tread structures will be controlled by the β plasma parameter and hence, FAC structure formation process will emerge at heights sufficiently close to the chromosphere, or just above it.

A simple physical analogue of magnetic flux tube immersed in plasma flow and generation of threads/filaments and their basic characteristics are examined, as well. The analogue consists of plasma flow velocity, velocity and magnetic field gradients, and sectors with non-zero azimuthal magnetic fields. This allows to making a 2-D model of the plasma circulation and the pitch angle evidence of the twisted magnetic field lines in flux rope models of the prominence threads. The velocity and the magnetic fields dependence on the distance x is of power law character. Structure organization and thread formation processes are studied on the basis of ideal MHD equation set. In the case of power law degree greater than one the vortices whose number depends on that degree are intermixed and various threads regimes along the axial magnetic field could be established. An example of one flux rope could be constructed in the limit of infinite series of odd power law-dependent velocity. A quadrupole flux rope which consists of four nearly distinct threads is formed in the simplest case of linear x-dependence. Field-aligned currents of the threads responsible for growing of the pitch-angle and for diminishing of the magnetic field tension are allowed for. The analytical and numerical results could be applied to solar prominence structure evolution and destabilization processes.

In summary, two factors responsible for structure formation processes in flux tubes: field-aligned current self-interaction and interaction with external plasma flow are taken into account. The following results would be inferred:

First, FAC filamentation process due to self-interaction process starts at certain threshold values of the FAC density in the prominence body. Changes of FAC intensity and the plasma parameter β along the magnetic field fluxes will control FAC structure formation processes. Preferable conditions for the examined FAC filament formation processes are met at heights approaching the chromosphere-corona transition region;

Second, flux tubes immersed in plasma flow are exposed to structure changes, as well. Factors responsible for structure formation of flux tube (visible in H_α line and other optical devices) are the flow velocity itself and velocity gradients;

Third, in the field of uniform plasma flow with velocity V_0 six-sheet FAC structures are easily formed. Note that such structures are evidenced in the Earth's magnetosphere (the FAC region 1 (R1) and 2 (R2) discovered by Iijima and Potemra (Iijima and Potemra, 1976);

Fourth, flow velocity gradient (a linear x-dependent velocity) causes quadrupole (octopole) structures. Their scales are of smaller sizes than the corresponding velocity gradient scale;

Fifth, flow velocity and the velocity shear results in structure complication. Transitional forms (e.g. modified dipole, three-cell, etc) could be generated.

Our theoretical results refer to the most simplified models of prominence structure destabilization. We consider FAC intensification process as basic QP destabilization mechanisms due to active chromosphere processes - the plasma density, velocity and/or conductivity enhancements.

The obtained criteria for QP destabilization are suitable for observational verification of the proposed mechanism of FAC filament and thread generation. Detailed examinations of all the plasma parameters and characteristics in the quiescent prominence body and experimental evidences are further needed.

4. Appendix A

1. *Characteristic time scales.* It is well known that the highest characteristic frequency associated with collective modes in non-magnetized plasmas is the electron plasma frequency

$$\omega_{pe} = (4\pi Ne^2/m_e)^{1/2}. \tag{A1}$$

In a system where the main effects come from the ambient magnetic field the Lorentz force is the dominant component. Should a flow change, say Δv appear particles of mass m and charge q in a first approximation respond to the Lorentz force

$$m\mathrm{d}\Delta \boldsymbol{v}/\mathrm{d}t = \Delta \boldsymbol{F} \approx q\, \Delta \boldsymbol{v}\times \boldsymbol{B}_0, \tag{A2}$$

where $\Delta \boldsymbol{v}$ is the disturbance flow velocity. From (A2) it follows that such a disturbance will grow with a growth rate

$$\gamma \approx qB_0/m = \omega_c. \tag{A3}$$

Thus, in an ensemble of charged particles immersed in strong ambient magnetic field $\boldsymbol{B}_0$ another characteristic time of collective processes equal to the cyclotron frequency $\omega_{c\alpha} = q_\alpha B_0/m_\alpha$ appears (α denotes the charged particle $\alpha = e,i$). The characteristic velocity of propagation of (electric) forces in magnetized plasmas is the Alfven velocity v_A expressed by $B_0/(\mu_0\rho_0)^{1/2}$, where ρ_0 is the plasma density. In term of above-mentioned characteristic frequencies, the Alfven velocity is equal to $(\omega_{ci}/\omega_{pi})c$, where c is the light speed, i.e. the characteristic velocity is modified by the ω_{ci}/ω_{pi} ratio. In plasmas where FAC structures of size d exist there is another time scale t_{FAC} equal to $d/v\mathbf{A}$. Therefore, *collective* plasma behavior is only observed on time-scales longer than the $\omega_{p\alpha}^{-1}$, $\omega_{c\alpha}^{-1}$ and/or t_{FAC}.

Let us now estimate time scales for FAC filament formation process in the chromospere-solar corona system. At heights close to the chromosphere the ambient magnetic field B_0 is approximately equal to 1-20 G, or less. The plasma number density is ~3-6x10^{17} m^{-3} and decreases slowly. The cross size d of ambient FACs is usually about 10^3 km and less and should increase in height. Taking it into account and using (A1) and (A3) we obtain that at chromospheric heights the greatest time scale equals to $t_{FAC} \approx 10^2$ s. Thus, we conclude that in the worst case the characteristic time of collective processes as FAC filament formation is less than of a minute. On the other hand, the FAC structures propagate to the chromosphere with the Alfven velocity v_A. Then the propagation time depends on the distance between the cromosphere and the region where the FAC filament formation appears. It equals from seconds to minutes. Fortunately, the time-scale for evolution of FAC structures treated by us is much less than that across the prominence body. Thus, the dynamics of Alfvén (shear and compressional) waves in such a dynamics can be neglected in zeroth order approximation.

2. *Condition for FAC closure through the chromosphere.* In the process of FAC structuring however the FAC intensity required for it increases inversely proportional to the FAC structure scale a. This means than FAC of smaller scales have higher intensities. The following question could arise: whether chromospheres can close prominence FAC structures whatever scale it is of.

The FAC intensity, or $j_{\|}$ is closed through the chromosphere by the Pedersen and Hall currents provided that

$$j_{\|} = j_{\|,\,\mathrm{chromo}} \equiv \nabla . \boldsymbol{J}_{\perp,\,\mathrm{chromo}} \tag{A4}$$

The perpendicular current $\boldsymbol{J}_{\perp,\,\mathrm{chromo}}$ is given by:

$$\boldsymbol{J}_{\perp,\,\mathrm{chromo}} = \Sigma_P \boldsymbol{E} + \Sigma_H(\boldsymbol{B}_0 \times \boldsymbol{E}) \tag{A5}$$

where Σ_P and Σ_H are the Pedersen and Hall conductances evaluated in height. They are assumed to be almost homogeneous in horizontal direction. Then (A4) becomes

$$j_{\|,\,\mathrm{chromo}} = \nabla .[\Sigma_P \boldsymbol{E} + \Sigma_H(\boldsymbol{B}_0 \times \boldsymbol{E})] \cong \Sigma_P \nabla . \boldsymbol{E} - \Sigma_H \boldsymbol{B}_0(\nabla \times \boldsymbol{E}) = -\Sigma_P \nabla .(v \times \boldsymbol{B}_0). \tag{A6}$$

where the $\nabla \times E$ part is considered to be zero (electrostatic field approximation). In the solar corona the (convection) electric field E is determined by the non-collisional Ohm's law $\boldsymbol{E} + v \times \boldsymbol{B}_0 = 0$. Under the condition of equipotentiality of the magnetic field tubes both the electric field and the corresponding fluid velocity v enter the chromosphere region. The following relationship between the fluid velocity v and the magnetic field disturbance b produced by FAC comes from the corona region:

$$b = (B_0/v_A)v, \tag{A7}$$

where v_A is the Alfven velocity. It is a consequence from the well-known Walen relation (in magnetohydrodynamics). Using the Maxwell equation

$$\nabla \times \boldsymbol{b} = \mu_0 \boldsymbol{j} \tag{A8}$$

where the displacement current $\partial E/\partial t$ is conventionally neglected we obtain for its parallel component:

$$(B_0/v_A)\Omega = \mu_0 j_{||} \tag{A9}$$

where factor (B_0/v_A) is assumed to vary along the ambient magnetic field direction $\boldsymbol{B}_0/B_0$ only, i.e. it is homogeneous in horizontal direction. By a comparison of (A4) and (A6) the condition for FAC closing through the chromosphere, $j_{||,\,\mathrm{chromo}} \geq j_{||}$, results in the following inequality

$$\Sigma_P \geq (\mu_0 v_A)^{-1} \equiv \Sigma_W. \tag{A10}$$

where Σ_W is the wave conductance. In practice, such magnitudes of the Pedersen conductance are frequently observed (Nenovski et al, 2001). Hence, irrespectively of the FAC intensity, the FAC closure condition through the chromosphere will be controlled by the Pedersen conductance magnitude.

The chromosphere itself will however not be simply a passive recipient of the field-aligned currents. The FAC can modify the chromospheric conductivity by ionising the neutral photosphere, producing additional electron-proton pairs due to particle fluxes permeating the chromospere-photosphere system. In the presence of a background electric field $\boldsymbol{E}$, the conductivity gradients produced by particles can lead to divergences in the perpendicular chromospheric currents, which in turn lead to additional parallel field-aligned currents.

5. References

Arshinkov, I., Bochev, A., Nenovski, P., Marinov, P., and Todorieva, L. (1985) *Adv. Space Res.* 5, 127-130.

Balescu, R. (1975) *Equilbrium and nonequilibrium statistical mechanics*, John Wiley, N.Y., ch. 6, 11.

Bellan, P. M.(2004) Final Technical Report for DOE Grant DE-FG03-97ER54438, Applied Physics, Caltech, Pasadena CA 91125, December 21.

Ballester, J. L., & Priest, E. R. 1989, A&A, 225, 213

Bashkirtsev, V., Kobanov, N., & Mashnich, G. 1983, Solar Phys., 82, 443

Bashkirtsev, V., & Mashnich, G. 1984, Solar Phys., 91, 93

Bennett, W.H. (1934) *Phys. Rev.* 45, 890-897.

Bommier, V., Leroy, J.-L., & Sahal-Brechot, S. 1986, A&A, 156, 79

Chmyrev, V.M., Bilichenko, S.V., Pokhotelov, O.A., Marchenko, V.A., Lazarev, V.I., Streltsov, A.V., and Stenflo, L. (1988) *Physica Scripta*, 38, 841-854.

Christon, S.P., Mitchell, D.G., Williams, D.J., Frank, L.A., Huang, C.Y. and Eastman, T.E. (1988) *J.Geophys. Res.* 93, 2562-2572.

Cui Shu, Hu Ju, Ji Guo-Ping, et al. 1985, Chin. A&A, 9, 49

Ecker, G. (1972) *Theory of fully ionized plasmas* (Academic press, New York)

Engvold, O. 1976, Solar Phys., 49, 238

Hirayama, T. 1986, in NASA CP-2442, Coronal and Prominence Plasmas, ed. A. I. Poland, 149

Hirayama, T., Nakagomi, Y., & Okamoto, T. 1979, in Phys. Solar Prominences, ed. E. Jensen, P. Maltby, & F. Q. Orrall, IAU Coll. 44, 48

Hirayama, T. 1971, Solar Phys., 17, 50

Hirayama, T. 1972, Solar Phys., 24, 310

Hirayama, T. 1985a, Solar Phys., 100, 415

Hirayama, T. 1985b, in Dyn. Quiescent Prominences, ed. V. Ruzdjak, & E. Tandberg-Hanssen, IAU Coll., 117, 187.

Iijima, T. and Potemra, T.A. (1976) *J. Geophys. Res.* 81, 2165-, 5971-.

Kaptsov, O.V. (1988) *Dokl. Akad.Nauk SSSR* 298, 597-600.

Knudsen, D.J. (1996) *J. Geophys. Res.* 101, 10 761-10772.

Kubot, J. 1980, in ed. A. F. Moriyama, & J. C. Henoux, Proc. Japan - France Seminar on Solar Physics, 178

Landman, D. A. 1984a, ApJ, 279, 183

Landman, D. A. 1984b, ApJ, 279, 438.

Leroy, J.-L. 1985, in NASA Conf. Publ. 2374, Measurements of Solar Vector Magnetic Fields, ed. M. J. Hagyard, 121

Lin, Y., Engvold, O., & Wiik, J. E. 2003, Sol. Phys., 216, 109;

Lin, Y. 2004, Ph.D. thesis, Univ. Oslo

Lin, Y., Engvold, O., Rouppe van der Voort, L. H. M., Wiik, J. E., & Berger, T. E. 2005, Sol. Phys., 226, 239

Lin, Y., Engvold, O., Rouppe van der Voort, L. H. M., & van Noort, M. 2007, Sol. Phys., 246, 65.

Malherbe, J. M., Schmieder, B., Ribes, E., & Mein, P. 1983, A&A, 119, 197

Martinov, N, Ourushev, D. and Georgiev, M. (1984) *J. Phys.* C 17, 5175-5184.

Martinov, N., Ourushev, D. and Chelebiev, E. (1986) J. *Phys. A: Math. Gen.* 19, 1327-1332.

Mein, P., & Mein, N. 1991, Solar Phys., 136, 317

Mein, P. 1977, Solar Phys., 54, 45

Ohtani, S., Zanetti, L.J., Potemra, T.A., Baker, K.B., Ruohoniemi. J.M., and Lui, A.T.Y. (1994) *Geophys. Res. Lett.* 21, 1879-1882.

Nenovski, P. (1996) *Phys. Scripta* 53, 345-350.

Nenovski, P., Dermendjiev, V. N., Detchev, M., Vial, J.-C and Bocchialini, K.(2001) On a mechanism of intensification of field-aligned currents at the soalr chromospere-quiescent prominence boundaries, Astronomy & Astrophysics, 375, p.1065-1074.

Nenovski, P.,Danov, D. and Bochev, A. (2003) On the field-aligned current filament formation in the magnetospere, J. Atm. Solar-Terr. Phys., 65, pp 1369-1383.

Nenovski, P. (2008) Comparison of simulated and observed large-scale, field-aligned current structures, Annales Geophysicae, 26, pp. 281-293

Nezlin, M.,Rulov A., Snezhkin E. N., Trubnikov A. S., (1987) Self-organizaion of spiral-vortex structures in shallow water with repid differenial rotation, Sov.Phys.JETP, v.65(1), pp.1-4.

Nezlin, M., Snezhkin E. N. (1993) Rossby Vortices, Spiral Structures, Solitons, Springer-Verlag, Berlin, Heidelberg.

Nezlin, M.V. (1994) Modeling of the Generation of Spiral Structure by Laboratory Experiments in Rotating Shallow Water, and Prediction of Interarm Anticyclones in Galaxies, In Proc. "Physics of the Gaseous and Stellar Disks of the Galaxy", SAO,

22-25 September 1993, Russia, ed. I.R.King, ASP Conference Series, 66, pp. 135-151, 1994.

Oliver, R. 1999, in ESA SP-448, Magnetic Fields and Solar Processes, ed. A. Wilson (Florence, Italy), 425

Northrup, E.F. (1907) Phys. Rev. 24 474 (1907).

Parker, E.N. (1979) *Cosmical magnetic fields. their origin and their activity*, Clarendon Press, Oxford, Vol.1, ch. 4

Sato, T. (1982) in *Magnetospheric Plasma Physics*, Reidel, Tokyo, ch. 4.

Schmieder, B., Malherbe, J. M., Mein, P., & Tandberg-Hanssen, E. 1984, A&A, 136, 81

Schmieder, B., Malherbe, J. M., Poland, A. I., & Simon, G. 1985, A&A, 153, 64

Shafranov, V.D. (1963) Ravnovecie plasmy v magnitnom pole, in *Voprocy teorii plasmy*(in russian), Atomizdat, Moscow, Vol.2, 92-176.

Shkarofsky, I.P., Johnston, T.W., and Bachynski, M.P. (1966) *The particle kinetics of plasmas*, Addison-Wesley, Reading, Massachusetts, ch.10.

Southwood, D.J. and Kivelson, M.G. (1993) *Adv. Space Res.* 13, No 4, (4)149-(4)157.

Speiser, T.W. (1965) *J. Geophys. Res.* 70, 4219-4226.

Speiser, T.W. (1967) *J. Geophys. Res.* 72, 3919-3932.

Stenflo J.O., (1976) Basic Mechanism of Solar Activity, Proc.IAU Symp. 7.1, Reidel, Dordrecht.

Strauss, H.R.(1977) Phys. Fluids, 19, 134, 1976; 20, 1354,

Streltsov, A.V., Chmyrev, V.M., Pokhotelov, O.A., Marchenko, V.A. and Stenflo, L (1990) *Physica Scripta*, 41, 686-692.

Tsubaki, T., & Takeuchi, A. 1986, Solar Phys., 104, 313

Tsubaki, T., Toyoda, M., Suematsu, Y., & Gamboa, G. 1988, PASJ, 40, 121

Tsubaki, T. 1989, in Solar and Stellar Coronal Structures and Dynamics, National Solar Observatory, ed. R. C. Altrock, 140

Vernazza, J. E., Avertt, E. H., & Loeser, R. 1981, ApJS, 45, 635

Vial, J.-C. 1982, ApJ, 253, 330

Vial, J.-C. 1986, in ed. A. I. Poland, NASA CP-2442, Coronal and Prominence Plasmas, 89

Vrsnak, B. 1993, Hvar Obs. Bull., 17, 23

Vrsnak, B., & Ruzdjak, V. 1994, in Solar Coronal Structures (VEDA Publ. House, Tatranska Lomnica), ed. V. Rusin, P. Heinzel, & J.-C. Vial, IAU Coll., 144,

Yi, Z., & Engvold, O. 1991, Sol. Phys., 134, 275

Yi, Z., Engvold, O., & Keil, S. 1991, Solar Phys., 132, 63

Wu, S.T., Hu, Y.Q., Nakagawa, Y., and Tandberg-Hanssen, E., Induced mass and wave motions in the solar atmosphere. I. Effects of shear motion of flux tubes, The Astrophys. J., 266 (1983) 866-881.

Zhang, Q. Z., Livingston, W. C., Hu, J., & Fang, C. 1987, Solar Phys., 114, 245 Zirin, H., & Severny, A. B. 1961, Observatory, 81, 155

Zirker, J. B., Engvold, O., & Martin, S. F. 1998, Nature, 396, 440

9

Small Scale Processes in the Solar Wind

Antonella Greco, Francesco Valentini and Sergio Servidio
Physics Department, University of Calabria, Rende (CS)
Italy

1. Introduction

The solar wind provides a fascinating laboratory for the investigation of a wide range of plasma physical nonlinear processes, such as, e.g., turbulence, intermittency, magnetic reconnection and plasma heating. One of the key aspects for a deep understanding of these phenomena is the plasma behaviour at small scales. This chapter is intended as a discussion forum on the role played by small scales in solar wind plasma dynamics and/or evolution. Processes occurring at large scales are anyhow responsible for the generation of small scale kinetic fluctuations and structures that in turn have important feedback on the global system evolution. In particular, we will focus our attention on two topics, namely magnetic reconnection and kinetic effects at short spatial scales.

For instance, magnetic reconnection occurring at non-MHD scales is linked to the small scale solar wind discontinuities. In particular, recent studies have shown that current sheets produced by turbulence cascade and discontinuities observed in the solar wind have very similar statistical properties and they are connected to intermittency.

Furthermore, the solar wind offers the best opportunity to study directly collisionless plasma phenomena and to attempt to address fundamental questions on how energy is transferred from fluid-large to small scales and how it is eventually dissipated. The processes by which energy is transferred from the fluid-scale inertial range into, ultimately, heating of ions and electrons are not well understood yet: there is growing evidence that multiple processes operate in the solar wind, either simultaneously or in different regimes. Kinetic effects (such as, for example, wave-particle resonant interaction) that presumably govern the short-scale dynamics are considered the best candidates to replace collisional processes in "dissipating" the energy at small wavelengths and in heating the plasma.

The aim of this chapter is to review the state of the art on these topics and their possible implications on space weather, both under theoretical and numerical standpoints, and comparing theoretical results with recent observations.

The chapter consists of two sections, at the end of which we give our conclusions.

2. Magnetic reconnection as an element of turbulence

Magnetic reconnection is a process that occurs in many astrophysical and laboratory plasmas (Moffatt, 1978). Systems like the solar surface (Parker, 1983), the magnetosphere (Sonnerup et al., 1981), the solar wind (Gosling & Szabo, 2008), and the magnetosheath (Retinò et al., 2007; Sundkvist et al., 2007) represent just some of the classical systems in which magnetic reconnection occurs. Another underlying common feature of the above systems is the presence of turbulence (Bruno & Carbone, 2005), so a simultaneous description of both reconnection and turbulence is needed.

In the past 60 years, most of the theoretical effort has been addressed to the study of the basic physics of reconnection, concentrating on idealized two-dimensional (2D) geometries. Generally, these 2D models are characterized by a strong current density peak, where a magnetic X-type neutral point is found (Dungey, 1958). A well-known description of this process was provided by Sweet (Sweet, 1958) and Parker (Parker, 1957). In their work, employing conservation of mass, pressure balance and constancy of the electric field, the essential large scale dynamics of magnetic reconnection was described. In this configuration, a narrow layer called the "diffusion region" forms, and here the field-lines break and reconnect. This process produces a plasma flow into the layer, accompanied by an outflow along the neutral sheet.

In many cases the reconnecting system has been idealized as occurring in a limited spatial region, employing a "rigid-box" topology in which the magnetic field is often arbitrarily chosen to be straight at the inflow-side boundaries. Moreover, simplified "outflow" boundaries are employed. However, such idealized conditions rarely occur in nature, since plasmas may frequently experience turbulence (Bruno & Carbone, 2005). In turbulence, magnetic reconnection may behave in a less predictable way, departing considerably from rigid-box models.

We view reconnection as an element of turbulence itself: it would be difficult to envision a turbulent cascade that proceeds without change of magnetic topology. Furthermore, turbulence provides a natural boundary condition, as opposed to arbitrary (imposed) conditions. Although some suggestions have been made regarding both the general role of reconnection in magnetohydrodynamic (MHD) turbulence (Carbone et al., 1990; Dmitruk & Matthaeus, 2006; Matthaeus & Lamkin, 1986; Veltri, 1999)) and the impact of small scale turbulence on reconnection of large structures (Lapenta, 2008; Malara et al., 1992; Matthaeus & Montgomery, 1980; Matthaeus & Lamkin, 1986; Veltri, 1999)), only recently a quantitative study of reconnection in turbulence has been presented (Servidio et al., 2009; 2010a). In the scenario proposed in these papers, multiple-reconnection events are present in turbulence. The properties of these events depend on the local topology of the magnetic field and the local turbulence conditions.

Our ideas on magnetic reconnection have broad applications, and one of them is the turbulent solar wind. In the free solar wind, in fact, strong magnetic discontinuities are commonly observed (Burlaga, 1968; Tsurutani & Smith, 1979). These consist of rapid changes of the magnetic field, across narrow layers. It is natural to ask whether these discontinuities are related to the process of reconnection. In recent works by Greco et al. (2008; 2009); Servidio et al. (2011) a link between these rapid changes of the magnetic field and the presence of intermittent current sheets was proposed. In the present book chapter we retrace these ideas providing evidence that reconnection and discontinuities may be different faces of the same coin.

2.1 Overview on 2D MHD turbulence

The investigations described here are carried out in the limited context of incompressible 2D MHD, for which the turbulence problem, as well as the well-resolved reconnection problem, are already very demanding.

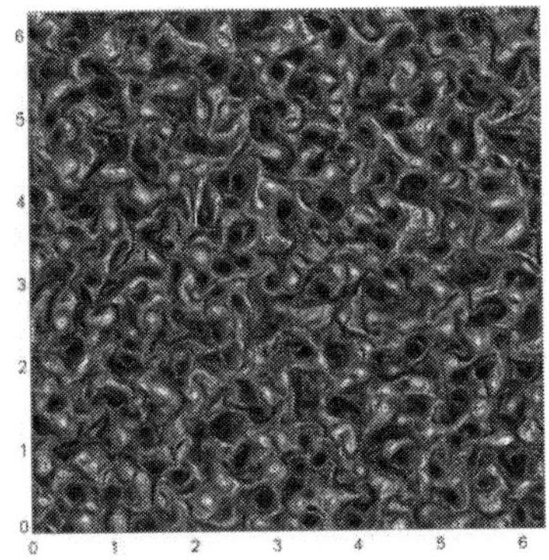

Fig. 1. Shaded contour of the current density j together with the line contours of the magnetic potential a at $t^* = 0.3$.

The 2D incompressible MHD equations can be written in terms of the magnetic potential $a(x,y)$ and the stream function $\psi(x,y)$. By choosing a uniform mass density $\rho = 1$, the equations read:

$$\frac{\partial \omega}{\partial t} = -(\boldsymbol{v} \cdot \boldsymbol{\nabla})\,\omega + (\boldsymbol{b} \cdot \boldsymbol{\nabla})\,j + R_\nu^{-1} \boldsymbol{\nabla}^2 \omega, \tag{1}$$

$$\frac{\partial a}{\partial t} = -(\boldsymbol{v} \cdot \boldsymbol{\nabla})\,a + R_\mu^{-1} \boldsymbol{\nabla}^2 a, \tag{2}$$

where the magnetic field is $\boldsymbol{b} = \boldsymbol{\nabla} a \times \hat{\boldsymbol{z}}$, the velocity $\boldsymbol{v} = \boldsymbol{\nabla}\psi \times \hat{\boldsymbol{z}}$, the current density $j = -\boldsymbol{\nabla}^2 a$, and the vorticity $\omega = -\boldsymbol{\nabla}^2 \psi$. Eqs. (1)-(2) are written in Alfvén units with lengths scaled to L_0. The latter is a typical large scale length such that the box size is set to $2\pi L_0$. Velocities and magnetic fields are normalized to the root mean square Alfvén speed V_A and time is scaled to L_0/V_A. R_μ and R_ν are the magnetic and kinetic Reynolds numbers, respectively (at scale L_0.) The latter coefficients are reciprocals of kinematic viscosity and resistivity.

Eqs. (1)-(2) are solved in a periodic Cartesian geometry (x,y), using a well tested dealiased (2/3 rule) pseudo-spectral code. We employ a standard Laplacian dissipation term with constant dissipation coefficients. The latter are chosen to achieve both high Reynolds numbers and to ensure adequate spatial resolution. A detailed discussion of these issues has been given by Wan et al. (2010). We report on runs with resolution from 4096^2 up to 16384^2 grid points, reaching Reynolds numbers $R_\nu = R_\mu \sim 10000$. Time integration is second order Runge-Kutta and double precision is employed.

Considering a representation of the fields in the Fourier space, for a particular run, the energy is initially concentrated in the shell $5 \le k \le 30$ (wavenumber k in units of $1/L_0$), with mean value $E = \frac{1}{2}\langle |\boldsymbol{v}|^2 + |\boldsymbol{b}|^2 \rangle \simeq 1$, $\langle ... \rangle$ denoting a spatial average. Random uncorrelated phases are employed for the initial Fourier coefficients. The latter implies that the cross helicity, defined as $H_c = 1/2\langle \boldsymbol{v} \cdot \boldsymbol{b} \rangle$, is negligible. The kinetic and the magnetic energy at the beginning of the simulation are chosen to be equal.

We consider for the statistical analysis the state of the system at which the mean square current density $\langle j^2 \rangle$ is very near to its peak value. At this instant of time the peak of small scale turbulent activity is achieved.

When turbulence is fully developed, coherent structures appear. They can be identified as magnetic islands (or vortices). A typical complex pattern of 2D MHD turbulence is shown in Fig. 1, at high Reynolds numbers . In the figure is represented a contour plot of the current j, together with the in-plane magnetic field (line contour of a). The current density j becomes very high in narrow layers between islands.

In Fig. 2-(a), a zoom into the turbulent field is represented, showing that the current is bursty in space. This behavior of the current is related to the intermittent nature of the magnetic field (Mininni and Pouquet, 2009) and can be interpreted as a consequence of fast and local relaxation processes (Servidio et al., 2008). The probability distribution function (PDF) of the current density strongly departs from a Gaussian, as shown in Fig. 2-(b). These coherent structures interact non-linearly, merge, stretch, connect, attract and repulse each other. Reconnection is a major element of this complex interaction.

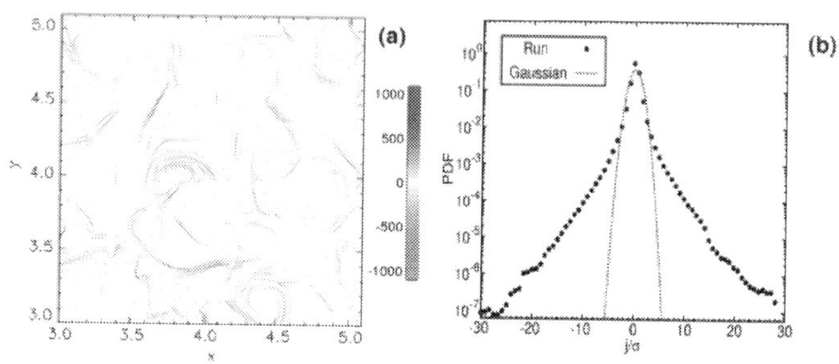

Fig. 2. (a) Shaded-contour of the current density j in a sub-region of the simulation box; (b) PDF of the current density, normalized to its variance, for high Reynolds number turbulence (black bullets). The Gaussian distribution is the red-dashed line.

2.2 Local reconnection events in turbulence

The reconnection rate of two islands is given by the electric field at the X-point. This is related to the fact that the magnetic flux in a closed 2D island is computed as the integrated magnetic field normal to any contour connecting the central O-point (maximum or minimum of a) with any other specified point. Choosing that point to be an X-point bounding the island, we find that the flux in the island is just $a(\mathrm{O-point})$, $-a(\mathrm{X-point})$. Flux is always lost at the O-point in a dissipative system, so the time rate of change of the flux due to activity at the X-point is

$$\frac{\partial a}{\partial t} = -E_{\times} = (R_{\mu}^{-1} j)_{\times}, \tag{3}$$

where $E_{\times}$ is an abbreviation for the electric field measured at the X-point (analogously for the current $j_{\times}$). Eq. (3) follows from the Ohm's law

$$\boldsymbol{E} = -\boldsymbol{v} \times \boldsymbol{b} + R_{\mu}^{-1} \mathbf{j}, \tag{4}$$

which in 2D involves only the out of plane component $E_z = -(\boldsymbol{v} \times \boldsymbol{b})_z + R_{\mu}^{-1} j$. Therefore, in order to describe the local processes of reconnection that spontaneously develop in turbulence

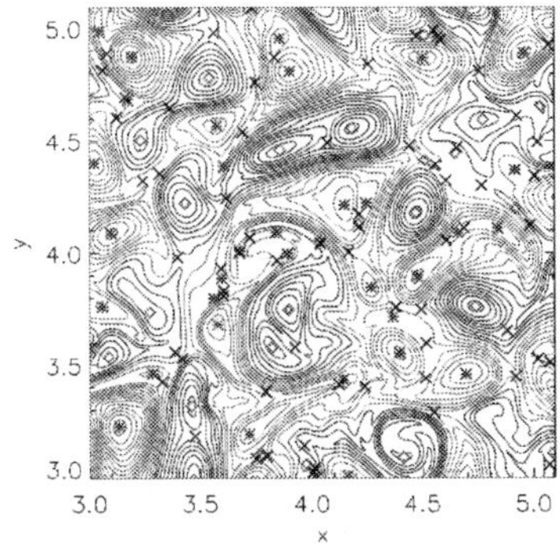

Fig. 3. Contour plot of the magnetic potential a with the position of all the critical points: O-points (blue stars for the maxima and red open-diamonds for the minima) and X-points (black $\times$).

we examine the topology of the magnetic potential studying the *Hessian matrix* of a, defined as

$$H^{a}_{i,j}(\boldsymbol{x}) = \frac{\partial^2 a}{\partial x_i \partial x_j}, \tag{5}$$

which we evaluate at the neutral points of the magnetic field. Further details on the methodology are provided in Servidio et al. (2010a). Here we briefly summarize the main steps of the analysis:

1. Identify critical points at $\boldsymbol{x}^*$, where $\nabla a = 0$
2. Compute the Hessian matrix, given by Eq. (5), at $\boldsymbol{x}^*$
3. Compute eigenvalues λ_1 and λ_2 of $H^{a}_{i,j}(\boldsymbol{x}^*)$, with $\lambda_1 > \lambda_2$
4. Classify the critical point as maximum (both $\lambda_i < 0$), minimum (both $\lambda_i > 0$) and saddle points (or X-points) ($\lambda_1 \lambda_2 < 0$).
5. Compute eigenvectors at each X-point. The associated unit eigenvectors are $\hat{e}_s$ and $\hat{e}_l$, where coordinate s is associated with the minimum thickness δ of the current sheet, while l is associated with the elongation ℓ. Note that the local geometry of the diffusion region near each X-point is related to the Hessian eigenvalues $\lambda_1 = \frac{\partial^2 a}{\partial s^2}$ and $\lambda_2 = \frac{\partial^2 a}{\partial l^2}$.
6. According to Eq. (3), the reconnection rates are given by the electric field at the X-points. These rates are then normalized to the mean square fluctuation δb^2_{rms}, appropriate for Alfvènic turbulence.

In Fig. 3 we show the magnetic potential a with the critical point locations, obtained with the above procedure. In this complex picture the X-points link islands with different size and energy.

From a scaling analysis $\frac{\ell}{\delta} \simeq \sqrt{\lambda_R}$, where $\lambda_R = \left|\frac{\lambda_1}{\lambda_2}\right|$. In the case in which the reconnection is in a stationary state, the rate depends on the above aspect ratio λ_R, satisfying the scaling $E_\times \sim \frac{\ell}{\delta} \sim \sqrt{\lambda_R}$. In Fig. 4, a scatter plot of the reconnection rates against the aspect ratio λ_R is shown. There is a clear trend in this figure, showing that the expression for $E_\times$ is satisfied. This suggests that locally the reconnection processes depend on the geometry and that they therefore are in a quasi steady-state regime.

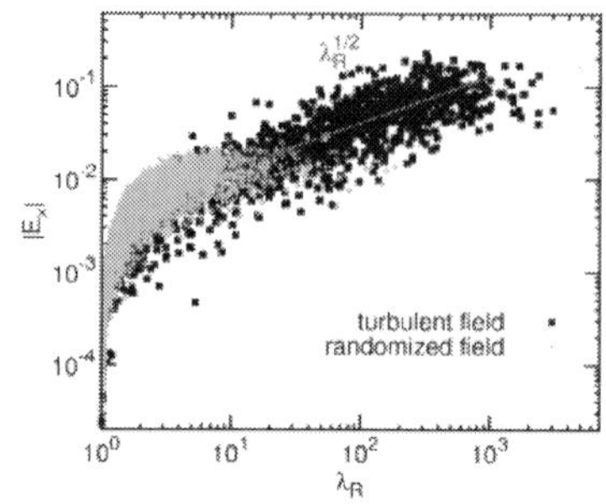

Fig. 4. Scatter plot (black full squares) of the reconnection rates *vs* the geometry of the reconnection region (ratio of the eigenvalues λ_R). The presence of a power-law fit (red line) demonstrates that there is a relation between the reconnection rate and the geometry of the diffusion region. The relative "randomized" reconnection rate is reported with (gray) crosses.

The approximate power-law scaling seen in Fig. 4 at larger values of λ_R suggests that the expression for $E_\times$ holds for the fastest reconnection events. The weaker reconnection events evidently follow a different scaling. We now show that the collection of slowly reconnecting (or even non-reconnecting) X-point regions is associated with a distribution of magnetic fields that is Gaussian. As described by Servidio et al. (2009; 2010a), we now employ a *phase-randomizing procedure*: the original turbulent field is compared with a hybrid field that has the same spectrum but random phases. The coherency of a turbulent pattern is, in fact, hidden in the phases of the Fourier expansion. Using this technique, one can distinguish between slow (Gaussian) and fast (non-Gaussian) reconnection events (Servidio et al., 2010a). As it can be seen from Fig. 4, the reconnection rates of the incoherent randomized magnetic field are on average much weaker than for the original case and they do not manifest any dependence on the aspect ratio of the eigenvalues. In fact the part of the distribution where we found the strongest reconnection sites and the scaling relation with aspect ratio is completely absent in the Gaussianized case. We would like to stress that phase-coherency analysis are widely used in the literature, and they are generally adopted to identify coherent structures (Hada et al., 2003; Sahraoui & Goldstein, 2010).

2.3 The link between magnetic reconnection and turbulence

Now we will take a closer look at the reconnection sites, trying to link them to the characteristic scales of MHD turbulence. Because of the complexity of the geometry we will focus only on the X-lines with higher reconnection rates, identified as described above. We need at this point to find a methodology to quantitatively characterize every reconnection region and extrapolate important information such as δ and ℓ. Since we know the ratio of the eigenvalues obtained from the Hessian matrix analysis, using $\frac{\ell}{\delta} \simeq \sqrt{\lambda_R}$, the problem reduces to find just one of these lengths, say δ.

We call $b_t(s)$ and $b_n(s)$ the normal and the tangential component of the magnetic field, respectively. These components are obtained by projecting the in-plane magnetic field into the system of reference given by $\{\hat{\mathbf{e}}_l, \hat{\mathbf{e}}_s\}$, that is $b_t = \hat{e}_l \cdot \boldsymbol{b}, b_n = \hat{e}_s \cdot \boldsymbol{b}$. Using the eigensystem of the Hessian matrix (λ_i and $\hat{\mathbf{e}}_i$), together with local fit-functions, the up-stream magnetic field can be estimated, locally, for each reconnection region. Note that, the process of reconnection

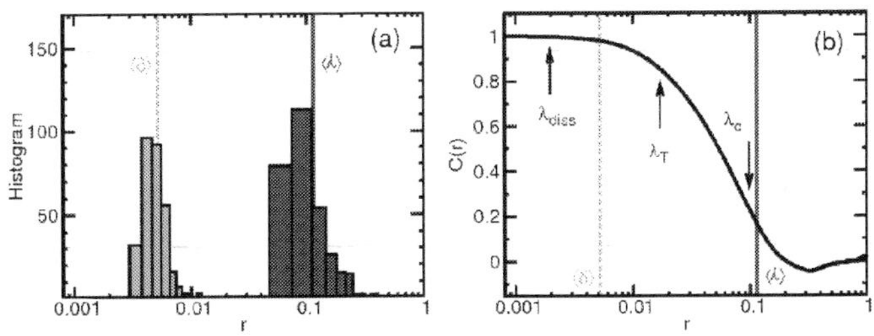

Fig. 5. (a) histograms of thicknesses (δ, gray bars) and elongations (ℓ, blue bars). Vertical lines are average values $\langle\delta\rangle$ (dashed gray) and $\langle\ell\rangle$ (full blue). (b) the magnetic field auto-correlation function (solid black line) is represented. The arrows (left to right) represent respectively: dissipation scale λ_{diss}, Taylor micro-scale λ_T and correlation length λ_C.

in turbulence is often asymmetric, so we define two upstream magnetic fields b_1 and b_2 (we suppressed subscript t).

The PDFs of δ and ℓ are reported in Fig. 5-(a), showing that they are well separated.

The present goal is to look for possible links between the reconnection geometry and the statistical properties of turbulence. In order to get more information about these associations we computed the auto-correlation function of the magnetic field. The correlation length is defined as $\lambda_C = \int_0^* C(r)dr$, where

$$C(\mathbf{r}) = \frac{\langle \boldsymbol{b}(\boldsymbol{x}+\mathbf{r}) \cdot \boldsymbol{b}(\boldsymbol{x})\rangle}{\langle b^2\rangle}, \tag{6}$$

where the direction of displacement r is arbitrary for isotropic turbulence in the plane, and the upper limit is unimportant if the distant eddies are uncorrelated. The correlation length λ_C is a measure of the size of the energy containing islands. The auto-correlation function is illustrated in Fig. 5-(b). In the same figure $\langle\delta\rangle$, $\langle\ell\rangle$ are reported as vertical lines for comparison. The dissipation length, at which the turbulence is critically damped, is defined as $\lambda_{diss} = R_\mu^{-\frac{1}{2}} \langle\omega^2 + j^2\rangle^{-\frac{1}{4}}$, while the Taylor micro-scale, a measure of mean-square gradients, is $\lambda_T = \sqrt{\frac{\langle|\boldsymbol{b}|^2\rangle}{\langle j^2\rangle}}$. The above lengths are represented in Fig. 5-(b).

It appears that the average elongation ℓ is strongly related to the correlation length where $C(r) \to 0$. For all simulations, we found that the values of diffusion layer thickness δ is distributed in the range between the Taylor scale and the dissipation scale, while the length ℓ, though broadly scattered, scales with λ_C (c.f. Fig. 5). The main features of this ensemble of reconnecting events, including the key length scales, are evidently controlled by the statistical properties of turbulence, setting the range of values of length and thickness of the diffusion regions according to the correlation length and the dissipation scale. Note that a correlation between diffusion width and dissipation was discussed experimentally by Sundkvist et al. (2007).

2.4 Applications to the turbulent solar wind

The statistical properties of reconnection have been investigated in the previous sections, leading to the conclusion that strong reconnection events can locally occur in 2D MHD

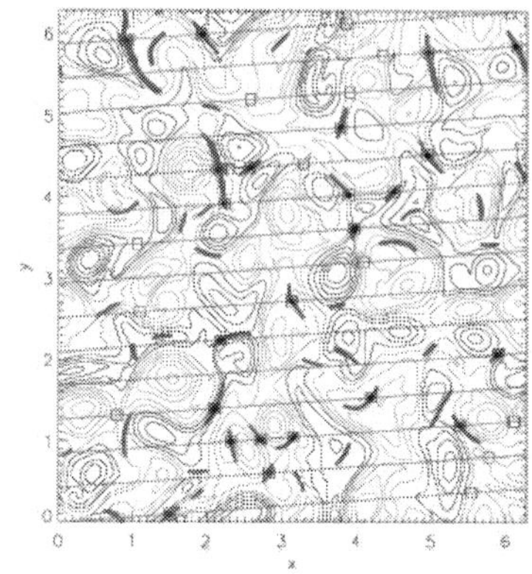

Fig. 6. Contour lines of the magnetic field (or line contour of a) together with the diffusion regions (blue shaded map), and with the one-dimensional path s (green solid line). On the same plot, the discontinuities identified by PVI technique with a threshold $\theta = 5$ in Eq. (8) (open magenta squares) are represented. Bullets (black) are discontinuities which correspond to reconnection sites.

turbulence. In this section we will review some of the main results about the link between solar wind discontinuities and local magnetic reconnection processes.

A well-known feature of solar wind observations is, in fact, the appearance of sudden changes in the magnetic field vector, defined as directional discontinuities (DDs), which are detected throughout the heliosphere (Burlaga, 1968; Ness & Burlaga, 2001; Neugebauer, 2006; Tsurutani & Smith, 1979). These changes are often seen at time-scales of 3 to 5 minutes, although similar discontinuities are seen at smaller time scales (Vasquez et al., 2007). In addition to identification based on characterization of discontinuities, coherent structures have also been identified using other approaches, such as wavelets (Bruno et al., 2001; Veltri & Mangeney, 1999) or phase coherency analysis (Hada et al., 2003; Koga et al., 2007; Koga & Hada, 2003).

One interpretation of magnetic discontinuities is that they are the walls between filamentary structures of a discontinuous solar wind plasma (Borovsky, 2006; Bruno et al., 2001; Burlaga,

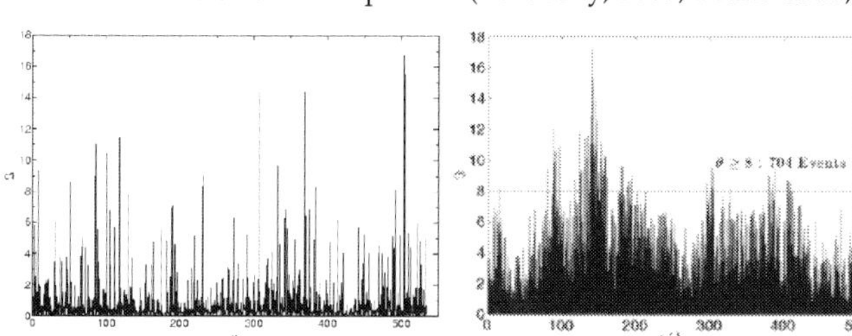

Fig. 7. Left: Spatial signal $\Im(\Delta s, \ell, s)$ (PVI) obtained from the simulation by sampling along the trajectory s in the simulation box, with $\Delta s \simeq 0.67\lambda_{diss}$ and $\ell \simeq 535\lambda_C$. Right: Same quantity obtained from solar wind data, with $\Delta s = 20$ s and $\ell \simeq 500\lambda_C$.

1969), while another is that some strong discontinuities are fossils from the birth of the solar wind (Borovsky, 2008; Burlaga, 1968). An alternative possibility is that the observed discontinuities are the current sheets that form as a consequence of the MHD turbulent cascade (Matthaeus & Montgomery, 1980; Veltri, 1999)). Recent studies on magnetic discontinuities show that their statistical properties are very similar to distributions obtained from simulations of MHD turbulence (Greco et al., 2008; 2009). This line of reasoning argues that thin current sheets are characteristic coherent structures expected in active intermittent MHD turbulence (Mininni and Pouquet, 2009), and which are therefore integral to the dynamical couplings across scales. Therefore, solar wind discontinuities are one of the best applications of our theory of reconnection-in-turbulence.

In this perspective, one is led naturally to suspect that at least some of the current sheets that are a common feature of the solar wind at 1 AU may be participating in small-scale magnetic reconnection (Gosling & Szabo, 2008; Phan et al., 2010; Sundkvist et al., 2007), as well as inhomogeneous interplanetary plasma dissipation and heating (Leamon et al., 2000; Osman et al., 2011). To further establish the relationship between current sheets and small scale reconnection in turbulence, some quantitative connection is needed.

We have in mind the particular question: If one identifies a current sheet in turbulence, how likely is it to be also an active reconnection site? Here we show, using MHD simulation data, that methods for identifying intermittent current sheet-like structures, when quantified properly, can identify sets of structures that are likely to be active reconnection regions.

For the present statistical analysis we will consider a 4096^2 run. Anticipating possible applications to spacecraft data, we focus on properties of discontinuities that are recorded by magnetic field measurements at a single spacecraft in interplanetary space. We adopt a spacecraft-like sampling through the simulation domain [see Fig. 6 and Greco et al. (2008)], and we call s this trajectory. In particular, we can define a set whose elements consist of the segments of a trajectory that passes through any reconnection zone, identified by the *cellular automaton method* (Servidio et al., 2010a;b). In this way we can build a set of strong reconnection site encounters (RS) associated with a trajectory. Fig. 6 shows an example of reconnection sites together with the one-dimensional path s.

Interpolating the magnetic field data along the one-dimensional path s Greco et al. (2008), we can identify discontinuities (TDs) with the following procedure:

1. First, to describe rapid changes in the magnetic field, we look at the increments

$$\Delta \boldsymbol{b}(s, \Delta s) = \boldsymbol{b}(s + \Delta s) - \boldsymbol{b}(s), \tag{7}$$

   where Δs the spatial separation or lag. For this simulation we choose a small scale lag, $\Delta s \simeq 0.67\lambda_{diss}$, which is comparable to the turbulence dissipation scales (see previous sections).

2. Second, employing only the sequence of magnetic increments, we compute the normalized magnitude

$$\Im(\Delta s, \ell, s) = \frac{|\Delta \mathbf{b}(s, \Delta s)|}{\sqrt{\langle |\Delta \boldsymbol{b}(s, \Delta s)|^2 \rangle_\ell}}, \tag{8}$$

   where $\langle \bullet \rangle_\ell = (1/\ell) \int_\ell \bullet ds$ denotes a spatial average over an interval of length ℓ, and Δs is the spatial lag in Eq. (7). The square of the above quantity has been called the *Partial Variance of Increments* (PVI) (Greco et al., 2008) and the method abbreviated as the PVI

Method	θ	# ITD	# IRS	efficiency (%)	goodness (%)
$\Im_1$	1	378	37	100	9.8
$\Im_5$	5	40	23	62.2	57.5
$\Im_8$	8	13	13	35.1	100

Table 1. First column: label of the method $\Im_\theta$. Second column: threshold θ imposed on PVI, cf., Eq.(9). Third column: $\#ITD$, number of discontinuities identified by the method. Fourth column: $\#IRS$, number of reconnection sites found by the method. Fifth column: $\#IRS/\#RS$, the relative efficiency of the method, identified reconnection sites as percent of all the reconnection sites present along the path. Last column: $\#IRS/\#ITD$, the relative goodness of the method, percent of identified reconnection events in set of identified discontinuities.

method. For the numerical analysis performed here $\ell \simeq 535\lambda_C$, where $\lambda_C = 0.18$ is the turbulence correlation length - a natural scale for computing averages.

The PVI time series, evaluated using Eqs. (7)-(8) is reported in Fig. 7. The illustration spans more than 500 correlation lengths. This spatial signal has been compared to a time signal measured by a ACE solar wind spacecraft, near 1 AU, over a period of about 20 days (right panel of the figure). In order to facilitate the comparison, we converted the time signal to a spatial signal, using the average velocity of the flow, and then normalized to a solar wind magnetic correlation length of 1.2×10^6 km.

The PVI increment time series is bursty, suggesting the presence of sharp gradients and localized coherent structures in the magnetic field, that represent the spatial intermittency of turbulence. These events may correspond to what are qualitatively called "tangential discontinuities" and, possibly, to reconnection events.

Imposing a threshold θ on Eq. (8), a collection of stronger discontinuities along the path s can be identified. That is, we select portions of the trajectory in which the condition

$$\Im(\Delta s, \ell, s) > \theta \tag{9}$$

is satisfied, and we will employ this condition to identify candidate reconnection sites. In Fig. 6, an example of the location of discontinuities along s, selected by the PVI method with a particular threshold θ, is shown. One can immediately see in Fig. 6 that there is an association, but not an identity, between the set of "events" identified using Eq. (9), and the encounters of the trajectory with reconnection regions. We will now study this association quantitatively using different values of threshold θ. To understand the physical meaning of the threshold θ, we recall Greco et al. (2008; 2009) that the probability distribution of the PVI statistic derived from a nonGaussian turbulent signal is empirically found to strongly deviate from the pdf of PVI computed from a Gaussian signal, for values of PVI greater than about 3. As PVI increases to values of 4 or more, the recorded "events" are extremely likely to be associated with coherent structures and therefore inconsistent with a signal having random phases. Thus, as θ is increased, stronger and more rare events are identified, associated with highly nonGaussian coherent structures.

We now adopt a procedure to count how many of the identified TDs [from Eq. (9)] are also reconnection sites (i.e., elements of the set RS), as follows: Every discontinuity is characterized by a starting and an ending point along the synthetic trajectory s. A set of discontinuities is identified, and a certain number of these discontinuities intersect reconnection regions. To automate the determination of the reconnection regions, we make use of a map (Servidio et al.,

2010b) that is generated using the cellular automaton procedure. The latter, in summary, is a 2D matrix that has 0 values in all cells outside of the diffusion regions, or values of 1 inside the diffusion regions. For this simulation, and for the selected trajectory (see Fig. 6), there are 37 reconnection sites along the path s. When at least one point of the identified candidate discontinuity overlaps with one point of the identified reconnection region, the event is counted as a "success". Otherwise the TD is not identified as an RS, and is a "failure". In the latter case the method is detecting a non-reconnecting, high-stress, magnetic field structure. However, such points are not associated with a region of strong reconnection, and therefore are not of interest in this analysis.

As an example, using $\theta = 5$ in Eq. (9), 40 discontinuities have been identified and 23 overlap a reconnection site and correspond to successful identification of a reconnection region. The goodness (quality) of this method can be defined as the number of the successes over the total number of identified discontinuities. For this example, the goodness is $\simeq 57.5\%$. An example of discontinuities, together with the reconnecting regions, is shown in Fig. 6.

Following the above procedure summarized by Eq. (9), we impose different threshold θ for the PVI signal. Each threshold characterizes a different set of discontinuities or "events", and we can label each algorithm as $\Im_\theta$. The parameters of different PVI-based algorithms are listed in Table 1, all of which use $\Im(\Delta s = 0.76\lambda_{diss}, \ell = 535\lambda_C)$. It can be seen that for higher values of θ an increasing fraction of the identified TDs corresponds to a reconnection site. That is, the goodness increases as the threshold θ is increased (Servidio et al., 2011).

For high θ, all the TDs correspond to reconnection sites. Once each reconnection site has been identified, the characteristic width δ' can be measured, as described by Servidio et al. (2011). For each TD captured by $\Im$, we measured each δ', and taking the average we obtained $\langle\delta'\rangle = 1.45 \times 10^{-2}$. From the 2D simulation, the average diffusion region thickness is $\langle\delta\rangle = 1.44 \times 10^{-2}$. The estimation $\langle\delta'\rangle$ is therefore in very good agreement with the average size of the diffusion region $\langle\delta\rangle$.

Other information such as the direction or orientation of each TD can be estimated. Using the assumption that the structures are one dimensional, in fact, there is a way to determine the normal vector to the discontinuity surface if single point measurements are used, namely the minimum variance analysis (MVA) technique (Sonnerup & Cahill, 1967). We will now test this technique, making use of the fact that we have a fully 2D picture of each RS from the simulation (see Fig. 6). In each TD detected with the PVI and expanded using the W-field, we compute the matrix $S_{ij} = \langle b_i b_j\rangle - \langle b_i\rangle\langle b_j\rangle$, where here $\langle \dots \rangle$ denotes an average on the trajectory within the TD. Then we compute the eigenvalues (λ_1, λ_2) and the normalized eigenvectors $(\hat{\mathbf{n}}, \hat{\mathbf{t}})$, where λ_1 is the maximum eigenvalue and $\hat{\mathbf{n}}$ $(\hat{\mathbf{t}})$ is the normal (tangential) eigenvector. The values of the ratio λ_1/λ_2 is very large for all the discontinuities selected by $\Im_8$, that is $100 < \lambda_1/\lambda_2 < 10^7$. Another feature is that the normal component b_n is almost null and constant, while b_t is strongly changing sign.

2.4.1 An example from solar wind

We have computed the PVI time series using ACE 1 second resolution magnetic field data from the interval 2004 May 1 to 18 (Osman et al., 2011). The increment (Δs) is 20 seconds and the averaging interval in the denominator in Eq. (8) is the entire data period. The average velocity was around 400 km/s. In Fig. 7, the PVI time series is shown. In order to facilitate the comparison with simulation, we converted the time signal to a spatial signal, using the

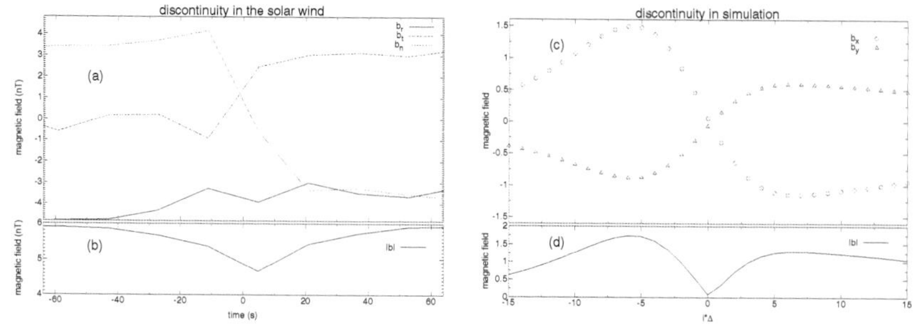

Fig. 8. Examples of discontinuities selected by the PVI method. Panel a: the three components of the magnetic field vector in solar wind data in the RTN reference frame; Panel b: magnitude of the magnetic field vector in solar wind data. The discontinuity, centered around zero, lasts few tens seconds. Panel c: the two components of the magnetic field vector in simulation data; Panel d: magnitude of the magnetic field vector in simulation data. Δ is the resolution data.

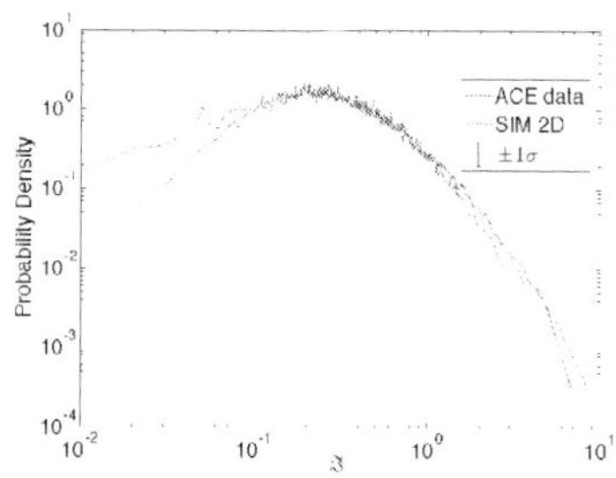

Fig. 9. Probability density function of the spatial signal $\Im$ (PVI) obtained from ACE measurements (blue line) and simulation (red line). The error bar $\pm\sigma$ is displayed in the legend and the value of σ is the expected fractional error in the PDF due to counting statistics.

average velocity of the flow, and then normalized to a solar wind magnetic correlation length of 1.2×10^6 km. Imposing a threshold $\theta > 8$ on Eq. (8), 704 events are identified. One of these TDs is illustrated in Fig. 8 along with an example of TD from the 2D MHD simulation. Finally, in Fig. 9, we show the probability distribution functions of the PVI signal for both the observational and simulation data. The comparison tells us that there is a great similarity within the errors. In general, we suggest that the methods developed here may have many applications to the the solar wind data, where the coexistence of turbulence and magnetic reconnection cannot be discarded.

2.5 Conclusion

In these sections we have assembled a digest of recent works that has examined magnetic reconnection, not as an isolated process that occurs in idealized, controlled conditions, but as a necessary ingredient in the complex nonlinear dynamical process that we call turbulence. Much of the progress in three dimensional (3D) non-steady or turbulent reconnection has been either experimental (e.g., Ren et al., 2005) or in a 3D numerical setup that is in effect nearly 2D (Daughton et al., 2011). It is noteworthy that the fully 3D case is substantial more complex and less understood, both theoretically (Priest and Pontin, 2009), and in numerical simulations (Dmitruk & Matthaeus, 2006). However, for weakly 3D setups, it has been amply confirmed that turbulence effects (Matthaeus & Lamkin, 1986; Servidio et al., 2009) persist (Daughton et al., 2011). While important aspects of the physics of reconnection revealed in the 2D paradigm can carry over to 3D, it is likely also that there are essential physical effects that occur only in a strongly 3D system or with kinetic effects at the small scales.

Most of the progress reviewed here has been in the context of nonlinear dynamics of magnetic reconnection in turbulence, investigated through direct numerical simulations of decaying 2D MHD. The reconnection is spontaneous but locally driven by the fields and boundary conditions provided by the turbulence.

The turbulent cascade produces a distribution of reconnecting islands. Computing the electric field at the X-points, we see that turbulence produces a broad range of reconnection rates. In addition, the strongest reconnection rates vary in proportion to ℓ/δ, the aspect ratio of the reconnection sites. This scaling appears superficially to differ greatly from classical laminar theories (Parker, 1957; Sweet, 1958). These results explain how rapid reconnection occurs in MHD turbulence in association with the most intermittent non-Gaussian current structures, and also how turbulence generates a very large number of reconnection sites that have very small rates.

In contrast to laminar reconnection models that provide a single predicted reconnection rate for the system, turbulent resistive MHD gives rise to a broad range of reconnection rates that depend on local turbulence parameters. Many potential reconnection sites are present, but only a few are selected by the turbulence, at a given time, to display robust reconnection electric fields. In this way, the present problem differs greatly from studies of reconnection that assume that it occurs in isolation or as a spontaneous process.

We have seen that reconnection becomes an integral part of turbulence, as suggested previously (Carbone et al., 1990; Matthaeus & Montgomery, 1980). This perspective on reconnection in turbulence that we have reviewed here seems to be potentially very relevant to space and astrophysical applications such as the turbulent solar wind (Gosling & Szabo, 2008; Sundkvist et al., 2007). On the basis of the current results, we would expect to find in the turbulent corona and solar wind a broad distribution of size of interacting islands, with a concomitantly broad distribution of reconnection rates. Furthermore a useful extension will be to employ models that are suited to low collisionality plasmas, where for example anomalous resistivity, or kinetic effects, may be important.

3. The electrostatic character of the high-frequency energy spectra in the solar wind

The interplanetary medium, the bubble of plasma that is generated by the Sun and that fills the Heliosphere, is known to be hotter than expected in an expanding plasma. Understanding how energy from the Sun can be dissipated into heat in such a collision-free system represents a top priority in space physics. The Sun injects energy into the Heliosphere through large wavelength fluctuations (Alfvén waves). This energy is then channeled towards short scales through a turbulent cascade until it can be transferred to the plasma particles in the form of heat.

The study of the short-wavelength region of the solar-wind turbulent cascade represents nowadays a subject of active interest in space plasma physics. Many experimental works (Alexandrova et al., 2009; Bale et al., 2005; Sahraoui et al., 2009), focused on the analysis of the solar-wind data from spacecraft, aim to investigate how the energy of the large-scale Alfvénic fluctuations can be transferred towards short scales and eventually turned into heat. Within this scenario a crucial point is the identification of the fluctuations that channel the energy from large to short wavelengths along the turbulent cascade.

Long ago it has been shown (Matthaeus et al., 1986) that in the solar wind the Magnetohydrodynamics fluctuations are mainly composed by two populations: the first one with wavevectors predominantly perpendicular to the ambient magnetic field (2D turbulence) and the second one with wavector aligned to the background field (slab turbulence). As recently discussed, for example, by means of Gyrokinetics simulations (Howes et al., 2008)), 2D turbulence seems to give rise, at length scales below the ion-gyro scale, to transverse electromagnetic fluctuations whose features are consistent with the so-called kinetic Alfvén waves. These results provide a significant interpretation to solar-wind observations from the Cluster spacecraft (Sahraoui et al., 2009) in which a quasi-two-dimensional cascade into kinetic Alfvén waves seems to be identified. This cascade represents then a channel available to bring energy from large to small scales.

The second population (slab turbulence) can produce a second channel, in the form of electrostatic fluctuations, for the transport of energy towards small scales. The first insights into the nature of this kind of phenomenon date back to the late seventies, when solar wind measurements from the Helios spacecraft (Gurnett & Andreson, 1977; Gurnett & Frank, 1978; Gurnett et al., 1979) have shown that the high-frequency (few kHz) range of the solar wind turbulent cascade is characterized by the presence of a significant level of electrostatic activity identified as ion-acoustic waves propagating parallel to the ambient magnetic field. The energy level of these fluctuations shows a certain correlation to the electron to proton temperature ratio T_e/T_p and surprisingly survives even for small values of T_e/T_p, for which linear Vlasov theory (Krall & Trivelpiece, 1986) predicts strong Landau dissipation. The propagation of these fluctuations seems to be correlated to the generation of non-Maxwellian proton velocity distributions that display the presence of beams of accelerated particles in the direction of the ambient magnetic field, moving with mean velocity close to the local Alfvén speed. More recent data from the WIND (Lacombe et al., 2002; Mangeney et al., 1999) and the CLUSTER (Pickett et al., 2004) spacecraft allowed to analyze in more detail the features of this electrostatic activity at high frequency in the solar wind. Subsequent experimental space observations confirmed that the particle velocity distributions show a general tendency to depart from the Maxwellian equilibrium configuration, displaying temperature anisotropy (Hellinger at al., 2006; Holloweg & Isenberg, 2002; Marsch et al., 2004) and generation of field-aligned accelerated beams (Heuer & Marsch, 2007; Marsch et al., 1982; Tu et al., 2004)).

These experimental results support the idea that kinetic effects are at work in the solar wind plasmas at short spatial scale lengths, but many aspects of the experimental evidences discussed above still need a convincing physical interpretation: (i) how electrostatic fluctuations of the ion-acoustic type can survive against damping in the case of cold electrons ($T_e \simeq T_p$), (ii) why the mean velocity of the field-aligned beam of accelerated protons is commonly observed to be of the order of the local Alfvén speed.

Recently, many authors used kinetic numerical simulations to reproduce the solar-wind phenomenology described above, that is the generation of longitudinal proton-beam velocity distributions associated with the propagation of electrostatic fluctuations. In particular, Araneda et al. (2008) presented one-dimensional hybrid Particle In Cell (PIC) simulations in which ion-acoustic (IA) fluctuations, generated through parametric instability of monochromatic Alfvén-cyclotron waves, produce field-aligned proton beams during the saturation phase of the wave-particle interaction process. More recently, Matteini et al. (2010) analyzed in detail the relationship between the kinetic aspects of the parametric instability of Alfvén waves (in the case of monochromatic pump waves and of a spectrum of waves) and the evolution of the proton distribution functions, again making use of numerical PIC simulations in hybrid regime.

The parametric instability of left-handed polarized Alfvén waves, considered in the papers referenced above, is efficient in producing IA fluctuations in regimes of low values of the proton plasma beta β_p (Longtin & Sonnerup, 1986) and for large values of the electron to proton temperature ratio, since IA waves are heavily Landau damped for small T_e/T_p (Krall & Trivelpiece, 1986). >From the solar wind observations, the mean velocity of the longitudinal proton beam is typically of the order of the local Alfvén speed. As discussed by Araneda et al. (2008); Matteini et al. (2010), the IA fluctuations, produced through parametric instability, trap resonant protons and dig the particle velocity distribution in the vicinity of the wave phase speed, thus creating the field-aligned beam; it follows that, in order to generate a beam with a mean velocity close to V_A through this mechanism, the phase velocity $v_\phi^{(IA)}$ of the IA fluctuations must be of the same order of V_A [$v_\phi^{(IA)} \simeq V_A$]. Taking into account that the phase speed of the IA waves is $v_\phi^{(IA)} \simeq \sqrt{T_e/m_p}$, the condition necessary to produce a beam with mean velocity of the order of V_A is $T_e/m_p \simeq V_A^2 \Rightarrow T_e/T_p \simeq m_p V_A^2/T_p \simeq 1/\beta_p$, or, equivalently, $(T_e/T_p)\beta_p \simeq 1$. Large values of T_e/T_p, needed for IA fluctuations to survive against Landau damping, require low values of β_p to keep this condition valid. This range of parameters is unusual for the solar wind plasma, where the electron to proton temperature ratio varies in the range $0.5 < T_e/T_p < 4$ (Schwenn & Marsch, 1991), while β_p is typically of order unity. Moreover, it is not clear why the electrostatic activity in the high frequency region of the solar-wind energy spectra is observed even at low values of T_e/T_p and why the secondary proton beam has a mean velocity always of the order of the local Alfvén speed.

Beside the numerical simulations described above, a newly developed Eulerian hybrid Vlasov-Maxwell code (Valentini et al., 2007)) has been used to propose a different mechanism for the generation of the proton-beam distributions associated to the short-scale electrostatic activity in the solar wind. This code solves numerically the Vlasov equation for the protons, while the electrons are considered as a fluid; a generalized Ohm's equation for the electric field, where the Hall term and the electron inertia terms are retained, is integrated. The Faraday equation, the equation for the curl of the magnetic field (where the displacement current is neglected) and an isothermal equation of state for the electron pressure close the set of equations. The quasi-neutrality assumption is considered.

These hybrid Vlasov-Maxwell simulations in 1D-3V phase space configuration (one dimension in physical space and three dimensions in velocity space) (Valentini et al., 2008); Valentini & Veltri, 2009)) were focused on a physical situation where Magnetohydrodynamics turbulence evolves to a state where a significant amount of energy is stored in longitudinal wavevector modes (slab turbulence) (Carbone et al., 1995; Dobrowolny et al., 1980; Matthaeus et al., 1986). The turbulent energy cascade is triggered by nonlinear wave-wave interaction of large scale ion-cyclotron (IC) waves. The numerical results from these simulations gave evidence that for large values of the electron to proton temperature ratio ($T_e/T_p = 10$) the tail of the turbulent cascade at short scales is characterized by the presence of electrostatic fluctuations, propagating in the direction of the mean magnetic field. The Fourier $k-\omega$ spectrum of the numerical signals revealed that, beside the IA branch, a new branch of waves with phase speed close to the proton thermal speed and with acoustic dispersion relation appears. These new waves have been dubbed ion-bulk (IBk) waves. It has been shown that the diffusive plateaus, created in the longitudinal proton velocity distribution through resonant interaction of protons with IC waves (Heuer & Marsch, 2007; Kennel & Engelmann, 1966), are responsible for the excitation of the IBk waves. This phenomenology leads to the generation of a beam of accelerated protons in the direction of the ambient field with mean velocity close to V_A. These results have been confirmed through hybrid-Vlasov simulations in 2D-3V phase space configuration (Valentini et al., 2010)). Moreover, in 2011 Valentini et al. (2011)) the existence of the IBk waves has also been demonstrated by means of electrostatic kinetic simulations, in which an external driver electric field is used to create a longitudinal plateau in the proton velocity distribution.

Here, we review the main results of a series of 1D-3V hybrid Vlasov-Maxwell simulations (in physical situation of slab turbulence), in which the development of the turbulent cascade towards short spatial lengths is investigated in terms of the electron to proton temperature ratio. This analysis allows to demonstrate that the electrostatic fluctuations at short wavelengths, generated as the result of the turbulent cascade, can last in typical conditions of the solar-wind plasma, that is even for low values of T_e/T_p. Moreover, through our numerical simulations we describe a physical mechanism leading to the generation of a field-aligned proton beam with mean velocity close to the Alfvén speed that works even for small values of T_e/T_p provided the proton plasma beta is of order unity.

3.1 Numerical results

As discussed in the Introduction, we numerically follow the kinetic dynamics of protons in 1D-3V phase space configuration (periodic boundary conditions are imposed in physical space). In the following, times are scaled by the proton cyclotron frequency Ω_{cp}, velocities by the Alfvén speed $V_A = B_0/\sqrt{4\pi\rho}$ (B_0 being the magnetic field and ρ the mass density), lengths by the proton skin depth $\lambda_p = V_A/\Omega_{cp}$ and masses by the proton mass m_p.

We assume that at $t = 0$ the plasma has uniform density and is embedded in a background magnetic field $\mathbf{B}_0 = B_0\mathbf{e}_x$, with superposed a set of Alfvén waves, circularly left-hand polarized in the plane perpendicular to the mean magnetic field and propagating along it. The explicit expressions for the velocity and magnetic perturbations [$\delta u_y(x)$, $\delta u_z(x)$, $\delta B_y(x)$ and $\delta B_z(x)$] were derived from the linearized two-fluid equations (Valentini et al., 2007)). The first three modes in the spectrum of velocity and magnetic perturbations are excited at $t = 0$, in such a way that the maximum perturbation amplitude is $A = 0.5$. No density disturbances are imposed at $t = 0$. The initial Maxwellian ion distribution is $f(x, \mathbf{v}, t = 0) = A(x)\exp\left[-(\mathbf{v}-\delta\mathbf{u})^2/\beta_p\right]$, where $\beta_p = 2v_{tp}^2/V_A^2$ ($v_{tp} = \sqrt{T_p/m_p}$ being the proton thermal speed and T_p the proton temperature); $A(x)$ is such that the velocity integral of f gives the equilibrium density $n_0 = 1$. The value of the proton plasma beta is fixed at

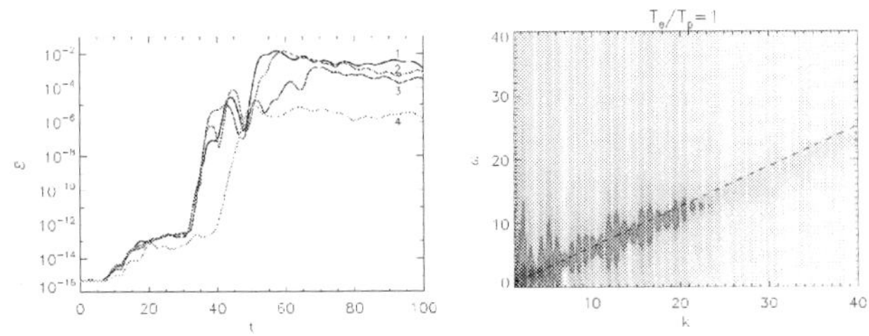

Fig. 10. Left: Time evolution of $\mathcal{E}$ for $T_e/T_p = 10$ [black line (1)], $T_e/T_p = 6$ [red line (2)], $T_e/T_p = 3$ [blue line (3)] and $T_e/T_p = 1$ [green line (4)]; right: k-ω spectrum of the parallel electric energy for the case $T_e/T_p = 1$.

$\beta_p = 0.5$ (the proton thermal velocity is $v_{tp} = 0.5$) and the system evolution is analyzed for different values of the electron to proton temperature ratio ($T_e/T_p = 1, 3, 6, 10$). The mass ratio is $m_e/m_p = 1/1836$. The length of the physical domain is $L_x \simeq 40.2$ (the fundamental wave number is $k_1 = 2\pi/L_x \simeq 0.156$), while the limits of the velocity domain in each direction are fixed at $v^{max} = 5v_{tp}$. We use 2048 gridpoints in physical space, and 51^3 in velocity space and a time step $\Delta t = 10^{-3}$. The simulation is carried up to $t = 400$.

The nonlinear three-wave interactions at large scales trigger the turbulent energy cascade. When the energy is carried at frequencies close to Ω_{cp}, the resonant interaction of the protons with the IC waves produces the formation of a diffusive plateau in the longitudinal velocity distribution in the vicinity of the phase speed $v_\phi^{(IC)}$ of the IC waves (Valentini et al., 2008)). For parallel propagating IC waves one has $v_\phi^{(IC)} \simeq V_A$ for frequencies smaller than Ω_{cp} (or, equivalently for small wave numbers) and $v_\phi^{(IC)} < V_A$ for frequencies close to Ω_{cp}. If β_p is of order unity one gets $v_\phi^{(IC)} \leq V_A \simeq v_{tp}$, this means that the diffusive plateau is created in the vicinity of the proton thermal speed, or, equivalently, in the bulk of the proton velocity distribution.

When the proton velocity distribution is flattened in the vicinity of v_{tp}, the IBk waves can be excited (Valentini et al., 2008); Valentini & Veltri, 2009); Valentini et al., 2011)) and the energy is transferred from large to short wavelengths along the IBk channel. From the analysis of the numerical results, one realizes that in the range of large wavenumbers, say $k > 10\lambda_p^{-1}$, the parallel electric energy is the dominant component of the energy spectrum, this meaning that the tail at short wavelengths of the turbulent cascade is characterized by the presence of electrostatic activity. On the left in Fig. 10, we report in semi-logarithmic plot the early time evolution ($0 < t < 100$) of the longitudinal electric energy at small scales evaluated as $\mathcal{E}(t) = \sum_k |E_{k_x}|^2$ with $k > 10\lambda_p^{-1}$. The black line (1) corresponds to a simulation with $T_e/T_p = 10$, the red line (2) to $T_e/T_p = 6$, the blue line (3) to $T_e/T_p = 3$ and the green line (4) to $T_e/T_p = 1$. It is clear from this figure that, during the system evolution, $\mathcal{E}$ displays a sudden exponential growth and then a saturation phase. We notice that for $T_e/T_p = 1$ the exponential growing phase is somewhat delayed with respect to the case with $T_e/T_p = 10, 6, 3$.

The physical mechanism responsible for this exponential growth consists in an instability process of the beam-plasma type (Valentini et al., 2011)): the resonant interaction of protons with IC waves of large amplitude creates regions of positive slope (small bumps) instead of flat plateaus in the longitudinal proton velocity distribution at v_{tp}; this triggers the growth of high wavenumbers electric field components in parallel propagation with phase speed comparable to v_{tp}. As recently shown by Valentini et al. (2011)), the IBk waves can be excited only when a plateau in the longitudinal proton velocity distribution is generated in the vicinity of v_{tp}. In the present hybrid Vlasov-Maxwell simulations, a diffusive longitudinal plateau is generated at $v_\phi^{(IC)} \leq V_A$, as the result of the resonant interaction between IC waves and protons; when β_p is of order unity one gets $v_\phi^{(IC)} \leq V_A \simeq v_{tp}$, this meaning that the plateau can be produced at v_{tp} and the IBk waves can be excited. We considered different simulations with $0.5 \leq \beta_p \leq 2$ obtaining the same qualitative system evolution. On the other hand, the mechanism described above cannot work for small values of β_p.

The growth of $\mathcal{E}$ corresponds to the excitation of electrostatic fluctuations at high wavenumbers. The energetic level of these fluctuations after the saturation of the exponential growth depends on T_e/T_p; the largest saturation value of $\mathcal{E}$ is found for $T_e/T_p = 10$, but even at $T_e/T_p = 1$ a significant level of fluctuations is recovered. On the right of Fig. 10 we show the k-ω spectrum of the parallel electric energy for the simulation with $T_e/T_p = 1$. The acoustic branch visible in this spectrum is the branch of the IBk waves with phase speed $v_\phi^{(IBk)} \sim 1.2 v_{tp}$ (black dashed line in the figure). As shown by Valentini et al. (2008)); Valentini & Veltri (2009)), for a simulation with $T_e/T_p = 10$ the branch of IA waves also appears in the k-ω spectrum of the parallel electric energy beside the IBk branch. For such a large value of T_e/T_p, the IA fluctuations generated in the early stage of the system evolution by ponderomotive effects can survive against Landau damping up to the end of the simulation. Nevertheless, here we show that, when decreasing the value of T_e/T_p these IA waves are Landau damped quite soon and disappear from the k-ω spectrum, as it is clear from the right plot of Fig. 10 for $T_e/T_p = 1$.

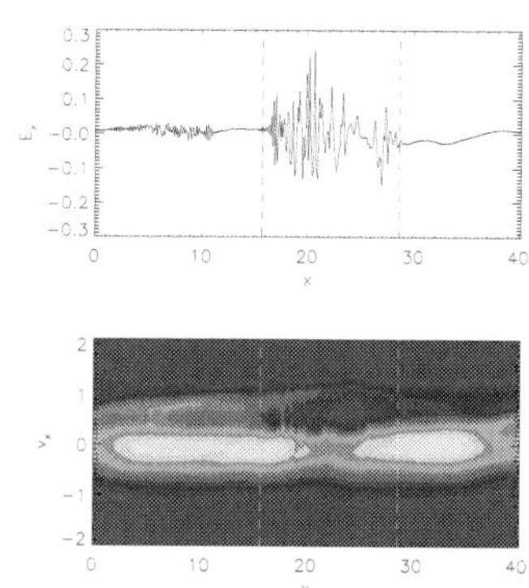

Fig. 11. At the top: parallel electric field E_x versus x at $t = 100$; at the bottom: longitudinal phase space level lines of the proton distribution function at $t = 100$.

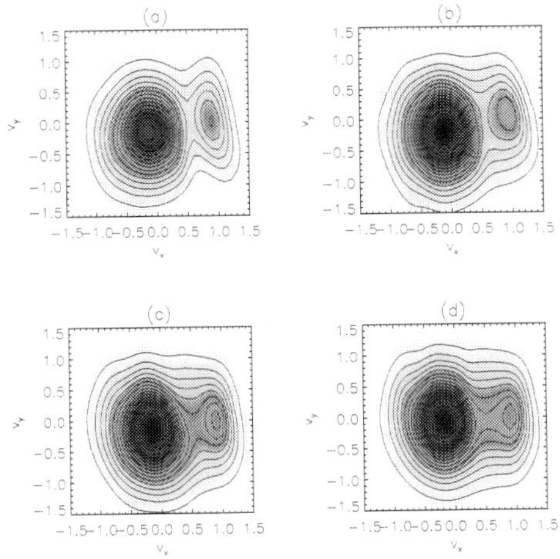

Fig. 12. v_x-v_y contour lines of the proton distribution function integrated over v_z and averaged over x in the region corresponding to the trapped particle population, for $T_e/T_p = 1$ (a), $T_e/T_p = 3$ (b), $T_e/T_p = 6$ (c) and $T_e/T_p = 10$ (d).

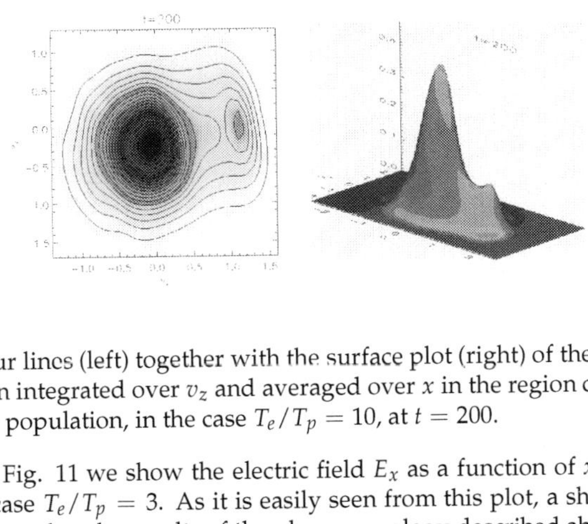

Fig. 13. v_x-v_y contour lines (left) together with the surface plot (right) of the proton distribution function integrated over v_z and averaged over x in the region corresponding to the trapped particle population, in the case $T_e/T_p = 10$, at $t = 200$.

In the top frame of Fig. 11 we show the electric field E_x as a function of x at the end of the simulation for the case $T_e/T_p = 3$. As it is easily seen from this plot, a short-scale localized wavepacket is generated as the results of the phenomenology described above. We point out that for simulations with $T_e/T_p = 1, 6, 10$ we observed the generation of similar structures, with amplitude that depends on T_e/T_p (the largest amplitude is found for $T_e/T_p = 10$). These electrostatic signals propagate with phase velocity $v_\phi \simeq v_{tp}$ (independent on T_e/T_p) and trap resonant protons moving with velocity close to v_{tp}. This is shown in the bottom frame of Fig. 11, where the contour lines of the longitudinal phase space proton distribution function are represented at $t = 100$ for $T_e/T_p = 3$; the region of trapped particles is delimited in space by the vertical white dashed lines and moves with mean velocity close to $v_{tp} = 0.5$.

In order to show how the generation of a trapped particle population affects the proton velocity distribution, in Fig 12 we report the v_x-v_y level curves of f, integrated over v_z and averaged over x in the interval corresponding to the trapping region (see the white dashed lines at the bottom in Fig. 11), for $T_e/T_p = 1$ (a), $T_e/T_p = 3$ (b), $T_e/T_p = 6$ (c) and $T_e/T_p = 10$ (d), at $t = 100$. In each plot of Fig. 12 a beam of accelerated protons is generated in the direction of the ambient magnetic field. We point out that the mean velocity of this secondary beam (of the order of $V_A = 1$) as well as its height is independent on the value of T_e/T_p. We emphasize that this numerical evidence provides a reliable interpretation of the physical mechanism leading to the generation of field-aligned beams of protons in the solar wind velocity distributions. The beam of accelerated particle in the direction of the mean magnetic field is very stable and long lived structure in time as wine show in Fig. 13, where the v_x-v_y contour lines of the proton distribution (left) together with a surface plot of the same distribution are displayed for the case $T_e/T_p = 10$ at $t = 200$.

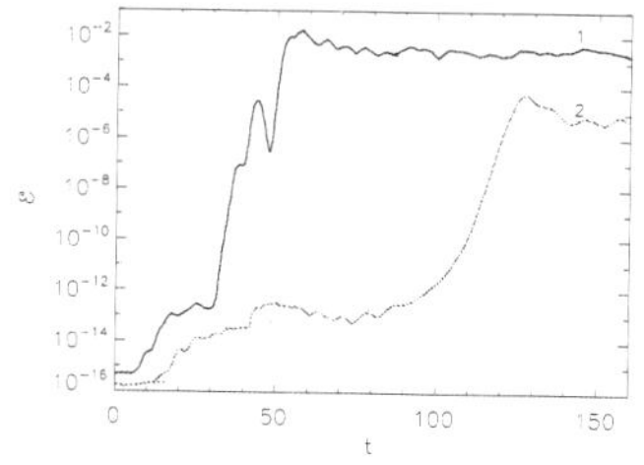

Fig. 14. Time evolution of $\mathcal{E}$ for a simulation with $T_e/T_p = 10$, for $A = 0.5$ [black line (1)], $A = 0.2$ [red line (2)].

As a next step, we consider a new simulation with $T_e/T_p = 10$ in which we decreased the amplitude of the initial Alfvénic perturbations to the value $A = 0.2$ and compare the results of this new simulation with those of the old simulation with $A = 0.5$. In Fig. 14 we report the time evolution of $\mathcal{E}$ (as defined above) for the case with $A = 0.5$ [black line, (1)] and $A = 0.2$ [red line (2)] up to $t = 160$. We point out that decreasing the amplitude of the initial perturbations corresponds to a delay in the exponential growth of the electrostatic fluctuations

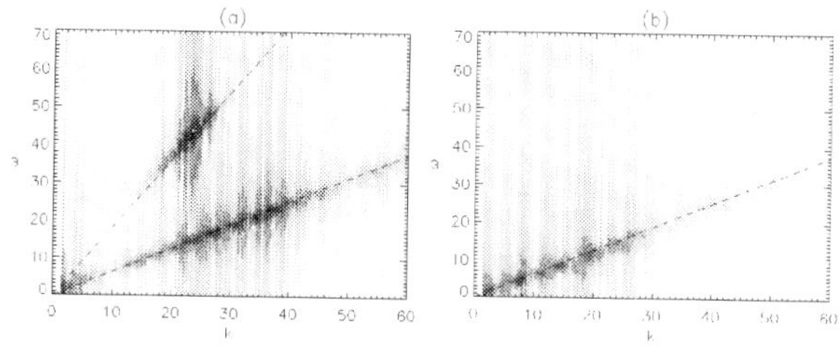

Fig. 15. k-ω spectrum of the parallel electric energy for a simulation with $T_e/T_p = 10$, for $A = 0.5$ (a) and $A = 0.2$ (b).

at small scales and also to a decrease of the growth rate of $\mathcal{E}$. Moreover, the saturation level of $\mathcal{E}$ results about two orders of magnitude lower for $A = 0.2$ with respect to $A = 0.5$. Except for the differences discussed above, Fig. 14 would suggest that the system dynamics is qualitatively similar for both cases $A = 0.2$ and $A = 0.5$; nevertheless, analyzing in more detail the numerical results, we realized that this is not true.

In Fig. 15 we show the k-ω spectra of the longitudinal electric energy for the two simulations with $T_e/T_p = 10$ and $A = 0.5$ (a) and $A = 0.2$ (b). As already shown by Valentini et al. (2008)), for $A = 0.5$ we distinguish two different branches of acoustic waves, the IA waves (upper branch) and the IBk waves (lower branch). The upper dashed line in the top plot of Fig. 15 represents the theoretical prediction for the ion-sound speed c_s (Krall & Trivelpiece, 1986), while the lower dashed line represents the IBk waves phase speed $v_\phi^{(IBk)} \sim 1.2v_{tp}$. As we discussed earlier, the IA waves are produced by ponderomotive effects. On the other hand, it is clear from the bottom plot of Fig. 15 that, when decreasing the amplitude of the initial perturbations to the value $A = 0.2$, the branch of the IA waves disappears. This is due to the fact that decreasing the amplitude of the perturbation produces a decrease in the density fluctuations generated through ponderomotive effects, thus making the IA fluctuations too weak to survive against Landau damping.

The numerical results shown in Fig. 15 allow to conclude that the IBk waves represent the main component of the longitudinal electric energy spectrum at short scales. The IA waves can represent an additional ingredient when the amplitude of the perturbations is large enough to allow ponderomotive effects to produce high level density fluctuations, but they disappear in the weak perturbation limit.

3.2 Conclusion

In these sections, we discussed numerical results of hybrid Vlasov-Maxwell simulations of turbulence at short scale in the solar wind. The system dynamics in the tail at short wavelengths of the turbulent cascade is analyzed in terms of the electron to proton temperature ratio. Our numerical results show that the electrostatic activity in the termination at small spatial scalelengths of the energy spectra, also recovered in the solar-wind data from spacecraft, mainly consists of a novel branch of waves, called ion-bulk waves, that propagate with phase speed comparable to the proton thermal velocity along the direction of the ambient magnetic field. The peculiarity of these electrostatic fluctuations is that, at variance with the usual ion-acoustic waves, they do not undergo Landau damping even at low values of the electron to proton temperature ratio, since they are sustained by the presence of diffusive plateaus in the longitudinal proton velocity distribution. We emphasize that this numerical evidence can be of strong impact for the case of the solar-wind plasma, where the ratio between electron and proton temperature is typically of order unity.

From the analysis of the numerical results of our hybrid Vlasov-Maxwell simulations, we also found that in correspondence with the propagation of the ion-bulk waves the proton velocity distribution displays the generation of a field-aligned beam of accelerated particles with mean velocity close to the local Alfvén speed. We point out that the mean velocity of this beam does not depend on the electron to proton temperature ratio and for β_p of order unity, appropriate value for the case of the solar wind, it always remains close to V_A, in agreement with the experimental data from observations. We emphasize that previously proposed mechanisms (Araneda et al., 2008; Matteini et al., 2010), based on the excitation of

IA fluctuations by parametric instability of large scale Alfvén waves, succeeds in reproducing the generation of the field-aligned proton beam at $v \simeq V_A$ for a different range of plasma parameters [large T_e/T_p, low β_p and $(T_e/T_p)\beta_p \simeq 1$].

On the other hand, the mechanism discussed in the present paper, based on the excitation of the electrostatic IBk branch, naturally works in the physical conditions of the interplanetary medium, even for small values of the electron to proton temperature ratio, provided the proton plasma beta is of order unity. These numerical results describe a reliable mechanism to explain the complex phenomenology detected in many solar wind measurements from spacecraft and thus can be of relevant importance in the study of the evolution of solar-wind turbulence towards short wavelengths.

4. References

Alexandrova, O.; Saur, J.; Lacombe, C.; Mangeney, A.; Mitchell, J.; Schwartz, S. J. & Robert, P. (2009). Universality of Solar-Wind Turbulent Spectrum from MHD to Electron Scales. *Phys. Rev. Lett.*, Vol.103, 165003.

Araneda, J. A.; Marsch, E. & Vinas, A. (2008). Proton Core Heating and Beam Formation via Parametrically Unstable AlfvÃl'n-Cyclotron Waves. *Phys. Rev. Lett.*, Vol. 100, 125003.

Bale, S. D.; Kellogg, P. J.; Mozer, F. S.; Horbury, T. S. & Reme, H. (2005). Measurement of the Electric Fluctuation Spectrum of Magnetohydrodynamic Turbulence. *Phys. Rev. Lett.*, Vol. 94, 215002.

Borovsky, J. (2006). Eddy viscosity and flow properties of the solar wind: Co-rotating interaction regions, coronal-mass-ejection sheaths, and solar-wind/magnetosphere coupling. *Phys. Plasmas*, Vol. 13, 056505.

Borovsky, J. (2008). Flux tube texture of the solar wind: Strands of the magnetic carpet at 1 AU? *J. Geophys. Res.*, Vol. 113, 08110.

Bruno, R. & Carbone, V. (2005). *Living Rev. Solar Phys.*, Vol. 2, 4.

Bruno, R.; Carbone, V.; Veltri, P.; Pietropaolo, E. & Bavassano, B.(2001). Identifying intermittency events in the solar wind. *Planet. Space Sci.*, Vol. 49, 1201.

Burlaga, L. F. (1968). Micro-Scale Structures in the Interplanetary Medium. *Solar Phys.*, Vol. 4, 67.

Burlaga, L. F. (1969). Directional Discontinuities in the Interplanetary Magnetic Field. *Solar Phys.*, Vol. 7, 54.

Carbone, V.; Veltri, P. & Mageney, A.(1990). Coherent structure formation and magnetic field line reconnection in magnetohydrodynamic turbulence. *Phys. Fluids A*, Vol. 2, 1487.

Carbone, V.; Malara, F. & Veltri, P. (1995). A Model for the Three-Dimensional Magnetic Field Correlation Spectra of Low-Frequency Solar Wind Fluctuations During AlfvÃl'nic Periods. *J. Geophys. Res.*, Vol. 100, 1763.

Cassak, P. A. & Shay, M. A. (2007). Scaling of asymmetric magnetic reconnection: General theory and collisional simulations. *Phys. Plasmas*, Vol. 14, 102114.

Daughton, W.; Roytershteyn, V.; Karimabadi, H.; Yin, L.; Albright, B. J.; Bergen, B. & Bowers, K. J. (2011). Role of electron physics in the development of turbulent magnetic reconnection in collisionless plasmas. *Nature Phys.*, Vol. 7, 539.

Dmitruk, P. & Matthaeus, W. H. (2006). Structure of the electromagnetic field in three-dimensional Hall magnetohydrodynamic turbulence. *Phys. Plasmas*, Vol. 13, 042307.

Dobrowolny, M.; Mangeney, A. & Veltri, P. (1980). Properties of magnetohydrodynamic turbulence in the solar wind. *Astron. Astrophys.*, Vol. 83, 26.

Dungey, J. W. (1958). *Cosmic Electrodynamics*, Cambridge University Press, England.

Gosling, J. T. & Szabo, A. (2008). Bifurcated current sheets produced by magnetic reconnection in the solar wind, *J. Geophys. Res.*, Vol. 113, A10103.

Greco, A.; Chuychai, P.; Matthaeus, W. H.; Servidio, S. & Dmitruk, P. (2008). Intermittent MHD structures and classical discontinuities, *Geophys. Res. Lett.*, Vol. 35, L19111.

Greco, A.; Matthaeus, W. H.; Servidio, S.; Chuychai, P. & Dmitruk, P. (2009). Statistical Analysis of Discontinuities in Solar Wind ACE Data and Comparison with Intermittent MHD Turbulence. *Astrophys. J.*, Vol. 691, L111.

Gurnett, D. A. & Andreson, R. R. (1977). Plasma Wave Electric Fields in the Solar Wind: Initial Results From Helios 1. *J. Geophys. Res.*, Vol. 82, 632.

Gurnett, D. A. & Frank, L. A. (1978). Ion Acoustic Waves in the Solar Wind, *J. Geophys. Res.*, Vol. 83, 58.

Gurnett; D. A., Marsch, E.; Pilipp, W.; Schwenn, R. & Rosenbauer, H. (1979). Ion Acoustic Waves and Related Plasma Observations in the Solar Wind. *J. Geophys. Res.*, Vol. 84, 2029.

Hada, T.; Koga, D. & Yamamoto, E. (2003). Phase coherence of MHD waves in the solar wind. *Space Sci. Rev.*, Vol. 107, 463.

Hellinger, P.; TravnÄšcek, P.; Kasper, J. C. & Lazarus, A. J. (2006). Solar wind proton temperature anisotropy: Linear theory and WIND/SWE observations. *Geophys. Res. Lett.*, Vol. 33, L09101.

Heuer, M. & Marsch, E. (2007). Diffusion plateaus in the velocity distributions of fast solar wind protons. *J. Geophys. Res.*, Vol. 112, A03102.

Holloweg, J. V. & Isenberg, P. A. (2002). Generation of the fast solar wind: A review with emphasis on the resonant cyclotron interaction. *J. Geophys. Res.*, Vol. 107, 1147.

Howes, G. G; Dorland, W.; Cowley, S. C.; Mammet, G. W.; Quataert, E.; Schekochihin, A. A. & Tatsuno, T. (2008). Kinetic Simulations of Magnetized Turbulence in Astrophysical Plasmas. *Phys. Rev. Lett.*, Vol. 100, 065004.

Koga, D.; Chian, A. C.-L.; Miranda, R. A. & Rempel, E. L. (2007). Intermittent nature of solar wind turbulence near the Earth's bow shock: Phase coherence and non-Gaussianity. *Phys. Rev. E*, Vol. 75, 046401.

Koga, D. & Hada, T. (2003). Phase coherence of foreshock MHD waves: wavelet analysis. *Space Sci. Rev.*, Vol. 107, 495.

Kennel, C. F. & Engelmann, F. (1966). *Phys. Fluids*, Vol. 9, 2377.

Krall, N. A. & Trivelpiece, A. W. (1986). *Principles of plasma physics*, San Francisco Press, San Francisco.

Lacombe C.; Salem, C.; Mangeney, A.; Hubert, D.; Perche, C.; Bougeret, J.-L.; Kellogg, P. J. & Bosqued, J.-M. (2002). Evidence for the interplanetary electric potential? WIND observations of electrostatic fluctuations. *Ann. Geophys.*, Vol. 20, 609.

Lapenta, G. (2008). Self-Feeding Turbulent Magnetic Reconnection on Macroscopic Scales. *Phys. Rev. Lett.*, Vol. 100, 235001.

Leamon, R. J.; Matthaeus, W. H.; Smith, C. W.; Zank, G.P.; Mullan, D. J. & and Oughton, S. (2000). MHD-driven Kinetic Dissipation in the Solar Wind and Corona. *Astrophys. J.*, Vol. 537, 1054.

Longtin, M. & Sonnerup, B. U.O. (1986). Modulation Instability of Circularly Polarized AlfvÃl'n Waves. *J. Geophys. Res.*, Vol. 91, 6816.

Malara, F.; Veltri, P. & Carbone, V. (1992). Competition among nonlinear effects in tearing instability saturation. *Phys. Fluids B*, Vol. 4, 3070.

Mangeney A.; Salem, C.; Lacombe, C.; Bougeret, J.-L.; Perche, C.; Manning, R.; Kellogg, P. J.; Goetz, K.; Monson, S. J. & Bosqued, J.-M. (1999). WIND observations of coherent electrostatic waves in the solar wind. *Ann. Geophys.*, Vol. 17, 307.

Marsch, E.; Muhlhauser, K.-H.; Schwenn, R.; Rosenbauer, H.; Pilipp, W. & Neubauer, F. (1982). Solar Wind Protons: Three-Dimensional Velocity Distributions and Derived Plasma Parameters Measured Between 0.3 and 1 AU. *J. Geophys. Res.*, Vol. 87, 52.

Marsch, E.; Ao, X.-Z. & Tu, C.-Y. (2004). On the temperature anisotropy of the core part of the proton velocity distribution function in the solar wind. *J. Geophys. Res.*, 109, A04102.

Matteini, L., Landi, S., Velli, M., & Hellinger, P. (2010), Kinetics of parametric instabilities of AlfvÃl'n waves: Evolution of ion distribution functions, *J. Geophys. Res.*, Vol. 115, A09106.

Matthaeus, W. H. & and Montgomery, D. (1980). Selective decay hypothesis at high mechanical and magnetic Reynolds numbers. *Ann. N.Y. Acad. Sci.*, Vol. 357, 203.

Matthaeus, W. H. & Lamkin, S. L. (1986). Turbulent magnetic reconnection. *Phys. Fluids*, Vol. 29, 2513.

Matthaeus, W. H.; Goldstein, M. L. & King, J. H. (1986). An Interplanetary Magnetic Field Ensemble at 1 AU. *J. Geophys. Res.*, Vol. 91, 59.

Mininni, P. D. & Pouquet, A. (2009). Finite dissipation and intermittency in magnetohydrodynamics. *Phys. Rev. E*, Vol. 80, 025401.

Moffatt, H. K. (1978). *Magnetic field generation in electrically conducting fluids*, Cambridge U. Press, Cambridge, England.

Ness, N. F. & Burlaga, L. F. (2001). Spacecraft studies of the interplanetary magnetic field. *J. Geophys. Res.*, Vol. 106, 15803.

Neugebauer, M. (2006). Comment on the abundances of rotational and tangential discontinuities in the solar wind. *J. Geophys. Res.*, Vol. 111, A04103.

Osman, K. T.; Matthaeus, W. H.; Greco, A. & Servidio, S. (2011). Evidence for Inhomogeneous Heating in the Solar Wind, *Astrophys. J.*, Vol. 727, L11.

Parker, E. N. (1957). Sweet's Mechanism for Merging Magnetic Fields in Conducting Fluids. *J. Geophys. Res.*, Vol. 62, 509.

Parker, E. N. (1983). Magnetic neutral sheets in evolving fields. Formation of the solar corona. *Astrophys. J.*, Vol. 264, 642.

Phan, T. D.; Gosling, J. T.; Paschmann, G.; Pasma, C.; Drake, J. F.; Oieroset, M.; Larson, D.; Lin, R. P. & Davis, M. S. (2010). The Dependence of Magnetic Reconnection on Plasma Îš and Magnetic Shear: Evidence from Solar Wind Observations. *Astrophys. J.*, Vol. 719, L199.

Pickett J. S.; Chen, L.-J.; Kahler, S. W.; Santolík, O.; Gurnett, D. A.; Tsurutani, B. T. & Balogh, A. (2004). Isolated electrostatic structures observed throughout the Cluster orbit: relationship to magnetic field strength. *Ann. Geophys.*, 22, 2515.

Priest, E. R. & Pontin, D. I. (2009). Three-dimensional null point reconnection regimes. *Phys. Plasmas*, Vol. 16, 122101.

Ren, Y.; Yamada, M.; Gerhardt, S.; Ji, H.; Kulsrud, R & Kuritsyn, A. (2005). Experimental Verification of the Hall Effect during Magnetic Reconnection in a Laboratory Plasma. *Phys. Rev. Lett.*, Vol. 95, 055003.

Retinò, A.; Sundkvist, D.; Vaivads, A.; Mozer, F.; André, M. & Owen, C. J. (2007). In situ evidence of magnetic reconnection in turbulent plasma, *Nature Phys.*, Vol. 3, 236.

Sahraoui, F.; Goldstein, M. L.; Robert, P.; & Khotyaintsev, Yu. V. (2009). Evidence of a Cascade and Dissipation of Solar-Wind Turbulence at the Electron Gyroscale. *Phys. Rev. Lett.*, Vol. 102, 231102.

Sahraoui, F. & and Goldstein, M. (2010). Structures and Intermittency in Small Scales Solar Wind Turbulence, *Twelfth International Solar Wind Conference, AIP Conference Proceedings*, Vol. 1216, 140, Saint-Malo, France, 21-26 June 2009.

Schwenn, R. & Marsch, E. (1991). *Physics of the Inner Heliosphere II. Particles, Waves and Turbulence*, Vol. 2, Springer.

Servidio, S.; Matthaeus; W. H. & Dmitruk, P. (2008). Depression of Nonlinearity in Decaying Isotropic MHD Turbulence. *Phys. Rev. Lett.*, Vol. 100, 095005.

Servidio, S.; Matthaeus, W. H.; Shay, M. A.; Cassak, P. A. & Dmitruk, P. (2009). Magnetic Reconnection in Two-Dimensional Magnetohydrodynamic Turbulence. *Phys. Rev. Lett.*, Vol. 102, 115003.

Servidio, S.; Matthaeus, W. H.; Shay, M. A.; Dmitruk, P.; Cassak, P. A. & Wan, M. (2010). Statistics of magnetic reconnection in two-dimensional magnetohydrodynamic turbulence. *Phys. Plasmas*, Vol. 17, 032315.

Servidio, S.; Wan, M.; Matthaeus, W. H. & Carbone, V. (2010). Local relaxation and maximum entropy in two-dimensional turbulence. *Phys. Fluids*, Vol. 22, 125107.

Servidio, S.; Greco, A.; Matthaeus, W. H.; Osman, K. T. & Dmitruk, P. (2011). Statistical association of discontinuities and reconnection in magnetohydrodynamic turbulence. *J. Geophys. Res.*, Vol. 116, A09102.

Sonnerup, B. U. O. & Cahill, L. J. (1967). Magnetopause Structure and Attitude from Explorer 12 Observations. *J. Geophys. Res.*, Vol. 72, 171.

Sonnerup, B. U. O.; Paschmann, G.; Papamastorakis, I.; Sckopke, N.; Haerendel, G.; Bame, S. J.; Asbridge, J. R.; Gosling, J. T & Russel, C. T. (1981). Evidence for Magnetic Field Reconnection at the Earth's Magnetopause. *J. Geophys. Res.*, Vol. 86, 10049.

Sweet, P. A. .(1958). *Electromagnetic Phenomena in Cosmical Physics*, Cambridge University Press, New York.

Sundkvist, D.; Retino, A.; Vaivads A. & Bale, S. D. (2007). Dissipation in Turbulent Plasma due to Reconnection in Thin Current Sheets. *Phys. Rev. Lett.*, Vol. 99, 025004.

Tsurutani, B. T. & Smith, E. J. (1979). Interplanetary discontinuities - Temporal variations and the radial gradient from 1 to 8.5 AU. *J. Geophys. Res.*, Vol. 84, 2773.

Tu, C. -Y; Marsch, E. & Qin, Z.-R. (2004). Dependence of the proton beam drift velocity on the proton core plasma beta in the solar wind. *J. Geophys. Res.*, Vol. 109, A05101.

Valentini F.; Travnicek, P.; Califano, F.; Hellinger, P. & Mangeney, A. (2007). A hybrid-Vlasov model based on the current advance method for the simulation of collisionless magnetized plasma. *J. Comput. Phys. 225*, Vol. 753.

Valentini, F.; Veltri, P.; Califano, F. & Mangeney, A. (2008). Cross-Scale Effects in Solar-Wind Turbulence. *Phys. Rev. Lett.*, Vol. 101, 025006.

Valentini, F. & Veltri, P. (2009). Electrostatic Short-Scale Termination of Solar-Wind Turbulence. *Phys. Rev. Lett.*, Vol. 102, 225001.

Valentini, F.; Califano, F. & Veltri, P. (2010). Two-Dimensional Kinetic Turbulence in the Solar Wind. *Phys. Rev. Lett.*, Vol. 104, 205002.

Valentini, F.; Califano, F.; Perrone, D.; Pegoraro, F. & Veltri, P. (2011). New Ion-Wave Path in the Energy Cascade. *Phys. Rev. Lett.*, Vol. 106, 165002.

Vasquez, B. J.; Abramenko, V. I.; Haggerty, D. K & Smith, C. W. (2007). Numerous small magnetic field discontinuities of Bartels rotation 2286 and the potential role of AlfvÃl'nic turbulence. *J. Geophys. Res.*, Vol. 112, 11102.

Veltri, P. (1999). MHD turbulence in the solar wind: self-similarity, intermittency and coherent structures, *Plasma Phys. Control. Fusion*, Vol. 41, A787.

Veltri, P. & and Mangeney, A. (1999). Scaling laws and intermittent structures in solar wind MHD turbulence, *Solar Wind IX, AIP Conference Proceedings*, Vol. 471, 543. Nantucket Island (Massachusetts, USA), 5-9 October, 1999.

Wan, M.; Oughton, S.; Servidio, S. & Matthaeus, W. H. (2010). On the accuracy of simulations of turbulence, *Phys. Plasmas*, Vol. 17, 082308.

10

Kinetic Models of Solar Wind Electrons, Protons and Heavy Ions

Viviane Pierrard
Belgian Institute for Space Aeronomy and Université Catholique de Louvain
Belgium

1. Introduction

In the present chapter, we describe the recent results obtained in the development of solar wind models using the kinetic approach. We show how the solution of the evolution equation is used to determine the velocity distribution function (VDF) of the solar wind particles and their moments. The solutions depend on the approximations and assumptions made in the development of the models. We describe in particular the results obtained with the collisionless exospheric approximation, the effects of Coulomb collisions obtained by using a Fokker-Planck term in the evolution equation, the effects of wave turbulence for the electrons and those of kinetic Alfven waves for the protons.

2. The kinetic approach

Solar wind models have been developed on the basis of the magnetohydrodynamic (MHD) and of the kinetic approaches. These two theories are complementary, but we will here emphasize the recent advances in the kinetic solar wind representation

Kinetic models provide the velocity distribution functions f ($\vec{r}$,$\vec{v}$,t) of the particles as a solution of the evolution equation:

$$\frac{\partial f}{\partial t}+\vec{v}.\frac{\partial f}{\partial \vec{r}}+\vec{a}.\frac{\partial f}{\partial \vec{v}}=\left(\frac{df}{dt}\right)_c \tag{1}$$

where the first term represents the time dependence of the VDF, the second term corresponds to the spatial diffusion ($\vec{r}$ is the position and $\vec{v}$ the velocity vector of the particles), the third term takes into account the effects of the external forces $\vec{F}$ ($\vec{a}=\vec{F}/m$ where $\vec{a}$ is the acceleration and m is the mass of the particles), and the term in the right hand side of the equation represents the effects of collisions and other interactions.

The calculation of the VDF moments gives the macroscopic quantities such as

the number density:

$$n(\vec{r})=\int_{-\infty}^{\infty} f(\vec{r},\vec{v})\, d\vec{v} \tag{2}$$

the particle flux:

$$\vec{F}(\vec{r}) = \int_{-\infty}^{\infty} f(\vec{r},\vec{v})\,\vec{v}\,d\vec{v} \tag{3}$$

the bulk velocity:

$$\vec{u}(\vec{r}) = \frac{\vec{F}(\vec{r})}{n(\vec{r})} \tag{4}$$

the pressure:

$$\vec{P}(\vec{r}) = m\int_{-\infty}^{\infty} f(\vec{r},\vec{v})(\vec{v}-\vec{u})(\vec{v}-\vec{u})d\vec{v} \tag{5}$$

the temperature:

$$T(\vec{r}) = \frac{m}{3kn(\vec{r})}\int_{-\infty}^{\infty} f(\vec{r},\vec{v})\left|v-u\right|^2 d\vec{v} \tag{6}$$

the energy flux.

$$\vec{E}(\vec{r}) = \frac{m}{2}\int_{-\infty}^{\infty} f(\vec{r},\vec{v})\left|v-u\right|^2 (\vec{v}-\vec{u})d\vec{v} \tag{7}$$

We are interested by the steady state solutions of the evolution equation, i.e., we consider $\partial f / \partial t = 0$.

3. Exospheric models

The simplest approximation in low density plasmas like the solar corona and the solar wind where kinetic processes prevail is to consider that there are no interactions at all between the particles above a certain radial distance called the exobase. In this collisionless region located already at low radial distances in the solar corona (the exobase is typically located between 1.1 Rs and 6 Rs), one considers the exospheric approximation: the right hand side term in eq. (1) is neglected, i.e., we assume $(df/dt)_c=0$. The particles are only submitted to the effects of the external forces, i.e., the gravitational force, the electric force and the Lorentz force due to the presence of the magnetic field $\vec{B}$:

$$\vec{a} = \left(\vec{g} + \frac{Ze\vec{E}}{m}\right) + \frac{Ze}{m}(\vec{v}\times\vec{B}) \tag{8}$$

The Vlasov equation can then be solved analytically assuming an electric force that ensures the equality of electron and ion fluxes. If the VDF of the particles is known at the reference level, the solution of the Vlasov equation determines the VDF of the particles at the other radial distances. Such a model was developed by Pierrard & Lemaire (1996) for the ion-exosphere and adapted by Maksimovic et al. (1997a) and later by Lamy et al. (2003) for the solar wind.

3.1 The electrons

In exospheric models, it is assumed that the energy and the magnetic moment are conserved. The trajectories of the particles depend on their velocity and pitch angle. The electrons are in an attractive potential due to the attractive electric and the gravitational potentials, so that 4 different orbits are then possible:

- Ballistic particles: these electrons have not enough energy to escape, so they come back to the Sun. The low energy part of the electron VDF is thus symmetric.
- Escaping particles: these electrons have enough energy to escape.
- Trapped particles: these electrons are trapped by the magnetic force. They can not escape and give also a symmetric part in the VDF.
- Incoming particles: these electrons correspond to particles coming from the interplanetary space to the Sun.

These different orbits are illustrated on Fig. 1 in the velocity plane parallel and perpendicular to the magnetic field.

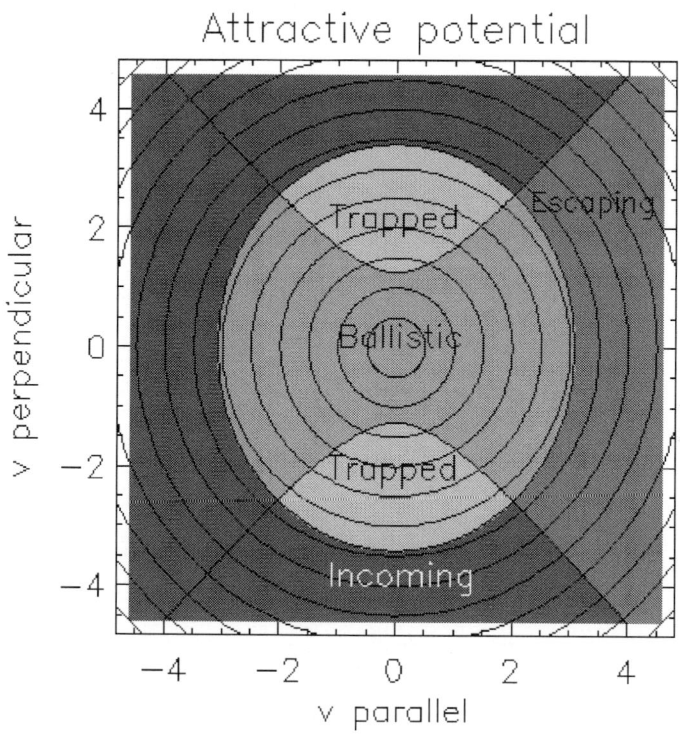

Fig. 1. Isocontours of the VDF of particles in an attractive potential like the solar wind electrons (circles). The 4 different classes of the particles (ballistic, escaping, trapped and incoming) determined by the energy and magnetic moment conservation laws are illustrated in different colours.

In exospheric models, it is assumed that there are no particles incoming from the interplanetary space so that the distribution is anisotropic. The escape flux is thus only contributed by energetic electrons. Indeed, only energetic electrons can pass over the potential barrier. The VDF is thus a truncated anisotropic distribution that leads to non-zero external flux and heat flux.

3.2 The protons

The protons are on the contrary in a repulsive potential, because the repulsive electrostatic potential exceeds the attractive gravitational potential at sufficiently large radial distances. There are then neither ballistic nor trapped orbits. Assuming again an empty population of incoming particles flying from the interplanetary space to the Sun, there are only escaping protons that are accelerated outwards in the solar wind by the electric force. When the exobase is located at very low radial distance like in the coronal holes, the potential can first be attractive at low radial distances and then become repulsive above a radial distance called r_{max} that is located at a few solar radii. The global proton potential shows then a maximum at low radial distance (Lamy et al., 2003). The different orbits of the solar wind protons in a repulsive potential located after r_{max} are illustrated on Fig. 2.

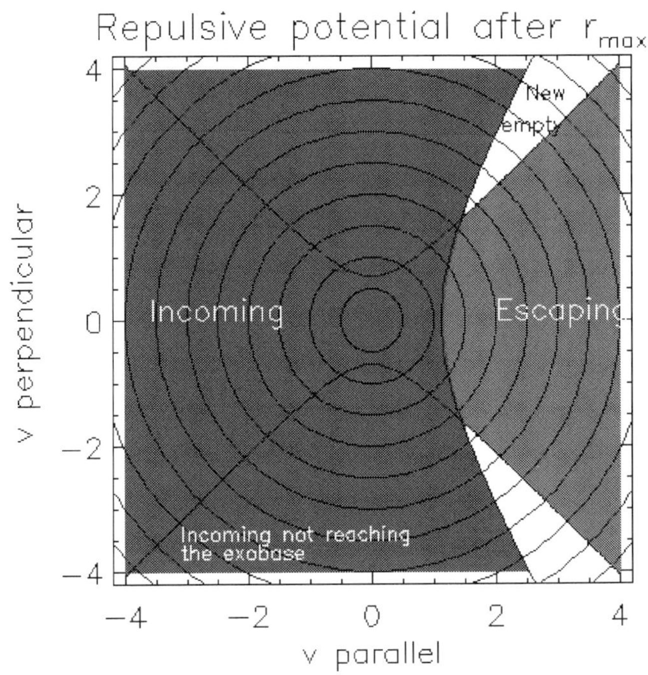

Fig. 2. Isocontours of the VDF of particles (circles) in a repulsive potential like the solar wind protons at large radial distances. The 2 different classes of particles (escaping and incoming) determined by the energy and magnetic moment conservation laws are illustrated in different colours.

3.3 The profiles of the moments

When the VDF of the particles is known at any radial distances, one can calculate the moments by eq. (2) to (6). Fig. 3 shows typical radial profiles of the solar wind obtained with an exospheric model assuming that at the reference exobase level at 1.1 Rs, the temperature of the electrons T_e= 8 10^5 K and of the protons T_p=10^6 K. The maximum of the proton potential at r_{max} is well visible on panel 7 of Fig. 3.

The exospheric models have established that the physical process implicated in the solar wind acceleration is the electric field that pushes the positive ions outwards (Lemaire & Pierrard, 2001). This electric field is increased when an enhanced population of suprathermal electrons is assumed to be present in the solar corona (Pierrard & Lemaire, 1996; Maksimovic et al., 2001). This can be modeled by assuming for instance a Kappa distribution for the electrons. This distribution decreases as a power law of the energy.

Electron VDFs measured in situ in the solar wind are characterized by a thermal core population and a halo of suprathermal electrons (Pierrard et al., 2001b). Such distributions with suprathermal tails are well fitted by the so-called Kappa or Lorentzian distributions. The value of the kappa index determines the slope of the energy spectrum of the suprathermal electrons forming the tail of the VDF. In the limit $\kappa \to \infty$, the Kappa function degenerates into a Maxwellian. The kappa values obtained by fitting the observed electron VDF from Ulysses in the high speed solar wind range between 2 and 7 (Maksimovic et al., 1997b). The Kappa fit parameters obtained in the slow speed solar wind are a little bit larger than in the high speed solar wind, confirming empirically a link between the velocity and the suprathermal electrons. The presence of similar power law distributions in many different space plasmas suggests a universal mechanism for the creation of such suprathermal tails (Pierrard & Lazar, 2010 for a review).

The electrostatic potential found in this example with an assumed kappa value of κ=5 for the electron VDF is illustrated on Panel 2 of Fig. 3. Note that contrary to the electrons, the proton distribution is assumed to be Maxwellian in the corona. Anyway, suprathermal protons have no significant influence on the final bulk velocity of the solar wind, contrary to the suprathermal electrons that are the only one to be able to escape and are thus crucial in the determination of the electric potential.

Figure 3 shows typical radial profiles of the solar wind obtained with the exospheric model assuming a moderate value of Kappa (κ=5) for the electrons at the exobase level 1.1 Rs where the temperature of the electrons is assumed to be T_e= 8 10^5 K and that of the protons T_p=10^6 K. This case gives a final bulk velocity at 215 Rs of 301 km/s (see panel 4) and a potential difference of 1250 V (see panel 2). This corresponds well to the slow speed solar wind that originates from equatorial streamers during minimum solar activity.

Smaller values of kappa (thus more suprathermal electrons) give larger potential differences and thus larger final bulk velocities at 1 AU, as observed in the high speed solar wind where the average velocity is around 800 km/s. With kappa=2 and the same other values for the temperatures at the exobase as in the example given Fig. 3, a bulk velocity as high as 1419 km/s is obtained at 1 AU. On the contrary, when κ tends to infinity, a simple Maxwellian distribution is recovered for the electron distribution and the bulk velocity is limited to 163 km/s. The most realistic velocities in the model, compared to what is observed in the high

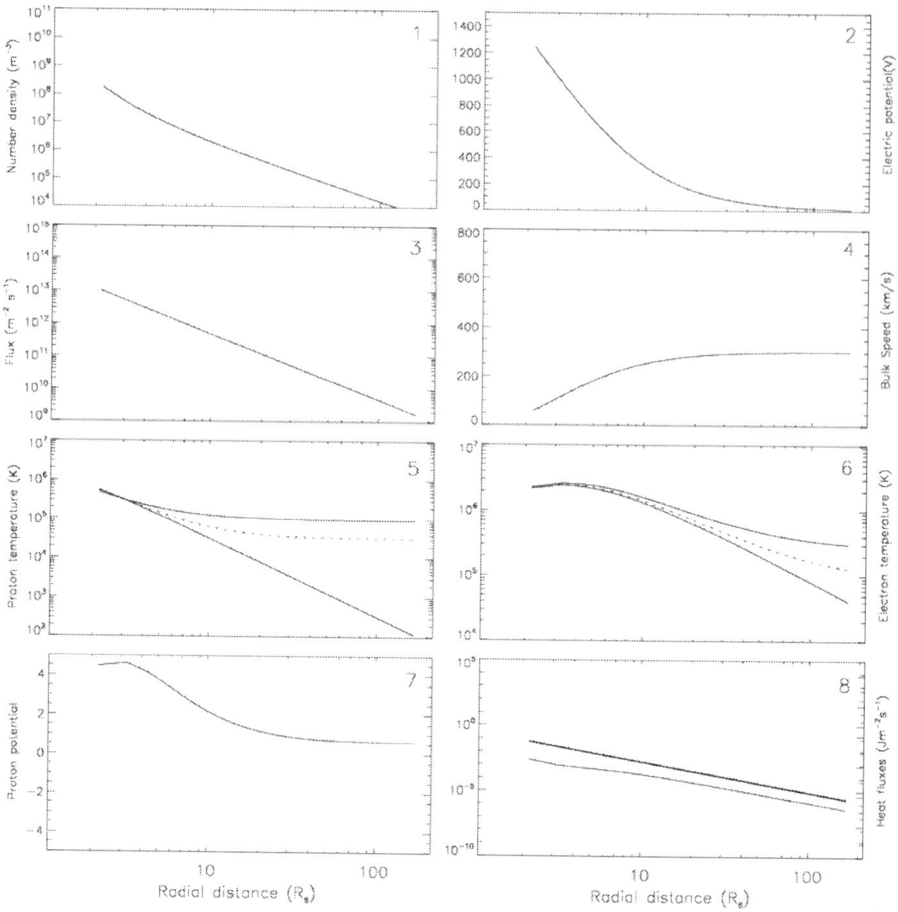

Fig. 3. Profiles of the FDV moments obtained with an exospheric model. Panels 1 to 8 correspond respectively to the density, electric potential, flux, bulk speed, proton temperatures (parallel: upper line, perpendicular: bottom line, average: dotted line), electron temperatures (idem), proton potential and heat flux (electrons: upper line, protons, bottom line).

speed solar wind originating from the coronal holes, are obtained with Kappa between 3 and 4. The position of the exobase plays also an important role in the solar wind acceleration: the bulk velocity at 1 AU is higher when the exobase is low.

Panels 1 and 3 show the fast decrease of the number density and of the particle flux as a function of the radial distance. Panel 5 and 6 show respectively the proton and electron parallel (upper solid line), perpendicular (bottom solid line) and total (dotted line) temperatures. The electron temperature shows a peak at low radial distance (see panel 6) that is also deduced from coronal observations made during solar eclipses. Simulating a coronal hole with a lower value of kappa in the exospheric model, the temperature peak obtained in the solar corona is slightly displaced at higher altitudes, in good agreement again with eclipse observations (Lemaire, 2012).

Note that the profiles of the different moments calculated by the exospheric model are in good agreement with the solar wind observations (Issautier et al., 2001), except for the temperature anisotropy $T_{//}/T_{perp}$ which is too high in the model for the electrons as well as for the protons. This is due to the exospheric assumption of the magnetic moment conservation and of the absence of interaction between the particles. The inclusion of a spiral magnetic field reduces the temperature anisotropies at low latitudes (Pierrard et al., 2001a).

The heat flow carried by Kappa distributions in the solar corona is especially high when kappa is small. It does not correspond to the classical Spitzer & Härm (1953) expression, except when kappa tends to infinity corresponding to the Maxwellian case. Dorelli & Scudder (1999) demonstrated that a weak power law tail in the electron VDF can allow heat to flow up a radially directed temperature gradient, contrary to the classical heat conduction law. For Kappa distributions, the heat flux is not necessarily proportional to -grad T (Pierrard, 2011a).

3.4 The other ions

Other ions than protons are also present in very low concentration in the solar wind. Protons represent around 90 % of the ions, Helium around 9%, and all the heavier ions less than 1% all together. They were included in the exospheric models (Pierrard et al., 2004). Their study is very interesting in such models including only the external forces because the ions have different masses and different charges leading to different bulk velocities for each species.

In the exospheric models, the electrostatic potential is determined by the dominant species (electrons and protons). Minor ions have no significant effect on the electrostatic potential, but they are accelerated by this potential at large radial distance since they have a positive charge (like the protons). So they have also a non monotonic potential energy that shows a maximum at low radial distances. An example is illustrated on Fig. 4 illustrating the results of an exospheric model including Helium He^{2+} and Oxygen ions O^{6+}.

Figure 4 illustrates the profiles of the number density, bulk speed, temperatures (parallel and perpendicular) and the total potential (gravitational and electrostatic) obtained with an exospheric model assuming an exobase at 2 Rs, κ=3 for the electrons, T_p=T_e=10^6 K, T_{He}=5.10^7 K and T_O=$2\ 10^8$ K. The H^+ ions are illustrated by the solid black line, He^{2+} by the magenta dashed-dotted line and O^{6+} by the green three dots-dashed line.

The maximum of the total potential is located at slightly larger radial distances for heavier ions since their mass on charge ratio is larger than that of the protons (see panel 4 of Fig. 4). The mass of the particles directly influences the gravitational attraction while the charge determines the electric force. Because the mass and level of ionization is different for each ion species, the bulk velocity is also different for each ion species.

The mass of the minor ions is very high so that they are more difficult to accelerate at high bulk velocities. Nevertheless, these heavy ions can reach high bulk velocities at 1 AU (see panel 2) if their temperature at the exobase is assumed to be sufficiently high (Pierrard et al., 2004). Very high ion temperatures in the solar corona are not unrealistic: their observations

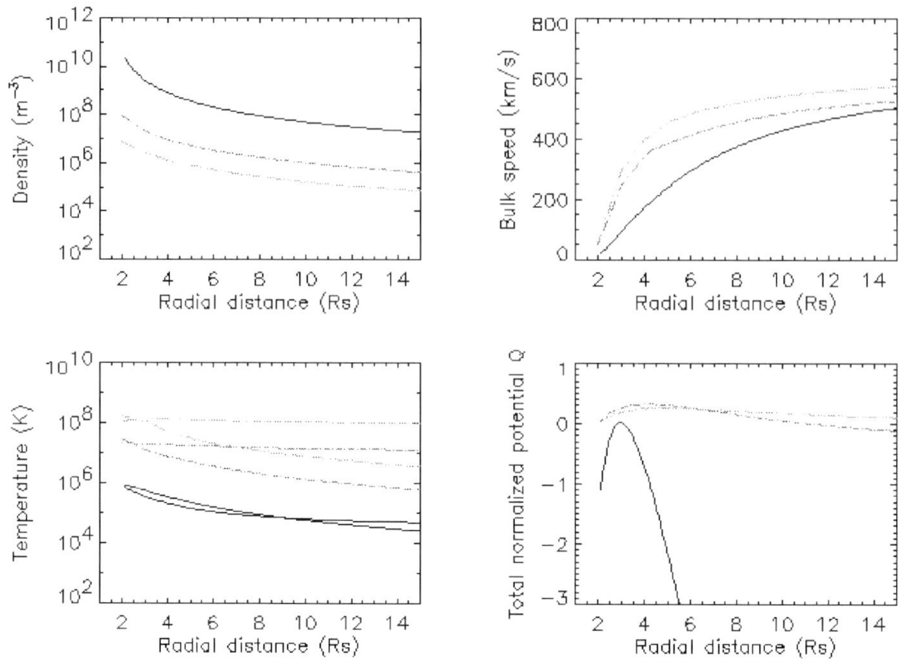

Fig. 4. Profiles of the FDV moments obtained with an exospheric model assuming an exobase at 2 Rs, κ=3 for the electrons, $T_p=T_e=10^6$ K, $T_{He}=5.10^7$ K and $T_O=2\ 10^8$ K. The moments of the H^+ ions are represented by the solid black line, He^{2+} by the magenta dashed-dotted line and O^{6+} by the green three dots-dashed line.

deduced from spectroscopy give indeed very high values. Only some ion species have temperatures that can be observed in the solar corona. It is the case of O^{5+} for which a temperature of the order of 10^8 K at 2 Rs is estimated from SUMER on SOHO (Esser & Edgar, 2000).

The number density decreases in the same proportion for the different particle species independently on their mass or charge, so that the composition of the solar wind remains almost the same in the solar wind as in the corona in the exospheric models (see panel 1).

The perpendicular temperature becomes always lower than the parallel temperature at large radial distances (see panel 3), but this anisotropy can be reduced by assuming large $T_{perp}/T_{//}$ in the solar corona as it is observed by SUMER at least for oxygen ions and protons (Esser & Edgar, 2000). The inclusion of the particle interactions such as the Coulomb collisions can also lead to a lower anisotropy. The effects of particle interactions are discussed in the next section.

Different processes have been suggested to explain the very high ion temperatures observed in the solar wind and the corona. Among others, the velocity filtration effect, as suggested initially by Scudder (1992), is an interesting possibility: the presence of an enhanced population of suprathermal particles leads to ion temperatures more than proportional to their mass and to an anticorrelation between the temperature and the density of the plasma. Pierrard & Lamy (2003) showed with a model of the solar corona in hydrostatic equilibrium that VDFs of particles characterized by an enhancement of suprathermal particles lead to a filtration effect that predicts a large increase of the electron and ion coronal temperatures. The temperatures of the ions are more than proportional to their mass, with a small correction depending on their charge state.

Solar wind ion distribution functions have been measured by WIND for Ne, O and He (Collier et al., 1996). Such ion distributions are observed to be characterized by suprathermal tails like in many space plasmas. They have been fitted by Kappa functions with the best kappa parameters as low as $\kappa=2$. Moreover, the temperatures of the ions measured in situ in the high speed solar wind are observed to be more than proportional to their mass, in good agreement with the assumptions made in the model.

4. The interaction term

In exospheric models, we considered only the effects of the external forces. We have shown that the averaged values of the moments correspond quite well to what is observed in the solar wind. Nevertheless, the temperature anisotropies are too high and the truncated VDFs are quite different from the solar wind VDF observed in situ. The inclusion of the different interaction processes can help to better model the solar wind and obtain more realistic VDF for the solar particles.

4.1 The Fokker-Planck Coulomb collision term

The solar wind is neither a collision-dominated plasma nor a completely collisionless one. This means that neither the hydrodynamic approach, based on the Euler or Navier-Stokes approximations, nor the exospheric or pure collisionless approach are truly appropriate to model the global expansion of the solar wind.

To describe the expansion of the plasma flow out of the hot solar corona, the most appropriate way is to use the kinetic approach since the kinetic processes prevail in the solar corona and solar wind low density plasmas (Marsch, 2006). One has to solve the Fokker-Planck equation which describes the evolution of the VDFs of the particles with the radial distance. The solution of the Fokker-Planck equation is especially significant in the collisional transition region where the Coulomb collisions effects become less and less important with increasing height.

Different models have been developed to study the steady state electron VDF in the corona and at larger radial distances in the solar wind by solving the Fokker-Planck equation (Echim et al., 2011 for a review). Lie-Svendsen et al. (1997) and Lie-Svendsen & Leer (2000) solved the Fokker-Planck equation using a finite-difference scheme. Landi & Pantellini (2001, 2003) developed direct particle simulations with binary collisions but the high computational load constrained the mass ratio to be very high ($m_i/m_e=400$). Recent results of electron VDF simulated using this model are provided in Landi et al. (2010) to study the effects of Coulomb collisions beyond 0.3 AU.

The most sophisticated collisional model regarding the self-consistent collision term was developed by Pierrard et al. (1999) and Pierrard et al. (2001c). They considered binary Coulomb collisions with electrons and protons using the Fokker-Planck collision operator given by Hinton (1983):

$$\left(\frac{df}{dt}\right)_c = -\frac{\partial}{\partial \vec{v}} \cdot \left[\vec{A}f - \frac{1}{2}\frac{\partial}{\partial \vec{v}} \cdot (\ddot{D}f)\right] \tag{9}$$

where $\vec{A}$ is the dynamic friction vector and $\ddot{D}$ is the velocity diffusion tensor. These terms are given by Rosenbluth et al. (1957) or Delcroix & Pers (1994) in spherical coordinates.

Pierrard et al. (1999) solved the Fokker-Planck equation using a spectral method where the VDF is expanded in polynomials in velocity, pitch angle and space:

$$f(x,\mu,z) = \exp(-x^2)\left(\sum_{l=0}^{n-1}\sum_{s=0}^{N-1}\sum_{m=0}^{M-1} a_{lsm} P_l(\mu) S_s(x) L_m(z)\right) \tag{10}$$

where x is the dimensionless velocity normalized by the thermal velocity, μ is the cosinus of the pitch angle between the velocity vector and the magnetic field direction, z is the radial distance, $P_l(\mu)$ are Legendre polynomials, $S_s(x)$ are speed polynomials and $L_m(z)$ are displaced Legendre polynomials.

This spectral method is described in detail in Pierrard (2011b). It was developed and applied previously for the polar wind escaping from the terrestrial atmosphere (Pierrard et al., 1998).

The test electrons are submitted to the influence of the external forces (gravitation, electric and Lorentz forces) and collide with background particles. The VDF of the background electrons was assumed to be the same as the VDF of the test electrons with an iterative process. So, the self collisions are treated consistently by matching the velocity distribution functions of the test and background proton distributions using this iterative numerical method.

The Fokker-Planck model shows the transformation of the electron VDFs in the transition region between the collision-dominated region in the corona and the collisionless region at larger radial distances. While VDF observed by WIND at 1 AU was used as boundary conditions in Pierrard et al. (1999), Pierrard et al. (2001c) have chosen to focus on the region between 2 and 14 solar radii where the Coulomb collisions have the most important effects. The upper limit (14 Rs) corresponds to the region where the mean free path of the particles becomes larger than the density scale height. This range of distance is also interesting because both fluid and exospheric models place the acceleration region of the solar wind at low radial distances from the Sun.

Figure 5 illustrates the electron VDF obtained at 13 Rs with the Fokker-Planck model assuming exospheric conditions at the top of the transition region, i.e., that electrons above a certain escape velocity do not return. Due to this condition, the VDF obtained with the Fokker-Planck model is anisotropic and different from a displaced Maxwellian as used in the Euler (five moments) fluid approximation. The VDF is close to an isotropic distribution at low radial distances and become more and more anisotropic in the transition region. The solar wind model based on the solution of the Fokker-Planck equation produces a core close to a Maxwellian due to Coulomb collisions. The halo component due to the enhancement of suprathermal electrons is reproduced only if it is assumed in the boundary conditions. In this case, the halo component increases with the radial distance (Pierrard et al., 1999). The VDF is also aligned to the magnetic field due to the mirror force, forming the strahl component (Pierrard et al., 2001b).

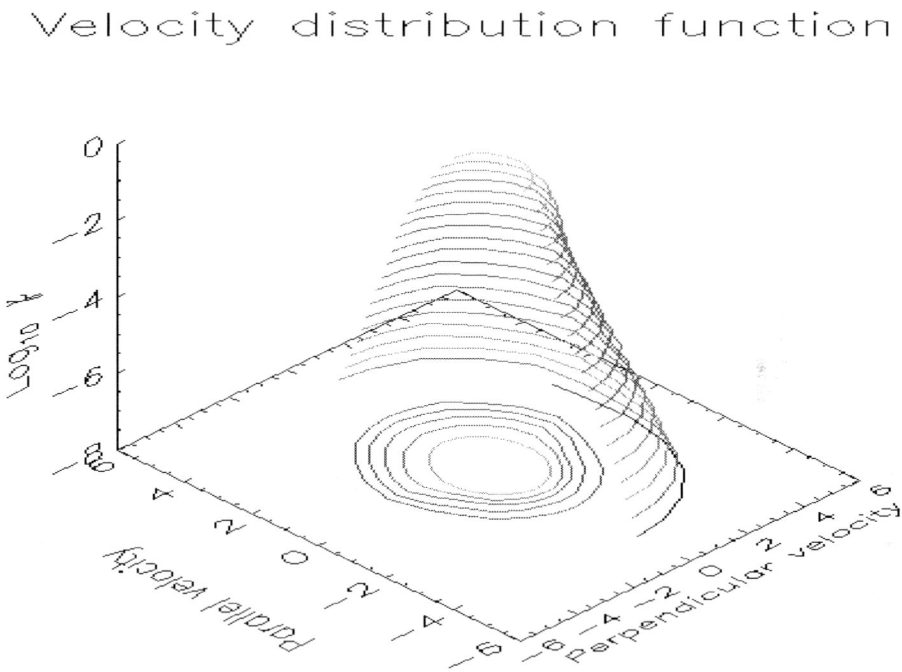

Fig. 5. Electron VDF obtained at 13 Rs with the Fokker-Planck model assuming exospheric conditions at the top of the transition region between collision-dominated and collisionless regions.

It is found that Coulomb collisions have important effects on angular scattering (i.e., on the pitch angle distribution of the electrons) without changing their average density and mean temperatures radial distributions. The boundary conditions have important effects on the solution of the Fokker-Planck equation.

Effects of the Coulomb collisions on the ion and proton VDF were also investigated by Marsch & Goldstein (1983) and Livi & Marsch, (1987). These authors showed that in the collisional regions like low-speed wind near the heliospheric current sheet, the Coulomb collisions can maintain an isotropic core, but an escaping tail is obtained in the major part of the high speed solar wind that can be considered as collisionless. It has been proposed that the formation of proton beams in the acceleration region at heliocentric distances 1.1-3 solar radii can be explained by the proton collisional runaway, or by the mirror force acting on the forward part of the proton velocity distribution.

Even if the inclusion of Coulomb collisions improves the temperature anisotropies and the global shape of the solar particles distributions, significant differences are still obtained between the VDF resulting from the collisional models and the VDF observed in the solar wind. The presence of halo electrons and of proton VDF with an anisotropic core and a beam aligned to the magnetic field remain difficult to explain considering only the external forces and the Coulomb collisions. This shows that other mechanisms exist and have to be taken into account.

4.2 The whistler turbulence for the electrons

Resonant interaction with whistler waves in the solar corona and the solar wind was suggested by Vocks & Mann (2003) and Vocks et al. (2008) to explain the generation of suprathermal electrons. Introducing antisunward-propagating whistler waves into a kinetic model in order to provide diffusion, Vocks et al. (2005) have shown that the whistler waves are capable of influencing the solar wind electron. Vocks (2011) developed a kinetic model for whistler wave scattering of electrons in the solar corona and wind.

Pierrard et al. (2011) have also recently considered the wave-particle resonant interactions in the plasma wave turbulence. These authors showed that the turbulent scattering mean free path is lower than the Coulomb collision mean free path in the solar wind, so that the turbulence mechanism is dominant compared to the Coulomb collisions at sufficiently large radial distances. The quasi-linear wave-particle scattering is again described by the Fokker-Planck equation, using the appropriate diffusion coefficients. They used the Fokker-Planck term determined by Schlickeiser (1989) in the presence of wave turbulence:

$$\left(\frac{df}{dt}\right)_{wp}=\frac{\partial}{\partial\mu}\left(D_{\mu\mu}\frac{\partial f}{\partial\mu}+D_{\mu p}\frac{\partial f}{\partial p}\right)+\frac{1}{p^2}\frac{\partial}{\partial p}p^2\left(D_{p\mu}\frac{\partial f}{\partial\mu}+D_{pp}\frac{\partial f}{\partial p}\right) \tag{11}$$

where p is the particle's momentum and the diffusion coefficients are given by Steinacker & Miller (1992). The electrons are assumed interacting with right-handed polarized waves in the whistler regime. They are considered in uniform fields and a superposed turbulent whistler wave spectrum. Only the slab modes propagating parallel to the interplanetary magnetic field are invoked because the energy exchange with oblique waves is expected to be less significant.

Figure 6 illustrates the electron VDF obtained at 190 Rs with the model including the whistler turbulence and using a typical electron VDF observed at 1 AU by WIND as boundary condition. An increase of the parallel temperature shows that turbulence can reduce the temperature anisotropy found in previous models neglecting this effect. Pierrard et al. (2011) found also that the acceleration of electrons in the solar wind remains mainly due to the electrostatic potential like in exospheric models. Nevertheless, wave turbulence determines the electron pitch-angle diffusion and can be responsible of the formation of the suprathermal tails observed in the solar wind. This is in good agreement with the results of Shizgal (2007) who showed that the VDF tends to Maxwellian only in presence of Coulomb collisions and in the absence of wave-particle interactions. When wave-particle interactions are included, an initial distribution tends to a steady state VDF with nonequilibrium (non-Maxwellian) tails and associated to an increase of the entropy (Leubner, 2004). The role of parallel whistlers can also extend to small altitudes in the acceleration region of the outer corona, where they may explain the energization and the presence of suprathermal electrons.

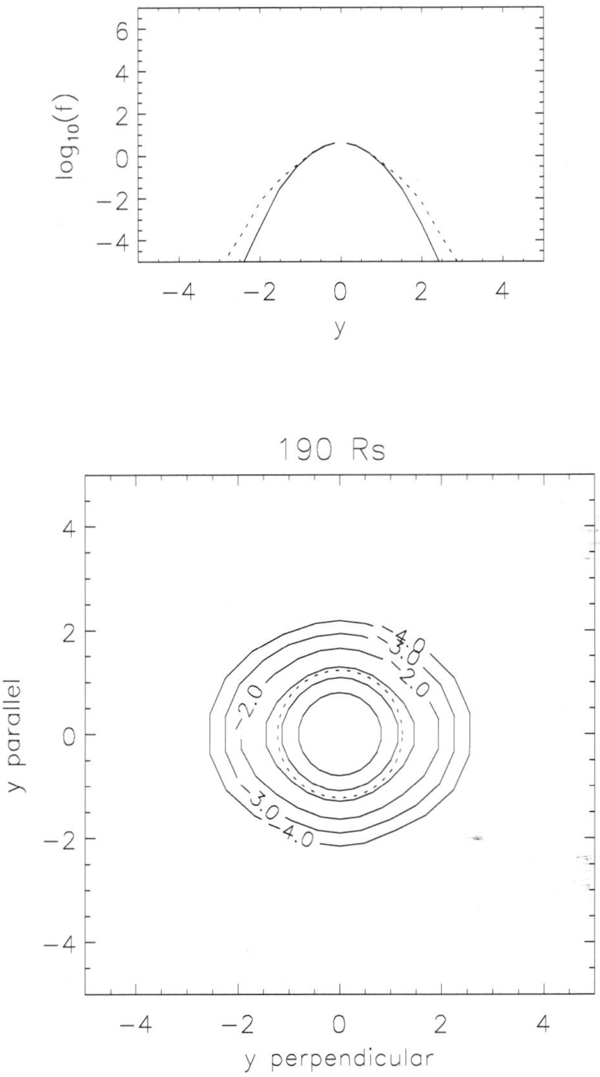

Fig. 6. Electron VDF obtained at 190 Rs with the model including the whistler turbulence (Pierrard et al., 2011).

4.3 Kinetic Alfven waves for the formation of the proton beam

Typical proton VDFs observed in the solar wind are characterized by an anisotropic core with $T_{perp} > T_{\backslash\backslash}$ and a proton beam aligned with the magnetic field direction (Marsch et al. 1982). Neither exospheric nor collisional Fokker-Planck models can fully explain these characteristics. In view of kinetic Alfven waves (KAW) activity observed in the solar wind, Pierrard & Voitenko (2010) proposed KAW to play a crucial role for the proton beam formation.

Recent investigations suggest that the wave-particle interactions in the solar wind can be dominated by the waves with short cross-field wavelengths, i.e. in the wavelength range of kinetic Alfven waves (Howes 2008, and references therein).

The generation mechanism for proton beams based on the proton trapping and acceleration by KAWs implies a flux of MHD Alfven waves converting into KAWs linearly and/or nonlinearly. When the proton gyroradius/cross-field wavelengths ratio becomes sufficiently small, the parallel potential well carried by the KAWs traps a fraction of core protons and accelerates them along the background magnetic field. This is illustrated in Fig. 7. The acceleration of trapped protons is caused by the accelerated KAW propagation under the condition that the normalized cross-field wave vector increases. There are several mechanisms in the solar wind, which can increase perpendicular wave number, like phase mixing in shear plasma flows, or in cross-field plasma inhomogeneities or turbulent cascades.

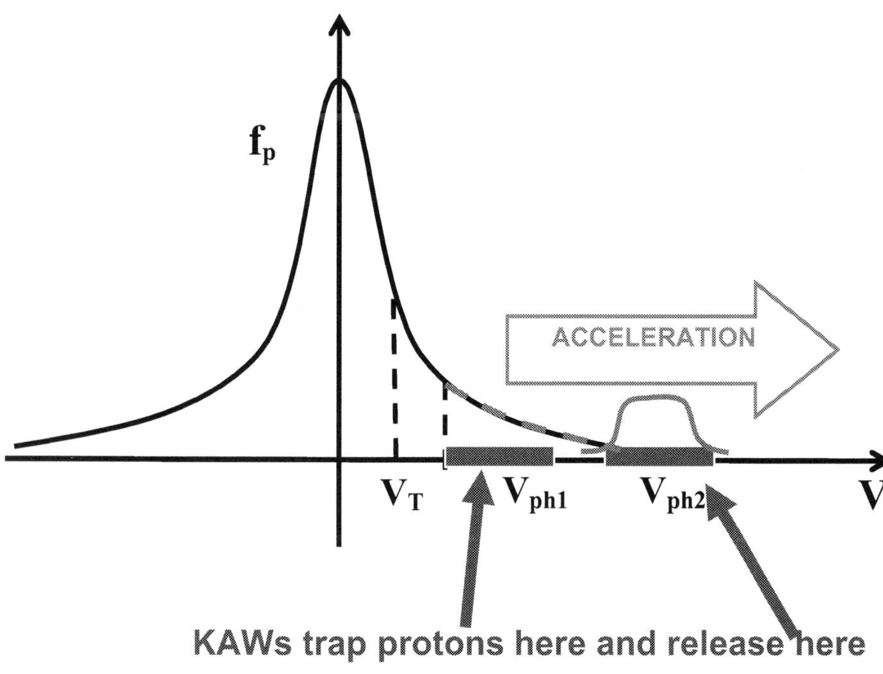

Fig. 7. Illustration of proton VDF obtained with proton trapping and acceleration by KAWs (Pierrard & Voitenko, 2010).

The interaction term that is used to simulate the quasi-linear Fokker-Planck diffusion due to kinetic Alfven wave turbulence is given by (Pierrard & Voitenko 2012):

$$\left(\frac{df}{dt}\right)_{QL} = \frac{\partial}{\partial v}(D_{2D}^{QL})\frac{\partial f}{\partial v} \tag{12}$$

where the proton diffusion coefficient is

$$D_{2D}^{QL} = \frac{\pi q^2}{2m^2}\sum_k \delta(\omega_k - k_{//}v)J_0\left(k_\perp \rho \frac{v_\perp}{v_T}\right)|E|^2 \tag{13}$$

where q and m are respectively the charge and the mass of the proton, $k_{//}$ and $k_\perp$ are the parallel and perpendicular components of the wave vector, ρ is the gyroradius, v_T is the thermal velocity, δ is the Dirac's delta function due to the resonant character of the wave-particle interaction, J_0 is the zero-order Bessel function and E is the parallel component of the electric field.

The model implies a flux of MHD Alfven waves converting into KAWs linearly and/or nonlinearly. When the proton gyroradius/cross-field wavelengths ratio becomes sufficiently small, the parallel potential well carried by the KAWs traps a fraction of core protons and accelerates them along the background magnetic field.

4.4 Other interactions

There are also several alternative possibilities to explain the origin of the proton beams (Tu & Marsch, 2002 and references therein). They can be injected in the solar wind by magnetic reconnection events at the coronal base. The beams can also be produced gradually by the evolution of the proton velocity distributions under the action of wave-particle interactions in the extended region in the solar wind from the acceleration region to r > 0.3 AU where they are observed. Tu & Marsch (2002) proposed also a mechanism for the proton beam formation based on the proton-cyclotron resonant interaction with cyclotron modes that can exist in the presence of alpha particles drifting with Alfven velocity.

Araneda et al. (2009) have studied parametric instabilities driven by Alfven-cyclotron waves and their influence on solar wind ions. It was shown that product waves generated by these instabilities can lead to a selective heating and acceleration of different ion species, generating in particular a proton beam with drift speed of about the Alfven speed. However, as there are pros and contras against all mentioned above mechanisms, there is no definite consensus about the physical mechanism producing proton beams.

KAW were also suggested as a possible source for the strong heating of ions across the magnetic field in the solar corona (Voitenko & Goossens, 2004). Cranmer (2002) has also studied different processes suggested to explain the very high ion temperatures in the corona, and specifically the ion-cyclotron waves with short wavelengths along the background magnetic field. Isenberg et al. (2010b) explained the energization of minor ions in the coronal holes by multiple cyclotron resonances in the presence of dispersive ion cyclotron waves. Solar coronal heating by Alfven wave turbulence was proposed by Bigot et al. (2010).

Wave-particles interactions are in any case supposed to be important in the evolution of solar wind ion distribution functions (Matteini et al., 2010). The quasilinear effects of resonant wave-particle interaction under condition of imbalanced turbulent heating in the collisionless coronal hole were investigated by Isenberg et al. (2010a). Other instabilities can also be taken into account, like the interplay between Weibel and firehose instabilities in coronal and solar wind outflows (Lazar et al., 2010).

5. Conclusion

The acceleration of the solar wind particles can be modeled using the hydrodynamic or the kinetic representations (Parker, 2010). The kinetic approach permits to analyze the effect of each physical process separately. The exospheric models emphasize the effects of the external forces, and especially the effect of the electrostatic potential that accelerates the solar wind particles outwards. Even if such models give already good average values of the solar wind lower order moments, the variety of particle velocity distributions observed in the solar wind cannot be explained by only one mechanism. By including different interaction terms such as Coulomb collisions and wave-particle interactions in the kinetic evolution equation, we can determine their effects on the VDF of the solar wind particles.

Kinetic plasma models including Coulomb collisions with proper boundary conditions are capable in reproducing solar wind speeds, number densities and temperatures compatible with observations. Whistler turbulence can explain the halo suprathermal population of electrons observed in the solar wind VDF. Salient features such as core temperature anisotropies and ion beams propagating with super-Alfven velocities require an additional energy source. An obvious source for that is provided by Alfven waves carrying energy fluxes enough for additional cross-field ion heating and beam production. A new mechanism for the proton beam production via proton trapping and acceleration by KAWs was proposed.

6. Acknowledgment

The research leading to these results has received funding from the European Commission's Seventh Framework Program (FP7/2007-2013) inside the grant agreement SWIFF (project n°2633430, www.swiff.eu). V. Pierrard thanks the STCE (Solar-Terrestrial Center of Excellence) and BISA for their support.

7. References

Araneda, J. A.; Maneva, Y. & Marsch, E. (2009). Preferential heating and acceleration of alpha particles by Alfven-cyclotron waves. *Phys Rev Lett.*, 102(17):175001, PMID: 19518788

Bigot, B.; Galtier, S. & Politano, H. (2010). Solar coronal heating via Alvfen wave turbulence. American institute of Physics, *AIP Conference Proceedings Solar Wind 12*, Volume 1216, pp.48-51

Collier, M. D.; Hamilton, D. C.; Gloeckler, G.; Boschler, P. & Sheldon, R. B. (1996). Neon-20, Oxygen-16, and Helium 4 densities, temperatures, and suprathermal tails in the solar wind determined by WIND/MASS. *Geophys. Res. Lett.*, 23, 1191-1194

Cranmer, S. R. (2002). Coronal holes and the high-speed solar wind. *Space Sci. Rev.*, 101, 229-294

Delcroix, J.-L. & Pers, A. (1994). Physique des plasmas 2, Savoirs Actuels, InterEditions/CNRS Editions, Paris, 499p.

Dorelli, J. C. & Scudder, J. D. (1999). Electron heat flow carried by Kappa distributions in the solar corona. *Geophys. Res. Lett.*, 23, 3537-3540

Echim, M.; Lemaire, J.& Lie-Svendsen, O. (2011). A Review on Solar Wind Modeling: Kinetic and Fluid Aspects. *Surveys in Geophysics*, 32 (1), pp. 1-70

Esser, R. & Edgar, R. J. (2000). Reconciling spectroscopic electron temperature measurements in the solar corona with in-situ charge state observations. *Astroph. J. Lett.*, 532, L71-74

Hinton, F. L. (1983). Collisional transport in plasma, in *Basic Plasma Physics I and II*, A. A. Galeev and R. N. Sudan (Ed.), North Holland, New York, pp. 148-200

Howes, G. G.; Cowley, S. C.; Dorland, W.; Hammett, G. W; Quataert, E. & Schekochihin, A. A. (2008). A Model of Turbulence in Magnetized Plasmas: Implications for the Dissipation Range in the Solar Wind. *J. Geophys. Res.*, 113, CiteID A05103

Isenberg, P. A.; Vasquez, B. J.; Chandran B. D. G. & Pongkitiwanichakul, P. (2010a). Quasilinear wave "reflection" due to proton heating by an imbalanced turbulent cascade. American institute of Physics, *AIP Conference Proceedings Solar Wind 12*, Volume 1216, pp. 64-67

Isenberg, P. A.; Vasquez, B. J. & Cranmer, S. R. (2010b). Modeling the preferential acceleration and heating of coronal holes O5+ as measured by UVCS/SOHO, American institute of Physics, *AIP Conference Proceedings Solar Wind 12*, Volume 1216, pp. 56-59

Issautier, K.; Meyer-Vernet, N.; Pierrard, V. & Lemaire, J. (2001). Electron temperature in the solar wind from a kinetic collisionless model: application to high-latitude Ulysses observations, *Astrophys. Space Sci.*, 277, 2, 189-193

Lamy, H.; Pierrard, V.; Maksimovic, M. & Lemaire, J. (2003). A kinetic exospheric model of the solar wind with a non monotonic potential energy for the protons. *J. Geophys. Res.*, 108, 1047-1057

Landi, S., & Pantellini, F. G. E. (2001). On the temperature profile and heat flux in the solar corona: Kinetic simulations, *Astronomy and Astrophys.*, 372, 686-701

Landi, S., & Pantellini, F. G. E. (2003) Kinetic simulations of the solar wind from the subsonic to the supersonic regime., *Astronomy and Astrophys.*, 400, 769-778

Landi, S.; Pantellini, F. & Matteini, L. (2010). Radial evolution of the electron velocity distribution in the heliosphere: role of the collisions. American institute of Physics, *AIP Conference Proceedings Solar Wind 12*, Volume 1216, pp.218-221

Lazar, M.; Poedts, S. & Schlickeiser, R. (2010). Nonresonant electromagnetic instabilities in space plasmas: interplay of Weibel and firehose instability. American institute of Physics, *AIP Conference Proceedings Solar Wind 12*, Volume 1216, pp. 280-283

Lemaire, J. F. (2012). Determination of coronal temperatures from electron density profiles. Submitted to *Solar Physics*

Lemaire, J. & Pierrard, V. (2001). Kinetic models of solar and polar winds. *Astrophys. Space Sci.*, 277, 2, 169-180

Leubner, M. P. (2004). Core-halo distribution functions: a natural equilibrium state in generalized thermostatistics. *The Astrophys. J.*, 604, 469-478

Lie-Svendsen, O.; Hansteen, V. H. & Leer, E. (1997). Kinetic electrons in high-speed solar wind streams: formation of high-energy tails. *J. Geophys. Res.*, 102, 4701

Lie-Svendsen, O. & Leer, E. (2000). The electron velocity distribution in the high-speed solar wind: Modeling the effects of protons. *J. Geophys. Res.*, 105, 35-46

Livi, S. & Marsch, E. (1987). Generation of solar wind proton tails and double beams by Coulomb collisions. *J. Geophys. Res*, 92, 7255

Maksimovic, M.; Pierrard, V. & Lemaire, J. (1997a). A kinetic model of the solar wind with Kappa distributions in the corona. *Astron. Astrophys.*, 324, 725-734

Maksimovic, M.; Pierrard, V. & Lemaire, J. (2001). On the exospheric approach for the solar wind acceleration. *Astrophys. Space Sci.*, 277, 2, 181-187

Maksimovic, M., Pierrard, V. & Riley, P. (1997b) Ulysses electron distributions fitted with Kappa functions. *Geophys. Res. Let.*, 24, 9, 1151-1154

Marsch, E. (2006). Kinetic physics of the solar corona and the solar wind, *Living Rev. Solar Phys.*, 3, 1 (www.livingreviews.org/lrsp-2006-1)

Marsch, E. & Goldstein, H. (1983). The effects of Coulomb collisions on solar wind ion velocity distributions. *J. Geophys. Res.*, 88, A12, 9933-9940

Marsch, E.; Muehlhauser, K.-H.; Schwenn, R.; Rosenbauer, H.; Pilipp, W. & Neubauer, F. (1982). Solar wind protons: Three-dimensional velocity distributions and derived plasma parameters. *J. Geophys. Res.*, 87, 52-72.

Matteini, L.; Lansi, S.; Velli, M. & Hellinger, P. (2010). On the role of wave-particle interactions in the evolution of solar wind ion distribution functions. American institute of Physics, *AIP Conference Proceedings Solar Wind 12*, Volume 1216, pp.223-226

Parker, E. N. (2010). Kinetic and hydrodynamic representations of coronal expansion and the solar wind. American institute of Physics, *AIP Conference Proceedings Solar Wind 12*, Volume 1216, pp.3-7

Pierrard, V. (2011a). Solar wind electron transport: interplanetary electric field and heat conduction. *Space Science Review (solar wind)*, doi: 10.1007/s11214-011-9743-6

Pierrard, V. (2011b). A numerical method to determine the particle velocity distribution functions in space. in *Astronum2010 Proceedings*, Numerical Modeling of Space Plasma Flows, Astronomical Society of the Pacific Conference series, Edited by N. V. Pogorelov, E. Audit and G. P. Zank, vol. 444, 166-176

Pierrard, V.; Issautier, K.; Meyer-Vernet, N. & Lemaire, J. (2001a). Collisionless solar wind in a spiral magnetic field. *Geophys. Res. Lett.*, 28, 2, 223-226

Pierrard, V. & Lamy, H. (2003). The effects of the velocity filtration mechanism on the minor ions of the corona. *Solar Physics*, 216, 47-58

Pierrard, V.; Lamy, H. & Lemaire, J. (2004). Exospheric distributions of minor ions in the solar wind. *J. Geophys. Res.,* vol. 109, A2, A02118, p.1-13, doi: 10.1029/2003JA010069

Pierrard, V. & Lazar, M. (2010). Kappa distributions: theory and applications in space plasmas. *Solar Physics*, vol. 287, N° 1, 153-174, doi: 10.1007/s11207-010-9640-2

Pierrard, V.; Lazar, M. & Schlickeiser, R. (2011). Evolution of the electron distribution function in the wave turbulence of the solar wind. *Solar Phys.* 269, 2, 421-438, DOI 10.1007/s11207-010-9700-7

Pierrard, V. & Lemaire, J. (1996). Lorentzian ion exosphere model. *J. Geophys. Res.*, 101, 7923-7934

Pierrard, V. & Lemaire, J. (1998) A collisional kinetic model of the polar wind. *J. Geophys. Res.*, 103, A6, 11701-11709

Pierrard, V.; Maksimovic, M. & Lemaire, J. (1999). Electron velocity distribution function from the solar wind to the corona. *J. Geophys. Res.*, 104, 17021-17032

Pierrard, V.; Maksimovic, M. & Lemaire, J. (2001b). Core, halo and strahl electrons in the solar wind. *Astrophys. Space Sci.*, 277, 2, 195-200

Pierrard,V.; Maksimovic, M. & Lemaire, J. (2001c). Self-consistent kinetic model of solar wind electrons, *J. Geophys. Res.*, 107, A12, 29.305-29.312

Pierrard, V. & Voitenko, Y. (2010). Velocity distributions and proton beam production in the solar wind. American institute of Physics, *AIP Conference Proceedings Solar Wind 12,* Volume 1216, pp. 102-105

Pierrard, V. & Voitenko, Y. (2012). Formation of proton beams due to kinetic Alfven waves in the solar wind. Submitted to Solar Phys.

Rosenbluth, M. N.; McDonald, W. & D. L. Judd, D. L. (1957). Fokker-Planck equation for an inverse-square force. *Phys. Rev.*, 107, 1

Schlickeiser, R. (1989). Cosmic-ray transport and acceleration. I - Derivation of the kinetic equation and application to cosmic rays in static cold media. II - Cosmic rays in moving cold media with application to diffusive shock wave acceleration. *The Astrophysical Journal,* Part 1 vol. 336, 243-293, ISSN 0004-637X

Scudder, J. D. (1992). Why all stars possess circumstellar temperature inversions. *The Astrophys. J.*, 398, 319-349

Shizgal, B. D. (2007). Suprathermal particle distributions in space physics: kappa distributions and entropy. *Astrophys. Space Sci.*, 312, 227-237

Spitzer, L. Jr. & Härm, R. (1953). Transport phenomena in a completely ionized gas. *Phys. Rev.*, vol. 89, 5, 977-981

Steinacker, J. & Miller, J. A. (1992). Stochastic gyroresonant electron acceleration in a low-beta plasma. 1. Interaction with parallel transverse cold plasma waves. *Astrophys. J.*, 393, 764-781

Tu, C. Y. & Marsch, E. (2002). Anisotropy regulation and plateau formation through pitch angle diffusion of solar wind protons in resonance with cyclotron waves. *Journal of Geophysical Research-Space Physics,* 107, (A9) pp. SSH 8-1, CiteID 1291, DOI 10.1029/2002JA009264

Vocks, C. (2011). Kinetic models for whistler wave scattering of electrons in the solar corona and wind. *Space Sci. Rev.*, doi 10.1007/s11214-011-9749-0

Vocks, C. & Mann, G. (2003). Generation of suprathermal electrons by resonant wave-particle interaction in the solar corona and wind. *The Astroph. J.*, 593, 1134-145

Vocks, C.; Salem, C.; Lin, R. P. & Mann, G. (2005). Electron halo and strahl formation in the solar wind by resonant interaction with whistler waves. *The Astrophys. Journal*, 627

Vocks, C. & Mann, G. (2008). Formation of suprathermal electron distributions in the quiet solar corona. *Astron. Astroph.*, 480, 527-536

Voitenko, Y. & Goossens, M. (2004). Cross-field heating of coronal ions by low-frequency kinetic Alfven waves. *The Astrophys. J.*, 605: L149-L152

11

Suprathermal Particle Populations in the Solar Wind and Corona

M. Lazar[1], R. Schlickeiser[1] and S. Poedts[2]
[1]*Institute for Theoretical Physics, Institute IV: Space and Astrophysics, Ruhr-University Bochum, Bochum*
[2]*Centre for Plasma Astrophysics, Leuven*
[1]*Germany*
[2]*Belgium*

1. Introduction

Understanding and predicting the transport of matter and radiation in the solar wind and terrestrial magnetosphere is one of the most challenging tasks facing space plasma scientists today. Space plasmas are hot ($T > 10^5$ K) and poor-collisional (mean free path ~ 1 AU), and contain ample free kinetic energy of plasma particles. Kinetic effects prevail leading to wave fluctuations, which transfer the energy to small scales: wave-particle interactions replace Coulomb collisions and enhance dispersive effects heating particles and producing suprathermal (non-Maxwellian) populations.

The existence of suprathermal distributions of charged particles has been regularly confirmed at any heliospheric distance in the solar wind and near Earth space by the spacecraft missions since the early 1960s (Feldman et al., 1975; Fisk & Gloeckler, 2006; Maksimovic et al., 1997a; 2005; Montgomery et al., 1968; Pilipp et al., 1987). Both deviations from isotropy and from thermal equilibrium are in general well explained by the action of plasma wave fluctuations, which seem to be the best agent in the process of conversion and transfer of the free energy to suprathermal populations. Even for the quiet times, the dilute plasma in the solar wind does not easily reach thermal equilibrium because the binary collisions of charged particles are sufficiently rare. Distributions with high energy suprathermal tails are therefore expected to be a characteristic feature for any low-density plasma in the Universe.

Processes by which suprathermal particles are produced and accelerated are of increasing interest in laboratory an fusion plasma devices, where they are known as the *runaway* particles decoupled from the thermal state of motion, and for a wide variety of applications in astrophysics. The astrophysical phenomena generally appear to involve an abundance of suprathermal ions and electrons observed to occur in the interplanetary medium, and which provide information about their source, whether it is the Sun or from the outer heliosphere. Accelerated particles (including electrons, protons and minor ions $^4He^{+2}$, $^{16}O^{+6(+7)}$, $^{20}Ne^{+8}$) are detected in the quiet solar wind and terrestrial magnetosphere, and in the solar energetic particle (SEP) events associated with flares and coronal mass ejections (CMEs) during intense solar activity (see discussions and references in Lin (1998) and Pierrard & Lazar (2010)). A steady-state suprathermal ion population is observed throughout the inner heliosphere with a velocity distribution function close to $\propto v^{-5}$ (Fisk & Gloeckler, 2006), and, on the largest scales, the relativistic cosmic-ray gas also plays such a dynamical role through the galaxy and its halo (Schlickeiser, 2002).

Solar wind distributions of charged particles comprise two different populations: a low energy thermal core and a suprathermal halo, and both are isotropically distributed at all pitch angles (Maksimovic et al., 2005; Montgomery et al., 1968). In the fast solar wind, the halo distribution can carry a magnetic field aligned strahl population, which is highly energetic and usually antisunward moving (Marsch, 2006; Pilipp et al., 1987). Suprathermal particles are present not only in electron distribution functions (Fig. 1) but in many ion species including H^+, He^{++}, and the heavier ions (Ne, N, O) present in the solar wind (Chotoo et al., 2000; Collier et al., 1996).

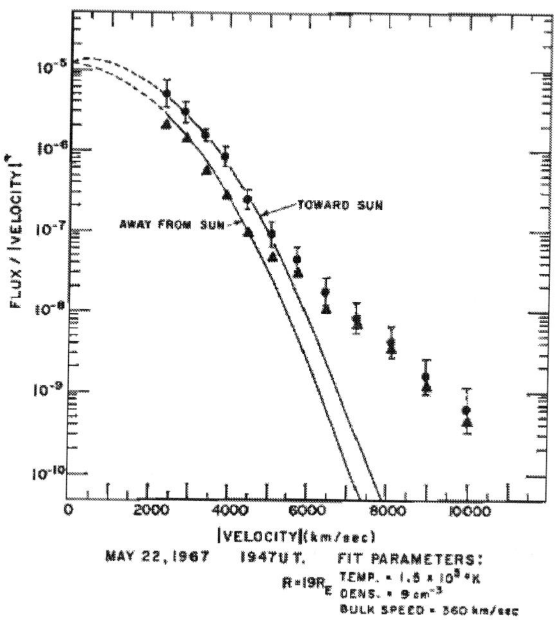

Fig. 1. Empirical distribution functions of the solar wind electrons measured at $\sim$ 1 AU by the Vela 4B satellite, and the Maxwellian fit (solid lines) to the low-energy data. Non-Maxwellian tails of the suprathermal electrons are clearly visible (after Montgomery et al. (1968)).

In this chapter we review the observational inventory of suprathermal populations, and present the theoretical and numerical techniques used to interpret and explain their origin in the solar wind, corona and some planetary environments. Recently, a short review of the theories and applications of Kappa particle distribution functions, has been provided by Pierrard & Lazar (2010), which only partly covered these aspects. The many spacecraft missions and observational reports of the last decades have greatly enhanced our knowledge about the existence of suprathermal populations in the solar wind. But their origin is still an open question because the multitude of models proposed to explain the acceleration of suprathermal particles are not always correlated with the observations. Understanding these mechanisms is essential for understanding key processes like heating in the corona and acceleration in the solar wind.

2. The observational evidence

The solar wind is the first, and up to now, the only stellar outflow to have been measured *in-situ* revealing important clues about the existence and the evolution of suprathermal particles, and their intimate connection to plasma wave turbulence and solar events.

2.1 The power-law Kappa model

The Kappa (or Lorentzian) function is a power-law generalization

$$f_\kappa(v) = \frac{1}{(\pi w_\kappa^2)^{3/2}} \frac{\Gamma(\kappa+1)}{\Gamma\left(\kappa-\frac{1}{2}\right)} \left(1+\frac{v^2}{\kappa w_\kappa^2}\right)^{-(\kappa+1)} \quad (1)$$

$$w_\kappa^2 = \left(1-\frac{3}{2\kappa}\right)\left(\frac{2k_B T}{m}\right) \quad (2)$$

introduced to describe suprathermal velocity distribution functions (Fig. 2) in plasmas out of thermal equilibrium (Maksimovic et al., 1997a; Vasyliunas, 1968). Here w_κ is the effective thermal velocity of the charged particles, m is the mass, n is the number density, T is the effective temperature, and $\Gamma(x)$ is the Gamma function. We call it a generalization because in the limit of a large power index, $\kappa \to \infty$ ($w_{\kappa\to\infty} = v_T = \sqrt{2k_B T/m}$), the power-law distribution function reduces to a Maxwellian, $f_\kappa(v) \to f_M(v)$ (red line in Fig. 2).

According to the form (1) of the Kappa distribution function, the power index κ must take values larger than a minimum critical value $\kappa > \kappa_c = 3/2$, for which the distribution function (1) collapses and the effective temperature is not defined. The distributions observed in the solar wind and terrestrial environments have been fitted with Kappas in this range. Furthermore, the velocity moments of the Kappa distribution functions can be defined only to the orders less than $\kappa + 1$, enabling a macroscopic description of the plasma with at least three moments of the Kappa distribution, the mean particle density, the streaming velocity and the effective temperature. The power index κ determines the slope of the energy spectrum of the suprathermal particles forming the tail of the velocity distribution function.

In collisionpoor or collisionless plasmas the independence of particles breaks down due to long-range correlations mediated by the plasma waves, and classical statistics cannot provide a basic derivation for Kappa distributions. Power-law distributions result instead from a new generalized statistics (Tsallis, 1995) for charged particles with long-range correlations supplied by the (quasi-stationary) field turbulence. Long-range correlations contribute to a super-extensive entropy (Leubner, 2002; Treumann & Jaroschek, 2008; Tsallis, 1995) implying generalization of the plasma temperature to non-thermal quasi-stationary states (Leubner, 2002; Livadiotis & McComas, 2009), and providing fundamental justification for Kappa distribution functions. Thus, the role of these distribution functions becomes of great significance not only in plasma physics and astrophysics, but generally, in electrodynamics and statistics.

When particle distributions measured in the solar wind are fitted with Kappa functions the power index κ for the electrons ranges from 2 to 5, and takes even larger values for protons

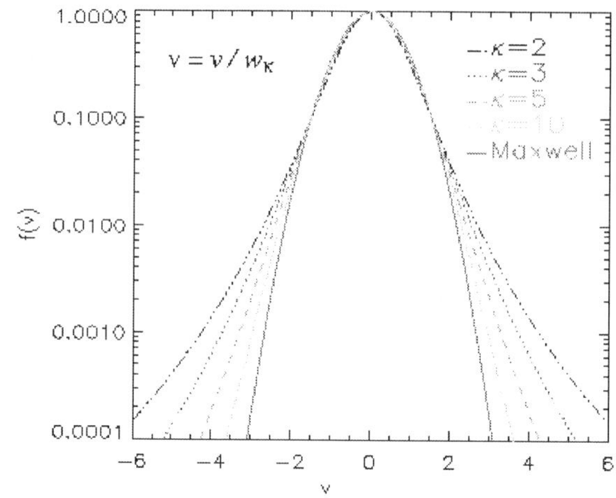

Fig. 2. Kappa distribution functions approach a thermal Maxwellian core at small velocities, less than thermal speed $v \ll w_\kappa$, but enhances showing superthermal tails at high energies, $v > w_\kappa$ (after Pierrard & Lazar (2010)).

and heavier ions (Pierrard & Lazar, 2010). As radial distance from the Sun increases and there are fewer collisions, the velocity distribution functions exhibit stronger suprathermal tails and values of the fitted power index κ decrease with distance, see Fig. 3 (Maksimovic et al., 2005). A sum of two Maxwellians has also been proposed to fit with the observations, but the best fit of the overall particle distribution including suprathermal tails is obtained using only one Kappa distribution function (Maksimovic et al., 1997a; Zouganelis et al., 2004). This needs, by comparison to two Maxwellians, a reduced number of macroscopic parameters (density, streaming velocity, temperature, and power index κ) to describe the plasma state.

Energy (or velocity) distribution functions of electrons and ions are frequently measured in space with electrostatic analyzers. These instruments measure three dimensional distributions, and can provide energy, mass and charge composition of the suprathermal populations. The quantity measured is the particle flux, namely the differential particle flux $J(W,\alpha,\vec{r})$ per unit area at given energy (W), pitch angle (α), and position ($\vec{r}$). This is the flux of particles measured in a certain energy interval, and which directly relates to the distribution function by

$$J(W,\alpha,\vec{r}) = \frac{v^2}{m} f(v_\parallel, v_\perp, \alpha, \vec{r}). \tag{3}$$

Measuring techniques have considerably been improved in the last decades. The quasi-thermal noise spectroscopy is a powerful tool for in situ space plasma diagnostics based on the analysis of the electrostatic field spectrum produced by the quasi-thermal fluctuations of electrons (Chateau & Meyer-Vernet, 1991; Le Chat et al., 2009; Zouganelis, 2008). The quasi-thermal noise is produced by the quasi-thermal fluctuations of the electrons and by the Doppler-shifted thermal fluctuations of the ions. The electron density and temperature are measured accurately since the quasi-thermal noise spectroscopy is less sensitive to the spacecraft perturbations than particle detectors. Measuring the noise peak just above the plasma frequency $\omega_p^2 = 4\pi ne^2/m$ allows an accurate evaluation of the plasma density n. In addition, since the peak shape strongly depends on the velocity distribution of electrons, the analysis of the spectrum reveals its properties, for instance, the temperature or the value of the power index κ when the fitting model is chosen to be a Kappa function. In contrast to a Maxwellian, for a power-law tail the peak is exactly at the plasma frequency. Despite the measuring constraints of the high-energy electron parameters (the plasma density fluctuations can broaden the spectral peak if the frequency/time resolution of the receiver is not high enough) the kappa parameter deduced with data from Ulysses is mainly in the range 2 - 5 indicating the presence of conspicuous suprathermal tails and agreeing with the results by Maksimovic et al. (1997a), who fitted the SWOOPS electron distribution functions on Ulysses and found a Kappa within the same range of values (Zouganelis, 2008).

2.2 Suprathermal electrons in the solar wind and terrestrial magnetosphere

Suprathermal electrons observed at different helisopheric distances in the solar wind and in the planetary environments have been fitted with suprathermal Kappa distributions with a power index $2 < \kappa < 6$ (Gloeckler & Hamilton, 1987; Maksimovic et al., 1997a). Helios I and II (before 1985) and the Wind missions (after 1994) provided measurements of electrons between 0.3 AU and 1 AU, and Ulysses (before 2009) from large heliocentric orbits over 1 AU to 5 AU. In terrestrial magnetosphere the first reports on the existence of suprathermal electrons came from the OGO missions in '60s-'70s, and then from the Cluster (after 2000) and Wind missions for all distinct regions including the plasmasheet, the magnetosheath and the radiation belts. Data provided by Helios, Ulysses, Cassini, Wind and the *Hubble Space Telescope* have also shown the existence of suprathermal electrons in the magnetosphere of other planets, like Mercury, Jupiter, Saturn, Uranus, Neptune, Titan (see Pierrard & Lazar (2010) and references therein).

The electron distributions are directly or indirectly measured in the solar wind and corona and show a global anticorrelation between the solar wind bulk speed and the value of the parameter κ: the high speed streams with important bulk velocities are emitted out of coronal regions where the plasma temperature is lower, and the low speed solar wind is originating in the hotter equatorial regions of the solar corona. Radial evolution of nonthermal electron populations in the low-latitude solar wind have been measured with Helios, Cluster, and Ulysses. The observations show a decrease of the relative number of strahl electrons with distance from the Sun, whereas the relative number of halo electrons is increasing (Maksimovic et al., 2005; Stverak et al., 2008). Further out in the solar wind, while the core density is roughly constant with radial distance, the halo and strahl densities vary in an opposite way (Maksimovic et al., 2005) indicating that suprathermal halo population consists partly of electrons scattered out of the strahl by broadband whistler fluctuations (Stverak et al., 2008).

Enhanced fluxes of suprathermal electrons have frequently been reported by the Ulysses mission upstream of co-rotating forward and reverse shocks in the solar wind at heliocentric distances beyond 2 AU (Gosling et al., 1993). The average duration of these events, which are most intense immediately upstream from the shocks and which fade with increasing distance from them, is ~ 2.4 days near 5 AU. The observations suggest that conservation of magnetic moment and scattering typically limit the sunward propagation of these electrons as beams to field-aligned distances of ~ 15 AU.

The solar wind halo includes electrons of energies over 100 eV to 1 keV, and is believed to originate from coronal thermal electrons which have a temperature of $\sim 10^6$ K (Lin, 1998). This spectrum referred to as the quiet time *primary* flux of suprathermal electrons appears to have a solar origin because it is measured in the absence of any solar particle events, and electrons of such energies are continuously escaping from the Sun being present in both the fast and slow solar winds. Primary fluxes of energetic ions are generally lower, possibly because at coronal temperatures ions are gravitationally bounded while electrons not.

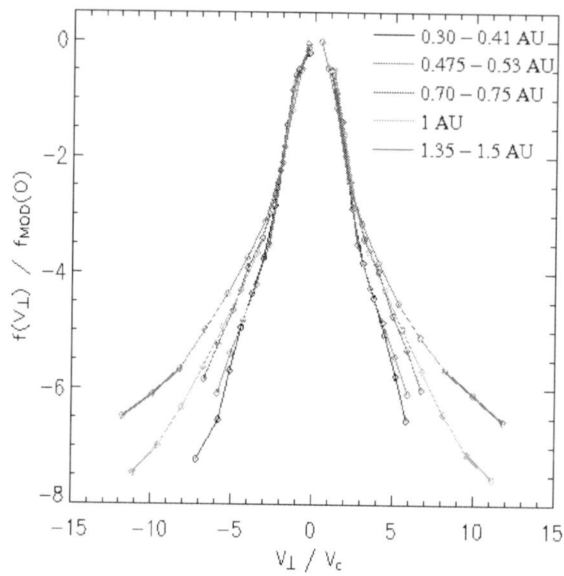

Fig. 3. Distribution function of the solar wind electrons for different heliospheric distances: the normalized core component remains unchanged at all radial distances while the relative number of halo electrons compared to the one of the core, increases with radial distance (after Maksimovic et al. (2005)).

The ion charge measurements stated by Ulysses in times of higher activity of the Sun, were found to be consistent with coronal Kappa distributions of electrons with kappa index ranging between 5 and 10. The presence of non-Maxwellian electron energy distributions in the solar transition region is suggested by the Si III line ratio from SUMER (Pierrard & Lazar (2010) and some references therein). Coronal origin of the energetic particles is also sustained by their traces in the emissions of solar flares and adjacent coronal sources (Kasparova & Karlicky, 2009). Observations in the corona suggest distributions having a nonthermal character that increases with altitude. Observations of electron suprathermal tails in the solar wind suggest their existence in the solar corona, since the electron mean free path in the solar wind is around 1 AU. Numerical computation of the velocity distribution function evolution from the profile measured at 1 AU back to the solar corona, supports a plausible coronal origin of the suprathermal populations (Pierrard et al., 1999).

There is also large observational evidence indicating that the formation of suprathermal fluxes is closely related to an enhanced activity of plasma waves and instabilities. The primary (coronal) flux of suprathermal particles is the source population for particles accelerated to higher energies by the impulsive solar energetic events at the site of flares, or by the gradual large SEP (LSEP) events in interplanetary CME-driven shocks (Ergun et al., 1998; Lin, 1998). Thus, a *secondary* flux of highly energetic called super-halo, electrons up to 2-100 keV and protons up to 10 MeV, will enhance the solar wind suprathermals which would have been originally accelerated in the corona and hence bear distinctive compositional and energetic features.

2.3 Suprathermal ions in the solar wind and terrestrial magnetosphere

The *Wind* mission has provided elemental and isotopic abundances for the solar wind ions near 1 AU, including temperatures, charge state distributions and reduced distribution functions, which extend well into the suprathermal tails (Collier et al., 1996). ^{4}He, ^{16}O, and ^{20}Ne distribution functions averaged over many days all appears well-fitted by Kappa functions (see Figure 4) with sufficiently small values of the power index $2.4 < \kappa < 4.7$. The average ^{4}He/^{20}Ne density ratio is 566 ± 87, but has significant variability with solar wind speed, and the average ^{16}O/^{20}Ne density ratio is 8.0 ± 0.6. The average ^{20}Ne/^{4}He and ^{16}O/^{4}He temperature ratios are close to unity at low solar wind speeds, but increase with the solar wind speed. Non-Maxwellian particles were also reported in H^+, He^{++}, and He^+

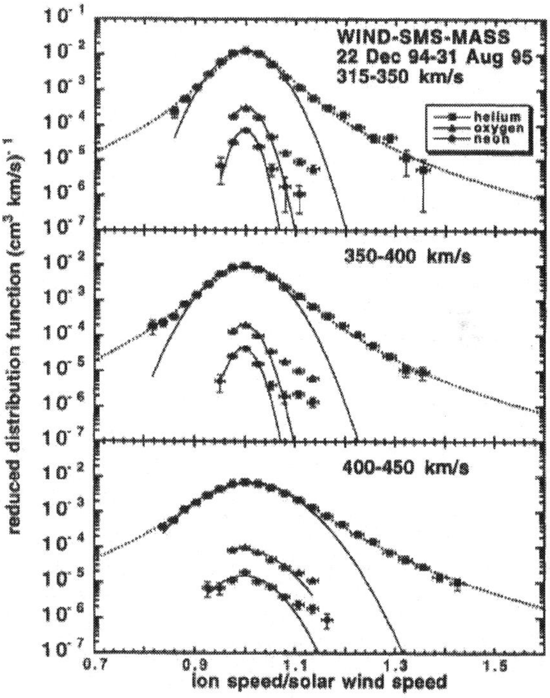

Fig. 4. Ion distribution functions measured at 1 AU in the solar wind (after Collier et al. (1996)).

distribution functions during co-rotating interaction region (CIR) events observed by *Wind* at 1 AU (Chotoo et al., 2000).

Primary fluxes of suprathermal ions recorded during a solar minimum are in general much lower than the electron fluxes (Lin, 1998), because at coronal temperatures the ions are gravitationally bounded while electrons not. For the same reason an ambipolar electric field (the Pannekoek-Rosseland field) is set up and extends into the solar wind accelerating protons outward and decelerating electrons. This potential varies inversely with distance from the Sun, with total drop of about 1 kV from the base of the corona to 1 AU (Lin, 1998).

Data from the plasma and magnetic field instruments on *Voyager* 2 indicate that non-thermal ion distributions could have key roles in mediating dynamical processes at the termination shock and in the heliosheath (Decker et al., 2008). The LECP (Low Energy Charged Particle) detectors on both *Voyager* 1 and 2 have established that a power law describes the heliosheath ions downstream of the termination shock in the energy range of $\sim$ 30 keV to 10 MeV (see Decker et al. (2008) and references therein). It is not known where exactly in the energy spectrum the power law tail begins, but the energy region below 30 keV (including the low range $\sim$ 0.01 - 6 keV) has been investigated by the IBEX (Interstellar Boundary Explorer) mission (Prested et al., 2008). IBEX was designed to make the first global image of the heliosheath beyond the heliospheric termination shock measuring the energetic neutral atoms (ENAs) created on the boundary of our solar system by charge exchange between downstream protons and interstellar hydrogen atoms, and that cannot be measured by conventional telescopes. The first simulated ENA maps of the heliosheath have used Kappa distributions of protons and calculate the ENAs that are traveling through the solar system to Earth. Considering suprathermal protons is well motivated by a significant increase of the ENA flux within the IBEX low-energy range $\sim$ 0.01 - 6 keV by more than an order of magnitude over the estimates using a Maxwellian (Prested et al., 2008).

The observations of the *ACE* and the *Ulysses* missions have revealed suprathermal tails that are always present within the integration times of the observations, and with intensities increasing in the presence of interplanetary shocks and other disturbed conditions. In the quiet-time conditions, suprathermal, power-law tails, known as quiet-time tails, are also present, and at 1 AU the tails include solar wind ions with a spectral index of -5 (see Fisk & Gloeckler (2006) and references therein). In the observations beyond 1 AU, the tails can be dominated by accelerated interstellar pickup ions.

For the time of an active Sun, the suprathermal flux of ions can be enhanced by the large solar energetic particle (LSEP) events, which produce significant fluxes of $\gtrsim$ 10 MeV protons. These events usually occur after an intense solar flare, and occasionally exhibit acceleration up to relativistic energies. In interplanetary space, LSEP events are associated with interplanetary CME-driven shocks. Electrons are also observed, but the fluxes of energetic protons dominate over electrons. Tens of LSEP events are detected per year near solar maximum Lin (1998). Direct observations confirm that, at least at some times, solar wind suprathermals can be augmented by suprathermals from LSEP events (Mason et al., 1995). One plausible scenario is that suprathermal ions (e.g., ^{3}He) remnant from impulsive events (at the flare site) may be a source population available for further acceleration by interplanetary shocks that accompany large SEP events, thereby leading to the ^{3}He enhancements in a significant fraction of large SEP events. There is also evidence of heavier suprathermal ions, e.g., Fe, remnant from flares and present in the source population of LSEP events (Tylka et al., 2001).

3. The existing models for the generation of suprathermal populations

Velocity distribution functions with suprathermal tails observed in the solar wind can be the result of their existence in the lower solar corona and the velocity filtration mechanism in gravitational and electrostatic (ambipolar) fields in the upper solar atmosphere (Scudder, 1992). Postulating that suprathermal particle distributions populate the transition region between the chromosphere and the corona, the temperature will increase with height through velocity "filtration" of the particle distributions in gravitational/electrostatic fields without invoking any local heating source. The temperature profile derived from the second order moment of such a Kappa distribution function is an increasing function of height (Pierrard & Lamy, 2003). The energy flux carried by high-energy electrons may be transferred into flow energy in the supersonic region of the flow leading to an enhancement of the energy density and the asymptotic flow speed of the solar wind. The velocity filtration model predicts the evolution of the electron velocity distribution function at higher altitudes in the solar wind with three distinct components, the core, halo, and strahl populations, similar to those observed in the interplanetary space (Viñas et al., 2000).

However, the velocity filtration mechanism does not address the question of how to generate and maintain the suprathermal populations in the solar wind and corona. In the chromosphere, the suprathermal electron distributions can be generated by the transit-time damping of the fast-magnetosonic waves (Roberts & Miller, 1998). This mechanism might operate reasonably well in the low-collisional region of the chromosphere, but it is a slow process and will be quenched rapidly in regions where collisional damping is important (Viñas et al., 2000). A realistic approach of the possible answers on the origin of suprathermal populations in the solar wind and corona requires a kinetic analysis of the coronal heating processes and the solar wind acceleration.

The anisotropy of the particle velocity distributions in stellar winds is basically controlled by the Chew-Goldberger-Low (CGL) mechanism: as the wind expands, the plasma density and magnetic field decrease radially, and if the particle motion is adiabatic and collisionless, the distributions become anisotropic in the sense that the pressure (or temperature) along the magnetic field exceeds the perpendicular pressure (or temperature, $T_\parallel > T_\perp$). In the case of violent outflows generating interplanetary shocks after solar flares or CMEs, injection of particle beams into the surrounding ionized interplanetary medium creates additional beam-plasma or temperature anisotropy.

These large deviations from isotropy quickly relax by the resulting wave instabilities, which act either to scatter particles back to isotropy or to accelerate them (Landau or cyclotron resonance) and maintain a superthermal abundance because thermalization is not effective at these time scales. The energy dissipation is a sustained interchange of energy between particles and plasma wave fluctuations, eventually lasting over sufficiently large time scales to be observed. The high rate of occurrence of an excess of perpendicular temperature ($T_\parallel < T_\perp$) at large radial distances in the heliosphere (Kasper at al., 2002; Stverak et al., 2008) is a proof that wave-particle interactions must be at work there dominating over the adiabatic expansion.

In low-collisional plasmas transport of matter and energy is governed by the selfconsistent correlation between particles and electromagnetic fields, which can, for instance, convect charged particles in phase space but are themselves created by these particles. The resulting Kappa distribution functions represent therefore not only a convenient mathematical tool, but a natural and quite general state of the plasma (Pierrard & Lazar, 2010). Suprathermal

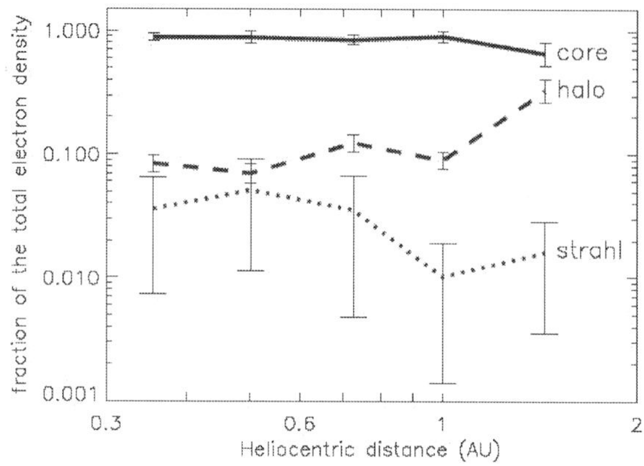

Fig. 5. Radial variations of the relative number density for the core (full line), halo (dashed line), and strahl (dotted line) populations, respectively, n_c/n_e, n_h/n_e, and n_s/n_e, where $n_e = n_c + n_h + n_s$. During the solar wind transport an important fraction of the strahl electrons may be diffused and transferred to the halo population (after Maksimovic et al. (2005)).

populations must therefore involve selfconsistently in both processes of wave turbulence generation and particle energization because the electromagnetic wave fields are the most efficient agent for charged particles energization. Stochastic acceleration of charged particles can account for both the momentum diffusion due to cyclotron or transit-time damping of the ambient electromagnetic fluctuations, and the momentum convection due to compression near shock waves in space (Schlickeiser, 2002). Unfortunately, none of these processes is well understood, mostly because these plasmas are poor-collisional and require progress in modeling the wave turbulence, going beyond MHD models to use a kinetic and selfconsistent description.

3.1 Models proposed for the acceleration of electrons

Looking to the observations reproduced in Fig. 5, the radial dependence in the solar wind shows an increasing number of the halo electrons, while the number of strahl electrons is decreasing with the distance from the Sun (Maksimovic et al., 2005). These observations suggest that the halo population consists partly of electrons scattered out of the strahl, and given the poor (particle/particle) collisionality of these populations, the mechanism that can explain their scattering is necessarily the interaction with plasma waves.

Cooling down of the suprathermal electrons originating in the corona due to a rapid adiabatic expansion in the interplanetary space would be also responsible for a decreasing in suprathermal populations, but in both components, the halo and the strahl, at the same time, and the effect should be an increasing of the core population. But such an evolution is not confirmed by the observations in Fig. 5, where the core relative number density remains roughly constant between 0.3 and 1.5 AU, that can be explained only by some some supplementary acceleration. Interactions with plasma waves would therefore be a plausible candidate to explain the scattering and isotropization of electrons from the strahl component (Pagel et al., 2007), as well as the acceleration of the low-energy electrons from the core (Ma &

Summers, 1999; Roberts & Miller, 1998), counterbalancing the cooling of suprathermals due to expansion.

Hasegawa et al. (1985) proposed the first wave-particle acceleration model showing that, although photons do not contribute directly to the velocity-space transport in a nonrelativistic plasma, an intense electromagnetic radiation can induce fluctuations in the Coulomb field enhancing the velocity space diffusion and producing a power-law distribution of electrons. The acceleration of plasma particles mechanism by wave-particle interaction is modeled within the Fokker-Planck formalism, where the velocity space diffusion is induced by the resonant Landau or cyclotron interactions.

Using the same kinetic formalism, but for a space plasma embedded into a stationary magnetic field, Ma & Summers (1999) have invoked the acceleration of electrons by the cyclotron resonance with whistlers. The turbulent spectrum of the whistlers has been assumed of Kolmogorov-type ($I \propto k_{\parallel}^{-q}$, $q = 5/3$) by extension of the wave-spectra observations from the inertial range. It was found that a very weak turbulence, with a very small power density of magnetic fluctuations $\delta B^2/B_0^2 < 10^{-7}$, can produce Kappa distributions of electrons with values of the index $\kappa = 2 - 8$ typically observed in the solar wind and magnetosphere.

Transit time damping of the low-frequency fast MHD oblique modes can also accelerate electrons and produce power law distributions in different conditions typical for solar flares and magnetosphere (Roberts & Miller, 1998; Summers & Ma, 2000). In the solar wind these waves arise in the acceleration region (the "reservoir") near the Sun, and interact and cascade to higher wavenumbers (the dissipation range) where eventually energize electrons out of an initially Maxwellian distribution. From the fast mode dispersion relation, the resonance condition can be written as $v_{\parallel} = v_A/\eta$ where v_A is the Alfvén speed, and $\eta = k_{\parallel}/k$. Hence, the threshold speed for the acceleration to occur is v_A, and electrons will therefore be preferentially energized, while only a negligible number of protons in the thermal tail can resonate (protons will, however, be accelerated less efficiently by Alfvén waves (Miller & Roberts, 1995)). Leubner (2000) has also shown that the main dynamics of the lower-hybrid and Alfvén waves-particle energy exchange due to Landau interaction, is regulated in a parallel direction with respect to the ambient magnetic field, providing the acceleration of electrons and the basis for the formation of (one-dimensional) Kappa distribution functions.

Such a model has systematically been invoked to explain the electron acceleration in a flaring solar coronal loop, where a spectrum of MHD waves is established by the primary energy release event, e.g., magnetic reconnection (De La Beaujardière & Zweibel, 1989; Miller, 1991). In such multiwave systems the stochastic acceleration leads to diffusion in the velocity space, and proceeds as long as the wave energy density exceeds the particle kinetic energy density. The behavior of a particle depends on whether or not it is in resonance with the wave spectrum, and the resonant particles follow chaotic trajectories and undergo diffusion in velocity space between wave effective potentials, thereby gaining energy (De La Beaujardière & Zweibel, 1989). The particle velocity spectrum can become very hard, with values of the power index κ below 4, in agreement with the observed non-thermal X-ray and radio emissions (Kasparova & Karlicky, 2009).

Primary fluxes of suprathermal populations originating in the corona and regularly observed in the quiet solar wind, can be reaccelerated to higher energies by the impulsive flares, or by the gradual large SEP events in the interplanetary CME fast shocks (Ergun et al., 1998; Lin, 1998) leading to the formation of a secondary (harder) flux of suprathermal populations. The electrostatic turbulence and the resulting electromagnetic decays (radio bursts) are produced by beams or counterstreaming plasmas and can be responsible for particle energization at these sites (Lin, 1998; Yoon et al., 2006). Marginally stable plateaued distributions coincide only occasionally with periods of local Langmuir emissions suggesting that competition of the electrostatic growing modes with whistlers and oblique mode instabilities may be important (Ergun et al., 1998).

Enhanced fluxes of suprathermal electrons reported by Ulysses beyond 2 AU seem to be caused by the leakage of shock heated electrons back into the upstream region of co-rotating forward and reverse shocks in the solar wind. These leaked electrons commonly counterstream relative to the normal solar wind electron heat flux. Although it seems unlikely that these shock-associated events are an important source of counterstreaming events near 1 AU, remnants of the backstreaming beams may contribute significantly to the diffuse solar wind halo electron population there (Gosling et al., 1993).

Strongly damped electrostatic modes are expected to be effective in processes of acceleration, but stochastic acceleration is typically slow due to the diffusive nature of the scattering process, and needs to be sustained. The nonlinear wave-wave and wave-particle interaction involving intense low-frequency Alfvén waves or electrostatic Langmuir and ion sound (weak) turbulence driven by the beam-plasma instabilities (Viñas et al., 2000; Yoon et al., 2006) can also be responsible for the acceleration in the corona, solar wind and magnetosphere. Instead of modeling it, Yoon et al. (2006) have proposed an advanced selfconsistent derivation of the wave intensity from a wave kinetic equation including spontaneous or induced emissions, wave-particle scattering and wave-decay processes for the electrostatic modes. In the upper regions of the solar atmosphere (in the presence of collisional damping) low-frequency, obliquely propagating electromagnetic waves can carry a substantial electric field component parallel to the mean magnetic field that can decay generating high-frequency plasma oscillations and low-frequency ion-acoustic waves. The resulting electrostatic modes would be rapidly damped out leading to the formation of suprathermal electron distributions near the Sun (Viñas et al., 2000). Due to large scale density fluctuations generated by the solar wind MHD turbulence, the Langmuir wave packets are trapped in the proton density holes leading to an efficient acceleration of electrons and formation of suprathermal tails (Califano & Mangeney, 2008; Yoon et al., 2006).

The nonlinear mechanisms of acceleration are usually investigated in numerical experiments, and the simulations indicate that the results of the quasilinear theory are applicable to the finite amplitude wave fluctuations typically observed in the solar wind and magnetosphere (Califano & Mangeney, 2008; Leubner, 2000; Roberts & Miller, 1998; Summers & Ma, 2000). It appears that the traditional quasi-linear theory provides a reasonable description of particle scattering and acceleration by a weak plasma turbulence. A quasilinear analysis is based on Boltzmann-Maxwell equations, and the resulting equations for the weak plasma turbulence include the kinetic equation for particle distributions (with Coulomb or wave-particle collisional terms) (Davidson, 1972; Yoon et al., 2006). While Coulomb collisions are not efficient in interplanetary space, the interaction with wave fields is the most promising leading to diffusion, pitch-angle scattering and acceleration of charged particles.

Two complementary kinetic models have been proposed to determine the heliospheric radial profiles of the distribution function and statistical moments starting either from coronal properties suggested by the observed emissions (Vocks & Mann, 2003; Vocks et al., 2005), or from more precise in-situ measurements at 1 AU in the solar wind (Pierrard et al.,

2011; 1999; 2001). The whistler waves eventually present in the corona are an important ingredient supporting coronal origin of the superthermal electrons observed in the solar wind. Based on this hypothesis the kinetic model developed by Vocks & Mann (2003) can explain the formation of the aligned antisunward moving strahl in the fast wind. If the whistler turbulence is further present in the solar wind, the antisunward waves reduce the anisotropy making superthermal halo present at all pitch angles, and the sunward moving waves scatter electrons out of the strahl into the halo (Vocks et al., 2005) as suggested by the observations (Maksimovic et al., 2005).

Otherwise, Pierrard et al. (2011) have shown that the solar wind plasma parameters measured at 1 AU, including the temperature and the anisotropy of the suprathermal electrons, can be a consequence of the whistler turbulence action at lower altitudes in the solar wind. Radial profiles of the electron-whistler scattering mean-free-path (mfp) suggest that at large heliospheric distances ($r > 0.5$ AU) whistler turbulence should in general be more efficient than Coulomb collisions leading to pitch-angle diffusion (temperature anisotropy) and electron acceleration (suprathermal tails). Collisions constrain at lower distances ($r \ll 0.5$ AU), but their role is usually limited to isotropization, while the same estimations of the wave-particle scattering mfp are still favorable to a possible contribution of whistlers to the electron energization in the active region of corona (Pierrard et al., 2011) giving valuable support to the complementary models (Kasparova & Karlicky, 2009; Vocks & Mann, 2003).

3.2 Models proposed for the acceleration of ions

Non-Maxwellian features of the solar wind proton distributions, including kinetic anisotropies and suprathermal tails, confirm expectations from resonant interaction with ion-cyclotron waves (see Marsch (2006) and references therein). In the initial impulsive phase of the solar flares, long-wavelength Alfvén waves are generated by restructuring of the magnetic field. These waves nonlinearly cascade to high wavenumbers and reach the dissipation range, where they can energize protons out the tail of the thermal distribution (Miller & Roberts, 1995). Then, these suprathermal protons are promptly accelerated to higher (relativistic) energies by the longer wavelength waves already present in the wave spectrum, and can produce the observed gamma-ray lines in impulsive solar flares.

Pierrard & Lamy (2003) have shown that the preferential heating of heavy ions relative to the protons, found in the empirical measurements, can be explained by the velocity filtration effect. The velocity filtration effect can account for the ion heating and the bulk acceleration of the solar wind particles without taking into account additional effect of wave-particle interactions. Moreover, the exospheric models show that with sufficiently high temperatures, the heavy ions are accelerated to high velocities in the low corona. Thus, the velocity filtration might contribute to the puzzling high temperatures observed in the corona and reduce the need for other heating mechanisms. While the source of coronal plasma waves is not always clear, the velocity filtration model needs suprathermal populations already present in the corona, an assumption that still waits for an observational confirmation and a theoretical explanation. A heuristic justification is that coronal suprathermal particles are collision poor due to the Coulomb collision cross section decreasing as $\propto v^{-4}$, and escape therefore more easily from the inner corona.

The first theoretical analyses of the quiet-time suprathermal ion population $\propto v^{-5}$ in the inner heliosphere have attributed their origin to the random compression and re-acceleration by the interplanetary wave turbulence (Fisk & Gloeckler, 2006). Recently Jokipii & Lee (2010) have reconsidered these theories showing that the compressive acceleration process does not produce power-law velocity spectra with indices less than (i.e., softer than) -3. Moreover, stochastic acceleration by a natural spectrum of Alfvén waves and oblique magnetosonic waves, yields comparable acceleration rates but also, do not produce power-law distributions with indices less than -3. Conversely, the process of diffusive shock acceleration, responsible for energetic storm particle events, co-rotating ion events and probably most large solar energetic particle (SEP) events, readily produces power-law velocity spectra with indices in a range including the observed -5. Consequently, the quiet-time suprathermal ion population will be composed predominantly of remnant ions from these events as well as a contribution from impulsive SEP events (Jokipii & Lee, 2010).

4. Some open questions and conclusions

Since the first reports on the existence of suprathermal populations in the solar wind and terrestrial environments, a significant progress has been made in many directions including the new modern techniques of observation and interpretation, and the large variety of theories and models proposed to explain particles acceleration and formation of suprathermal distributions. These are non-Maxwellian plasmas out of thermal equilibrium, and therefore expected to behave much different from the standard Maxwellian. Theoretically, the effects of suprathermal populations on the wave dispersion and stability properties have been extensively studied using Kappa distribution functions, isotropic or anisotropic, including or not drifts, and always making contrast with Maxwellian models (see (Pierrard & Lazar, 2010) and references therein).

Thus, establishing a realistic correlation between measurements of the suprathermal populations and the associated wave fluctuations detected at the same intervals of time and the same locations in the solar wind, would involve not only a technical progress, but will provide an important support for a correct understanding of the role of plasma waves in the process of acceleration. Distinction must be made between the wave fluctuations driven by various kinetic anisotropies of the suprathermal populations (like beams or temperature anisotropy), and those originating from other further sources in the solar wind, but passing through the same suprathermal sample at the time of observation.

The radio plasma imagers on board of satellites can stimulate such plasma emissions and echoes, known as plasma resonances, which are then reproduced on plasmagrams. Because these resonances are stimulated at the electron cyclotron frequency f_{ce}, the electron plasma frequency f_{pe}, and the upper-hybrid frequency $f_{uh} = (f_{pe}^2 + f_{ce}^2)^{1/2}$, they are measured to provide the local electron density and magnetic field strength. Calculations of these resonances using dispersion characteristics based upon a nonthermal Kappa distribution function appear to resolve the frequency discrepancy between these resonances in the magnetosphere and those predicted by a Maxwellian model (Viñas et al., 2005).

These calculations based upon an isotropic Kappa model simply suggest that departures from the standard Maxwellian provided by the dispersion/stability theory of the Kappa distributed plasmas should be introduced in the future techniques of evaluation and parametrization of the suprathermal populations and their effects in the solar wind. Moreover, accurate measurements of the kinetic anisotropies of plasma particles and the resulting wave fluctuations (Bale et al., 2009; Pagel et al., 2007; Stverak et al., 2008) can provide further support for theoretical modeling. The anisotropic Kappa distribution functions proposed to model kinetic anisotropies of the suprathermal particles are presently under debate because of the contradictory results provided by the existing models, namely, the bi-Kappa and the product-bi-Kappa functions (including or not drifts), with respect to the standard bi-Maxwellian (Lazar et al., 2010; 2011a). Correlating the empirical fitting data and the results from numerical experiments might be a valuable starting point for more comprehensive models developed to explain a full 3D formation of suprathermal populations in the solar wind.

The strahl component is the main contributor to the (electron) heat flux and the main contributor to the anisotropy of suprathermals, apparently a manifestation of the adiabatic focusing. Solar wind observations show that suprathermal strahl electrons have a width of the pitch angle distributions that decreases with increasing electron kinetic energy up to a few hundred eV (Pilipp et al., 1987), and becomes broader for more energetic electrons up to 1 keV (Pagel et al., 2007). Such a diversity implies a diversity of scattering mechanisms for the suprathermal strahl. The highly anisotropic strahl with $T_{\parallel} \gg T_{\perp}$ might be a driver of the firehose low-frequency wave instability, which can contribute to the electron scattering, but the firehose instability based on a bi-Maxwellian model requires a sufficiently large $\beta_{\parallel} > 1$, a condition rarely satisfied by the tenuous strahl. A simple bi-Kappa model incorporating all components, thermal and suprathermal, with the same averaged values for the particle density and the power index Kappa, requires even larger values of the plasma $\beta_{\parallel}$ (Lazar et al., 2011b). But a refined Kappa model accounting for a realistic, finite relative density of the halo and strahl components would probably provide better chances for this instability to develop confirming expectations from the observations, which indicate a constant presence of the corresponding magnetic field fluctuations in the solar wind (Bale et al., 2009; Stverak et al., 2008).

At smaller wavelengths, a second possible source of electron scattering are the whistler waves (Pagel et al., 2007), but their origin in the solar wind is not always understood. In the quiet solar wind conditions, the simulations show that enhanced whistler waves with finite damping lead to strahl pitch angle distributions which broaden in width with increasing kinetic energy, in agreement with observations, but at the same time, the strahl is broadened as a function of wave amplitude and relative strahl density (Saito & Gary, 2007). In addition, for the times the solar wind is disturbed by the electron radio bursts, the electrostatic electron/electron (two-stream) instability driven by a strahl with a large average speed leads to substantial strahl velocity scattering perpendicular to mean magnetic field. If the strahl speed is large compared to the halo thermal speed, after instability saturation, the width of the electron pitch angle distribution exhibits a maximum as a function of electron energy (Gary & Saito, 2007). Numerical experiments have also shown possibility to switch from focusing to scattering in the model of formation of the strahl/halo (including eventually, the superhalo) configuration due to a simple geometric effect of the Parker spiral magnetic field (Owens et al., 2008). Further out from the Sun, the pitch-angle scattering dominates because focusing of the field-aligned strahl is effectively weakened by the increasing angle between the magnetic field direction and intensity gradient, a result of the spiral field.

The results presented here emphasize the importance of studying these suprathermal plasma populations by not only the quantity of observations that attest their presence in all species of charged particles in the solar wind and terrestrial environments, but the exclusive and invaluable nature of informations about the natural plasmas out of thermal equilibrium. Understanding mechanisms by which suprathermal particles are produced and accelerated is essential for understanding key processes, like heating in the corona and acceleration in the solar wind. The central conclusion of our comparative analysis is that these topics clearly need further investigations to confront numerical simulations with the observations and make them consistent with dispersion and stability properties of the most appropriate suprathermal Kappa models.

5. References

Bale S.D., Kasper J.C, Howes G.G. et al. (2009). Magnetic fluctuation power near proton temperature anisotropy instability thresholds in the solar wind, *Phys. Rev. Lett.* 103, 211101.

Califano F. & Mangeney A., (2008). A one dimensional, electrostatic Vlasov model for the generation of suprathermal electron tails in solar wind conditions, *J. Geophys. Res.* 113, A06103.

Chateau Y.F. & Meyer-Vernet N., 1991. Electrostatic noise in non-Maxwellian plasmas: generic properties and "Kappa" distributions, *J. Geophys. Res.* 96, 5825.

Chotoo K, Schwadron N., Mason G., et al. (2000). The suprathermal seed population for corotating interaction region ions at 1 AU deduced from composition and spectra of H+, He++, and He+ observed on Wind, *J. Geophys. Res.* 105, 23107.

Collier M.R., Hamilton D.C., Gloeckler G. et al. (1996). Neon-20, oxygen-16, and helium-4 densities, temperatures, and suprathermal tails in the solar wind determined with WIND/MASS, *Geophys. Res. Lett.* 23, 1191.

Davidson R.C. (1972). *Methods in Nonlinear Plasma Theory*, Academic, New York.

Decker R.B., Krimigis S.M., Roelof E.C., et al. (2008). Mediation of the solar wind termination shock by non-thermal ions, *Nature* 454, 67.

De La Beaujardière J.-F. & Zweibel E.G. (1989). Magnetohydrodynamic waves and particle acceleration in a coronal loop, *Astrophys. J.* 336, 1059.

Ergun R.E., Larson D., Lin R.P. et al. (1998). Wind spacecraft observations of solar impulsive electron events associated with solar type-III radio bursts, *Astrophys. J.* 503, 435.

Feldman W.C., Asbridge J.R., Bame S.J. et al., (1975). Solar wind electrons, *J. Geophys. Res.* 80, 4181.

Fisk L.A. & Gloeckler, G. (2006). The common spectrum for accelerated ions in the quiet-time solar wind, *Astrophys. J.* 640, L79.

Gary S.P. & Saito S. (2007). Broadening of solar wind strahl pitch-angles by the electron/electron instability: Particle-in-cell simulations, *Geophys. Res. Lett.*, 34, L14111.

Gloeckler G. & Hamilton D.C. (1987). AMPTE ion composition results, *Phys. Scripta* T18, 73.

Gosling J.T., Bame S.J., Feldman W.C. et al. (1993). Counterstreaming suprathermal electron events upstream of corotating shocks in the solar wind beyond ~2 AU: Ulysses, *Geophys. Res. Lett.* 20, 335.

Hasegawa A., Mima K. & Duong-van N. (1985). Plasma distribution function in a superthermal radiation field, *Phys. Rev. Lett.* 54, 2608.

Hellinger, P.; Travnicek, P.; Kasper, J. C. & A. J. Lazarus (2006). Solar wind proton temperature anisotropy: Linear theory and WIND/SWE observations, *Geophys. Res. Lett.*, 33, L09101.

Jokipii J.R. & M.A. Lee (2010). Compression acceleration in astrophysical plasmas and the production of $f(v) \propto v^{-5}$ spectra in the heliosphere, *Astrophys. J.* 713, 475.

Kasparova J. & Karlicky M. (2009). Kappa distribution and hard X-ray emission of solar flares, *Astron. Astrophys.* 497, L13.

Kasper, J. C.; Lazarus, A. J. & Gary S. P. (2002). Wind/SWE observations of firehose constraint on solar wind proton temperature anisotropy, *Geophys. Res. Lett.*, 29, 1839.

Lazar. M., Schlickeiser, R., & Podts S. (2010). Is the Weibel instability enhanced by the suprathermal populations or not?, *Phys. Plasmas*, 17, 062112.

Lazar. M., Poedts, S. & Schlickeiser, R. (2011a). Instability of the parallel electromagnetic modes in Kappa distributed plasmas - I. Electron whistler-cyclotron modes, *Mon. Not. R. Astron. Soc.*, 410, 663.

Lazar. M., Poedts, S. & Schlickeiser, R. (2011b). Proton firehose instability in bi-Kappa distributed plasmas, *Astron. Astrophys.*, 534, A116.

Le Chat G., Issautier K., Meyer-Vernet N., et al. (2009). Quasi-thermal noise in space plasma: ŞkappaŤ distributions, *Phys. Plasmas*, 16, 102903.

Leubner M.P. (2000). Wave induced suprathermal tail generation of electron velocity space distributions, *Planet. Space Sci.* 48, 133.

Leubner M.P. (2002). A nonextensive entropy approach to kappa-distributions, *Astrophys. Space Sci.* 282, 573.

Lin R.P. (1998). Wind observations of suprathermal electrons in interplanetary medium, *Space Sci. Rev.* 86, 61.

Livadiotis G. & McComas D.J. (2009). Beyond kappa distributions: Exploiting Tsallis statistical mechanics in space plasmas, *J. Geophys. Res.* 114, A11105.

Ma C. & Summers D. (1999). Correction to "Formation of Power-law Energy Spectra in Space Plasmas by Stochastic Acceleration due to Whistler-Mode Waves", *Geophys. Res. Lett.* 26, 1121.

Maksimovic M., Pierrard V. & Riley, P. (1997). Ulysses electron distributions fitted with Kappa functions, *Geophys. Res. Let.* 24, 1151.

Maksimovic M., Pierrard V. & Lemaire J.F. (1997). A kinetic model of the solar wind with Kappa distribution functions in the corona, *Astron. Astrophys.* 324, 725.

Maksimovic M., Zouganelis I., J.-Y. Chaufray, et al. (2005). Radial evolution of the electron distribution functions in the fast solar wind between 0.3 and 1.5 AU, *J. Geophys. Res.* 110, A09104, doi:10.1029/2005JA011119.

Marsch E. (2006). Kinetic physics of the solar corona and solar wind, *Living Rev. Solar Phys.* 3.

Mason G.M., Mazur J.E. & Dwyer J.R. (1999). ^{3}He enhancements in large solar energetic particle events, *Astrophys. J.* 525, L133.

Miller J.A. (1991). Magnetohydrodynamic turbulence dissipation and stochastic proton acceleration in solar flares, *Astrophys. J.* 376, 342.

Miller J.A. & Roberts D.A. (1995). Stochastic proton acceleration by cascading Alfvén waves in impulsive solar flares, *Astrophys. J.* 452, 912.

Montgomery M.D., Bame S.J. & Hundhause A.J. (1968). Solar wind electrons: Vela 4 measurements, *J. Geophys. Res.* 73, 4999.

Owens J., Crooker N.U. & Schwadron N.A. (2008). Suprathermal electron evolution in a Parker spiral magnetic field, *J. Geophys. Res.* 113, A11104.

Pagel C., Gary S.P., de Koning C.A. et al. (2007). Scattering of suprathermal electrons in the solar wind: ACE observations, *J. Geophys. Res.* 112, A04103.

Pierrard V. & Lazar M. (2010). Kappa distributions: theory and applications in space plasmas, *Sol. Phys.* 267, 153.

Pierrard V., Lazar M. & Schlickeiser R. (2011). Evolution of the electron distribution function in the whistler wave turbulence of the solar wind, *Sol. Phys.* 269, 421.

Pierrard V., Maksimovic M. & Lemaire J.F. (1999). Electron velocity distribution functions from the solar wind to corona, *J. Geophys. Res.* 104, 17021.

Pierrard V., Maksimovic M. & Lemaire J.F. (2001). Self-consistent model of solar wind electrons, *J. Geophys. Res.* 106, 29,305.

Pierrard V. & Lamy H. (2003). The effects of the velocity filtration mechanism on the minor ions of the corona, *Solar Phys.* 216, 47.

Pilipp W.G., Miggenrieder H., Montgomery M.D., et al. (1987). Variations of electron distribution functions in the solar wind, *J. Geophys. Res.* 92, 1075.

Prested C., Schwadron N., Passuite J, et al. (2008). Implications of solar wind suprathermal tails for IBEX ENA images of the heliosheath, *J. Geophys. Res.* 113, A06102.

Roberts D.A. & Miller J.A. (1998). Generation of nonthermal electron distributions by turbulent waves near the Sun, *Geophys. Res. Lett.* 25, 607.

Saito S. & Gary S.P. (2007). Whistler scattering of suprathermal electrons in the solar wind: Particle-in-cell simulations, *J. Geophys. Res.*, 112, A06116.

Schlickeiser R. (2002). *Cosmic Ray Astrophysics*, Springer, Heidelberg.

Scudder J.D. (1992). On the causes of temperature change in inhomogenous low-density astrophysical plasmas, *Astrophys. J.* 398, 299.

Stverak, S.; Travnicek, P.; Maksimovic, M. et al. (2008). Electron temperature anisotropy constraints in the solar wind, *J. Geophys. Res.*, 113, A03103.

Summers D. & Ma C. (2000). Rapid acceleration of electrons in the magnetosphere by fast-mode MHD waves, *J. Geophys. Res.* 105, 15,887.

Treumann R.A. & Jaroschek C.H. (2008). Gibbsian theory of power-law distributions, *Phys. Rev. Lett.* 100, 155005.

Tsallis C. (1995). Non-extensive thermostatistics: brief review and comments, *Phys. A*, 221, 277.

Tylka A.J., Cohen C.M.S., Dietrich W.F., et al. (2001). Evidence for remnant flare suprathermals in the source population of solar energetic particles in the 2000 Bastille Day event, *Astrophys. J.* 558, L59.

Vasyliunas V.M. (1968). A Survey of low-energy electrons in the evening sector of the magnetosphere with OGO 1 and OGO 3, *J. Geophys. Res.* 73, 2839.

Viñas A.F., Wong H.K. & Klimas A.J. (2000). Generation of electron suprathermal tails in the upper solar atmosphere: implications for coronal heating, *Astrophys. J.* 528, 509.

Viñas A.F., Mace R.L. & Benson RF. (2005). Dispersion characteristics for plasma resonances of Maxwellian and Kappa distribution plasmas and their comparisons to the IMAGE/RPI observations, *J. Geophys. Res.* 110, A06202.

Vocks C. & Mann G. (2003). Generation of suprathermal electrons by resonant wave-particle interaction in the solar corona and wind, *Astroph. J.* 593, 1134.

Vocks C., Salem C., Lin R.P. & Mann G. (2005). Electron halo and strahl formation in the solar wind by resonant interaction with whistler waves, *Astroph. J.* 627, 540.

Yoon P.H., T. Rhee & C.-M. Ryu (2006). Self-consistent formation of electron κ distribution: 1. Theory, *J. Geophys. Res.* 111, A09106.

Zouganelis I., Maksimovic M., Meyer-Vernet N., et al. (2004). A transonic collisionless model of the solar wind, *Astrophys. J.* 606, 542.

Zouganelis I. (2008). Measuring suprathermal electron parameters in space plasmas: Implementation of the quasi-thermal noise spectroscopy with kappa distributions using in situ Ulysses/URAP radio measurements in the solar wind, *J. Geophys. Res.* 113, A08111.

Part 4

The Solar Wind Magnetic Field Powered by the Sun

12

Impact of the Large-Scale Solar Magnetic Field on the Solar Corona and Solar Wind

A.G. Tlatov[1] and B.P. Filippov[2]
[1]*Kislovodsk Mountain Station of the Central Astronomical Observatory of RAS at Pulkovo*
[2]*Pushkov Institute of Terrestrial Magnetism, Ionosphere and Radio Wave Propagation, Russian Academy of Sciences, Troitsk, Moscow Region*
Russia

1. Introduction

In 1955 Soviet astrophysicists Vsehsvyatskiy, Nikolskiy, Ponomarev and Cherednichenko (Vsekhsvyatskiy et al., 1955) showed that broad corona loses its energy for radiance, and can be in hydrodynamic equilibrium, and there should be a flow of materials and energy.

This process is a physical basis for the important phenomenon of "dynamic corona". The magnitude of the flow of materials was evaluated due to the following considerations: if the corona were in hydrodynamic equilibrium, then the altitudes of homogenous atmosphere for hydrogen and iron would correlate as 56/1. In other words, in such case iron ions must not be observed in the distant corona. But this is not so. In 1955 it was a considerable achievement, but nobody believed in the phenomenon of "dynamic corona".

Three years later Eugene N. Parker came to the conclusion that hot solar stream in Chapman model and particle flux, blowing away commentary tails in Birmann's hypothesis – these are manifestation of the same phenomenon, and Eugene N. Parker called it "solar wind".

Parker (Parker, 1958) showed - despite the fact that solar corona is greatly gravitated to the Sun, it is a strong heat conductor, it remains hot even at great distance. The farther the distance, the less the solar gravitation is, there is a supersonic discharge from the upper corona into interplanetary space.

Solar wind represents a flux of ionized particles, thrown out of the Sun in all the directions with the speed about 300-1200 km/sec. The source of the solar wind is solar corona. The temperature of the solar corona is so high, that gravitation force is not able to hold its substance near the surface, and part of this substance constantly moves to interplanetary space.

First direct gaging of the solar wind was carried out in 1959 by the automatic interplanetary station "Luna-1". The observations were made by means of a scintillometer and a gas ionization detector. Three years later the same gaging was implemented by the American scientists on board the station "Mariner-2".

First numerical models of solar wind in corona with using the equations of magnetofluid dynamics were created by Pneuman and Kopp in 1971 (Pneuman & Kopp, 1971).

In the end of the 90s there were observations of the areas of uprise of fast solar wind in solar poles, those observations were made on board of the satellite SOHO by means of Ultraviolet Coronal Spectrometer (UVCS). It turned out that acceleration of the wind is much higher than it was presupposed, judging from purely thermodynamic extension. Parker's model predicted that the wind speed becomes hypersonic in 4 solar radii altitude from photosphere, but the observations showed that this transition takes place significantly lower, approximately in 1 solar radius altitude, confirming the existence of the extra mechanism of acceleration of the solar wind.

Long-term observations from the orbit of the Earth (about 150 000 000 km distance from the Sun) showed that the solar wind is structured and can usually be classified as steady and perturbative (sporadic and recurrent).

Depending on the speed, sporadic streams of solar wind can be divided into slow (approximately 300—500 km/sec near the orbit of the Earth) and fast (500—800 km/sec near the orbit of the Earth) (Fig. 1).

Slow solar wind is generated by the "steady" part of the solar corona (area of coronal streamers) with its gas-dynamic extension: with the temperature of corona about 2×10^6 K - cjrona cannot be in the condition of hydrostatic equilibrium, and this extension under the present border conditions must lead to the acceleration of the corona substance up to hypersonic speeds.

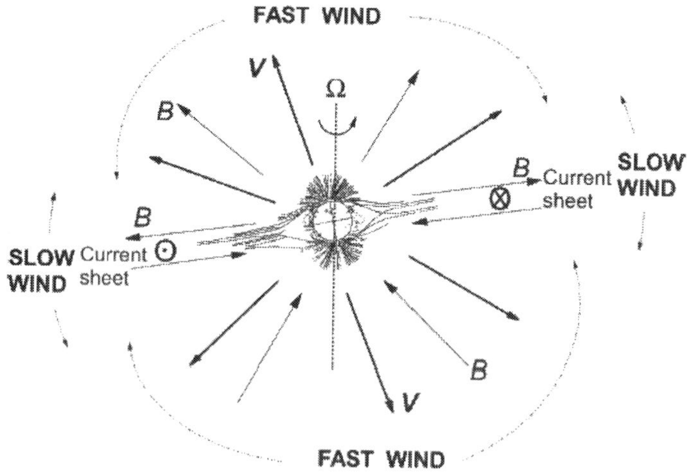

Fig. 1. Simplified picture of the large-scale structure of the solar wind near sunspot minimum, when the solar magnetic dipole makes a small angle with the spin axis (dotted line). The velocity and field lines are sketched in bold and thin lines respectively. The magnetic polarity is the one that existed when the WIND and Ulysses observations were acquired; this polarity reverses every 11 years (Meyer-Vernet, 2007).

Solar corona heating up to such temperatures takes place due to convectional nature of heat transfer in the solar photosphere: the development of the convective turbulence in plasma is accompanied by the generation of intensive magnetoacoustic waves. In its turn, during transmission towards solar atmosphere density reduction, sound waves are transforming into shock waves; shock waves are effectively absorbed by the corona substance and heat it till the temperature of $(1-3)\times10^6$ K. The streams of recurrent fast solar wind are emanated by the Sun during several months and have a return period of 27 days (the period of the rotation of the Sun) when being observed from the Earth. These streams are associated with coronal holes - the areas of the corona with relatively low temperature (approximately $0{,}8\times10^6$ K), low-density of plasma (that is all in all a quarter of density of the sporadic corona areas) and radial in relation to the Sun magnetic field. Owing to high conductivity of the solar wind plasma magnetic field of the Sun turns out to be frozen-in into the flowing out wind streams, and can be observed within the interplanetary atmosphere as an interplanetary magnetic field.

Because of the solar wind the Sun loses about one million tons of its substance every second. The solar wind consists basically of electrons, protons hellions (alpha-particles); nuclei of other elements and non-ionized particles (electrically neutral) are contained in rather inconspicuous quantity.

2. Long-term changes of coronal shape and geomagnetic disturbance

Solar wind parameters are changing during the solar activity. Their direct determination has begun relatively recently, with the beginning of space age. But the observations of the geomagnetic disturbance and coronal shape give us the opportunity to evaluate variations of the solar wind throughout the period of more than 100 years.

The large-scale solar corona structure corresponds to the large-scale configuration of solar magnetic fields. Since the magnetic field of the Sun is subjected to cyclic variations, the coronal shape also changes cyclically. Processing 12 photographs of the corona during solar eclipses, Ganskiy (Ganskiy, *1897*) classified 3 types of corona, i.e., maximum, intermediate, and minimum. In 1902, in the report concerning the solar eclipse of 1898, Naegamvala (*Naegamvala,* 1902) also gave a corona classification that depended on the sunspot activity. Description of the corona shape involves the use of characteristic features and the phase of solar activity that is given by $\Phi = \dfrac{T - T_{\min}}{|T_{\max} - T_{\min}|}$. The values of Φ are positive and negative at the rising and declining branches of the solar cycle. Vsekhsvyatskiy (Vsekhsvyatskiy et al., 1965) gave a somewhat different classification of the structure types. They are (1) a maximum type $\Phi > 0.85$ in which polar-ray structures are not seen, large streamers are observed at all heliolatitudes and are situated radially; (2) an intermediate premaximum or postmaximum type $0.85 > |\Phi| > 0.5$, in which polar-ray structures are observed at least in one hemisphere, and large coronal streamers situated almost radially are clearly seen at high latitudes; (3) an intermediate preminimum or postminimum type $0.5 > |\Phi| > 0.15$ in which polar-ray structures are clearly seen in both hemispheres and large coronal streamers strongly deviate toward the solar equator plane; (4) a minimum type $0.15 > |\Phi|$ in which polar-ray structures are clearly seen in both hemispheres and large coronal streamers are parallel to the equator plane; and (5) an ideally minimum type $0.05 > |\Phi|$ in which powerful structures of large coronal streamers are situated along the equator. Changes in the extent of the polar-ray structures, the degree of corona flattening, the average angle between large coronal streamers, and other characteristics of the corona that depend on the solar cycle phase have been widely studied (Loucif and Koutchmy, 1989, Vsekhsvyatskiy et al., 1965, Golub & Pasachoff, 2009).

Solar eclipses of 2006 Mar 26, 2008 Jul 22, and 2009 Aug 1 have enabled detailed examination of changes in the shape of the corona in the period of minimum solar activity in the modern era. However, the shape of the corona in the current minimum is slightly different from the ideal shape of the eclipses in the minimum of activity.

This section contains the comparative analysis of the solar corona structure during the minimums of activity cycles 12-24, and their changes throughout the centuries are discussed.

2.1 Processing method and results

The initial data in the analysis were drawings of the corona shape taken from the catalogs of (Vsekhsvyatskiy et al., 1965) (see Fig. 3, 1-8), Loucif and Koutchmy, 1989) (Fig. 3, 9) and also drawings of the eclipses at minima of cycles 21 and 23 taken from Waldmeier (Waldmeier, 1976) (Fig. 3, 10), Gulyaev (Gulyaev, 1997) (Fig. 2, 11) (Eclipse photo of 01.08.2008 with permission of M. Druckmüller), (Fig. 2, 12). (Vsekhsvyatskiy et al., 1965) separated out an ideally minimum corona. It is supposed that the cycle phase must be $|\Phi| < 0.05$. In fact, the ideally minimum corona type was observed on in 1954 (see Fig. 3). This happened at the solar activity minimum before the largest one in the history of observation cycle 19. Most probably, such a corona type did not occur for 50 years before and after this event. The corona shape of 1954 was close to a dipole one. This means that large coronal streamers rapidly approach the solar equator plane. However, during other eclipses, such as 1889 Jan 1 ($\Phi = -0.18$), 1889 Dec 21 ($\Phi = 0.03$), 1901 May 17 ($\Phi = -0.07$), 1923 Sep 10 ($\Phi = 0.04$), and 1965 May 30 ($\Phi = 0.14$), the large coronal streamers at distances more than 2 solar radii do not lie on the equator plane. They expand from mid-latitudes in parallel or at a slight angle to the equatorial plane (see Fig. 3). To analyze the corona shape at the eclipses during a solar activity minimum epoch, a corresponding index should be chosen. We need the index that characterizes the shape of the corona of the minimal type and is applicable to images and drawings of different qualities. The corona of the solar minimum is characterized by pronounced polar ray structures and large coronal streamers (Pasachoff et al., 2008). Let us introduce the parameter γ that characterizes the angle between high-latitude boundaries of the large coronal streamers at a distance of $2 \cdot R_O$. The γ parameter is the sum of the angles at the eastern and western limbs: $\gamma=180-(\gamma_W+\gamma_E)$. Figure 2 gives the scheme showing how parameter γ is determined from the shapes of the corona in Fig. 3, and the error does not exceed 4°.

The parameter γ is close to the index of the extent of the polar regions (Nikolskiy 1955; Loucif & Koutchmy, 1989; Golub, 2009). But in comparison with the polar index (Loucif & Koutchmy, 1989), it is not calculated close to the limb, but at a height R_O above the solar limb. This allows taking the compression of coronal rays to the plane of which is the equator into account, possibly linked with non-radial spreading of the coronal streamers. This is especially significant for the minimum activity epoch (Tlatov, 2010a).

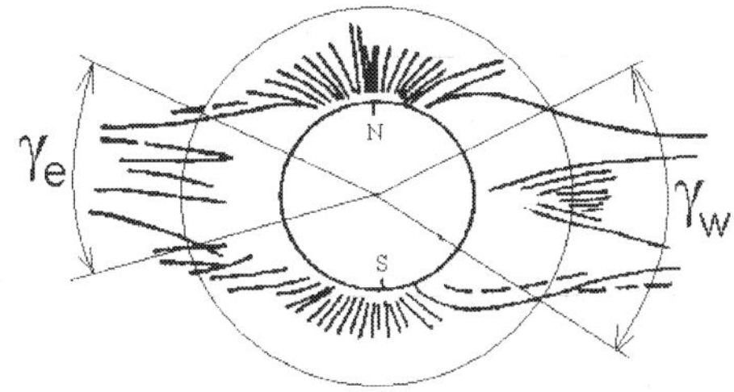

Fig. 2. Scheme showing how the angles defining parameter γ =180-(γ_W +γ_E) are found for the eclipse of 1923. The outer thin circle has a radius of $2R_o$

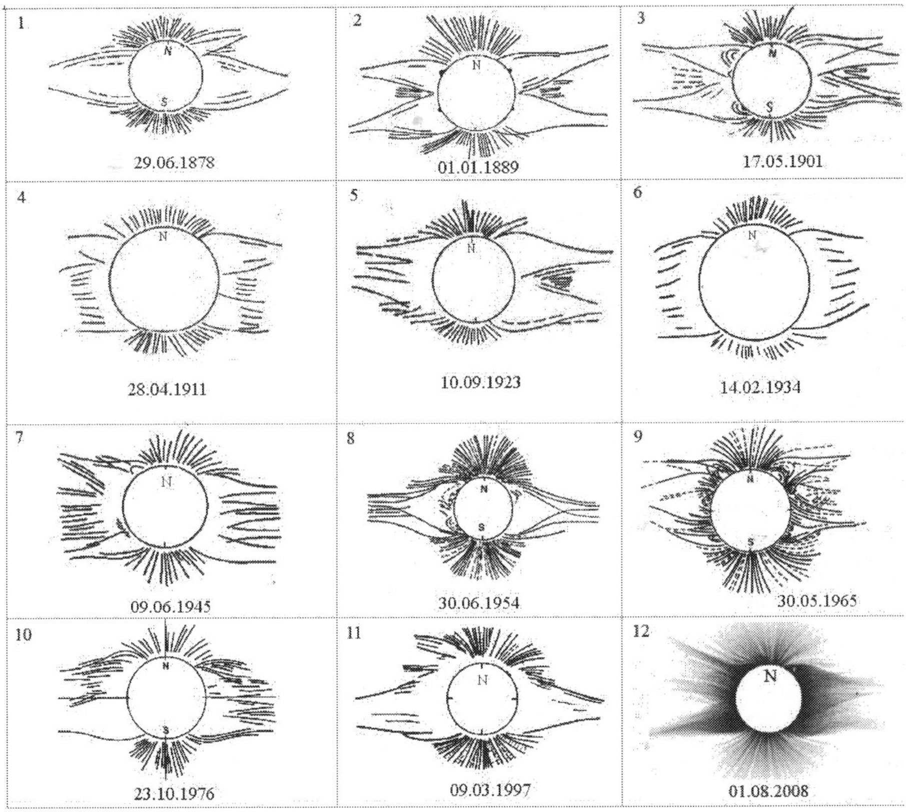

Fig. 3. Eclipses close to the minima of cycles 12--24.

Figure 3 presents the shapes of eclipses for the epochs close to the solar activity minima of cycles 12-24. The calculated values of the parameter γ for these eclipses are listed in Table 1. The γ parameter varies within 40 ÷ 100 degrees. Table 1 also gives the solar cycle phase Φ (see Eq. 1). One can see that the highest magnitudes of the parameter γ occurred during the

Cycle No	Date	W	Φ	γ
12	29 June 1878	75	-0.06	65
13	21 December 1889	88	0.03	60
14	17 May 1901	63	-0.07	80
15	28 April 1911	103	-0.18	80
16	10 September 1923	77	0.04	85
17	14 February 1934	114	0.14	97
18	09 June 1945	151	0.28	88*
19	30 June 1954	190	0	98
20	30 May 1965	106	0.14	90
21	23 October 1976	155	0,08	67
22	22 November 1984	158	-0.35	69*
23	09 March 1997	125	0.1	60
24	01 August 2008	--	~0.1	54

Table 1. Parameters γ for the eclipses of cycles 12 - 24. W - the amplitude of the Wolf number at the sunspot maximum, Φ- the phase of solar activity for eclipse date. For cycle 18 and 22 γ was modified taking the phase of the cycle into account

period 1934--1955. The remaining magnitudes fit the enveloping curve fairly well with a maximum during cycles 17--19 (see Fig. 4). No information on the solar corona structure during the eclipses at the minima of cycles 18 and 22 has been found in the literature. To fill the gaps, the eclipses during the phases of growth or decline of the solar cycle can be used.

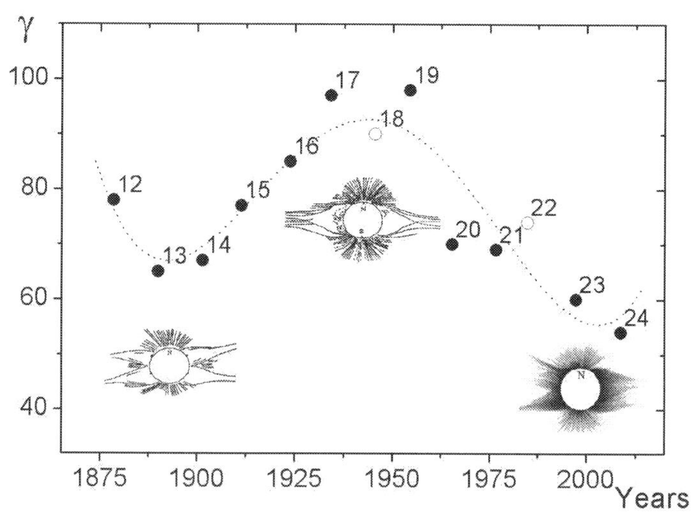

Fig. 4. Distribution of parameter γ for the structure of the corona of the minimal type. Numbers of activity cycles are given. The shape of the solar corona to cycles 13,19 and 24 are present.

One can see in Table 1 that the eclipses of 1945 and 1984 are rather far from the minimum phase in solar activity. A modified parameter $\gamma^{*}=180-(\gamma_W+\gamma_E)\cdot(1-|\Phi|)$ can be introduced for these eclipses. This parameter reduces parameter γ to the minimum phase. The application of this procedure is effective for recognizing the shape of the corona close to the minimum activity with the phase $\Phi<0.4$. Figure 4 presents variations in parameter γ during the last 13 activity cycles.

2.2 Long-term changes of coronal shape and geomagnetic disturbance

The presence of long-term trends in the solar corona structure can be caused by changes in the configuration of the global magnetic field of the Sun. The role of active region formation during a solar activity minimum is not significant. It has long been known that large coronal streamers typically lie above the polarity-inversion lines of the large-scale magnetic field marked by filaments and prominences (Vsekhsvyatskiy et al., 1965). For this reason, investigations of the large-scale corona shape give valuable information on the structure of the large-scale fields during a long time interval. During the activity minimum, the properties of the global magnetic field of the Sun manifest themselves in the most pronounced way.

The magnetic field of the Sun is determined by large-scale structures. The northern and southern hemispheres of the Sun have magnetic fields of opposite polarity. The strength of the polar magnetic field is significantly higher than the fields in middle and low latitudes in the activity minimum period. Along with this, one can conclude from the analysis that assuming that the global solar field configuration is in the form of a dipole structure is probably incorrect. The corona configurations for the eclipses of 1889, 1901, and others correspond instead to the quadrupole form, or to an octopole form if different polarities at the poles are taken into account.

Thus, long-term variations in parameter γ should manifest changes in the dipole component during solar activity minima. This hypothesis can be checked using the data on configurations of the large-scale magnetic fields. Figure 5 shows changes in the dipole moment and the envelope drawn through solar activity minima. The data were obtained from the analysis of synoptic H-alpha charts of patterns of polarity inversion lines from (Makarov & Sivaraman, 1989; Vasil'eva, 2002), and the Kislovodsk Astronomical Mountain Station (http://www.solarstation.ru). The greatest dipole moment corresponded to the minimum of cycle 19 in 1954. These data were obtained from *H-alpha* synoptic maps. The amplitude of the dipole contributions $l=1$, which determine parameter A_1 depend on the topology of the large-scale magnetic fields. The method of decomposition is given in the articles (Makarov & Tlatov, 2000; Tlatov, 2009). On the whole, the envelope of the dipole moment shows similar trends in the changes in the corona shape parameter γ (see Figs. 4 and 5).

The growth in the strength of the radial component of the interplanetary magnetic field (Cliver, *2002*) determined from the geomagnetic activity index *aa* is another important problem that has been widely discussed recently. Figure 6 shows variations in the geomagnetic index *aa*. Data were taken from the National Geophysical Data Center (http://www.ngdc.noaa.gov). The first half of the 20th century was characterized by a growth in this index; i.e., the slowly varying component that was especially pronounced

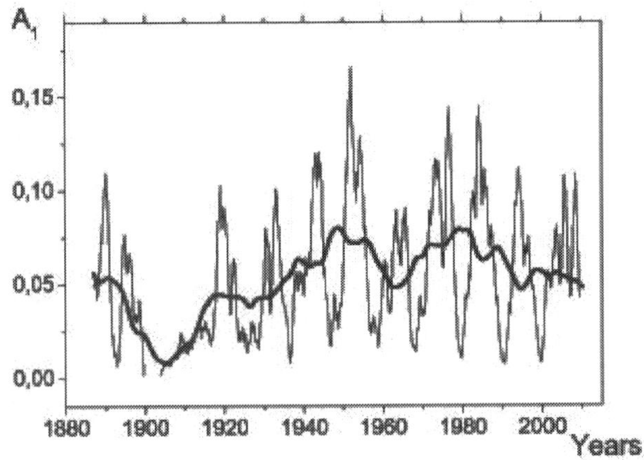

Fig. 5. Variations in the dipole moment derived from synoptic H-alpha charts of the Sun $A(t) = \sum_{m,l=1} (g_l^m g_l^m + h_l^m h_l^m)$, l=1, m=0,1, where g_l^m and h_l^m are the coefficients of the spherical function expansion. The thin line indicates monthly values, and the thick line are 11-year running means.

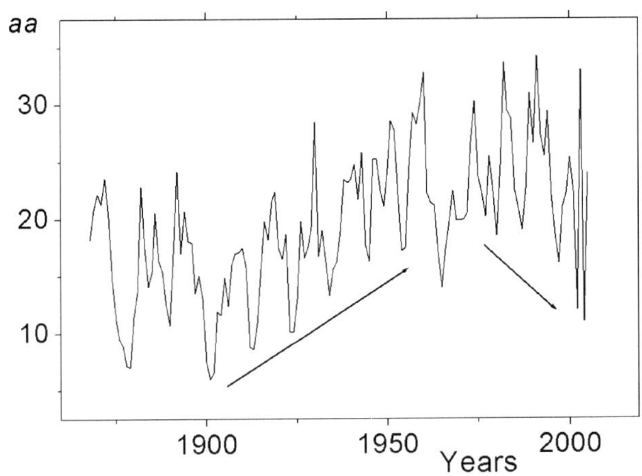

Fig. 6. Annual mean of the geomagnetic indices *aa* from 1868 (according to the National Geophysical Data Center, NGDC), smoothed by 2 years. The arrows mark the index growth in the first half of the 20th century and the index decrease during the last decades for the solar activity minimum epochs.

during the activity minimum epochs grew. During the past decades, a decrease in the *aa* index during the activity minimum epochs was observed. Probably, this is due to rearrangement of the global magnetic field of the Sun accompanied by changes in the solar corona structure during the minimum epochs. Variations in the geomagnetic index and the dipole moment of the large-scale magnetic field of the Sun during the activity minimum epoch are almost identical (see Fig. 7).

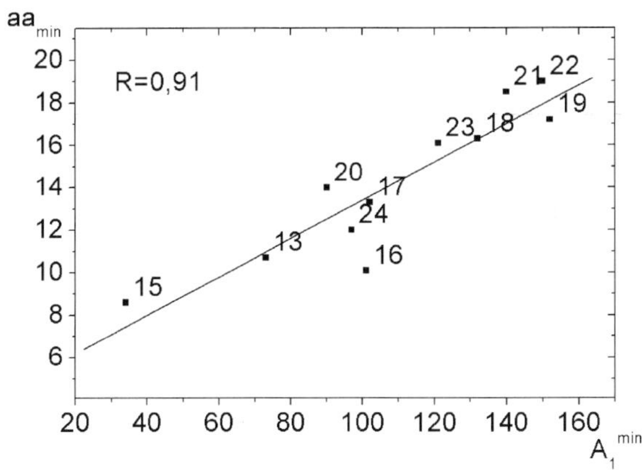

Fig. 7. Relation between the geomagnetic index aa and the magnitude of the dipole moment of the large-scale magnetic field A_1 according to Figs. 4 and 5 during the solar activity minimum epoch. Numbers of activity cycles and linear regression are also shown.

We can also test changes deviations of coronal streamers from the radial direction for over 100 years. You can use the catalogs of solar eclipses (Loucif and Koutchmy, 1989; Naegamvala, 1902; Vsekhsvyatskiy et al., 1965).

The corona of the solar minimum is characterized by pronounced polar ray structures and large coronal streamers. Let us introduce index γ that characterizes the angle between high-latitude boundaries of the large coronal streamers at a distance of 2R. The γ index is the sum of the angles at the eastern and western limbs: $\gamma = \gamma_N + \gamma_S$. In fact, the γ index is a simpler version of the corona flattening indices (Nikolskiy, 1955) but it is calculated at height R above solar limb. Figure 4 presents variations in parameter γ during the last 13 activity cycles (Tlatov, 2010a).

Thus, analysis of the corona shape has revealed a long-term modulation of the global magnetic field of the Sun. Possibly, there exists a secular modulation of the global solar magnetic field which is most pronounced during the solar activity minimum epoch. During the secular cycle of the global magnetic field of the Sun the relation between the dipole and quadrupole components of the magnetic field changes. The largest amplitude of the dipole component occurred during the interval 1944--1955. At the boundary between the 20th and 21st centuries the solar corona shape and, possibly, the global magnetic field correspond to the configuration close to the octupole one (Fig. 8).

3. Non-radial spreading of the solar corona structures in solar cycle and variations of the solar wind parameters

In this section we will consider in detail the physical mechanism linked with variations of the coronal shape and changes of the solar wind parameters.

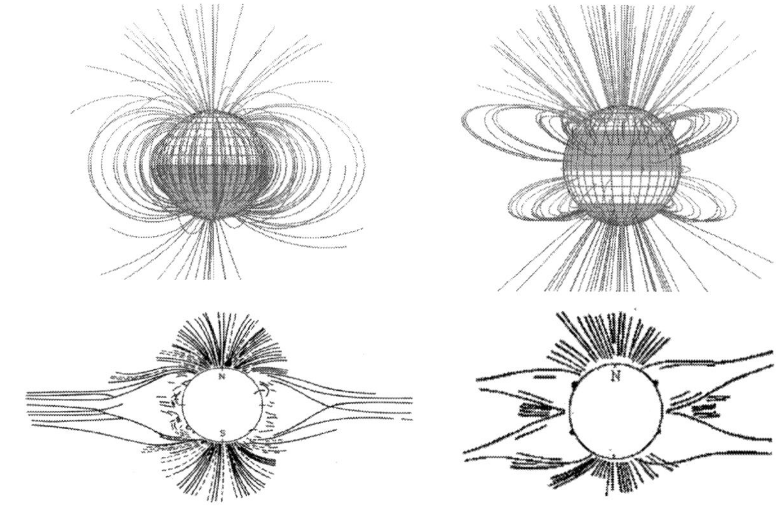

Fig. 8. Configuration of magnetic force lines for the corona of the dipole type (left) and quadrupole or octopole type if different polarities at the poles are considered (right). Regions of negative polarity are darkened. The structure of the corona for the epoch of minimum 13 and 19 activity cycles are also shown.

The solar corona structure corresponds to the configuration of solar magnetic fields. Since the magnetic field of the Sun is subjected to cyclic variations, the corona shape also changes cyclically. The coronal rays are distinctive structures in the solar corona, which propagate at a small angle to the radial direction from the Sun and display the electron density in K-corona enhanced by the factor of 3 to 10. The angle $\Delta\theta$ that describes the deviation of the rays from the radial position varies with the phase of the solar cycle and the latitude (Eselevich and Eselevich, 2002; Tlatov 2010b).

The regular observations with the SOHO/LASCO and Mark-III/IV coronagraphs at the Mauna Loa solar observatory make it possible to analyze the structure of the solar corona for the time comparable with the duration of the solar cycle. These data substantially complement extended series of observations of the corona in spectral lines carried out with extra-eclipse coronagraphs, since they make it possible to analyze coronal structures at sufficiently large distances from the solar limb, and also occasional observations of the "white light" corona during total eclipses. The coronagraph -- polarimeter Mark III detected the structure of the solar corona at the heights ~ 1,15 ÷ 2,45 R_{Θ} in 1980-1999. In 1998, at Mauna Loa observatory the new low-noise coronagraph Mark-IV, with a liquid-crystal modulator of polarization and a CCD, was mounted. To decrease the radial gradient and consequently to increase the contrast, we applied to the Mark data an artificial vignetting function. The LASCO-2 coronagraphic telescope on board of SOHO satellite has been working since 1996 and covers the distance 1.5 ÷ 6 R_{Θ} above the solar limb. Thereby, here we have analyzed the structure of the corona for 1980-2008 on the basis of the data obtained at ground-based observations with Mark-III/IV coronagraphs and for 1996-2009 with the SOHO/LASCO-2 data.

3.1 The identification for the deviation angles of coronal rays

In order to determine the deviation of coronal rays, we developed a technique of the identification of coronal streamers in two-dimensional images of the corona obtained with SOHO/LASCO-2 and Mark-III/IV in automatic mode. The analysis is based on discrimination of central parts of bright coronal structures propagating, as a rule, at some angle to the radial direction, discrimination of the points of the local maxima, and determination of the parameters of the approximating line section (Figure 9). The procedure included the following stages. Initially, the coordinates of the center and the radius of the Sun were measured in pixels. Further on, we calculated the average limb brightness of the corona for different heights above the solar limb $I(r)$. We considered the regions in which the brightness at a given distance from the limb was no smaller than $0.3 \cdot I(r)$. Then we selected all the points with the brightness exceeding that of the points apart of it by the angle ±1,5° along the limb. The obtained collections of points, as a rule, represented regions close to the center of brightness of the coronal rays, extended from the limb of the Sun at some angle to the radial direction. For these regions, we inscribed the least-square approximating straight lines. From the parameters of these straight lines we determined the polar angle of the base of a coronal ray θ and the deviation from the radial direction $\Delta\theta$. Figure 10 presents an example of such identification for a Soho/Lasco-2 image obtained on 2007.03.16. We used for the analysis the deviation angle for these straight lines with respect to the radial propagation. We took into account line sections with the length no smaller than $1.0 \cdot R$ for SOHO/Lasco-2 images and than $0.5R$ for Mark-III/IV images. The selection of rays, the linearity of which can be traced to a fairly large distance, allows to filter out a considerable number of coronal structures associated with eruptive processes. Thereby, we have processed about $4.2 \cdot 10^3$ images for the time interval 1996-2009 and discriminated approximately 10^5 coronal rays from the Soho/Lasco-2 data and $7 \cdot 10^3$ days for the Mark-III/IV data for 1980-2008.

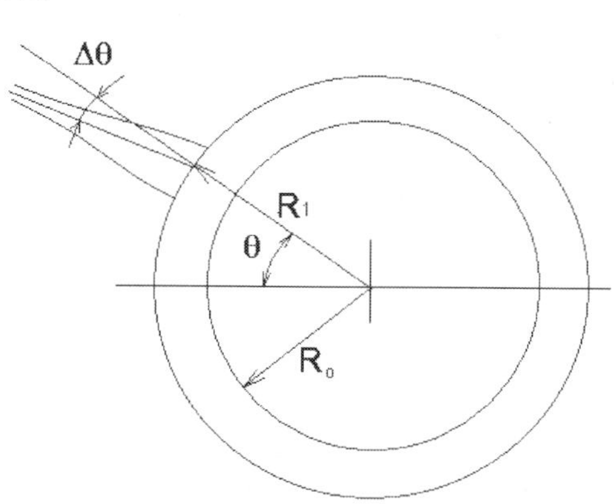

Fig. 9. The determination of deviation angles for coronal rays.

In this analysis, we recorded the coronal beams of various types. Among the rays belonging to the helmet-type, and chains of streamers, rays belonging to the low-latitude and polar regions (*Saito et al. 2000, Eselevich & Eselevich, 2002*). These rays can be formed over the bipolar and unipolar magnetic structures. There are also dynamic coronal rays, but we

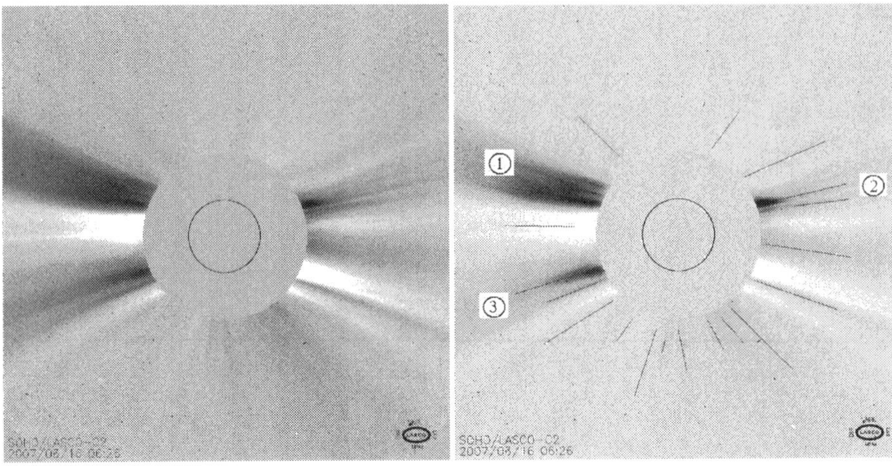

Fig. 10. An example of the identification of the direction of the propagation of coronal streamers for an image of the corona obtained with Soho/Lasco-2 on 2007.03.16.

decided that their number is small enough to that would affect the comparison. Thus Fig. 11 shows that the footpoints rays are usually evident in a few days. In addition to the geometry of the rays in the process of selection, we also recorded the absolute and relative brightness of the rays. This allowed to make a comparative analysis nonradially rays and define their types. In order to define the brightness of rays we calculated the average intensity within the function from the height over the solar limb $I_{avr}(r) = 1/2\pi \int_0^{2\pi} I_{avr}(r,\alpha)d\alpha$, where α- is a polar angle. Then we defined the average intensity of a ray at the beginning and ending of a segment, approximating the ray $\overline{I} = (I_{beg} + I_{end})/2$. The rays with the intensity $\overline{I} > 2 \cdot I_{avr}$, where I_{avr} was calculated at a height $(R_{beg} + R_{end})/2$, were related to bright rays. The rays with the intensity $\overline{I} < 1.3 \cdot I_{avr}$ were conditionally related to non-bright rays. Figure 10 shows the systems of bright rays, that are marked with numbers 1-3. The rays of lower intensity are normally located within the area of high latitudes (4, Fig. 10).

Footpoints of bright coronal rays, obtained under the assumption of linear distribution, as a rule, close to the neutral line (Fig. 11).

Fig. 12 shows the change of nonradiality parameter for bright rays $\Delta\theta_{br}$ associated with helmet-rays and soft rays $\Delta\theta_{low}$, which are typical for the chain of coronal streamers from the unipolar regions. Rays of varying brightness show close cyclical course of the parameter $\Delta\theta$, although the degree of nonradiality for bright rays is slightly less: $\Delta\theta_{br}=1.46+0.47\cdot\Delta\theta_{low}$, are correlated with $r=0.79$. Rays of different latitudinal zones also show a close behavior (Fig. 13). Relationship between the parameter $\Delta\theta$, for rays of the equatorial and polar zones is the following: $\Delta\theta_{pol}=4.1+0.84\cdot\Delta\theta_{eq}$; $r=0.89$. Thus, the nonradial parameter $\Delta\theta$ cycle exists for the various types of coronal rays.

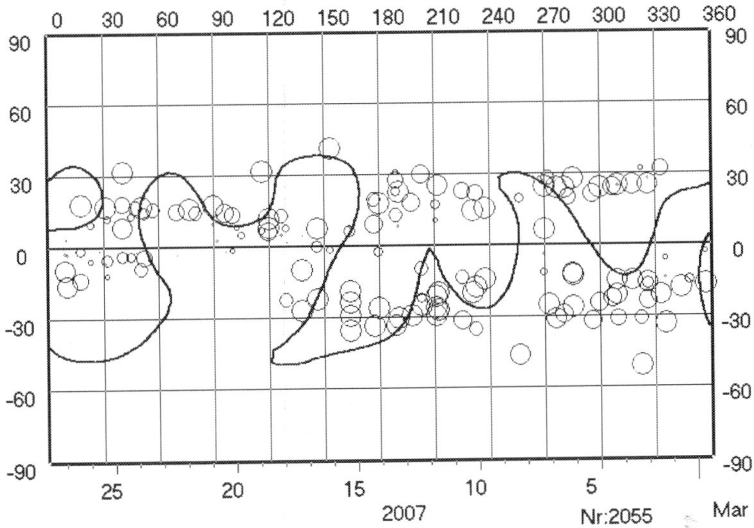

Fig. 11. Example of synoptic maps for Carrington rotation N2055 which marked the neutral line, calculated at an altitude R = 1.9 · R$_\odot$ according to the Wilcox Solar Observatory. Presents the footpoints of bright coronal rays.

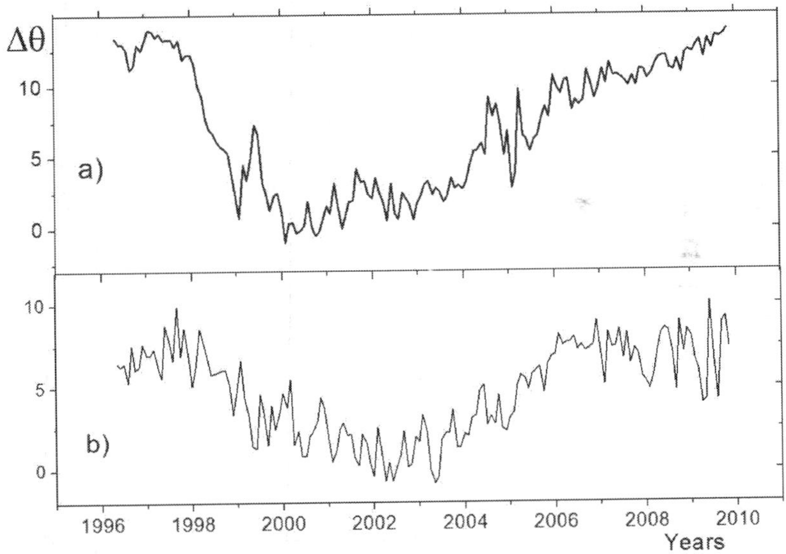

Fig. 12. The monthly-averaged deviation from the radial direction Δθ derived from the SOHO/Lasco-2 data a) for coronal rays of low intensity $I<1.3 \cdot I_{avr}$. b) for bright coronal rays ($I>2 \cdot I_{avr}$).

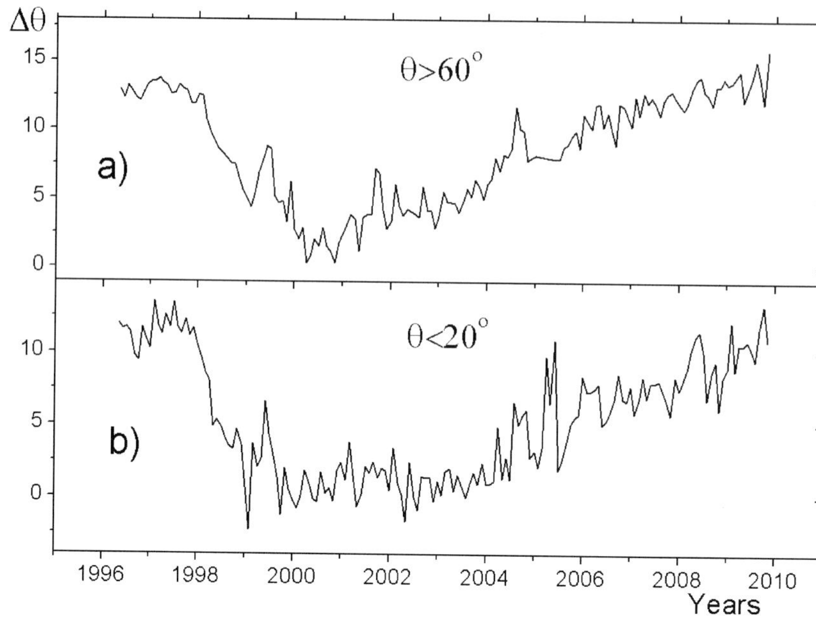

Fig. 13. The monthly-averaged deviation from the radial direction Δθ derived from the SOHO/Lasco-2 data a) for the polar rays b) for the rays of the equatorial zone.

3.2 Connection between the non-radial coronal shape and the solar wind parameters

The analysis of the SOHO/Lasco-2 data for 1996-2009 indicates that the deviation angle Δθ varies with the 11-year cycle of the activity, reaching the maximum values at the minimum of the activity. Figure 14 presents the time-latitude diagram for Δθ variations. In the vicinity of the equator, the deviation of the rays from the radial direction reaches the minimum. Negative angles were seen in the northern hemisphere in 2002-2004 and in the southern hemisphere in 2000-2003, at the latitudes smaller than 30°. The coronal rays at the minimum of the activity and the phases of the decline are, as a rule, turned towards the solar equator. Only rather low-latitude coronal rays at the maximum of the activity slightly deviate towards the poles (Fig. 14). The largest deviation from the radial direction is seen at the minimum of the activity at the latitudes 30 ÷ 60°. At the time of the maximum of the activity and the polarity reversal of the solar magnetic field the rays are directed either parallel to the equator or slightly deviate towards the poles. Individual fluctuations of the angles of deviation of the rays are seen in different latitude zones (Figure 13), which indicates the general type of the perturbations of coronal structures.

Figure 15 presents the monthly averages for the deviation, averaged along the entire limb for all types of coronal rays. For comparison, the graph also shows the variation of the angle of tilt angle τ of the heliospheric current sheet (HCS) (Hoeksema & Scherrer, *1986*). Tilt angle HCS is calculated in a potential assumption of this photospheric magnetic field for $R=3.25 \cdot R_0$. Between these parameters there is the ratio: $\Delta\theta = 0.49 + 0.2 \cdot \tau;\ r = 0.91$.

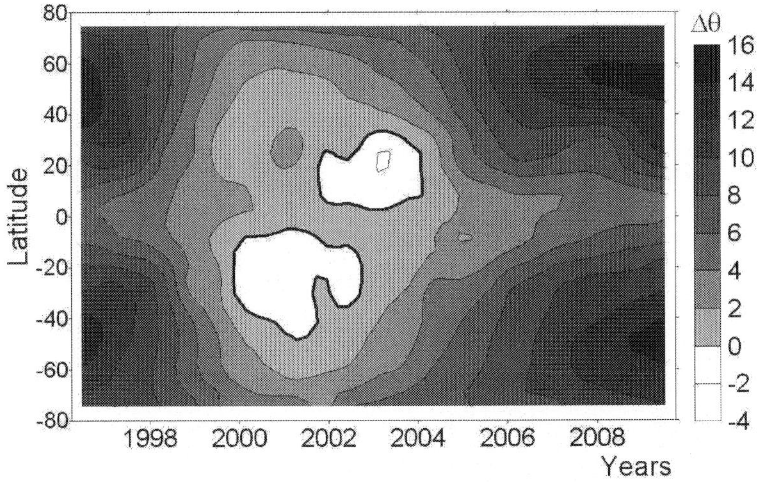

Fig. 14. The latitude and time distribution of the angles of the deviation of coronal rays from the radial direction in the time interval 1996-2009 derived from the SOHO/Lasco-2 data. The regions of the deviation of the rays towards the equator are darkened. The levels are indicated at the intervals of 2 degrees; the level corresponding to $\Delta\theta = 0^{\circ}$ is shown by the thick line.

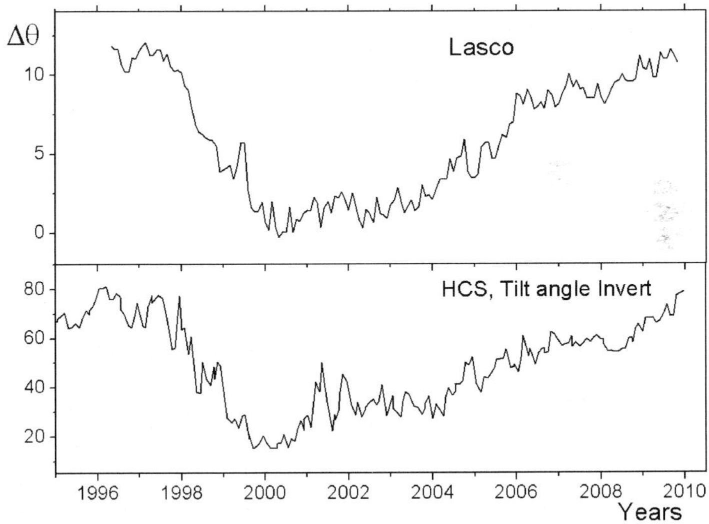

Fig. 15. Comparison between the deviation $\Delta\theta$ of coronal rays from SOHO/Lasco-2 (top panel) and the inverted tilt angle of the heliospheric current sheet according to WSO (http://wso.stanford.edu).

These conclusions are also confirmed by the analysis of the structure of the solar corona at the heights 1.2 ÷ 2.5R based on the MLSO Mark-III/IV data. Figure 16 presents the angle Δθ for the latitude zone ±30° in the time interval 1980-2008. For the period 1996-2008 angle Δθ according to the coronagraph Mark was slightly lower than according to Lasco/C2: $\Delta\theta_{Mark}=2.55+0.66\cdot\Delta\theta_{Lasco}$; $r=0.95$. The nonradial variation Δθ according to Mark telescope during 1980-2008 change of the parameter also has a good correlation with the angle of tilt angle of HCS $r\sim0.8$ (Fig. 16).

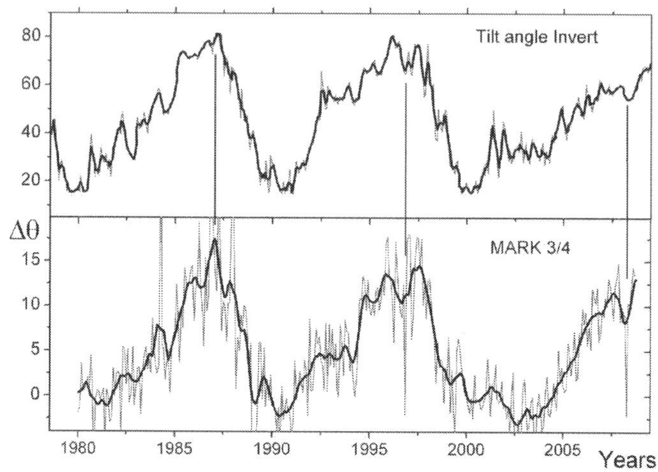

Fig. 16. The comparison between the deviations Δθ of coronal rays located in the middle-latitude zone ±30°, from the MLSO- MarkIII/IV data (bottom panel) and the inverted tilt angle of the heliospheric current sheet according to the WSO (top panel). The values are smoothed for 6 months.

Figure 17 presents the latitude distribution for the Δθ angle for the time of the minimum of the 23-rd cycle. The maximum deviation is seen at the latitudes ~40-50°, which corresponds to the data obtained with the Lasco telescope (Eselevich & Eselevich, 2002).

The deviation of coronal rays from the radial direction Δθ related to the solar activity cycle may affect substantially the formation of the solar wind and geomagnetic perturbations. From geometrical consideration, in Figure 1 the flux of the solar wind at the distance r is related to that at the distance R_1, where the current lines become straight, as follows: $nv / n_1 v_1 = 1 + (\theta - \Delta\theta)\cdot(r - R_1) / \theta \cdot R_1$, for Δθ<θ. Assuming that at the years of the minimum activity θ =40° and Δθ =20° (Figure 9), we obtain that the ratio $nv / n_1 v_1$ at the distance 1 AU increases roughly twice compared to the case of radial expansion.

Figure 18 presents the graph of the deviation angle in comparison with the indexes connected with parameters of a solar wind. In the years of the maximum activity, the influence of active regions and flares is substantial. However during a minimum of activity the deviation Δθ to a plane of solar equator can be one of the reasons of delay of a solar wind speed (Figure 18c), and to modulate indexes of geomagnetic activity aa (Figure 18d) and cosmic rays intensity (CRI) (Figure 18b).

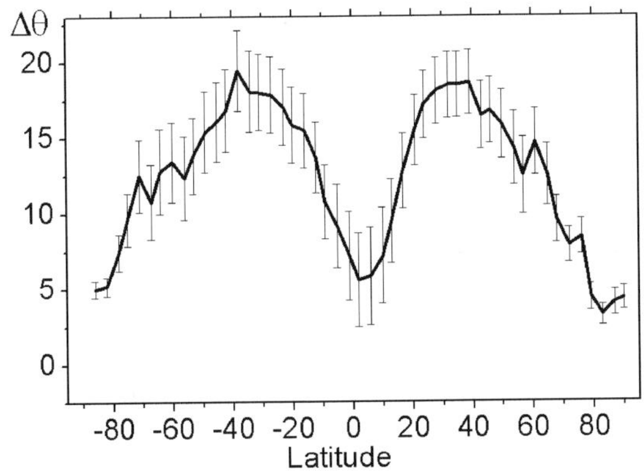

Fig. 17. The variations of the angle $\Delta\theta$ as a function of the latitude, obtained from the MLSO-MarkIII data at the minimum of the 23-rd cycle of the solar activity (1996-1997).

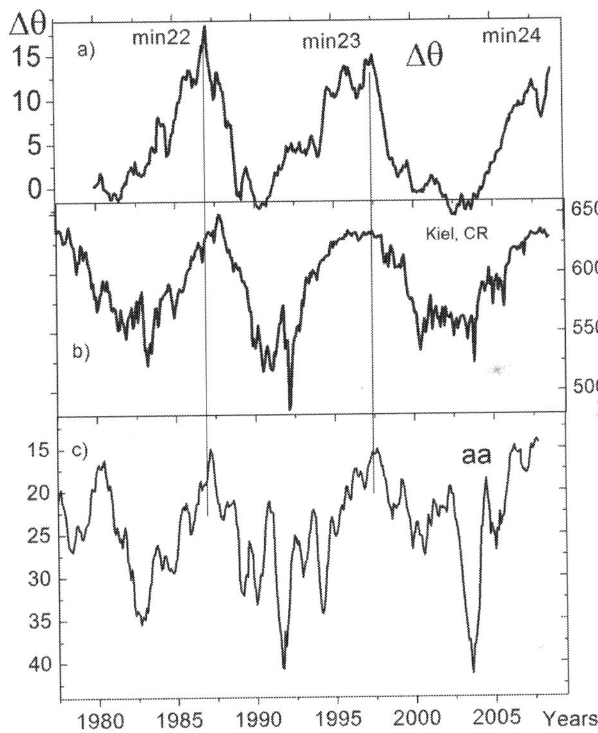

Fig. 18. The comparison between a) the deviations of coronal rays $\Delta\theta$; b) data of Kiel cosmic-ray intensity (arbitrary units, monthly means); d) the geomagnetic index *aa*. The data are smoothed with a sliding window technique for 6 months.

The similarity of the behavior of the shape of the coronal rays with the indices of geomagnetic activity and the radial flux give evidence concerning the connection between the large-scale organization of the corona and solar wind parameters. But is it possible to establish the link directly between the coronal shape and solar wind parameters, measured from the Earth? Data bases OMNI1 and OMNI2 (http://nssdc.gfc.nasa.gov/omniweb) contain the information concerning hourly average value of key parameters of the solar activity, interplanetary atmosphere and geomagnetic disturbance since 1964. We collated changes of Δθ parameter with solar wind parameter $\beta = 8\pi nkT / B^2$ and magnetic Mach number $M_a = v / v_a = v\sqrt{\mu\mu_0\rho} / B$ (Fig. 19). In minimal activity epoch there is a correspondence between the non-radial parameter Δθ and solar wind parameters. Comparing parameters β и M_a we can make a conclusion that non-radial corona influences the relation ρ / B^2 of the solar wind, which grows in minimum solar activity and therefore magnetic field in minimal activity squeezes the solar wind flux towards helio-equator. These results complement the findings of in comparison with the previous three minima, this solar minimum has the slowest, least dense, and coolest solar wind, and the weakest magnetic field (Jian et al., 2011).

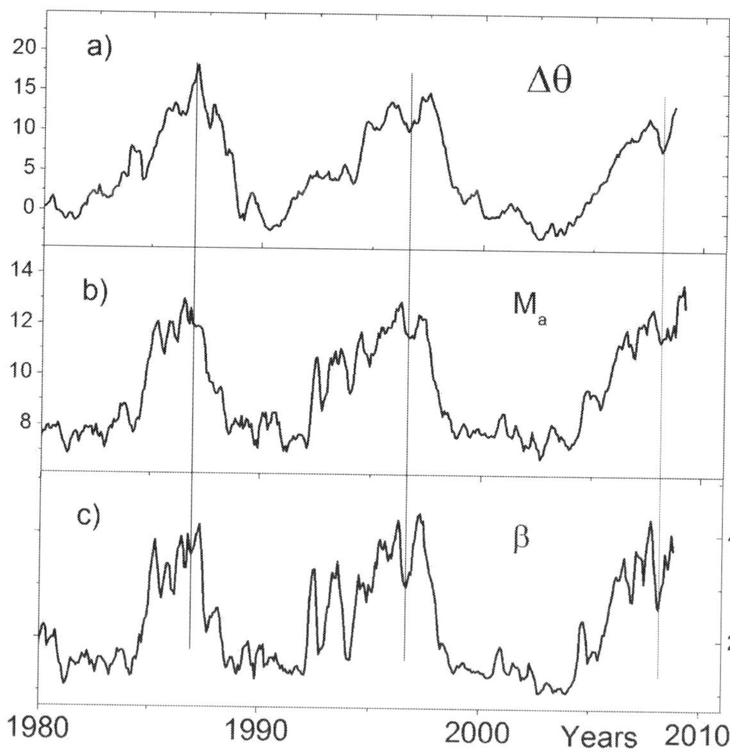

Fig. 19. The comparison between a) the deviations of coronal rays Δθ with the solar wind parameters according to OMNI2 database b) Alfven mach number Ma and c) Plasma beta β. Data of the solar wind parameters were are smoothed on 20 Bartels rotations.

4. Conclusion

Nowadays to prognose the solar wind parameters empirical models are used (Wang, Y.-M., & Sheeley, 1990; Arge & Pizzo, 2000) , they are based on the partial understanding of the physical processes taking place on the Sun. At the heart of these ideas there is a hypothesis about the radial character of the solar wind spreading. The details of the streams formation of the solar wind are complicated and are the subject of many researches. Nowadays there aren't any exact mathematical models of the physical processes influencing the formation of the solar wind flux.

In our research we pay attention to the fact that in minimal activity the conditions of forming the solar wind can depend on the configuration of the large-scale field and the shape of the solar corona. The configuration of the global magnetic field in minimal activity corresponds to the low harmonics of decomposition according to the Legendre polynomials $l=1,2,3$.

The link between the coronal shape, geomagnetic disturbances and solar wind parameters can be clearly seen in 11-year activity cycle. Global magnetic field can lead to non-radial spreading of the coronal streamers and probably, to the spreading of the solar wind. In minimal activity when the value of the global field reaches its maximum, coronal rays are deflected aside from the helio-equator.

Thus, analysis of the corona shape has revealed a long-term modulation of the global magnetic field of the Sun. Possibly, a secular modulation exists of the global solar magnetic field that is most pronounced during the solar activity minimum epoch. During the secular cycle of the global magnetic field of the Sun, the relation between the dipole and octopole components of the magnetic field changes. The largest amplitude of the dipole component occurred during the interval 1944--1955. At the turn of the 19th to 20st and 20th to 21st centuries the solar corona shape and, possibly, the global magnetic field correspond to the configuration close to the octopole one (see Fig. 8).

The period of variation in the corona's shape during the epoch of minimal activity is about 100 ÷ 120 years (see Fig. 4), which is close to the Gleissberg cycle for the sunspots, but probably precedes it to some extent in phase (*Hathaway*, 2010, Fig. 34).. The maximum of the secular variation in the global magnetic field of the Sun occurred before cycle 19 and preceded the sunspot activity maximum. This allows us to put forward the hypothesis that secular variations in the solar activity are caused by secular modulation of the global magnetic field of the Sun. Another conclusion of this work is the supposition that the slowly changing component of the geomagnetic activity derived from the data on the *aa* index from changes in the dipole component of the large-scale field of the Sun (Figure 7).

The link between the coronal shape, geomagnetic disturbances and solar wind parameters can be clearly seen in 11-year activity cycle. Global magnetic field can lead to non-radial spreading of the coronal streamers and probably, to the spreading of the solar wind. In minimal activity when the value of the global field reaches its maximum, coronal rays are deflected aside from the helio-equator. Here with lines of force influence the elapsing flux of the solar wind, squeezing it towards the ecliptic plane (Fig. 19). It is possible that long-term changes of geomagnetic disturbances that do not tend to zero even in the minimum activity are not conditioned this very deflection of the solar wind.

The awareness of the variations of the angle $\Delta\theta$ with the phase of a cycle is important for theoretical models describing the structure of the corona and the geometry of the magnetic field above the solar limb (Wang, 1996). The applied aspect of the studies for the deviations of coronal streamers, along with the plasma flows in the solar wind, is also of great importance, since the deviation may affect the geo-efficiency of the solar wind impact. The presence of long-term trends in the solar corona structure can be caused by changes in the configuration of the global magnetic field of the Sun. The non-radial propagation of the solar wind may explain the relationship between geomagnetic indices detected during the minima of activity and the amplitude of a subsequent solar cycle (Ohl, 1966). Indeed, the relation between the amplitude of the angles $\Delta\theta$ at the minima of the 22-th - 24-th cycles is close to that for the index of the large-scale magnetic field of the Sun in these times (Tlatov, 2009), and this field is one of prognostic indices of the solar activity. Thereby, we may suggest a link between the large-scale magnetic field and the deviation of coronal rays towards the equator, which in turn affects the level of geomagnetic indices at a solar minimum as well as solar-terrestrial relations.

The deviation coronal streamers $\Delta\theta$ occur as the cyclic process appreciable at all solar latitudes (Figure 13,14). Probably, it is connected with global processes, for example hemispheric current layer (Filippov, *2009*).

In the analysis we have considered the various structures, including possibly producing a unipolar coronal streamer belt (called' 'chains of streamers" or "streamer belt without a neutral line" by (Eselevich et al., *1999*) as well as the bipolar streamer belt (Zaho & Webb, *2003*). These streamers vary in brightness and location of footpoint. Our analysis of separation of their brightness and latitude showed that cyclic changes in the nonradial parameter $\Delta\theta$ exist for various types of coronal streamers. The good agreement between the change in time for parameter $\Delta\theta$ change the angle of inclination of HCS, suggest that the parameter of nonradiality, is possibly conditioned by the existence of current in the heliospheric current layer. The current in the HCS can be estimated from the parameter $\Delta\theta$.

In the epoch of the 18-19 activity cycles in the middle of the last century the shape of the solar corona in its minimal activity mostly corresponded to the corona of the minimal type, that is - coronal radiance spreaded along helio-equator (Fig. 3). Just in that period the highest level of geomagnetic activity was observed (Fig. 6), what is more, not only during the years of solar spots maximum, but also in the minimal activity. We can assume that during that period global magnetic field of the Sun most effectively squeezed the solar wind flux towards the heio-equator plane, which led to such an effect.

Variation changes of the low and high harmonics order are illustrated on figure 20. Here the amplitudes of the harmonics were smoothed by the "running window" of 11 years long. One introduces amplitudes of axisymmetric mode for dipole and quadruple ($l=1,2$; $m=0$) and for harmonics, characterizing sector structure of the higher harmonics ($l>3$; $m>0$). One can notice that changes in time within the amplitudes of these two types occur in antiphase. One should also notice that axisymmetric modes ($m=0$) show that magnetic structures of the opposite polarities are situated on different sides of the equator. On the contrary, the amplitude growth of the sector harmonics shows that opposite signs structures are drawn along meridians. The amplitudes of the axisymmetric harmonics reach their maximum in

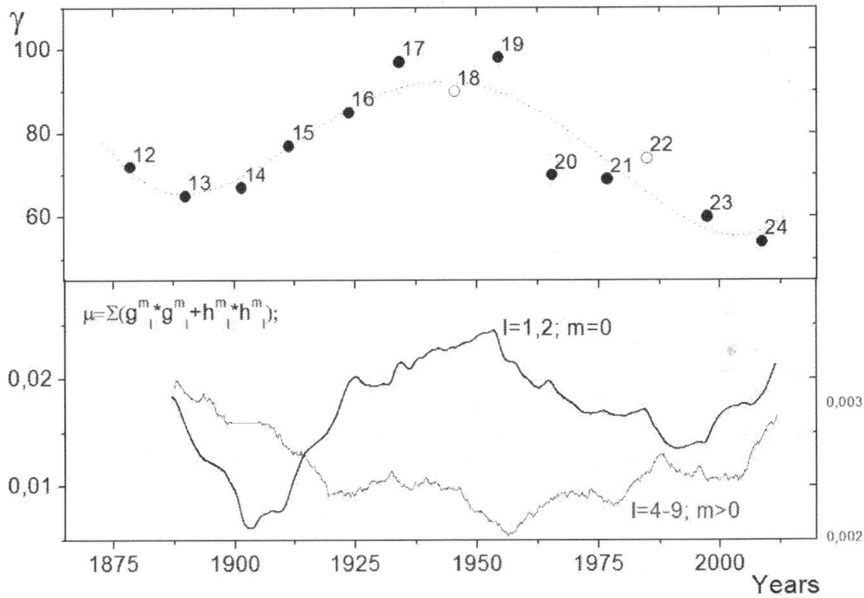

Fig. 20. Variation changes of the low and high harmonics order are illustrated on figure 20. Here the amplitudes of the harmonics were smoothed by the "running window" of 11 years long. Bottom panel introduces amplitudes of axisymmetric mode for dipole and quadruple (l=1,2; m=0) and for harmonics, characterizing sector structure of the higher harmonics (l>3; m>0). At the top changes of parameter γ form the solar corona.

the epoch of years 1940-1950, while sector harmonics were minimal at that moment. Such allocation corresponds to the Babcock hypothesis concerning magnetic fields generation. In fact, reconnection of the opposite through equator fields causes mutual destruction of the fields of the leading polarity, and it should cause the solar activity growth. But on the contrary, reconnection of the magnetic field along the meridians does not promote the creation of global field of dipole type and leads to the decline of solar activity level.

Consequently, in the epoch of big solar activity cycles the growth of geomagnetic activity takes place not only due to the result of active processes increase, but also under the influence of solar wind's formation by the global magnetic field of the Sun in activity minimums. The following fact contributes to this - before the high cycles of activity the intensity of global magnetic field is also high (Fig. 5). It is possible that age-long cycle of changes of solar wind parameter and geomagnetic activity, close to Gleissberg cycle is caused not only by the level of activity, but also by long-term changes of the large-scale magnetic field.

Thereby, we may suggest a link between the large-scale magnetic field and the deviation of coronal rays towards the equator, which in turn affects the level of geomagnetic indices at a solar minimum as well as solar-terrestrial relations.

5. Acknowledgements

This paper was supported by the Russian Fund of Basic Researches and Program of the Russian Academy of Science.

6. References

Arge, C. N., & Pizzo, V. J. (2000). Improvement in the prediction of solar wind conditions using near-real time solar magnetic field updates, *J. Geophys. Res.*,Vol. 105,pp. 10465-10480.

Cliver, E. W. & Ling, A. G. (2002). Secular change in geomagnetic indices and the solar open magnetic flux during the first half of the twentieth century, *J. Geophys. Res.*, SSH 11-1, 107.

Eselevich, V. G., & Eselevich, M. V. (2002). Study of the nonradial directional property of the rays of the streamer belt and chains in the solar corona, *Solar Phys.*, Vol. 208, pp.

Eselevich, V. G., Fainshtein, V. G. & Rudenko, G. V., (1999). Study of the structure of streamer belts and chains in the solar corona, *Solar Phys.* , Vol. 188,pp. 277-297.

Filippov, B. P. (2009), Non-radial coronal streamers in the course of the solar cycle, *Astronomy Reports*, Vol. 53,pp. 564-568.

Ganskiy, A. P. (1897). *Die totale sonnenfinsterniss am 8. august 1896*, Proc. Royal. Akad. Sci. (in Russian), Vol. 6, 251.

Golub, L. & Pasachoff, J. M. (2009). *The Solar Corona, second edition*, (Cambridge University Press).

Gulyaev, R. A. (1998). New results of observations of eclipsing the solar corona, *in Proceedings of the Conference ''A New Cycle of Solar Activity: Observations and Theoretical Aspects'' (in Russian)*, ed. A. V. Stepanov, (Pulkovo, St.Petersburg), pp. 61-66.

Hathaway, D. H. (2010). The Solar Cycle, *Living Rev. Solar Phys.*, 7, (http://solarscience.msfc.nasa.gov).

Hoeksema, J. T. & Scherrer, P. H. (1986). An atlas of photospheric magnetic field observations and computed coronal magnetic fields: 1976-1985, *Sol.Phys.*, Vol. 105,pp. 205-211.

Jian, L. K.; Russell, C. T. & Luhmann, J. G. (2011). Comparing Solar Minimum 23/24 with Historical Solar Wind Records at 1 AU, *Solar Phys.*, Doi 10.1007/s11207-011-9737-2.

Loucif, M. L. & Koutchmy S. (1989). Solar cycle variations of coronal structures, *Astronomy and Astrophysics Supplement Series*, Vol. 77, pp. 45-66.

Makarov, V. I. & Sivaraman, K.R. (1989). Evolution of latitude zonal structure of the large-scale magnetic field in solar cycles, *Solar Phys.*, 119, 35-44.

Makarov, V. I. & Tlatov A. G. (2000). The Large-Scale Solar Magnetic Field and 11-Year Activity Cycles, *Astron. Rep.*, Vol. 44, pp. 759-764.

Meyer-Vernet N. (2007). *Basics of the Solar Wind*, ISBN 978-052-1814-20 Cambridge University Press

Naegamvala, K. D. (1902). Report on the total solar eclipse of January 21-22, 1898, (Bombay), 49.

Nikolskiy, G. M. (1955). Forecast form the solar corona on June 20, 1955, *Astron. Circular*, 160, 11-12.

Ohl, A. I. (1966). *Soln. Danie*, N 12, 84.

Parker E. (1958). Dynamics of the Interplanetary Gas and Magnetic Fields, *The Astrophysical Journal*, Vol. 128, pp. 664-676

Pasachoff, J. M., Ružin, V., Druckmüller, M., Druckmülerovä, H., Belik, M., Saniga, M., Minarovjech, M., Markovä, E., Babcock, B. A., Souza, S. P. & Levitt, J. S. (2008). Polar Plume Brightening During the 2006 March 29 Total Eclipse, *ApJ*, Vol. 682, pp.

Pneuman, G. W. & Kopp, Roger A. (1971). Gas-Magnetic Field Interactions in the Solar Corona , *Solar Phys.*, Vol. 18, pp.258-270

Saito, T., Shibata, K., Dere, K.P. & Numazawa, S. (2000). Non-Radial Unipolar Coronal Streamers in Magnetically High Latitudes and Radial Bipolar Streamers at the Magnetic Equator of the Sun, *Adv. Space Res.*, Vol. 26,pp. 807-810.

Tlatov, A. G. (2009). The Minimum Activity Epoch as a Precursor of the Solar Activity, *Solar Phys.*, Vol. 260, pp. 465-477.

Tlatov, A. G. (2010a). The centenary variations in the solar corona shape in accordance with the observations during the minimal activity epoch, *Astronomy and Astrophysics*, Vol. 522, id.A27

Tlatov, A. G. (2010b). The Non-radial Propagation of Coronal Streamers within a Solar Cycle, *ApJ*, Vol. 714,pp. 805-809.

Vasil'eva, V. V. (1998). Recovery of large-scale synoptic maps of magnetic fields for the period 1880-1894, *in Proceedings of the Conference "A New Cycle of Solar Activity: Observations and Theoretical Aspects" (in Russian)*, ed. A. V. Stepanov, (Pulkovo, St.Petersburg), 213-216.

Vsekhsvyatskiy, S. K., G. M. Nikolskiy, V. I. Ivanchuk, A. T. Nesmyanovich, et al. (1965). *Solar Corona and Corpuscular Radiation in the Interplanetary Space*, (Kiev, Naukova Dumka), 293.

Vsekhsvyatskiy, S. K., G. M. Nikolskiy, V. I. Ponamarev E.A. & Cherednichenko V. I. (1955). On the corpuscular radiation from the sun, *Astronomy Rep.*, Vol. 32, pp. 165-176

Waldmeier, M. (1976). *Astron. Mitt. Eidgen. Sternw. Zurich*, Vol. 351,pp. 13.

Wang Y.-M. (1996). Nonradial Coronal Streamers, *ApJ*, Vol. 456, pp. L119-L121.
Wang, Y.-M., & Sheeley, N. R., Jr. (1990). Solar wind speed and coronal flux-tube expansion, *ApJ*, Vol. 355, pp. 726-732
Zhao, X. P. & Webb, D. F. (2003). Source regions and storm effectiveness of frontside full halo coronal mass ejections *JGR*, Vol. 108, pp. SSH 4-1.

13

Variability of Low Energy Cosmic Rays Near Earth

Karel Kudela
Institute of Experimental Physics, Slovak Academy of Sciences, Kosice Slovakia

1. Introduction

Almost a century ago Victor Hess in his balloon experiments discovered cosmic rays (Hess, 1912). The history in understanding the nature, physical mechanisms affecting the temporal and spatial variability of cosmic rays (CR) observed on the ground and within the atmosphere, as well as of the anisotropy of secondary particles due to geomagnetic effects and due to the state of atmosphere is reviewed e.g. in the monographs (Hillas, 1972; Dorman, 2004; 2009; Rossi and Olbert, 1970 and references therein). Progress in clarifying mechanisms controlling the secondary CR flux required the design and construction of new detectors. Production of secondary CRs is reviewed in the books (Grieder, 2001; 2010). Until half of the past century the elementary particle physics was driven mainly by CR studies. With the improvement of the acceleration technique, the domain of CR for subnuclear physics research, was shifted to high energies. Even today the high energy CRs observed by the large ground based detectors remain the only one source of information about the particles of extremal energy. The astroparticle physics aspect of CR is described e.g. in the books (Grupen, 2005; Gaisser, 1990; Hayakawa, 1969). Relations of high energy astrophysics to CR physics include the book (Longair, 1981 and the newer versions).

The important step in CR studies represented the International Geophysical Year in 1957-1958. The new detectors and its networks were built (Simpson 1958; Hatton, 1971; Stoker, 2009 among others) and CR research attracted many scientists, engineers and students. Most important stimulus for the progress in low energy cosmic rays (LECR) was the launch of the first satellites of Earth which meant beginning of the Space Era. Shortly after measurements of radiation on the first satellites, the new populations of energetic particles in space (not providing secondary „signal" on the Earth) have been found. That direction of research was important also for plasma physics. Processes of particle acceleration, transport and losses in the plasma regions in space, where conditions are inimitable for laboratory plasma experiments, brought new knowledge into plasma physics. There are several books and papers summarizing/illustrating in detail CR physics in its history. At low energies to which this chapter is devoted, much more informations can be found e.g. in (Dorman, 1974; Bieber et al., 2001; Vainio et al., 2009; rapporteur papers in the proceedings of ICRCs). Here we mention selected results in the experimental studies related to LECR with empasis to the papers published in the past 2-3 years.

2. LECR variations

Two populations of LECR can be assigned, namely (1) at the energies above, and (2) below the atmospheric threshold (~ 400 MeV for protons). Above that energy the ground based measurements provide the direct information about changes in the flux of primaries. For lower energies the measurements of particles on satellites, space probes, rockets and balloons are most relevant. Two types of experimental devices on the ground are very important for the detection of LECR flux variability above the atmospheric threshold, namely neutron monitors (NM) and muon telescopes (MT) sensitive to different ranges of energy spectra of primaries.

Galactic CRs entering the heliosphere are affected by the interplanetary magnetic field (IMF) which, especially in the inner heliosphere, is controlled by the solar wind plasma having higher energy density than IMF. Concept of field lines frozen in the high conductivity solar wind plasma is often used. CR in the inner heliosphere having lower energy density than the IMF, can be assumed as a specific "autonomous" population of particles. In the outer heliosphere, however, the relations are changing. The heliosphere via the IMF is modulating the CR flux at the low edge of its energy spectra. In addition, it contributes itself to energetic particle populations by acceleration at the Sun as well as at plasma discontinuities in the interplanetary space, and within the magnetospheres of planets with strong magnetic field. Additionally, heliosphere is transparent for access of neutral atoms from outer space. They can be ionized in the heliosphere and subsequently accelerated, which contributes to the suprathermal particle population known as anomalous cosmic rays (Garcia-Munoz et al., 1973). Modulation of CR depends on its primary energy. Solar wind with the embedded IMF flowing outward from the Sun screens the access of primary CR into the heliosphere. The physical framework in which the energetic particles and CR propagate in the heliosphere is also denoted as heliospheric magnetic field, HMS (recent review e.g. by Balogh and Erdös, 2011). Below few hundreds MeV practically no galactic CR enter the inner heliosphere (Jokipii, 1998). The modulation below ~ 10 GeV is present even during solar activity minimum. The main feature of a long term variation of low energy galactic CR (GCR) near Earth is the anticorrelation of the flux with solar activity having about 11 year cyclicity. In addition to ~11 year periodicity, the ~ 22 year modulation cycle in CR flux due to solar magnetic field polarity reversals, is observed (e.g. Webber and Lockwood, 1988). Around the epoch of solar activity maxima there is observed a double-peak structure in many solar activity factors. A distinctive minimum between the peaks is called Gnevyshev gap (Storini et al., 2003 and references therein). In CR this minimum is connected e.g. with decrease of ~27 day quasi-periodicity.

Theory of CR transport, used basically until present with several small modifications, was described first by E.N. Parker (1965). Energetic particles in the interplanetary space walk randomly in irregularities of the large-scale IMF when irregularities are moving with the solar wind velocity. The distribution function determined by a Fokker-Planck equation describes the time evolution of the probability density function of the position and momentum of particle. In addition to the convection and diffusion, CRs experience two additional effects. One of them is the acceleration or deceleration. Solar wind plasma is expanding in free space and compressing at the shocks near the planets or in the interplanetary space. The inhomogenities with different IMF become mutually more distant or more close to each other. This leads to the adiabatic cooling or heating due to multiple interactions of CR with inhomogenities. Modulation of GCR in the inner heliosphere is controlled by convection in solar wind, diffusion, particle drifts and adiabatic energy losses. Using the experimental data from various space missions, an overview of the current understanding of the modulation over the past decade is given by (Potgieter, 2011). At low energies, 10–100 MeV, where the adiabatic cooling plays the primary role, its effect can be best seen in the framework of the force-field approximation (Gleeson & Axford 1968; Fisk et al. 1973). In addition, the curvature and gradB drifts in IMF play role in the modulation. Gyration of particle around the field line is faster than scattering. Thus particles are subject of drift due to large scale spatial structure of the IMF. CR transport and modulation was studied in many papers (e.g. Jokipii, 1966; 1967; Potgieter and Le Roux, 1994; Zank et al., 1998; Cane et al., 1999 among others) and it is examined further along with the new knowledge of CR flux at various points in the heliosphere and near its boundaries. For the transport equation the determination of the coefficients deduced from the measurements at various positions and energies is important. Reviews of transport coefficients for LECR is e.g. in (Palmer, 1982; Valdés-Galicia, 1993). Recently (Kecskeméty et al., 2011) examined the energy spectra of 0.3–100 MeV protons and found, that at the lower energies, the galactic particle spectra are significantly steeper than the $J(E) \sim E$ one, predicted by analytical approximations, such as the force-field model of modulation. Usoskin et al. (2005) provide a long series of a parameter allowing for a quantitative estimate of the average monthly differential energy spectrum of CR near the Earth for long time interval.

CR, measured directly by its secondaries at Earth over more than half of century, indicate complicated structure in its temporal behavior. The nucleonic component of primary CR is appropriate to study solar modulation from ground based measurements (e.g. Storini, 1990). CR modulation as observed from the Earth is discussed and reviewed e.g. by (Belov, 2000). Single point measurements are influenced by the large scale structure of IMF over the heliosphere, which, in the given time, in addition to the measured IMF and solar wind e.g. at L1, includes also "memory effect" of solar activity due to relatively slow outward motion of solar wind with the IMF inhomogenieties, if compared to the speed of CR particles. Long term modulation is observed as a series of steplike decreases. Outward propagating diffusion barriers were identified as merged interaction regions, MIRs (Burlaga et al., 1985). A comprehensive review of various CR intensity variations over different time scales that have been conducted over 1970s and 1980s can be found in paper (Venkatesan and Badruddin, 1990) and in references therein. Recently (Strauss et al., 2011) review some of the most prominent CR observations made near Earth, and indicate how these observations can be modelled and what main insights are gained from the modelling approach. Also, discussion on drifts, as one of the main modulation processes, is given as well as how the drift effects manifest in near Earth observations. Specifically, discussion on explanation of the observations during past unusual solar minimum, is included. Siluszyk et al. (2011) developed a two-dimensional (2-D) time dependent model of the long period variation of GCR intensity, where they included the slope of power spectrum density of IMF fluctuations, magnitude of B, tilt angle of heliocentric current sheet and effect of particle drift depending on solar magnetic field polarity. Solar modulation of GCR over the past solar cycles is described and discussed in detail e.g. by (Chowdhury et al., 2011). Recent solar minimum was specific one, unusually long and deep (e.g. Badruddin, 2011). The modulation of CR was minimal for the more than 70-year-long period of direct measurements (Bazilevskaya et al., 2011). In 2009 (Stozhkov et al., 2011) recorded the highest

CR fluxes (particles with energy > 0.2 GeV) in the history of the CR measurements in the stratosphere. An increase of the flux on NMs during that minima was also reported (e.g. Moraal and Stoker, 2010). GCR modulation for solar cycle 24 began at Earth's orbit in January 2010 (Ahluwalia and Ygbuhay, 2011). That paper reports that some NMs are undergoing long-term drifts of unknown origin. Such effects have to be examined in detail because of correct understanding the long term CR variations from the direct measurements. Corrections of data in the NM network are discussed also e.g. by (Dvornikov and Sdobnov, 2008). There are several sources of the data from CR measurements by NMs (recently e.g. NMDB data base at http://nmdb.eu).

Modulation of the low energy component of GCR inside the heliosphere gives us insight on the relevance of solar phenomena that determine the structure and evolution of the heliosphere. Reviews of the advancement in the comprehension of the phenomena controlling the transport of LECR in heliosphere can be found e.g. in (Valdés-Galicia, 2005). Time profiles of CR observed at a given position of NM on the ground are result of superposition of many transitional effects due to the temporary and spatially changing structures of IMF irregularities within the heliosphere, by the rotation of the Earth with detector, by the effects of magnetosphere and by the variable state of the atmosphere. In addition to the network of NMs, the information about the variability of CR at higher energies provide the detectors in relatively new installations. The decription of few of such installations along with the first results are reported e.g. in papers (Augusto et al., 2011; De Mendonca et al., 2011; Maghrabi et al., 2011).

Charge dependent modulation is important point in the CR drift models (e.g. Kóta and Jokipii, 1983; Potgieter and Moraal, 1985). The drift effects have to be marked differently for CR particles with the opposite sign of electric charges during epochs with opposite polarities of the solar magnetic field. Thus the interest to the measurements of positrons and antiprotons is increasing. Important finding was done by PAMELA experiment (Adriani et al., 2009; 2010). The authors present data on the positron abundance in the CR in the energy range 1.5 - 100 GeV with high statistics. The data deviate significantly from predictions of secondary production models, and the authors stress that it may constitute the first indirect evidence of dark matter particle annihilations, or the first observation of positron production from near-by pulsars. The evidence of solar activity affecting the abundance of positrons at low energies is reported too. This result was recently confirmed independently by the observations of FERMI (Ackermann et al., 2011), where the authors report first time measurement of absolute CR positron spectrum above 50 GeV, and indicate that the fraction has been determined above 100 GeV. Increase of positron fraction with energy between 20 and 200 GeV is found. The future measurements with greater sensitivity and energy resolution, such as those by AMS-02, are necessary to distinguish between many possible explanations of this increase. Serpico (2011) summarizes the global picture emerging from the data and recapitulates the main features of different types of explanations proposed. Testing of different scenarios and inferring some astrophysical diagnostics from current/near future experiments is also discussed.

2.1 Irregular variations

Coronal Mass Ejections (CME) are causing changes in CR intensity measured at Earth. Although the decrease in CR coincident with geomagnetic field depression (horizontal component) was first observed by Forbush already in 1937 (in the book Van Allen, 1993, p. 117), this phenomena attracts the CR physicists until present. Forbush decreases are generally correlated with co-rotating interaction regions (CIRs) or with the Earth-directed CMEs from the Sun (e.g. Prasad Subramanian, 2009). The characteristics of CMEs in the inner heliosphere are discussed in detail by (Gopalswamy, 2004). One of the first papers suggesting that the solar cycle dependent modulation of GCR can be explained by CMEs and by related IMF inhomegeneities in the heliosphere was that by (Newkirk et al., 1981). Correlation of CR intensity and CME ocurrence at a single NM with high geomagnetic cut-off is reported e.g. by Mishra et al. (2011) for different solar magnetic field polarities.

The Forbush decrease (FD) or cosmic ray storm is produced as a result of the transient diffusion-convection of CR caused by the passage of IMF shock wave. Nagashima et al. (1992) show that the storm is frequently accompanied by non-diffusion-convection-type phenomena, depending on local time, as precursory decrease of CR at the front of the shock in the morning hours of local time, and a post-shock increase different from the diurnal variation. FDs are analyzed by global spectrographic method using worldwide network of NMs (e.g. Sdobnov, 2011 and references therein). Recently (Dumbovic et al., 2011) performed statistical study of the relationship between characteristics of solar wind disturbances, caused by interplanetary CMEs (ICMEs) and corotating interaction regions, as well as with properties of FDs. It was found that the amplitudes of the CR depression are primarily influenced by the increase in magnetic field strength and fluctuations, and the recovery phase also depends on the magnetic field strength and size of the disturbance. The use of FDs for space weather studies is discussed by (Chauhan et al., 2011). For this application the complexity of relations between the geomagnetic storms and FDs must be assumed. Mustajab and Badruddin (2011) critically analyze the differences in geoeffectiveness due to different structures and features, with distinct plasma/field characteristics. Distinct relations of FDs and geomagnetic activity measured by Dst in different events is reported by (Kudela and Brenkus, 2004; Kane, 2010). Richardson and Cane (2011) analyzed large number of ICMEs and their associated shocks passing the Earth during years 1995 - 2009. They found that magnetic clouds are more likely to participate in the deepest GCR decreases than ICMEs that are not magnetic clouds. Examining simultaneous observations of FD events by different CR stations remains a subject of interest. Variability in the manifestations of FDs demonstrates that there are still open questions in this field (e.g. Okike and Collier, 2011; Pintér et al., 2011). Study (Oh and Yi, 2009) may support the hypothesis that the simultaneous FDs occur when stronger magnetic barriers pass by the Earth, and in contrast that the nonsimultaneous FDs occur only if the less strong magnetic barriers pass the Earth on the dusk side of the magnetosphere. The FDs are observed not only by NMs but also at higher energies of primary CRs (e,g. Braun et al., 2009; Abbrescia et al., 2011; Bertou, 2011) or by a lead free NM (Mufti et al., 2011). Rigidity spectra of FDs and their relations to the index of power spectra density (PSD) of IMF fluctuations in the frequency range in which the interaction of IMF with CR can be efficient, is studied in papers (Alania and Wawrzynczak, 2008; 2011). The theoretically derived relationship between rigidity spectrum exponent (γ) and exponent (ν) of IMF PSD is confirmed. Figure 1 illustrates different relations of FDs to geomagnetic activity.

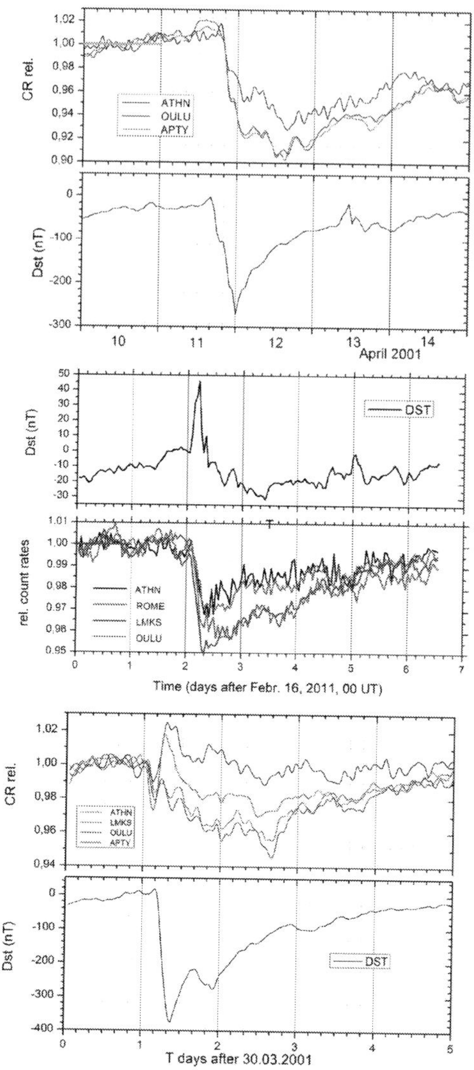

Fig. 1. Three decreases of GCR as measured by NMs (Athens, Oulu, Apatity, Rome, Lomnicky stit) with different relations to geomagnetic activity. Upper panel: CR storm accompanied by Dst depression. FD is better pronounced at low cut-off rigidity NMs (Apatity, Oulu) than at higher one (Athens). Middle panel: FD observed at several european NMs not accompanied by any Dst depression (IMF Bz>0). Lower panel: At low cut-off NMs CR decrease is seen (geomagnetic cut-offs near or below the atmospheric ones), while at middle and high cut-off positions (Lomnicky stit, Athens) an increase is seen due to the improvement of magnetospheric transmissivity.

Analysis of FDs is important for better understanding of the magnetic field structure related to shock waves and fast streams originating at the Sun. Quenby et al. (2008) examined the temporal history of the integral GCR fluence (≥100 MeV) measured by the high-sensitivity telescope (HIST) aboard the Polar spacecraft, along with the solar wind magnetic field and plasma data from the ACE spacecraft during a 40-day period encompassing September 25, 1998 FD. The authors also analyzed FD and energetic storm particle event on October 28, 2003, one of the largests in the past decades. Short-scale GCR depressions during a test period in September through October 1998 did not show correlation with changes in magnetic scattering power or fluctuations in solar wind speed or plasma density. However, IMF and solar wind data during the test period of FD suggest the presence of ICME. Mulligan et al. (2009), using the high resolution energetic particle data from ACE SIS, the Polar high-sensitivity telescope, and INTEGRAL's Ge detector measuring GCR background with a threshold of 200 MeV, show similar, short-period GCR variations in and around the FD. NMs have lower statistics. Earlier paper (Kudela et al., 1995) indicated that the Dst decreases are correlated with the „prehistory" of CR fluctuations on time scales longer than tens of minutes especially during the years with high solar activity. In the future high temporal resolution data on CR are needed, and the analysis based on combination of NM data with the satellite measurements by detectors having large geometrical factors is important (e.g. Grimani et al., 2011).

Another class of irregular variations of CR intensity observed in the vicinity of Earth are high energy particle populations accelerated in the solar flares or at the discontinuities in interplanetary space. Two types of solar energetic particle (SEP) events, namely impulsive and gradual ones, have been recognized since 1980s (e.g. Lin, 1987). Reames (1999) summarized the knowledge about energetic particle populations coming from solar flares, from shock waves driven outward by CMEs, from magnetospheres of the planets and bow shocks and reviewed various acceleration processes throughout the heliosphere. Miroshnichenko (2001) surveyed in detail the results of solar CR investigations since 1942, with including a large amount of data, obtained during long time period of observations of SEP. The book also covers theoretical models and gives an extensive bibliography. Recently Hudson (2011) reviews the knowledge of solar flares with the focus on their global properties. Flare radiation and CME kinetic energy can have comparable magnitudes, of order 10^{25} J each for an X-class event, with the bulk of the radiant energy in the visible-UV continuum. The author argues that the impulsive phase of the flare dominates the energetics of all of these manifestations, and also points out that energy and momentum in this phase largely reside in the electromagnetic field, not in the observable plasma. Barnard and Lockwood (2011) constructed a database of gradual SEP events for 1976-2006 using mainly data of > 60 MeV protons. Although number of events decreases when solar activity is low, the events during solar minimum are observed with higher fluence. Thus, very strong flares may be more likely at lower solar activity. Ground level events (GLE) are observed by NMs when the energy of accelerated particles in the flare or in interplanetary space exceeds the atmospheric threshold and geomagnetic cut-off rigidity. Characteristics of GLEs for the past solar cycles are summarized and discussed e.g. by (Gopalswamy et al., 2010; Andriopoulou et al., 2011). Moraal and McCracken (2011) analyzed all GLEs for the cycle 23. Three of the 16 GLEs have a double-pulse structure. They are associated with western flares and have good magnetic connection to the Earth. All have fast anisotropic first pulse followed by a smaller, gradual, less anisotropic second pulse. Vashenyuk et al. (2011) present a GLE modeling technique applied for 35 large GLEs for the period 1956 - 2006 and obtained features of prompt and delayed components of relativistic solar particles. Kurt et al. (2011) studied signatures of protons with energy above several hundred of MeV associated with major solar flares and observed by NMs during GLEs. The authors revealed that the delay of the earliest arrival time of high-energy protons at 1 AU with respect to the observed peak time of the solar bursts did not exceed 8 min in 28 events. This indicates that efficient acceleration of protons responsible for the GLE onset is close to the time of the main flare energy release. For the GLE observations are important high altitude NMs due to their high count rate and high statistics. List of GLEs observed at one high mountain NM can be found in (Kudela and Langer, 2008).

Important source of the information about protons accelerated near the Sun and about their interactions with residual solar atmosphere are solar gamma-rays and neutrons. Their observation near Earth is not affected by magnetic field as it is in the case of protons GLEs, SEP events. Production of γ -rays and neutrons results from convolution of the nuclear cross-sections with the ion distribution functions in the atmosphere. Recently Vilmer et al. (2011) reviewed the γ-ray and neutron observations with the emphasis on the very detailed RHESSI measurements, namely the high spectral resolution revealing line shapes and fluences, and gamma-ray imaging technique. The authors point out also still open question for the study of high energy neutral emissions from the Sun. Chupp et al. (1973) reported first observations of gamma ray lines from solar flares in August 1972 using data from OSO-7 satellite. Ramaty et al. (1979) reviewed the gamma-ray line emission from the Sun due to nuclear deexcitation of ambient nuclei following the interactions of accelerated particles. Although Biermann et al. (1951) long time ago proposed that high energy neutrons created by nuclear interactions of protons with the atmosphere of the Sun can be observed on Earth's orbit, first direct indication of solar neutrons on the ground was reported from high mountain NMs Jungfraujoch and Lomnický štít in the flare event June 3, 1982 (Debrunner et al., 1983; Efimov et al., 1983) and on satellite by electrons from the neutron decay (Evenson et al., 1983). After that event the interest to detection of solar neutrons increased. Several NMs started to measure with better temporal resolution, and new experimental devices for detection of gamma rays and neutrons both on the ground as well as on the satellites were constructed. High energy gamma rays and neutrons during several solar flares have been observed in the past decade also e.g. on low altitude polar orbiting satellite Coronas-F (Kuznetsov et al., 2006; 2011; Kurt et al., 2010). Review on experimental and theoretical works related to solar neutrons is in (Dorman, 2010a).

2.2 Periodic and quasi-periodic variations

The quasi-periodic and periodic variations in CR intensity observed at Earth are studied for rather long time. The solar diurnal wave, being the fixed one, was checked starting from papers (Forbush, 1937; Singer, 1952; Thompson, 1938; Brunberg and Dattner, 1954; Ahluwalia and Dessler, 1962). Studies of its higher harmonics, namely of semi-diurnal one, can be found in papers since (Ahluwalia, 1962; Nicolson and Sarabhai, 1948) and on the tri-diurnal starting probably from (Mori et al., 1971; Ahluwalia and Singh, 1973). At longer quasi-periodicities first attention was paid to ~ 11 yr, ~22 yr and ~27 d variabilities. The detailed review is in the books (Dorman, 1975; 2004) and references therein.

For the shape of the PSD we examined the long time series of Climax NM, skipping the data influenced by the GLEs, interpolating linearly the gaps, and applying the FFT method. PSD obtained is plotted in Figure 2.

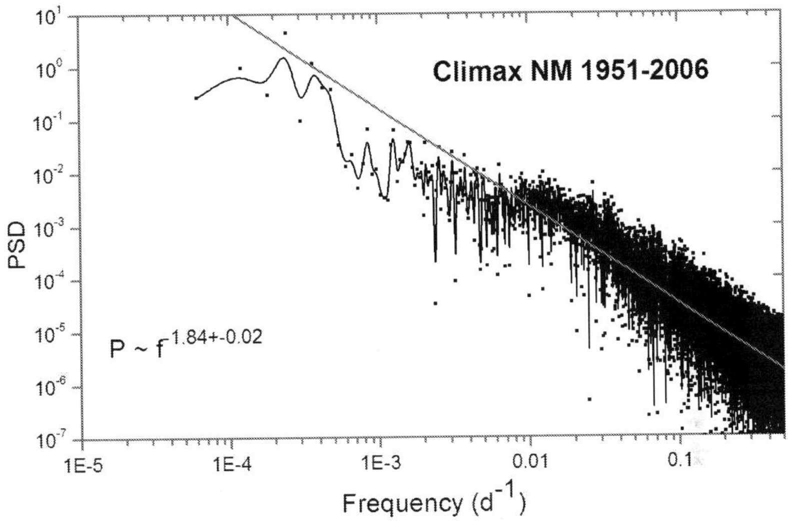

Fig. 2. The power spectrum density (PSD by FFT method) of daily averages of count rate of Climax NM (dots). Data downloaded from ftp://ulysses.sr.unh.edu/NeutronMonitor/DailyAverages.1951-.txt. Line is for the cubic spline connection.

The spectral density at higher frequency ranges is plotted in Figure 3.

At frequencies $f < 5.8 \times 10^{-6}$ Hz (upper panel of Figure 3) the slope is consistent with the theory (Jokipii and Owens, 1974b) where the authors indicate that including the effects of non-field-aligned diffusion, which dominates the power spectrum of NMs at low frequencies ($< 5 \times 10^{-6}$ Hz) produces a spectrum of f^{-2}. The lower panel clearly indicates the presence of the fixed frequencies of the diurnal, semi-diurnal and tri-diurnal waves. Probably the fourth harmonic is present too. Index of the spectra is lower, around -1.5. At $f > 5.10^{-6}$ Hz the theory with field-aligned diffusion is satisfactory for explanation of the shape (Jokipii and Owens, 1974a).

The solar diurnal anisotropy fixed at PSD as a single periodicity is resulting from the co-rotational streaming of particles past Earth (Duldig, 2001). Kudela et al. (2011) indicate the difference in the slopes of PSD at NM Lomnický štít (data at http://neutronmonitor.ta3.sk) for different phases of a solar activity cycle, namely the hardening of the PSD at solar minimum in comparison with the maximum. Similar behavior is found in the spectral analysis of the IMF magnitude B. Such picture is in qualitative agreement with the slopes of IMF measured in different solar cycle phases at large distances (Burlaga and Ness, 1998). El-Borie and Al Thoyaib (2002) studied power spectra of CR in the range 2 - 500 days and found significant differences in the individual spectra of solar maxima for different cycles.

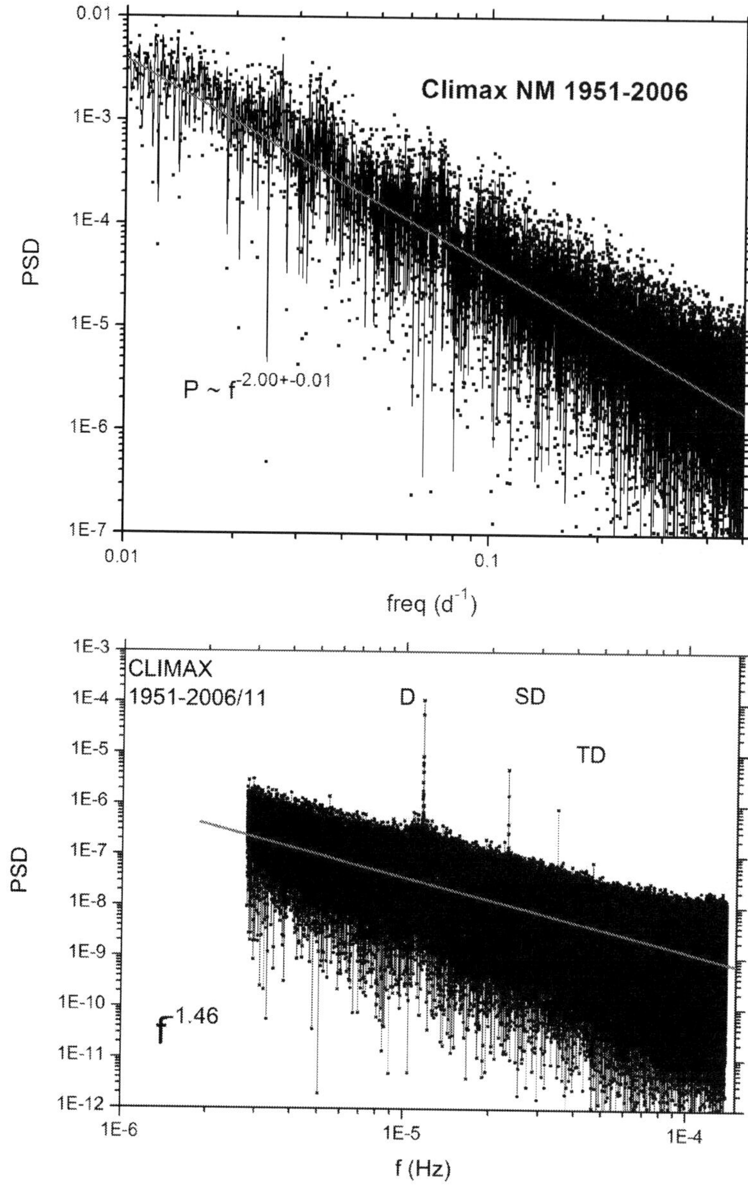

Fig. 3. Power spectrum density of Climax NM data at two ranges of frequencies. For the upper panel the daily means were used, for the lower one the hourly corrected ones. D, SD and TD is foor diurnal, semi-diurnal and tri-diurnal variation. While the amplitude of D has ~11 yr variation, the phase has ~22 yr one.

Spectra in range 2.7×10^{-7} - 1.4×10^{-4} Hz measured by muon underground detector have been examined by (Sabbah and Duldig, 2007). Flatter spectra having lower power when the interplanetary magnetic field (IMF) is directed away from the Sun above the heliosphercic current sheet ($A > 0$) than when the IMF is directed toward the Sun above the current sheet ($A < 0$), are reported.

For the description of the ~27 day variability related to the solar rotation, the daily count rate means of NM Climax was filtered in the frequency range corresponding to the time scale 25 - 33.3 days and the wavelet transform was applied on the data (Figure 4).

The double-peak structure, namely with maxima at ~ 27 days and another at ~ 31 days is found around solar activity maxima, similar to (Dunzlaff et al., 2008) based on data of GCR, EPHIN on SOHO. The structure is more complex during the long time period. Transport models (Gil et al., 2005) and measurements analyzed in paper (Richardson, 2004) suggest the dependence on solar magnetic field polarity. Vivek Gupta and Badruddin, (2009) found that the average behavior of GCR-oscillations during Carrington rotation is different in $A > 0$ from that in $A < 0$ epoch. Correlation of solar wind speed with GCR intensity during the course of Carrington rotation is stronger for $A > 0$ than for $A < 0$. The amplitudes of GCR-oscillations show somewhat weak dependence on the tilt angle of the heliospheric current sheet. Krymsky et al. (2008) indicate that temporal change of the power spectrum of ~13.5-day and ~27-day variations repeats the power spectrum change of the number of sunspots and tilt angle of the current sheet, and that the dependence of ~ 27-day variation on the polarity of general magnetic field of the Sun is not found. This feature has to be examined in future by wavelet technique in more detail. The ~27 day variation correlates with B, Bz, v, and B(v x B) - (Agarwal et al., 2011a). Similarly to ~27-day variability we examined the vicinity of ~13.5 and ~9 day periodicity contributions. At ~13.5 days we confirm the result by (Filisetti and Mussino, 1982) using ionisation chamber data, indicating the maximum contribution is correlated with the sunspot number. At ~9 days the results can be found in paper (Sabbah and Kudela, 2011).

Rieger et al. (1984) reported 154 day periodicity in solar X-ray and gamma ray flares. Pap et al. (1990) examined various periodicities in solar activity time series. There is no explanation for 150-157 day period found in several data sets. 154-day periodicities in the near-Earth IMF strength, in solar wind speed (Cane et al., 1998) and in solar proton events (Gabriel et al., 1990) have been reported. Hill et al. (2001) using Voyager 1 data, have shown that at anomalous CRs the quasi-periodic variations are in phase, with O, He having periods ~ 151 days, while protons exhibit a period ~ 146 days. Results about quasi-periodicity ~150 days in CR measured on the ground can be found e.g. in papers (Mavromichalaki et al., 2003; Kudela et al., 2010).

Another quasi-periodicity observed in CR is ~1.7 year. It was eported first by (Valdés-Galicia et al., 1996), analyzed by wavelet technique by (Kudela et al., 2002), found also in the outer heliosphere in Voyager data (Kato et al., 2003). Earlier, using NM data Calgary and Deep River, (Kudela et al., 1991) indicated that around ~20 month the change of the shape of PSD occurs. Recently Okhlopkov (2011) reports that length of the quasi-2 year periodicity in even and odd numbered cycles differs by ~2 months. Mendoza et al. (2006) by examining solar magnetic fluxes in the period 1971–1998 found that ~ 1.7 year is the dominant fluctuation for all the types of fluxes analyzed and that it has a strong tendency to appear

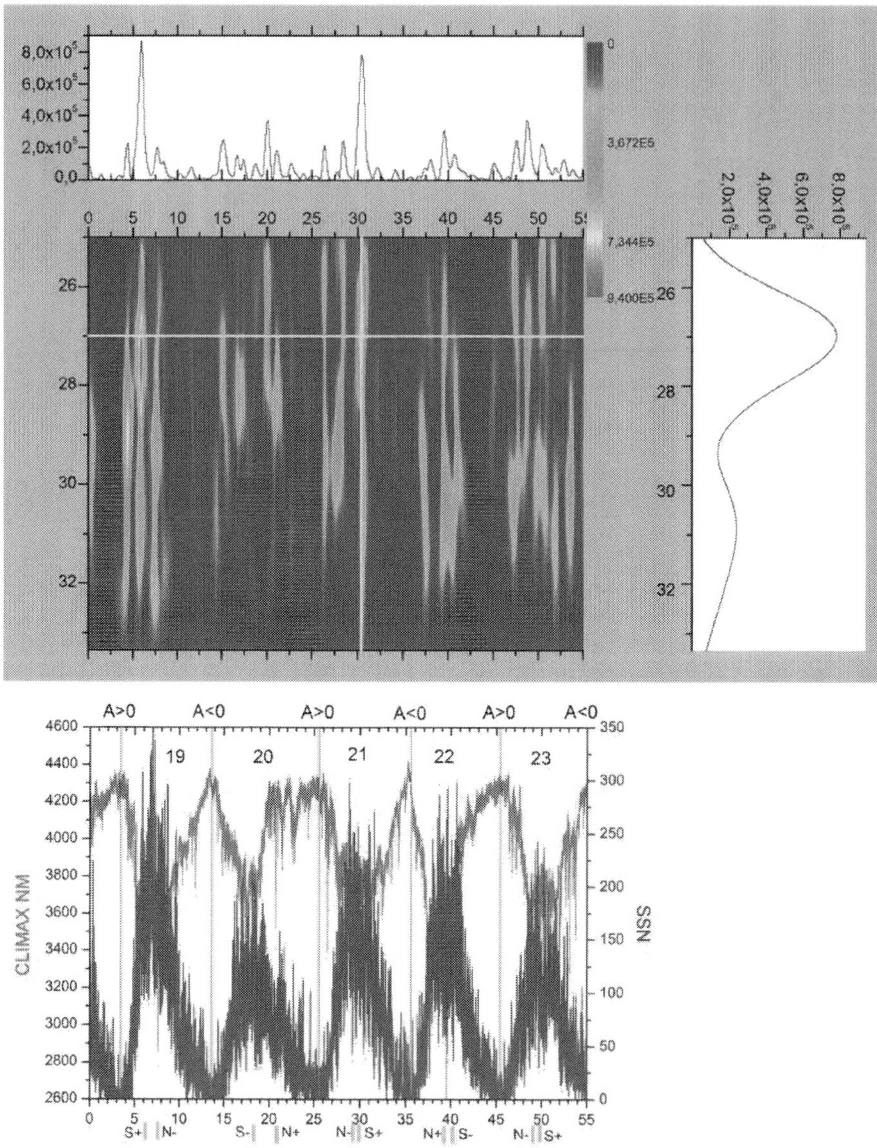

Fig. 4. The wavelet spectrum density (Morlet) of Climax NM daily means (middle panel). The upper panel is cross section of the density at 27 days over long time. Right panel is cross section over periods 25 - 33.3 d for the time of solar maxima ~ 1981. Low panel displays sunspot numbers and CR NM Climax count rates.

during the descending phase of solar activity. Quasi-periodicities of ~1.3 year (observed in solar wind) and ~1.7 years were seen neither often nor prominently in several solar activity indices (Kane, 2005). Rouillard and Lockwood (2004) relate a strong 1.68-year oscillation in GCR fluxes to a corresponding oscillation in the open solar magnetic flux and infer CR propagation paths confirming the predictions of theories in which drift is important in modulating CR flux. Charvátová (2007) indicated an interesting approach to the ~1.6 yr variation. The author calculated the solar motion due to the inner (terrestrial) planets (Mercury, Venus, Earth, Mars) for the years 1868–2030 and found that spectrum of periods shows the dominant periodicity of 1.6 years. Kane et al. (1949) reported that starting with Alfven's original suggestion, it is possible to develop a quantitative equilibrium theory for the trapping of CR in the magnetic field of the Sun, where in addition to the effect of scattering in the geomagnetic field there is also taken into account the direct absorption of the CR by five heavenly bodies Mars, Venus, the Earth, the Sun, and the Moon. This may be one of candidates for finding the link between the result (Charvátová, 2007) and ~ 1.7 year quasi-periodicity observed in CR.

At longer time scales McCracken et al. (2002) identified the presence of ~ 5 year variability in CR over epochs with low solar activity in the past. It is desirable to investigate whether a correlated 5-year signal exists in other geophysical and biological records, and if so, it could provide an additional source of data on the characteristics of the sun at times of low solar activity. El-Borie (2002) studied solar wind speed and density for quasi-periodic cyclicity and found some other long term peridicities. The 9.8, 3.8, and 1.7 - 2.2 year periods are the most significant found in the interplanetary proton flux at 190 - 440 MeV in IMP 8 data (Laurenza et al., 2009). Mavromichalaki et al (2005) reported ~2.3 years periodicity in coronal index calculated using Fe XIV 530.3 nm coronal emission line from ground-based measurements by the worldwide network of coronal stations (Rybanský, 1975). The wavelet analysis at various cut-offs for NM and for muon detector data is required to clarify whether that quasi-periodicity has a cut-off energy and how it is evolved over several tens of years. Figure 5 presents a brief summary of CR quasi-periodicities obtained in long time of measurements by NM.

Most of the quasi-periodicities identified in Figure 5 were reporteded by (Mavromichalaki et al., 2003) in the analysis of the data until 1996. For detailed analysis and for identification of contribution of various quasi-peridicities in the signal over the long time, the wavelet transform technique is suitable to be used in future, along with analysis of data from NMs at different geomagnetic cut-off positions and from muon telescopes sensitive to higher energies. The wavelet technique has been utilized also for checking the presence of other periodicities reported e.g. by (Chowdhury et al., 2010; Agarwal et al., 2011b; Zarrouk and Bennaceur, 2009) and, if used for the same intervals and applied on time series of CR as well as of solar, geomagnetic and interplanetary activity indices, may help in discriminating the links of CR to the atmospheric processes.

3. Geomagnetic effects

At low energies the trajectories of charged particles in the geomagnetic field are usually described with use of guiding center approximation (Roederer, 1970). Three adiabatic invariants connected with the three different cyclic motions, namely with the gyration, bounce and azimuthal (longitudinal) drift, are utilized. This is useful if the phases of the

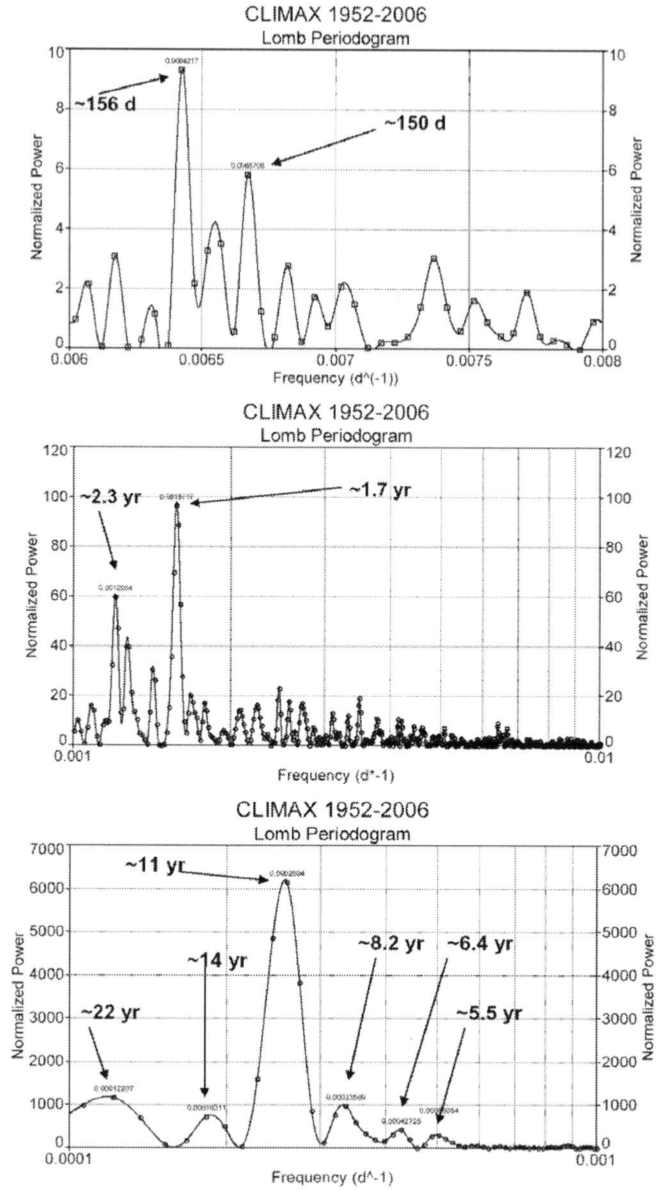

Fig. 5. Lomb-Scargle periodogram of Climax NM data at three intervals for which the filter was used. Upper panel shows the double structure with significant peaks at ~156 and ~150 days. Middle panel: in addition to ~1.7 yr variation also ~2.3 yr is statistically significant (0.99). For that level only ~11 yr variability is significant at lower plot.

motions are not of importance. Such approach is frequently used for trapped particle populations since the discovery of radiation belts. This approximation is valid if the three periodicities are very distinct. For given type of particle (proton, electron) the frequencies corresponding to the three types of motions in a dipole field depend on McIlwain's L parameter and kinetic energy (Schulz and Lanzerotti, 1974, Figure 6). The approximation becomes inapplicable when the periodicities are comparable. This is the case of higher energies in outer magnetosphere. Such particles from the point of view of magnetosphere can be assigned as CR. Detailed review of CR in the magnetospheres of planets is in (Dorman, 2009).

Trajectory description for CR particles in the Earth's magnetosphere is possible only by numerical solution of the equation of motion in given geomagnetic field model. The trajectory of particle with opposite sign of charge and velocity vector is traced starting from the point above the detector and continuing either up to the magnetosphere boundary (if the trajectory is allowed) or to the point on the ground (forbidden trajectory). For allowed trajectories the asymptotic directions are obtained. System of forbidden and allowed trajectories determines the geomagnetic cut-off rigidity. Cooke et al. (1991) summarize the definitions of the characteristics relevant for the cut-offs. Such procedure is used for long time, its history with relevant references is in the paper (Smart and Shea, 2009). Crucial for the results is the geomagnetic field model. Earlier the IGRF model was used, later the models with external current systems were introduced. Desorgher et al. (2009) discuss the geomagnetic field models used for the CR trajectory tracing. Usually for given position of NM the geomagnetic cut-off rigidity is computed for the vertical incidence of particles.

Due to geomagnetic field evolution on long time scale the geomagnetic cut-off rigidities at given point of the Earth's surface are changing (e.g. Smart and Shea, 2003; Kudela and Bobík, 2004). During the geomagnetic storms the contribution of external current systems in magnetosphere is important for the transmissivity function, asymptotic directions and geomagnetic cut-offs. Procedure that allows to determine the cut-off rigidity and asymptotic direction changes during geomagnetically active periods from measurements of magnetic field variations is presented in (Flückiger et al., 1986). Different transmissivity of CR through the Earth's magnetosphere for different empirical geomagnetic field models during strong geomagnetic storms is expected (Kudela et al., 2008). The correlations of cut-offs estimated from the global network of NMs and from trajectory tracing in one model of the field during a disturbed period, is discussed by Tyasto et al. (2011). Penetration boundary of SEP into magnetosphere is a specific tool for checking the validity of geomagnetic field models (e.g. Lazutin et al., 2011).

Recently, experiment PAMELA has shown that antiprotons produced due to nuclear interactions with the residual atmosphere are trapped in the geomagnetic field and observed at the altitude of several hundreds km (Adriani et al., 2011). Theoretical works published earlier have shown such possibility (Pugacheva et al., 2003; Gusev et al., 2008).

4. LECR, space weather and atmospheric effects

Space Weather is a relatively new discipline of science and CR play a role in this study in both aspects, namely in (a) direct one – irradiation of materials in space, in atmosphere and on the ground with various consequences for the technological systems and for the people, and (b) as one of the precursors due to changes of LECR anisotropy several hours before the onset of geomagnetic storm when CME arrives to the vicinity of Earth. LECR provide for that "remote" information about CME propagation in interplanetary space (alert for geoeffective events). There are several books and reviews on space weather effects and its possible forecasting (e.g. Song et al, 2001; Goodman, 2005; Bothmer and Daglis, 2007; Moldwin 2008; Lilensten et al., 2008), and on relations between cosmic rays, energetic particles in space and space weather (e.g. Kudela et al., 2000; 2009; Daglis (ed) 2004; Lilensten and Bornarel, 2006, Flückiger, 2007) as well as on physics behind (e.g. Scherer et al., 2005; Dorman 2010; Kallenrode, 2004; Hanslmeier, 2007; Kamide and Chian, 2007 among others). Singh et al. (2011) and Siingh (2011) reviewed CR effects on terrestrial processes such as electrical phenomena, lightning discharges cloud formation and cloud coverage, temperature variation, space weather phenomena, Earth's climate and the effects of GCRs on human health. The paper includes the new results and the authors point out many basic phenomena which require further study as well as new and long data sets.

Ions accelerated to several tens to hundreds of MeV are most important for the radiation hazard effects during solar radiation storms with electronic element failures on satellites, communication and biological consequences. Before their massive arrival, NM, if good temporal resolution and network by many stations is in real time operation, can provide useful alerts several minutes to tens minutes in advance. Kuwabara et al. (2006) report a system that detects count rate increases recorded in real time by eight NMs and triggers an alarm when GLE is detected. The GLE alert precedes the earliest indication from GOES (100 MeV or 10 MeV protons) by ~10–30 minutes. Oh el al. (2010) studied characteristics of SPE connected with GLEs.

Important point stressed by the recent papers is requirement of global detector network operating in real time with good statistical resolution is essential for space weather applications using ground based measurements. One of such systmes using neutron monitors is described by Mavromichalaki et al. (2006). At higher energies the Global Muon Detector Network is important source of the precursory information for geomagnetic storms (e.g. Rockenbach et al., 2011) and for sounding of CME geometry before its arrival to Earth (Kuwabara et al., 2004). Precursor signatures of SSC at the beginning of relatively small geomagnetic storm was also observed (Braga et al., 2011). Recently (Agarwal et al., 2011b) studied the cosmic ray, geomagnetic and interplanetary plasma/field data to understand the physical mechanism responsible for Forbush decrease and geomagnetic storm that can be used as a signature to forecast space weather and stressed the importance of change of geomagnetic cutoff rigidity.

Cosmic rays and energetic particles of lower energy interact with the material of the satellites, airplanes, atmosphere and may cause the failures. There is variety of effects with consequences on the reliability of the electronic elements. The energy deposition in materials resulting in permanent damage in silicon semiconductor devices and the single event effects due to the individual events caused by interaction of particles inside the active volume of silicon devices, along with the review of processes of electromagnetic interaction, nuclear interaction with matter is described in detail e.g. in the book (Leroy and Rancoita, 2009). The memory circuits are also partially affected by CR and its secondary products. Autran et al. (2010) review recent (2005–2010) experiments and modeling-simulation work dedicated to the evaluation of natural radiation-induced soft errors in advanced static memory (SRAM) technologies. The impact on the chip soft-error rate (SER) of both terrestrial neutrons induced by CR and alpha particle emitters, generated from traces of radioactive contaminants in CMOS process or packaging materials, has been experimentally investigated by life (i.e. real-time) testing performed at ground level on the Altitude Single-event Effect Test European Platform (ASTEP) and underground at the underground laboratory. Soft errors are caused by CR striking sensitive regions in electronic devices. Paper (Wang and Agrawal, 2010) illustrates how soft errors are a reliability concern for computer servers, and indicates a possible soft error rate (SER) reduction method that considers the CR striking angle to redesign the circuit board layout.

Miroshnichenko (2003) provides phenomenological picture of the radiation environment of Earth, summarizes observational data and theoretical findings related to main sources of energetic particles in space as well as surveys the methods of prediction of radiation risk on spacecraft. Dartnell (2011) reviews in detail the influence of ionizing radiation including CR on the emergence and persistence of life. Not only effects of ionizing radiation on organisms and the complex molecules of life are discussed, but also pointed out that ionizing radiation performs many crucial functions in the generation of habitable planetary environments and the origins of life. There are reports on the effects of short time increases of LECR on the dose within the atmosphere (airplanes, eg. Spurný et al., 2001; 2004; Felsberger et al., 2009) as well as in outer space (important for planned missions to the planets both for humans and for reliability of electronic systems); the changes of the status of the ionosphere with consequences on navigation. LECR and its measurement is important not only for monitoring radiation and its temporal and spatial variability (significant for preparing models), but its systematic measurement with good temporal resolution by many ground based devices has a potential to be one of the elements for schemes of space weather effects prediction.

Variability of CR with the aim of deducing the features useful in search of correlation between CR and atmospheric processes is described and discussed by (Bazilevskaya, 2000). Studies on relation of CR to the atmospheric processes (started probably from Svensmark and Friis-Christensen, 1997 and references therein) and references therein, recent paper reporting results of CLOUD experiment (Kirkby et al., 2011), as well as the availability of long term series of CR measurements from various NM and muon detectors until now, motivate to describe in detail LECR variability. Harrison et al. (2011) report cloud base height distributions for low cloud (<800 m) measured at the Lerwick Observatory, Shetland, UK, is varying with CR conditions. 27 day and 1.68 year periodicities characteristic of cosmic ray variations are present, weakly, in the cloud base height data of stratiform clouds, when such periodicities are present in neutron monitor CR data. Papers (Sloan and Wolfendale, 2008; Erlykin et al., 2009a,b) do not indicate that the large portion of the clouds is related to CR. No response of global cloud cover to Forbush decreases at any altitude nor latitude is reported by (Calogovic et al., 2010). Fichtner et al. (2006) point out that presence of a 22-year periodicity can not only be understood on well-known physical grounds, but must be expected if CR play a role in climate driving. The test of whether 22-year periodicities in climate indicators are present or not is a promising tool to bring the presently intensely led debate to a satisfactory conclusion. Discussion is continuing. For the purposes of checking long term variations of CR with atmospheric characteristics it is suitable to use indirectly estimated time profile of CR for the past. Usoskin et al. (2002) used the reconstructed magnetic flux as an input to a spherically symmetric quasi-steady state model of the heliosphere, and calculated the expected intensity of GCR at Earth position since 1610.

In recent decade the relations of CR to the atmospheric electricity has been studied extensively. When studying the intensity variations of secondary CR during thunderstorms (Lidvansky, 2003; Lidvansky and Khaerdinov, 2011) with the Carpet shower array of the Baksan Neutrino Observatory it was found that, in addition to regular variations correlating with the near-ground electric field, there existed considerable transient changes of the intensity. Chilingarian et al. (2010) presented the energy spectra of electrons and gamma rays from the particle avalanches produced in the thunderstorm atmosphere, reaching the Earth's surface. Paper by (Ermakov et al., 2009) shows that the main parameters of atmospheric electricity are related to CR. The mechanisms of solar forcing of the climate and long term climate change is summarized, and the role of energetic charged particles (including CR) on cloud formation and their effect on climate is discussed in (Siingh et al., 2010; 2011). Results of spectral analysis of surface atmospheric electricity data (42 years of Potential Gradient, PG at Nagycenk, Hungary) showed ~1.7 year quasi-periodicity (Harrison and Märcz, 2007). ~1.7 year periodicity in the PG data is present 1978 - 1990, but absent in 1963 - 1977. It is of interest to continue checking the occurrence of that quasi-periodicity in CR and in the data of atmospheric electricity after 1990.

Lightning is connected with the short time increases of the high energy photon flux in the atmosphere. Terrestrial Gamma-ray Flash (TGF) is a brief (<1ms) pulse of γ-rays with energies extending up to around 40 MeV, and average energy ~2 MeV (Smith et al., 2005) observed on low altitude satellites. TGFs exhibit both spatial and temporal correlations with lightning activity (e.g. Fishman et al., 1994). Study based on RHESSI data shows that on average the TGFs were found to precede the associated lightning events, with a mean delay of -0.77 ms (Collier et al., 2011). Spatial coincidence of the location of the lightning flashes with conjugate X-ray enhancements, and their simultaneity, was reported by (Bučík et al., 2006).

Cosmic ray characteristics along with the geomagnetic and solar activity are discussed also in connection with hurricanes (e.g. Kavlakov et al., 2008; Mendoza and Pazos, 2009 Perez-Peraza et al., 2008; Kane, 2006). This topic is reviewed in detail by (Mendoza, 2011).

5. Acknowledgments

We acknowledge the University of New Hampshire, "National Science Foundation Grant ATM-9912341" for Climax NM data. Author wishes to acknowledge support by the grant agency VEGA, project 2/0081/10 and by MVTS COST ES 0803 project of EU.

6. References

Abbrescia, M., S. Aiola, R. Antolini et al., Observation of the February 2011 Forbush decrease by the EEE telescopes, Eur. Phys. J. Plus, 126: 61, 2011.

Ackermann, M. et al., Measurement of separate cosmic-ray electron and positron spectra with the Fermi Large Area Telescope, arXiv:1109.0521v1, 2011.

Adriani, O. et al., PAMELA Collaboration, A statistical procedure for the identification of positrons in the PAMELA experiment, Astropart. Phys. 34, 1. arXiv:1001.3522, 2010.

Adriani, O. et al., PAMELA collaboration. Observation of an anomalous positron abundance in the cosmic Radiation, Nature, 458:607-609,2009; ArXiv:0810.4995v1, 2009.

Adriani, O. et al., The discovery of geomagnetically trapped cosmic ray antiprotons, Ap. J. Lett., 737, L29 (5pp), August 20, 2011.

Agarwal Rekha, R.K. Mishra, S.K. Pandey and P.K. Selot. 27 day variation of cosmic rays along with interplanetary parameters. Proc. 32nd ICRC, Beijing, paper 0132, 2011a.

Agarwal Rekha, Rajesh K. Mishra, M.P. Yadav and S.K. Pandey, Cosmic Rays and Space Weather Prediction, Proc. 32nd ICRC, Beijing, paper icrc0129, 2011b.

Ahluwahlia, H. S. and A.J. Dessler. Diurnal variation of cosmic radiation intensity produced by a solar wind. Planetary and Space Science, Vol. 9, p.195, 1962.

Ahluwalia, H.S. and R.C. Ygbuhay. The onset of sunspot cycle 24 and galactic cosmic ray modulation. Adv. Space Res., 48, 1, 61-64, 2011.

Ahluwalia, H.S. and S. Singh, On Higher Harmonics in Cosmic-Ray Solar Daily Variation. Proc. 13th ICRC, Denver, 2, 948, 1973.

Ahluwalia, H.S. Semidiurnal Variation of Cosmic Rays on Geomagnetically Disturbed Days. Proc. Phys. Soc. 80, 472, 1962.

Alania, M.V., Wawrzynczak, A., Energy dependence of the rigidity spectrum of Forbush decrease of galactic cosmic ray intensity, Advances in Space Research, online, 2011.

Alania, M.V., Wawrzynczak, A., Forbush decrease of the Galactic Cosmic Ray Intensity: Experimental Study and Theoretical Modeling, Astrophys. And Space Sci. Transactions, 4, 59-63, 2008.

Andriopoulou, M., H. Mavromichalaki, C. Plainaki, A. Belov and E. Eroshenko. Intense Ground-Level Enhancements of Solar Cosmic Rays During the Last Solar Cycles, Solar Phys., 269, 155-168, 2011.

Augusto, C.R.A., C. E. Navia, H. Shigueoka, and K. H. Tsui. Muon excess at sea level from solar flares in association with the Fermi GBM spacecraft detector. Phys. Rev. D 84, 042002, 2011.

Autran, J.L., D. Munteanu, P. Roche, G. Gasiot, S. Martinie, S. Uznanski, S. Sauze, S. Semikh, E. Yakushev, S. Rozov, P. Loaiza, G. Warot, M. Zampaolo. Soft-errors induced by terrestrial neutrons and natural alpha-particle emitters in advanced memory circuits at ground level. Microelectronics Reliability 50, 1822–1831, 2010.

Badruddin. Solar modulation during unusual minimum of solar cycle 23: Comparison with past three solar minima, Proc. 32nd ICRC, Beijing, paper icrc0116, 2011.

Balogh, A. and G. Erdös. The heliospheric magnetic field. Space Sci. Rev., published online, DOI 10.1007/s11214-011-9835-3, 2011.

Barnard, L. and M. Lockwood. A survey of gradual solar energetic particle events. JGR, 116, A05103, 2011.

Bazilevskaya, G.A. Observations of Variability in Cosmic Rays, Space Science Rev., 94, 25-38, 2000.

Bazilevskaya, G.A. , Krainev, M.B., Makhmutov, V.S., Svirzhevskaya, A.K., Svirzhevsky, N.S., Stozhkov, Y.L. , Features of cosmic ray variation at the phase of the minimum between the 23rd and 24th solar cycles, Bull. Russian Acad. Sci: Physics , 75, 6, 779-781, 2011.

Belov, A. Large Scale Modulation: View From the Earth. Space Sci. Rev., 93,1/2, 79-105, 2000.

Bertou, X. Background radiation measurement with water Cherenkov detectors, Nucl. Instruments and Methods in Physics research, section A, 639, 1, 73-76, 2011.

Bieber, J.W.; Eroshenko, E.; Evenson, P.; Flückiger, E.O.; Kallenbach, R. , Editors. Cosmic Rays and Earth. Proceedings of an ISSI Workshop 21-26 March 1999, Bern, Switzerland, Series: Space Sciences Series of ISSI, Vol. 10, 2001.

Biermann, L., O. Haxel, A. Schluter. Neutrale Ultrastrahlung von der Sonne. Z. Naturforsch. 6a, 47, 1951.

Bothmer, V. and I.A. Daglis. Space Weather - Physics and Effects. Springer, pp. 437, 2007.

Braga, C.R., A. Dal Lago, M. Rockenbach, N.J. Shuch, L.R. Vieira, K. Munakata, C. Kato, T. Kuwabara, P.A. Evenson, J. W. Bieber, M. Tokumaru, M.L. Duldig, J.E. Humble, I.S. Sabbah, H.K. Al Jassar, M.M. Sharma. Precursor signatures of the storm sudden commencement in 2008. Proc. 32nd ICRC, Beijing, paper icrc0717.

Braun, I., J. Engler, J.R. Horandel and J. Milke. Forbush decreases and solar events seen in the 10–20 GeV energy range by the Karlsruhe Muon Telescope, Advances in Space Research 43, 480–488, 2009.

Brunberg, E. A. and A. Dattner. On the Interpretation of the Diurnal Variation of Cosmic Rays. Tellus, 6, 1, 73-83, 1954.

Bučík, R., K. Kudela and S.N. Kuznetsov. Satellite observations of lightning-induced hard X-ray flux enhancements in the conjugate region, Ann. Geophys., 24, 1969–1976, 2006.

Burlaga, L.F. and N.F. Ness, Magnetic field strength distributions and spectra in the heliosphere and their significance for cosmic ray modulation: Voyager 1, 1980-1994, *JGR*, 103, A12, 29,719-29,732, 1998.

Burlaga, L.F., F.B. McDonald, M.L. Goldstein and A.J. Lazarus, Cosmic ray modulation and turbulent interaction regions near 11 AU, J. Geophys. Res., 90, A12, 12,027-12,039, 1985.

Calogovic, J., C. Albert, F. Arnold, J. Beer, L. Desorgher, and E. O. Flückiger, Sudden cosmic ray decreases: No change of global cloud cover, Geophys. Res. Lett., 37, L03802, doi:10.1029/2009GL041327, 2010.

Cane, H.V., G. Wibberenz, I.G. Richardson and T.T. von Rosenvinge, Cosmic ray modulation and the solar magnetic field, Geophys. Res. Lett., 26, 5, 565-568, 1999.

Cane, H.V., I.G. Richardson and T.T. von Rosenvinge, Interplanetary magnetic field periodicity of ~153 days, *GRL*, 25, 4437-4440, 1998.

Collier, A.B., T. Gjesteland and N. Østgaard, Assessing the power law distribution of TGFs. J. Geophys. Res., 116, A10320, doi:10.1029/2011JA016612, 2011.

Cooke, D.J., J.E. Humble, M.A. Shea, D.F. Smart, N. Lund, I.L. Rasmussen, B. Byrnak, P. Goret and N. Petrou. On cosmic-ray cut-off terminology. Il Nuovo Cimento 14, 213–234, 1991.

Daglis, I.A. (editor), Effects of Space Weather on Technology Infrastructure, Proc. of the NATO ARW on Effects of Space Weather on Technology Infrastructure, Rhodes, Greece, 2003, Kluwer Academic Publishers, 2004.

Dartnell, L.R. Ionizing Radiation and Life. Astrobiology, 11, 6, 551-582, 2011.

Debrunner, H., E.O. Flückiger, E.L. Chupp & D.J. Forrest. The solar cosmic ray neutron event on June 3, 1982. Proc. 18th ICRC,Bangalore, India, 4, 75–78, 1983.

DeMendonca, R.R.S., J.-P.Raulin, F.C.P.Bertoni, E.Echer, V.S.Makhmutov, G.Fernandez, Long-term and transient time variation of cosmic ray fluxes detected in Argentina by CARPET cosmic ray detector, J. Atmos. Solar Terr. Physics, 73, 1410-1416, 2011.

Desorgher, L., Kudela, K., Flückiger, E. O., Bütikofer, R., Storini, M., Kalegaev, V., Comparison of Earth's magnetospheric magnetic field models in the context of cosmic ray physics, Acta Geophysica, 57, 1, 75-87, 2009.

Dorman, L.I. Cosmic ray variations and space weather, Physics - Uspekhi, 53 (5), 496-503, 2010b.

Dorman, L.I. Variations of galactic cosmic rays, Moscow, MGU Publ. House, pp. 214, in Russian, 1975.

Dorman, Lev. Cosmic Rays in Magnetospheres of the Earth and other Planets, Astrophysics and Space Science Library, 358, Springer, pp. 770, 2009.

Dorman, Lev. Cosmic Rays in the Earth's Atmosphere and Underground. Astrophysics and Space Science Library, pp. 855, Kluwer, 2004.

Dorman, Lev. Cosmic Rays: Variations and Space Explorations. Elsevier Science Publishing Co Inc.,U.S. pp. 691, 1974.

Dorman, Lev. Solar Neutrons and Related Phenomena, Astrophysics and Space Science Library, 365, Springer, pp. 873, 2010a.

Duldig, M.L., Australian Cosmic Ray Modulation Research. Publ. Astron. Soc. Austr., 18, 12-40, 2001.

Dumbovic, M., B. Vrsnak, J. Calogovic, and M. Karlica, Cosmic ray modulation by solar wind disturbances, A&A 531, A91, 2011.

Dunzlaff, P., Heber, B., Kopp, A., Rother, O., Müller-Mellin, R., Klassen, A., Gómez-Herrero, R., Wimmer-Schweingruber, R. Observations of recurrent cosmic ray decreases during solar cycles 22 and 23, Ann. Geophys., 26, 3127–3138, 2008.

Dvornikov, V.M. and V. E. Sdobnov, Correction of Data from the Neutron Monitor Worldwide Network, Geomagnetism and Aeronomy, Vol. 48, No. 3, pp. 314–318, 2008.

Efimov, Yu. E.; Kocharov, G. E.; Kudela, K. On the solar neutrons observation on high mountain neutron monitor. Proc. 18th ICRC, Bangalore, India, 10, 276 - 278, 1983.

El-Borie, M.A. and S.S. Al-Thoyaib, Power Spectrum of Cosmic-ray Fluctuations During Consecutive Solar Minimum and Maximum Periods. Solar Phys., 209, 397–407, 2002.

El-Borie, M.A. On Long-Term Periodicities In The Solar-Wind Ion Density and Speed Measurements During The Period 1973-2000. Solar Phys., 208, 345–358, 2002.

Erlykin, A.D.; Gyalai, G.; Kudela, K.; Sloan, T.; Wolfendale, A.W., On the correlation between cosmic ray intensity and cloud cover, Journal of Atmospheric and Solar-Terrestrial Physics, Volume 71, Issue 17-18, 1794-1806, 2009a.

Erlykin, A.D., G Gyalai, K Kudela, T Sloan, A W Wolfendale. Some aspects of ionization and the cloud cover, cosmic ray correlation problem, Journal of Atmospheric and Solar-Terrestrial Physics, 71, 8-9, 823-829, 2009b.

Ermakov, V.I., V. P. Okhlopkov and Yu. I. Stozhkov, Influence of cosmic rays and cosmic dust on the atmosphere and Earth's climate, Bull. Rus. Acad. Sci., Physics, 73, 3, 416-418, DOI: 10.3103/S1062873809030411, 2009.

Evenson, P.; Meyer, P.; Pyle, K. R., Protons from the decay of solar flare neutrons, Astrophysical Journal, Part 1, 274, 875-882, 1983.

Felsberger, E., K. O'Brien, P. Kindl. : Iason-free: Theory and experimental comparisons. Radiation Protection Dosimetry, Vol. 136, Issue 4, 16 July 2009, Article number ncp128, 267-273, 2009.

Fichtner, H., K. Scherer and B. Heber. A criterion to discriminate between solar and cosmic ray forcing of the terrestrial climate. Atmos. Chem. Phys. Discuss., 6, 10811–10836, 2006.

Filisetti, O. and V. Mussino, Periodicity of about 13 days in the cosmic-ray intensity in the solar cycles no. 18, 19 and 20. Rev. Bras. Fis., Vol. 12, No. 4, p 599 - 610, 12/1982.

Fishman, G. J., et al. (1994), Discovery of intense gamma-ray flashes of atmospheric origin, Science, 264(5163), 1313–1316, doi:10.1126/science.264.5163.1313, 1994.

Fisk, L. A.; M.A. Forman and W.I. Axford. Solar modulation of galactic cosmic rays. 3. Implication of the Compton-Getting coefficient. JGR, 78, 995-1006, 1973.

Flückiger, E.O. Cosmic Rays and Space Weather, presentation at http://www.slidefinder.net/c/cosmic_rays_and/space_weather/1510213, 2007.

Flückiger, E. O., D.F. Smart, M.A. Shea. A procedure for estimating the changes in cosmic ray cutoff rigidities and asymptotic directions at low and middle latitudes during periods of enhanced geomagnetic activity. Journal of Geophysical Research 91, 7925-7930, 1986.

Gabriel, S., R. Evans, and J. Feynman, Periodicities in the occurrence rate of solar proton events, *Sol. Phys.*, 128, 415-422, 1990.

Gaisser, T.K. Cosmic Rays and Particle Physics. Cambridge University Press, pp. 279, 1990.

Garcia-Munoz, M., G.M. Mason and J.A. Simpson. A New Test for Solar Modulation Theory: the 1972 May-July Low-Energy Galactic Cosmic-Ray Proton and Helium Spectra. ApJ., 182, L81, 1973.

Gil, A., Iskra, K., Modzelewska, R., Alania, M. V., On the 27-day variations of the galactic cosmic ray anisotropy and intensity for different periods of solar magnetic cycle, Adv. Space Res., 35, 687-690, 2005.

Gleeson, L. J. and Axford, W.I. Solar modulation of galactic cosmic rays, The Ap. J., 154, 1011-1026, 1968.

Goodman, J.M. Space Weather & Telecommunications, Kluwer International Series in Engineering and Computer Science, Springer,pp.382, 2005.

Gopalswamy, N. A Global Picture of CMEs in the Inner Heliosphere. In The Sun and the heliosphere as an integrated system. Editors G. Poletto and S.T. Suess, Kluwer, 201-252, 2004.

Gopalswamy, N., H. Xie, S. Yashiro and I. Usoskin. Ground level enhancement events of solar cycle 23, Indian Journal of Radio & Space Physics, 39, 240-248, 2010.

Grieder, P.K.F. Cosmic Rays at Earth. Researcher's Reference Manual and Data Book, pp. 1093. Elsevier, 2001.

Grieder, P.K.F. Extensive Air Showers, vol. 1 and 2, pp. 1113, Springer, 2010.

Grimani, C., H.M. Araujo, M. Fabi, I. Mateos, D.N.A. Shaul, T.J. Sumner and P. Wass. Galactic cosmic-ray energy spectra and expected solar events at the time of future space missions. Classical and Quantum Gravity, 28, 9, Art. No. 094005, 2011.

Grupen, K. Astroparticle Physics. Springer Berlin Heidelberg New York, pp. 441, 2005.

Gusev, A.A., G. I. Pugacheva, V. Pankov, J. Bickford, W. N. Spjeldvik, U. B. Jayanthi and I. M. Martin, Antiparticle Content in the Magnetosphere, Advances in Space Research, Volume 42, Issue 9, 3 November 2008, Pages 1550-1555, 2008.

Hanslmeier, A, The Sun and Space Weather, Astrophysics and Space Physics Library, 347, Springer, pp. 315, 2007.

Harrison, R.G. and F. Märcz, Heliospheric timescale identified in surface atmospheric electricity, *GRL*, 34, L23816, 2007GL031714, 2007.

Harrison, R.G., M.H. P. Ambaum and M. Hapgood, Cloud base height and cosmic rays, Proc. R. Soc. A doi:10.1098/rspa.2011.0040, 2011.

Hatton, C.J. The Neutron Monitor, in J., G. Wilson and S.A. Wouthuysen (eds.), Progress in Elementary Particle and Cosmic-ray Physics, vol. 10, chapter 1, North Holland Publishing Co., Amsterdam, 1971.

Hayakawa, S. Cosmic ray physics: nuclear and astrophysical aspects, Wiley-Interscience, U. California, pp. 774, 1969.

Hess, V.F. Über beobachtungen der durchdringenden Strahlung bei sieben Freiballonfahrten, Phys. Ztschr., 13, 1084-1091, 1912.

Hill, M.E., D.C. Hamilton and S.M. Krimigis, Radial and Latitudinal Intensity Gradients of Anomalous Cosmic Rays During the Solar Cycle 22 Recovery Phase, *JGR*, 106, A5,8315, 2001.

Hillas. A.M. Cosmic Rays. Pergamon Press, pp. 297, 1972.

Hudson, H.S. Global properties of solar flares. Space Sci. Rev., 158, 5-41, 2011.

Charvátová, I., The prominent 1.6-year periodicity in solar motion due to the inner planets, Ann. Geophys., 25, 1227–1232, 2007.

Chauhan, M.L., Manjula Jain and S. K. Shrivastava. Space weather application of forbush decrease events. Proc. 32nd ICRC Beijing, paper icrc0155, 2011.

Chilingarian, A., A. Daryan, K. Arakelyan, A. Hovhannisyan, B. Mailyan, L. Melkumyan, G. Hovsepyan, S. Chilingaryan, A. Reymers, and L. Vanyan, Ground-based observations of thunderstorm-correlated fluxes of high-energy electrons, gamma rays, and neutrons. Phys. Rev. D 82, 043009, 2010.

Chowdhury, P., B.N. Dwivedi and P.C. Ray. Solar modulation of galactic cosmic rays during 19–23 solar cycles, New Astronomy, 16, 430-438, 2011.

Chowdhury, Partha; Khan, Manoranjan; Ray, P. C., Evaluation of the intermediate-term periodicities in solar and cosmic ray activities during cycle 23, *Astrophys. Space Sci.*, 326, 191-201, 2010.

Chupp, E. L., D.J. Forrest, P.R. Higbie, A.N. Suri, C.Tsai and P.P. Dunphy. Solar Gamma Ray Lines observed during the Solar Activity of August 2 to August 11, 1972. Nature, 241, 5388, 333-335, 1973.

Jokipii, J. R., Cosmic-ray propagation, 2, Diffusion in the interplanetary magnetic field, Astrophys. J., 149, 405, 1967.

Jokipii, J. R., Cosmic-ray propagation, 1, Charged particles in a random magnetic field, Astrophys. J., 146, 480, 1966.

Jokipii, J.R. and A.J. Owens, Cross correlation between cosmic-ray fluctuations and interplanetary magnetic-field fluctuations, GRL, 1,329, 1974b.

Jokipii, J.R. and A.J. Owens, Cosmic Ray Scintillations, 4. The Effects of Non-Field-Aligned Diffusion. JGR, 81, 13, 2094-2096, 1974a.

Jokipii, J.R. Cosmic Rays. In Auroras, Magnetic Storms, Solar Flares, Cosmic Rays. Ed. Suess, S.T. & Tsurutani, B., AGU, 123-13, 1998.

Kallenrode, May-Britt. Space Physics: An Introduction to Plasmas and Particles in the Heliosphere and Magnetospheres, Springer-Verlag Berlin, Heidelberg, pp. 482, 2004.

Kamide, Y. and A. Chian, editors. Handbook of the Solar-Terrestrial Environment. Springer, pp. 539, 2007.

Kane, E.O., J.B. Shanley and J.A. Wheeler,Influence on the Cosmic-Ray Spectrum of Five Heavenly Bodies, *Rev. Mod. Phys.*, 21, 1, 51-71, 1949.

Kane, R.P. Severe geomagnetic storms and Forbush decreases: interplanetary relationships reexamined, Ann. Geophys., 28, 479–489, 2010.

Kane, R.P. Spectral characteristics of Atlantic seasonal storm frequency. J. of India Meteor. Dept. (MAUSAM), 57, 597-608, 2006.

Kane, R.P., Short-Term Periodicities in Solar Indices, *Sol. Phys.*, 227, 155–175, 2005.

Kato, C.; Munakata, K.; Yasue, S.; Inoue, K.; McDonald, F. B., A ~1.7-year quasi-periodicity in cosmic ray intensity variation observed in the outer heliosphere. *JGR*, 108, A10, 1367, 2003.

Kavlakov, S., J. Perez-Peraza and J.B. Elsner, A statistical link between tropical cyclone intesification and major geomagnetic disturbances, Geofisica Internacional, 47, 207-213, 2008.

Kecskeméty, K., Yu. I. Logachev, M. A. Zeldovich and J. Kota, Modulation of the galactic low energy proton spectrum in the inner heliosphere, The Astrophysical Journal, 738:173 (10pp), 2011 September 10.

Kirkby, J. et al., Role of sulphuric acid, ammonia and galactic cosmic rays in atmospheric aerosol nucleation, Nature 476, 429–433 (25 August 2011).

Kóta, J. and J.R. Jokipii. Effects of drift on the transport of cosmic rays VI. A three dimensional model including diffusion. Astrophys. J. , 265, 573, 1983.

Krymsky, G.F., V.P. Mamrukova, P.A. Krivoshapkin, S.K. Gerasimova, S.A. Starodubtsev. Recurrent variations in the high-energy cosmic ray intensity. Proc. 30th ICRC, Mexico, v. 1, 381-384, 2008.

Kudela, K., A.G. Ananth and D. Venkatesan. The low-frequency spectral behavior of cosmic ray intensity. JGR, 96, 15,871-15,875, 1991.

Kudela, K. and P. Bobík. Long-Term variations of geomagnetic rigidity cutoffs, Sol. Phys., 224, 1-2, 423-431, 2004.

Kudela, K. and R. Brenkus. Cosmic ray decreases and geomagnetic activity: list of events 1982–2002. J. Atmos. Solar Terr. Phys., 66, 1121–1126, 2004.

Kudela, K. and R. Langer. Ground Level Events Recorded at Lomnicky Stit Neutron Monitor. Proc. 30th ICRC, Mérida, Mexico. 1, SH.1.8, 205-208, 2008.

Kudela, K., Cosmic Rays and Space Waether: Direct and Indirect Relations, Proc. 30th ICRC, Mexico, ed. R. Caballero, J.C. D'Olivo, G. Medina-Tanco and J.F. Valdés-Galicia, UNAM, 195-208, 2009.

Kudela, K., Venkatesan, D., Flückiger, E.O., Martin, I.M., Slivka, M. and H. Graumann, Cosmic Ray Variations: Periodicities at T<24 hours, Proc. 24th ICRC, Rome, vol. 4, p. 928-931, 1995.

Kudela, K., Mavromichalaki, H., Papaioannou, A., Gerontidou, M., On Mid-Term Periodicities in Cosmic Rays, Sol. Phys., 266, 173–180, 2010.

Kudela, K., Storini, M., Hofer, M.Y., Belov, A. Cosmic rays in relation to space weather. Space Sci Rev 93(1–2):153–174, 2000.

Kudela, K.; Rybák, J.; Antalová, A.; Storini, M., Time Evolution of low-Frequency Periodicities in Cosmic ray Intensity, *Sol. Phys.*, 205, 165 - 175, 2002.

Kudela, K. et al., On quasi-periodic variations in cosmic rays, submitted for Proc. 13th ICATPP, Como, Italy, 2011.

Kurt, Victoria, B. Yushkov, K. Kudela, and V. Galkin. High-Energy Gamma Radiation of Solar Flares as an Indicator of Acceleration of Energetic Protons. Cosmic Research, 48, 1, 70-79, 2010.

Kurt, Victoria, B. Yushkov, A. Belov, I. Chertok and V. Grechnev. A Relation between Solar Flare Manifestations and the GLE Onset, Proc. 32nd ICRC, Beijing, paper icrc0441, 2011.

Kuwabara, T., et al. Geometry of an interplanetary CME on October 29, 2003 deduced from cosmic rays, Geophys. Res. Lett., 31, L19803, doi:10.1029/2004GL020803, 2004.

Kuwabara, T., J. W. Bieber, J. Clem, P. Evenson, and R. Pyle . Development of a ground level enhancement alarm system based upon neutron monitors , Space Weather, 4, S10001, SW000223, 2006.

Kuznetsov, S.N., V.G. Kurt, B.Y. Yushkov, K. Kudela and V.I. Galkin. Gamma-Ray and High-Energy -Neutron Measurements on CORONAS-F during the Solar Flare of 28 October 2003. Sol. Phys., 268, 1, 175-193, 2011.

Kuznetsov, S.N., V.G. Kurt, I.N. Myagkova, B. Y. Yushkov and K. Kudela, Gamma-ray emission and neutrons from solar flares recorded by the SONG instrument in 2001–2004, Solar System Res., 40, 2, 104-110, 2006.

Laurenza, M.; Storini, M.; Giangravè, S.; Moreno, G., Search for periodicities in the IMP 8 Charged Particle Measurement Experiment proton fluxes for the energy bands 0.50-0.96 MeV and 190-440 MeV, *JGR*, 114, A01103, 2009.

Lazutin, L.L., E.A.Muraveva, K.Kudela and M.Slivka : Verification of Magnetic Field Models Based on Measurements of Solar Cosmic Rays Protons in the Magnetosphere, Geomagnetism and Aeronomy, 51, 2, 198-209, 2011.

Leroy Claude and Pier-Giorgio Rancoita. Principles of Radiation Interacction in Matter and Detection, 2nd Edition. World Scientific, pp. 930, 2009.

Lidvansky, A.S. and N.S. Khaerdinov, Cosmic Rays in Thunderstorm atmosphere: variations of different components and accompanying effects, in press, Proc. 13th ICATPP, Como, Italy, 2011.

Lidvansky, A.S. The Effect of the Electric Field of the Atmosphere on Cosmic Rays, J. Phys. G: Nucl. Part. Phys., vol. 29, pp. 925-937, 2003.

Lilensten, J. and J. Bornarel, Space Weather, Environment and Societies, Springer, Dordrecht, The Netherlands, pp. 241, 2006.

Lilensten, J., A. Belahaki, M. Messerotti, R. Vainio, J. Watermann and Stefaan Poedts, editors. Development the scientific basis for monitoring, modelling and predicting Space Waetaher. COST Office, Brussels, pp. 359, 2008.

Lin, R. P., Solar particle acceleration and propagation, Rev. Geophys., 25, 676, 1987.

Longair, M. High Energy Astrophysics. Cambridge University Press, 1981.

Maghrabi, A.H., Al Harbi, H., Al-Mostafa, Z.A., Kordi, M.N., Al-Shehri, S.M., The KACST muon detector and its application to cosmic-ray variations studies, Advances in Space Research, doi: 10.1016/j.asr.2011.10.011, 2011.

Mavromichalaki, H., G. Souvatzoglou, C. Sarlanis, G. Mariatos, C. Plainaki, M. Gerontidou, A. Belov, E. Eroshenko, V. Yanke, Space weather prediction by cosmic rays, Advances in Space Research, 37, 1141–1147, 2006.

Mavromichalaki, H., Preka-Papadema, P., Petropoulos, B., Tsagouri, I., Georgakopoulos, S., and Polygiannakis, J. Low- and high-frequency spectral behavior of cosmic-ray intensity for the period 1953-1996. Ann. Geophys., 21, 1681-1689, 2003.

Mavromichalaki, H.; Petropoulos, B.; Plainaki, C.; Dionatos, O.; Zouganelis, I., Coronal index as a solar activity index applied to space weather, Adv. Space Res. ,35, 410–415, 2005.

McCracken, K.G.,Beer, J. and McDonald, F.B., A five-year variability in the modulation of the galactic cosmic radiation over epochs of low solar activity, GRL, 29, NO. 24, 2161, 2002.

Mendoza, B. and M.A. Pazos. A 22-yrs hurricane cycle and its relation to geomagnetic activity, J. Atm. Solar-Terr. Phys., 71, 17-18, 2047-2054, 2009.

Mendoza, B. The effects of space weather on hurricane activity. INTECHopen, ed. A. Lupo, April 2011 (http://www.intechopen.com/articles/show/title/the-effects-of-space-weather-on-hurricane-activity).

Mendoza, B.V., V. M. Velasco and J. F. Valdés-Galicia, Mid-Term Periodicities in the Solar Magnetic Flux, Sol. Phys., 233, Issue 2, pp.319-330, 2006.

Miroshnichenko, L.I. Radiation Hazard in Space. Astrophys. And Space Science Library 207, Kluwer, Dordrecht, pp. 238, 2003.

Miroshnichenko, L.I. Solar Cosmic Rays. Astrophysics and Space Physics Library, 260, pp. 480, Kluwer, 2001.

Mishra, B.K., P. J. Shrivastava and R.K. Tiwari. A Study of the Role of the Coronal Mass Ejections in Cosmic Ray Modulation, J. Pure Appl. & Ind. Phys. Vol.1 (4), 222-226, 2011.

Moldwin, M. An Introduction to Space Weather. Cambridge U. Press, pp. 134, 2008.

Moraal, H. and K.G. McCracken. The Time Structure of Ground Level Enhancements in Solar Cycle 23, Space Sci Rev., DOI 10.1007/s11214-011-9742-7, online 2011.

Moraal, H. and Stoker, P. H.: Long-term neutron monitor observations and the 2009 cosmic ray maximum, Geophys. Res., 115, A12109, 2010.

Mori, S., S. Yasue and M. Ichinose, The Daily-Variation Third Harmonic of the Cosmic Radiation., paper MOD-37, Proc. 12th ICRC, Hobart, 2, 666, 1971.

Mufti, S., M.A.Darzi, P.M.Ishtiaq, T.A.Mir and G.N.Shah, Enhanced diurnal variation and Forbush decreases recorded with Lead-Free Gulmarg Neutron Monitor during the solar active period of late October 1989, Planet. Space Sci., 59, 394-401, 2011.

Mulligan, T., J.B. Blake, D. Shaul, J.J. Quenby, R.A. Leske, R.A. Mewaldt and M. Galamertz, Short-period variability in the galactic cosmic ray intensity: High statistical resolution observations and interpretation around the time of a Forbush decrease in August 2006, JGR, 114, A07105, 2009.

Mustajab, F. and Badruddin, Geoeffectiveness of the interplanetary manifestations of coronal mass ejections and solar-wind stream–stream interactions, Astrophys Space Sci (2011) 331: 91–104, 2011.

Nagashima, K., K. Fujimoto, S. Sakakibara, I. Morishita, R. Tatsuoka. Local-time-dependent pre-IMF-shock decrease and post-shock increase of cosmic rays, produced respectively by their IMF-collimated outward and inward flows across the shock responsible for forbush decrease, Planetary and Space Science, 40, 8, 1109-113, 1992.

Newkirk, G., Jr.; Hundhausen, A. J.; Pizzo, V., Solar cycle modulation of galactic cosmic rays - Speculation on the role of coronal transients, JGR, 86, 5387-5396, 1981.

Nicolson, P. and V. Sarabhai, The Semi-Diurnal Variation in Cosmic Ray Intensity, Proc. Phys. Soc. 60, 509, 1948.

Oh, S.Y. and Y. Yi. Statistical reality of globally nonsimultaneous Forbush decrease events. JGR, 114, A11102, 2009.

Oh, S.I., Y. Yi, J.W. Bieber, P. Evenson and Y.K. Kim, Characteristics of solar proton events associated with ground level enhancements. JGR, 115, A10107, 2010.

Okhlopkov, V.P., Distinctive properties of the frequency spectra of cosmic ray variations and parameters of solar activity and the interplanetary medium in solar cycles 20-23, Moscow U. Phys. Bull, 66, 1, 99–103, 2011.

Okike, O. and A.B. Collier. A multivariate study of Forbush decrease simultaneity. Journal of Atmospheric and Solar-Terrestrial Physics, 73, 7-8, 796-804, 2011.

Palmer, I.D. Transport coefficients of low energy cosmic rays in interplanetary space, Rev. Geophysics, 20, 2, 335-351, 1982.

Pap, J., W.K. Tobiska and S.D. Bouwer, Periodicities of solar irradiance and solar activity indices, Sol. Phys., 129, 165-189, 1990.

Parker, E.N. 1965. The passage of energetic charged particles through interplanetary space. Planet. Space Sci., 13, 1, 9-49, 1965.

Pérez-Peraza, J., Velasco, V. and S. Kavlakov. Wavelet coherence analysis of atlantic hurricanes and cosmic rays. Geofísica Internacional, 47, 231-244, 2008.

Pintér, T., M. Rybanský, K. Kudela, I. Dorotovič, Peculiarities in evolutions of cosmic radiation level after sudden decreases, in press, Sun and Geosphere, 2011.

Potgieter, M. S. & Le Roux, J. A, The Long-Term Heliospheric Modulation of Galactic Cosmic Rays according to a Time-dependent Drift Model with Merged Interaction Regions, Astrophysical Journal v.423, p.817-827, 1994.

Potgieter, M.S. and H. Moraal, A drift model for the modulation of galactic cosmic rays. Astrophys. J., 294, 425–440, 1985.

Potgieter, M.S. Cosmic Rays in the Inner Heliosphere: Insights from Observations, Theory and Models, Space Sci Rev., DOI 10.1007/s11214-011-9750-7, online, 2011.

Prasad Subramanian. Forbush decreases and Space weather. Available at http://www.iiap.res.in/ihy/school/prasad_lecture.pdf, 2009.

Pugacheva, G. I., A. A. Gusev, U. B. Jayanthi, N. G. Schuch, W. N. Spjeldvik, and K. T. Choque, Antiprotons Confined in the Earth's Inner Magnetosphere, Astroparticle Physics, 20, p.257-265, 2003.

Quenby, J.J., T. Mulligan, J.B. Blake, J.E. Mazur and D. Shaul. Local and nonlocal geometry of interplanetary coronal mass ejections: Galactic cosmic ray (GCR) short-period variations and magnetic field modeling, JGR, 113, A10102, 2008.

Ramaty, R., B. Kozlovsky and R.E. Lingenfelter. Nuclear gamma-rays from energetic particle interactions, Astrophys. J. Suppl. Ser., 40, 487-526, 1979.

Reames, D.V. Particle acceleration at the Sun and in the heliosphere. Space Science Reviews, 90, 3/4, 413-491, 1999.

Rieger, E., Kanbach, G., Reppin, C., Share, G. H., Forrest, D. J., Chupp, E. L., A 154-day periodicity in the occurrence of hard solar flares?, Nature, 312, 623-625, 1984.

Richardson, I.G. and H. V. Cane. Galactic Cosmic Ray Intensity Response to Interplanetary Coronal Mass Ejections/Magnetic Clouds in 1995-2009. Sol. Phys., 270, 2, 609-627, 2011.

Richardson, I.G. Energetic Particles and Corotating Interaction Regions in the Solar Wind. Space Sci. Rev., 111, 267-376, 2004.

Rockenbach, M., A. Dal Lago, W. D. Gonzalez, K. Munakata, C. Kato, T. Kuwabara, J. Bieber, N. J. Schuch, M. L. Duldig, J. E. Humble, H. K. Al Jassar, M. M. Sharma, and I. Sabbah, Geomagnetic storm's precursors observed from 2001 to 2007 with the Global Muon Detector Network (GMDN), GRL, 38, L16108, doi:10.1029/2011GL048556, 2011.

Roederer, J.G. Dynamics of Geomagnetically Trapped Radiation. Springer, pp. 166, 1970.

Rossi, B. and S. Olbert. Introduction to the Physics of Space. McGraw-Hill Book Co., pp. 454, 1970.

Rouillard, A. and M.A. Lockwood, Oscillations in the open solar magnetic flux with a period of 1.68 years: imprint on galactic cosmic rays and implications for heliospheric shielding, Ann. Geophys., 22, 4381–4395, 2004.

Rozelot, J.-P. Solar and Heliospheric Origins of Space Weather Phenomena. Lecture Notes in Physics 699, Springer, pp. 166, 2006.

Rybanský, M., Coronal index of solar activity. I - Line 5303 A, year 1971. II - Line 5303 A, years 1972 and 1973, Bull. Astron. Inst. Czech. 26, 367–377, 1975.

Sabbah, I. and K. Kudela, Third harmonic of the 27 day periodicity of galactic cosmic rays: Coupling with interplanetary parameter, JGR, 116, A04103, 2011.

Sabbah, I. and M.L. Duldig, Solar Polarity Dependence of Cosmic Ray Power Spectra Observed with Mawson Underground Muon Telescopes. Solar Phys., 243, 231–235, 2007.

Sdobnov, V.E., Analysis of Forbush effect in May 2005 by the method of spectrographic global survey, (in Russian), Izv. RAN, ser. Phys., 75, 6, 872-874, 2011.

Serpico, P.D.. Astrophysical models for the origin of the positron "excess". Astroparticle Physics, in press, arXiv:1108.4827v1, 2011.

Scherer, K., H. Fichtner, B. Heber, U. Mall (eds.), Space Weather: The Physics Behind a Slogan, Lecture Notes in Physics, Springer Berlin and Heidelberg, 2005, pp. 297, 2005.

Schulz, M. and L.J. Lanzerotti. Particle Diffusion in the Radiation Belts. Springer, pp. 215, 1974.

Siingh, D., Singh, R.P. Singh, A.K. , Kulkarni, M.N. , Gautam, A.S. , Singh, A.K,. Solar Activity, Lightning and Climate, Surveys in Geophysics, 32, 6, 659-703, 2011.

Siingh, Devendraa; Singh, R. P., The role of cosmic rays in the Earth's atmospheric processes, Pramana, vol. 74, issue 1, pp. 153-168, 2010.

Siluszyk, M., A. Wawrzynczak and M.V. Alania. A model of the long period galactic cosmic ray variations. Journal of Atmospheric and Solar-Terrestrial Physics, Volume 73, Issue 13, 1923-1929, 2011.

Simpson, J.A. Cosmic Radiation Neutron Intensity Monitor, Annals of the Int. Geophysical Year IV, Part VII, Pergamon Press, London, p. 351, 1958.

Singer, S.F. Cosmic Rays and the Sun's Magnetic Field: Diurnal Variation of Cosmic Rays and the Sun's Magnetic Field. Nature, 170, 4315, 63-64, 1952.

Singh, A.K., Devendraa Siingh, Singh, R.P. Impact of galactic cosmic rays on Earth's atmosphere and human health, Atmos. Environment, 3806-3818, 2011.

Sloan, T. and A.W. Wolfendale, Testing the proposed causal link between cosmic rays and cloud cover, Environ. Res. Lett. 3, 024001 (6pp), 2008.

Smart, D. F. and M.A. Shea. Geomagnetic Cutoff Rigidity Calculations at 50-Year intervals between 1600 and 2000, Proc. 28th ICRC, Tsukuba, Japan, 4201-4204, 2003.

Smart, D.F. and M.A. Shea. Fifty years of progress in geomagnetic cutoff rigidity determinations. Advances in Space Research, 44, 1107-1123, 2009.

Smith, D. M., L. I. Lopez, R. P. Lin, and C. P. Barrington-Leigh. Terrestrial gamma-ray flashes observed up to 20 MeV, Science, 307 (5712), 1085–1088, doi:10.1126/science.1107466, 2005.

Song, P., Howard J. Singer and George L. Siscoe, editors. Space Weather. AGU Monograph Series, 125, pp. 440, Washington, DC, 2001.

Spurný, F., and Ts. Dachev. Intense Solar Flare Measurements, April 15, 2001, Radiat. Prot. Dosim., 95, p. 273-275, 2001.

Spurný, F., Kudela, K., Dachev, T. Airplane radiation dose decrease during a strong Forbush decrease. Space Weather 2, S05001, 2004.

Stoker, P. The IGY and beyond: A brief history of ground-based cosmic-ray detectors. Advances in Space Research, Volume 44, Issue 10, 1081-1095, 2009.

Storini, M. Galactic cosmic-ray modulation and solar-terrestrial relationships. Nuovo Cimento C, Serie 1, vol. 13 C, 103-124,1990.

Storini, M., G.A. Bazilevskaya, E.O. Flückiger, M.B. Krainev, V.S. Makhmutov and A.I. Sladkova, The Gnevyshev gap: A review for space weather, Adv. Space Res., 31, 4, 895-900, 2003.

Stozhkov, Y.I., N. S. Svirzhevsky, G. A. Bazilevskaya, M. B. Krainev, A. K. Svirzhevskaya, V. S. Makhmutov, V. I. Logachev, and E. V. Vashenyuk. Cosmic rays in the stratosphere in 2008–2010. Astrophys. Space Sci. Trans., 7, 379-382, 2011.

Strauss, R.D., M.S. Potgieter and S.E.S. Ferreira. Modeling Ground and Space Based Cosmic Ray Observations, Advances in Space Research, accepted manuscript, online, doi:10.1016/j.asr.2011.10.006, 2011.

Svensmark, H. and E. Friis-Christensen, Variation of cosmic ray flux and global cloud coverage-a missing link in solar-climate relationships, J. Atmos. Sol. Terr. Phys., 59, 1225–1232, 1997.

Thompson, J.L. Solar Diurnal Variation of Cosmic-Ray Intensity as a Function of Latitude. Phys. Rev., 54, 2, 93-96, 1938.

Tyasto, M.I., O. A. Danilova and V. E. Sdobnov. Variations in the Geomagnetic Cutoff Rigidity of CR in the Period of Magnetospheric Disturbances of May 2005: Their Correlation with Interplanetary Parameters, Bull. Russian Acad. Sci., ser. Phys., 75, 6, 808-811, 2011.

Usoskin, I.G., K. Alanko-Huotari, G.A. Kovaltsov and K. Mursula. Heliospheric modulation of cosmic rays: Monthly reconstruction for 1951-2004, JGR, 110, A12, A12108, 2005.

Usoskin, I.G., K. Mursula, S.K. Solanki, M. Schüssler, M. and G.A. Kovaltsov. A physical reconstruction of cosmic ray intensity since 1610. JGR, 107, A11, 1374, 2002.

Vainio, R., L. Desorgher, D. Heynderickx, M. Storini, E. O. Flückiger, R.B. Horne, G.A. Kovaltsov, K. Kudela, M. Laurenza, S. McKenna-Lawlor, H. Rothkaehl and I. Usoskin. Dynamics of the Earth´s Particle Radiation Environment, Space Science Rev., 147, no. 3-4, 187-231. 2009.

Valdés-Galicia, J. F., et al., The Cosmic-Ray 1.68-Year Variation: a Clue to Understand the Nature of the Solar Cycle? Sol. Phys., 67, 409 – 417, 1996.

Valdés-Galicia, J.F. Energetic particle transport coefficients in the heliosphere, Space Science Reviews, 62, no. 1-2, 67-93, 1993.

Valdés-Galicia, J.F., Low energy galactic cosmic rays in the heliosphere, Adv. Space Res., 35, 755–767, 2005.

Van Allen, James A., Editor. Cosmic Rays, the Sun and Geomagnetism: The works of Scott E. Forbush, AGU, pp. 471, 1993.

Vashenyuk, E.V., Yu. V. Balabin, and B. B. Gvozdevsky. Features of relativistic solar proton spectra derived from ground level enhancement events (GLE) modeling, Astrophys. Space Sci. Trans., 7, 459–463, 2011.

Venkatesan, D. and Badruddin. Cosmic ray intensity variations in the 3-dimensional heliosphere, Space Sci. Rev., 52, 121-194, 1990.

Vilmer, N. A. L. MacKinnon, and G. J. Hurford. Properties of Energetic Ions in the Solar Atmosphere from γ -Ray and Neutron Observations, Space Sci. Rev., 159:167–224, 2011.

Vivek Gupta and Badruddin, Solar magnetic cycle dependence in corotating modulation of galactic cosmic rays, Astrophys Space Sci., 321: 185–195, 2009.

Wang, F. and Agrawal, V.D. Soft Error Considerations for Computer Web Servers, Proc. 42nd Southeastern Symposium on System Theory (SSST), 2010.

Webber, J.W. and J.A. Lockwood. Characteristics of the 22-year modulation of cosmic rays as seen by neutron monitors. JGR, 93, 8735-8740, 1988.

Zank, G.P. W.H. Matthaeus, J.W. Bieber and H. Moraal, The radial and latitudinal dependence of the cosmic ray diffusion tensor in the heliosphere, J. Geophys. Res., 103, A2, 2085-2097, 1998.

Zarrouk, N. and R. Bennaceur, Extrapolating cosmic ray variations and impacts on life: Morlet wavelet analysis, Acta Astronautica, 65, 1-2, 262-272, 2009.

Part 5

The Interaction of the Solar Wind with the Magnetosphere

14

Impact of Solar Wind on the Earth Magnetosphere: Recent Progress in the Modeling of Ring Current and Radiation Belts

Natalia Buzulukova[1], Mei-Ching Fok[2] and Alex Glocer[2]
[1]*NASA Goddard Space Flight Center/CRESST/University of Maryland College Park*
[2]*NASA Goddard Space Flight Center*
USA

1. Introduction

When solar wind interacts with the Earth's magnetosphere it causes disturbances in the near-Earth plasma environment. Large disturbances result in geomagnetic storms, and affect not only Earth magnetosphere but also space-borne and ground-based technological systems. The systematic studies of cause-effect relations between solar wind variations and resulting disturbances in near-Earth plasma environment as well as the construction of relevant numerical models are important subjects of *Space Weather*, which is currently a very active topic of research.

A most common reaction of the magnetosphere to prolonged solar wind disturbances are geomagnetic storms. Understanding geomagnetic storms and their solar wind drivers is one of the most important problems in geophysics and space weather. A geomagnetic storms is defined by a deviation of the H component of the Earth's magnetic field at low latitudes, i.e., the Dst index. It is believed that the main source of this deviation is the so called ring current which is comprised of plasma with energies 1–200 keV. Hence, to understand geomagnetic storms, we need to understand how the ring current is formed. Many previous studies have addressed the question of solar wind drivers for geomagnetic storms. Most of these were statistical in nature (Borovsky & Denton, 2006; Denton et al., 2006; Turner et al., 2009; Weigel, 2010; Yermolaev et al., 2010; Zhang, Richardson, Webb, Gopalswamy, Huttunen, Kasper, Nitta, Poomvises, Thompson, Wu, Yashiro & Zhukov, 2007). They focused upon the effectiveness of different types of solar wind structures in producing geomagnetic disturbances (e.g., a combinations of global activity indices and/or various coupling functions) and different types of storms.

In addition to statistical analysis, a great deal of effort has been put into the development of ring current models. Typically a kinetic approach is used in which modeling ring current plasma is described with a number of species. Each population has its own energy and drift velocity which is a sum of the gradient/curvature drift and E$\times$B drift. Plasma in this representation is generally anisotropic and is not in a thermodynamic equilibrium. The corresponding transport equations are usually written in terms of bounce-averaged quantities. This approach is sometimes referred to as 'drift physics' approach and has proven to be very useful in describing the plasma population of inner magnetosphere with energies from tens of eV to hundreds keV (e.g., Wolf et al. (2007)). Ring current models can also incorporate losses due to loss cone, charge exchange, Coulomb interactions, and interactions between different species due to wave activity. (For a detailed description of processes which control ring current losses, see Ebihara & Ejiri (2003) and Khazanov (2011)).

The electromagnetic field in the simulation region can be decomposed into a sum of 'external' field and 'internal' fields. The 'external' field is created outside of the simulation region while the 'internal' field is created by the ring current itself. For the sake of simplicity, the internal contribution is sometimes omitted. There exist a number of different codes which solve this problem with different assumptions, equations, and numerical methods (e.g., Ebihara & Ejiri (2003)) and many of them are widely used by the geophysical community. The Rice Convection Model (RCM) calculates internal electric fields self-consistently with the total pressure distribution (Harel et al., 1981; Jaggi & Wolf, 1973; Toffoletto et al., 2003). An externally prescribed magnetic field is assumed in the RCM. Each species is described by one adiabatic invariant (energy invariant), and the particle distribution function is assumed to be isotropic in pitch-angle. Self-consistency of the electric field with the plasma distribution is maintained by feedback from the ionosphere via field-aligned currents.

The Fok Ring Current (FokRC) model (Fok & Moore, 1997; Fok et al., 1995) and the Ring Current-Atmosphere Interaction Model (RAM) (Jordanova et al., 1994; 1997; Liemohn et al., 1999) solves the bounce-averaged Boltzmann equation for a number of species with given 'external' electric and magnetic fields. Each species is described by two adiabatic invariants μ, K (FokRC model) or, equivalently, energy and equatorial pitch angle (RAM model). The anisotropic pitch angle dependence of distribution function is calculated in both models.

The Comprehensive Ring Current Model (CRCM) (Fok, Wolf, Spiro & Moore, 2001) and the Self-Consistent RAM model (Ridley & Liemohn, 2002) calculate the internal electric fields self-consistently. The equilibrium RCM, RCM-E [Lemon et al., 2004] is a modification of RCM model with internal ring current magnetic field model that is calculated from 3-D force equilibrium assumption and treated as a correction to an external empirical magnetic field model. A modification of the RAM model with an internal ring current magnetic field also uses a 3-D force equilibrium assumption to calculate corrections to an external empirical magnetic field model (Zaharia et al., 2006). The Enhanced CRCM model (ECRCM) (Ebihara et al., 2008) is a modification of CRCM model with an internal ring current magnetic field that is calculated from the Biot-Savart law and included as a correction to an external empirical magnetic field model.

In addition to models of ring current plasma, there has been an increased interest in recent years in models of Earth's radiation belts. The Earth's radiation belts consist of energetic electrons ($\sim$100 keV to several MeV) and ions ($\sim$100 keV to several hundred MeV) trapped in the magnetosphere roughly from $1.2 < L < 8$. The energetic electrons reside in 2 distinct regions: the inner belt and the outer belt, which are usually separated by the slot region ($1.8 < L < 3$) of depleted particle populations. Pitch-angle diffusion loss of electrons by inter-acting with whistler mode plasmaspheric hiss is believed to be the cause of the slot region (Albert, 1994; Lyons et al., 1972; Meredith et al., 2007). The inner belt is relatively stable while the outer belt is highly variable with geomagnetic activity.

High energy electrons of radiation belts constitute a potential danger for Earth-orbiting satellites. High fluxes of energetic electrons cause harmful charging effects for sensitive electronics (Baker, 2002). The ability to predict how the radiation belts will react to solar wind drivers, and to geomagnetic storms in particular, is extremely important both for space weather applications and for geophysics in general. Approximately half of all moderate and intense storms cause a net increase of radiation belt fluxes by a factor of 2 or more (Reeves et al., 2003). At the same time, a quarter of all storms result in a net decrease of the fluxes by more than a factor of 2 (Reeves et al., 2003). There is an interplay between the number of processes that define the response of radiation belts to solar wind driving: acceleration, transport (convective and diffusive) and losses (Elkington et al., 1999; Fälthammar, 1965; Friedel et al., 2002; Horne, 2007; Schulz & Lanzerotti, 1974; Shprits et al., 2008; Summers et al., 2007; Ukhorskiy & Sitnov, 2008).

The main challenge to the community is how to describe and predict the variability of the radiation belts. There are three main approaches: (1) statistical approach based on analysis of relations between radiation belt fluxes and solar wind parameters like velocity and density, e.g. (Lyatsky & Khazanov, 2008; Reeves et al., 2011) and references therein; (2) modeling of electron phase space density with Fokker-Plank type equation (e.g. (Albert & Young, 2005; Li et al., 2001; Shprits et al., 2009; Varotsou et al., 2005)); (3) modeling of electron phase space density with combined Fokker-Plank and drift advection equation (Bourdarie et al., 1997; Fok et al., 2008; Miyoshi et al., 2006; Zheng et al., 2003). A necessity including advection terms (for gradient B drift and ExB drift) is dictated by the extension of a model to wider energy and L-shell ranges. In this case, a radiation belt model is similar to a ring current model with additional wave-diffusion terms. As a result both of these populations of near-Earth plasma, ring current and radiation belts, may be described within the same numerical model. This approach enables study of the complex and uneven relations between ring current and radiation belts. For example, when the ring current changes the electric field in the inner magnetosphere it influences radiation belts in (at least) two different ways: First, the electric field is important for convective transport. Second, electric field shapes the plasmasphere, a reservoir of cold and dense plasma of ionospheric origin. The plasmasphere directly drives radiation belt dynamics by excitation of waves; wave-particle interaction are an important acceleration mechanism for radiation belt electrons. Hence, we should describe the radiation belt dynamics together with the ring current and plasmasphere to ensure we do not miss important physics.

In this review paper, we will describe recent progress in the modeling of the ring current/radiation belt plasma with the Comprehensive Ring Current Model (CRCM) (Fok, Wolf, Spiro & Moore, 2001) and Radiation Belt Environment model (RBE) (Fok et al., 2005; Fok, Moore & Spjeldvik, 2001; Fok et al., 2008). Both models are based on a solution of the bounce-averaged transport equation for evolution of phase space density. We will describe the current progress of coupling the CRCM and the RBE with global MHD model of the Earth magnetosphere, modeling of ring current-plasmasphere interactions with the CRCM, and recent extensions to the RBE. In the discussion, we will describe an efforts we are going to make toward a combined ring current-radiation belt-plasmasphere model and putting this model inside a global MHD. The overall objective of this research can be described as follows: we are going to create a fully coupled model of ring current, radiation belts, plasmasphere and MHD with a significant level of self-consistency between all elements. This will be an excellent tool to study complex and non-linear response of Earth's inner magnetosphere to solar wind drivers.

2. Description of the Comprehensive Ring Current Model (CRCM)

The CRCM solves the distributions of ring current ions and electric potential at the ionosphere in a self-consistent manner (Fok, Wolf, Spiro & Moore, 2001). The temporal variation of the phase space density of ring current species is calculated by solving the bounce-averaged transport equation (Fok & Moore, 1997; Fok et al., 1996):

$$\frac{\partial \bar{f}_s}{\partial t} + \langle \dot{\lambda} \rangle \frac{\partial \bar{f}_s}{\partial \lambda} + \langle \dot{\phi} \rangle \frac{\partial \bar{f}_s}{\partial \phi} = -v\sigma_s \langle n_H \rangle \bar{f}_s - \left(\frac{\bar{f}_s}{0.5\tau_b} \right)_{loss\ cone} \tag{1}$$

where $\bar{f}_s = \bar{f}_s(\lambda, \phi, M, K)$ is the average distribution function of species s on the field line between mirror points. λ and ϕ are the magnetic latitude and local time, respectively, at the ionospheric foot point of the geomagnetic field line. M is the relativistic magnetic moment and $K = J/\sqrt{8m_0M}$, where J is the second adiabatic invariant. The motion of the particles is described by their drifts across field lines, which are labeled by their ionospheric foot points. The M range is chosen to cover the energy ranges from 1–200 keV. The K range is chosen tending to cover the loss cone so that particle precipitations can be calculated.

The left hand side of (1) represents the drifts of the particle population. The right hand side of (1) refers to losses. The calculation of the bounce-averaged drift velocities, $\langle \dot{\lambda} \rangle$ and $\langle \dot{\phi} \rangle$, are described in detail in Fok & Moore (1997). $\langle \dot{\lambda} \rangle$ represents the radial transport term and $\langle \dot{\phi} \rangle$ is the azimuthal drift velocity. These drifts include gradient-curvature drift and E × B drift from convection and corotation electric fields. The effects of inductive electric fields due to time-varying magnetic fields are also taken into account implicitly in the model. For this purpose, we assume that field lines are rooted at the ionosphere, so that the inductive electric field there is zero. However, the shapes of field lines at higher altitudes vary as a function of time according to the magnetic field model. If field lines are perfect conductors, the field line motion at high altitudes (e.g., at the equator) will generate an induction electric field with the form, $E_{ind} = -v_0 \times B_0$ where B_0 and v_0 are the field line velocity and magnetic field at the equator.

The right hand side of (1) includes two loss terms; one is for the loss-cone and the other for the charge-exchange. To solve Eq. (1), we specify the particle distribution on the nightside boundary, which is set at 8–10 R_E or at the last closed field line, which one comes first. We use the Tsyganenko & Mukai (2003) model for density/temperature at the polar boundary, or output from the BATSRUS MHD (two-way coupled with ring current model). To describe O^+ component, we use the standard relation of Young et al. (1982).

The ionospheric electric field is calculated self-consistently with the ring current distribution. Field-aligned currents are calculated from a current continuity equation between magnetosphere and ionosphere (Fok, Wolf, Spiro & Moore, 2001):

$$J_{\|i} = \frac{1}{r_i \cos^2 \lambda} \sum_j \left(\frac{\partial \eta_j}{\partial \lambda} \frac{\partial W_j}{\partial \phi} - \frac{\partial \eta_j}{\partial \phi} \frac{\partial W_j}{\partial \lambda} \right) \tag{2}$$

where summation is at a fixed (λ, ϕ) point and over all (M, K) points; $J_{\|i}$ is a sum of ionospheric field-aligned current density for two hemispheres; W_j is the kinetic energy of a particle with given (λ, ϕ, M, K); and, η_j is the number of particles per unit magnetic flux associated with ($\Delta M, \Delta K$): $\eta_j = 4\sqrt{2}\pi m_0^{3/2} \bar{f}_s(\lambda, \phi, M, K) M^{1/2} \Delta M \Delta K$. Using the distribution of field-aligned currents, the ionospheric potential is obtained from $\nabla \cdot (-\Sigma \cdot \nabla\Phi) = J_{\|i} \sin I$, where Φ is an electric field potential (the same for both hemispheres by definition); Σ is a tensor of ionospheric conductivities (field-line integrated, a sum for two hemispheres); and, I is a magnetic field inclination angle. The CRCM grid is assumed to be the same in both hemispheres. This is usually true with the exception of points near polar CRCM boundary (e.g. Buzulukova et al. (2010)). Assuming the source of field-aligned current is located well inside the polar boundary, we may expect that errors in the solution for the potential are not too large. The detailed analysis of the asymmetry problem, however, is beyond the scope of this work and needs to be addressed separately.

3. Coupling of the CRCM with BATSRUS MHD

A reason for coupling of a ring current model with a global MHD model is twofold. First, MHD models apparently fail to adequately describe the inner magnetosphere. It follows from MHD assumption that the E×B drift velocity is dominant. This assumption is not valid in the inner magnetosphere where magnetic fields and their gradients are strong. Under such conditions, plasma cannot be treated as single fluid with some flow velocity and temperature. Hence, a coupling between a ring current model and MHD model provides MHD the missing 'drift physics' for better representation of the inner magnetosphere. Second, ring current models need a self-consistent input. Usually, ring current models use several empirical models as input. There is no guarantee these empirical models will be self-consistent with each other. MHD model can provide a ring current model with fully self-consistent electromagnetic fields and plasma parameters.

Here we present the results of so-called 'one-way coupling' between the CRCM and the Block-Adaptive-Tree Solar-Wind Roe-Type Upwind Scheme (BATSRUS) MHD model (Powell et al., 1999). An approach we follow is similar to one described in (De Zeeuw et al., 2004; Toffoletto et al., 2004). A computational domain is divided into two regions: 'inner' magnetosphere and 'outer' magnetosphere. The boundary between these two domains is located near the open/closed magnetic field lines boundary. In the outer magnetosphere, only the MHD model is calculated. In the inner magnetosphere, the MHD model runs together with RC model. In 'two-way' coupling mode the pressure from ring current model is mapped back to the MHD grid providing a feedback. This approach was called two-way coupling to emphasize that ring current and MHD models exchange information between each other. A first step is to couple the CRCM and BATSRUS in 'one-way' mode without pressure feedback from ring current to MHD. It means BATSRUS model provides the CRCM with an 'external' B field, electric field potential near the polar cap, and plasma pressure / temperature at the boundary between the two domains (open/closed boundary). The CRCM calculates the particle distribution function and 'internal' subauroral electric field in the inner magnetosphere. Toffoletto et al. (2004) considered 'one-way' coupling of the RCM with the Lyon-Fedder-Mobarry (LFM) MHD code. De Zeeuw et al. (2004) considered 'two-way' coupling of the RCM with BATSRUS MHD code. Only quasistationary states were considered and a substorm due to northward-southward Bz turning was not included.

For the coupled CRCM-BATSRUS, we show here the results for the idealized case of southward-northward-southward Bz turning and substorm development. The details of coupling methodology and run setup can be found in Buzulukova et al. (2010). Fig.1 from Buzulukova et al. (2010) shows the Cross Polar Cap Potential (CPCP) and IMF Bz for the MHD BATSRUS run.

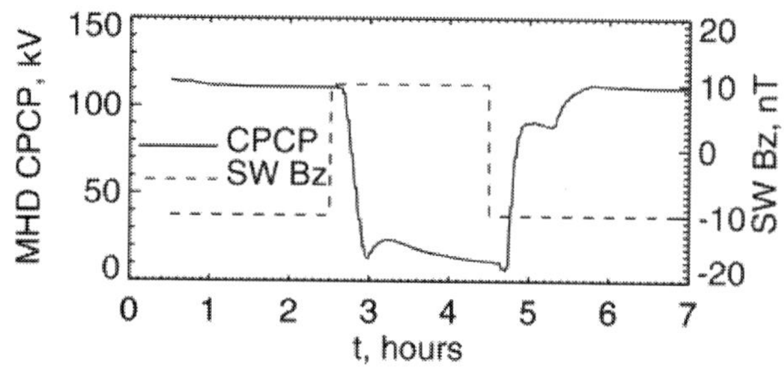

Fig. 1. MHD BATSRUS cross polar cap potential (CPCP) and solar wind Bz for an idealized case. Adopted from Buzulukova et al. (2010).

There is a characteristic structure of CPCP in response to the variation of Bz. A quasistationary CPCP ~110 kV at southward Bz corresponds to a strong convection. After the CPCP reaches the minimum at t = 3 h, northward IMF-associated (NBZ) current system (Iijima & Shibaji, 1987; Rasmussen & Schunk, 1987) begins to develop. At t~4.6 h, a southward Bz arrives at the magnetopause and cancels the NBZ current system. A typical two-cell convection system is restored. The substorm growth phase begins at ~5 h followed by substorm onset a t~5.3 h. At the end of the simulation, the profile reaches a plateau of 110 kV, the same as at the beginning of the run.

Fig.2 shows the dynamics of the inner magnetosphere for several snapshots during the CRCM-BATSRUS run. For each selected time, a plot of total RC pressure and a plot of field-aligned current are shown. All quantities are mapped to the equatorial plane. The plot of field-aligned currents is overlayed with equipotentials for the convection electric field without corotation. At the end of the Bz southward interval (t = 2.57 h), the RC is strongly asymmetric. There is a band of enhanced electric field in the dusk sector (MLT ~20) which can be associated with a Subauroral Ion Drift/Polarization Jet signature (SAID/PJ) (Galperin et al., 1974; Spiro et al., 1978). This feature is formed after ~1 h of simulations and can be interpreted as a result of shielding from enhanced electric field (Southwood & Wolf, 1978). Shielding is also manifested by the skewing of the equipotential lines in the morning sector (Brandt et al., 2002; Wolf et al., 2007). At t = 2.93 h, the polar cap potential drops substantially, and the RC-imposed electric field becomes apparent. This is an overshielding effect. At t = 4.64 (the end of the northward Bz interval) the RC becomes symmetric, as expected, and the RC-imposed electric field is almost zero.

At t = 4.7 h, the potential drop begins to increase; the RC becomes asymmetric again. As a result, a strong longitudinal pressure gradient and field-aligned current arises at t = 5.1 h. In the dusk sector (MLT ~20), field-aligned currents trigger the appearance of strong electric fields that resemble a SAID/PJ signature. This structure intensifies in the end of the growth phase. This is an effect of RC reconfiguration. The pressure gradients are formed by the preexisting population, so the time of formation is short, about ~15 min after the potential drop starts to increase. After the dipolarization starts at t=5.3 h, a strong convection and an induction electric field push particles into the inner magnetosphere, creating an injection. The injection is concentrated on the nightside and makes a quadrupole structure of field-aligned currents (t = 5.4, 5.53 h).

At t = 5.53 h, two populations persist: a preexisting one, and a newly injected population. Each population is spatially confined, and carries two layers of field-aligned currents. This layered structure is clearly seen in the dusk sector and interlaced in the dawn sector where the two populations coexist. There are two separated signatures of SAID/PJ in the dusk sector: one of them corresponds to the old population and the second corresponds to the newly injected population. Gradually the new population merges with the old one. At t = 6.8 h, a new quasi-stationary picture is established, similar to that at t = 2.5 h before the southward-northward Bz turning, with quasistationary SAID/PJ signature, shielding, and Region II currents.

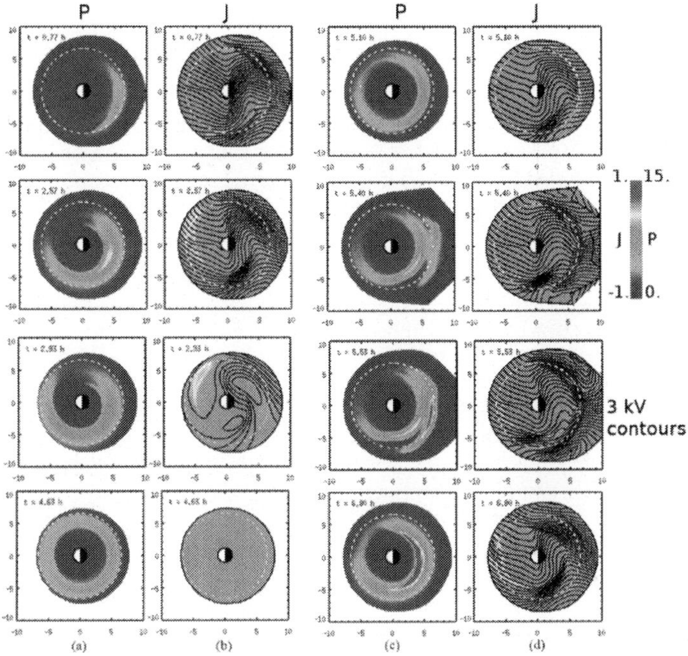

Fig. 2. Ring current dynamics for the CRCM-BATSRUS run. (a and c) Ring current (RC) pressure for 1–180 H+ (in color, logarithmic scale) at different times, t. (b and d) Snapshots of ionospheric Birkeland currents (BC) for the same times (in color) overplotted with convection equipotentials; corotation is excluded. All quantities are mapped onto the equatorial plane. BC and RC pressure are in mA/m^2 (from -1 to +1) and nPa (from 0 to 15), respectively. Equipotentials are 3 kV spacing. BC are positive down into the ionosphere. Sun is to the left. Dashed circles denote geostationary orbit. Adopted from Buzulukova et al. (2010).

For this idealized event the CRCM-BATSRUS coupled model successfully reproduces well known features of inner magnetosphere electrodynamics: strong convection under the southward Bz; the development of an electric field shielding and overshielding; a weak convection under prolonged interval of northward Bz; an induction electric field during substorm development; SAID/PJ signatures in the dusk sector during strong convection. Moreover, the stable structure of Region II field-aligned currents is formed only during prolonged intervals of strong quasi-stationary convection. During transient states when Bz changes polarity and/or substorm occurs, multilayered structures of field-aligned currents are formed. We interpret these as complex structures of partial ring current (including injections), created by spatial and temporal variations of electric fields (both convection and induction), and plasma sheet source at polar boundary of ring current model. In the next section we consider another case which demonstrate a complex structure of partial ring current and field-aligned currents during transient states, and associated plasmapause undulations.

4. CRCM with plasmasphere model and plasmapause undulations

On 17 April 2002, IMAGE EUV imager captured clear images of plasmaspheric undulations (Goldstein et al., 2005). They found shortly after a substorm onset at 1900 UT, the plasmasphere ripple was propagating westward along the plasmapause. Fok et al. (2005) used the combined models of the CRCM and the dynamic plasmasphere model of Ober et al. (1997) to simulate this event. They were able to reproduce the observed undulation and suggested that substorm injection, shielding and convection fields all play a role. We revisited this problem and examined in detail the effects of magnetosphere-ionosphere coupling and inductive electric field on the formation of plasmaspheric undulation with the CRCM model and Ober et al. (1997) model of plasmasphere (Buzulukova et al., 2008).

The dynamic plasmasphere model of Ober et al. (1997) calculates the plasma flux tube contents and equatorial plasma density distribution over time in the inner magnetosphere. It includes the influences of convection on the flux tube volumes; the daytime refilling; and, the nighttime draining of plasma.

Columns 1–3 in Fig. 3 show the simulated potential pattern, field aligned current mapped to the equator, plasmasphere density and ring current pressure at 3 times during the undulation event in a time-varying magnetic field of Tsyganenko (Tsyganenko, 1995). Column 4 displays results at t = t0 with magnetic field kept constant throughout the simulation. In Fig. 3, field-aligned current (top panels) in red represents downward current and blue is for upward current. The arrows in the plasmasphere plots (middle panels) indicate the locations of the ripples. As shown in Fig. 3 at t = t0 (column 1), particle injection on the nightside forms a bulge in the ring current pressure near the geostationary orbit in the dusk-midnight sector. Thirty minutes later when cross polar cap potential decreases due to northward turning of IMF Bz, the shielding field produced by the strong field aligned current dominates over the convection electric field (overshielding). A reverse flow region is formed near dusk local time outside the plasmapause and a ripple at the plasmapause is developed westward of this special flow pattern (column 2, middle panel). The ripple then propagates westward as the ring current-field aligned current move westward (column 3). The westward edge of the ring current pressure bulge in column 1 is potentially interchange unstable following the criterion of Xing & Wolf (2007). It is possible that interchange instability enhances the development of undulations for this particular event. Column 4 shows the configuration at t = t0 if magnetic field is kept constant. It can be seen that the particle injection is weaker without the inductive electric field associated with magnetic field fluctuations. As a result, no plasmaspheric undulation is developed at later times (results not shown). This example demonstrates the complexity and nonlinearity of ring current-ionosphere-plasmasphere. It also shows the ring current's relations to the plasmasphere undulation, overshielding, induction electric field and interchange instability.

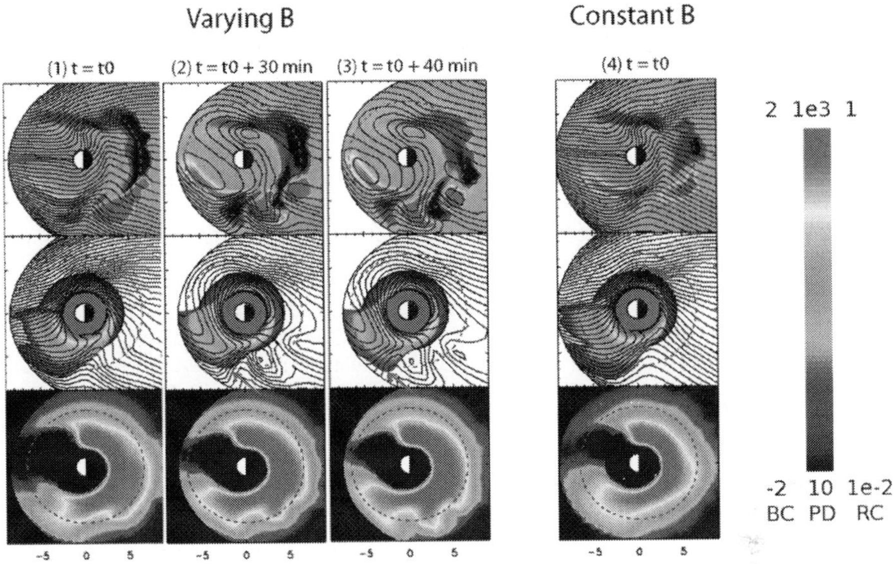

Fig. 3. Ring current-plasmasphere interaction. Columns 1-3 are simulated Region 2 Birkleland current (BC) mapped to the equator (top panels), plasmasphere density (PD) (middle panels) and ring current (RC) pressure for 16–27 keV H+ (bottom panels) at 3 times during the substorms on 17 April 2002. Column 4 is simulation result with constant magnetic field configuration. The dashed cycles are geosynchronous orbits. Potential contours without and with corotation are overlaid in top and middle panels, respectively. Adapted from Buzulukova et al. (2008).

5. The Radiation Belt Environment model (RBE)

The Radiation Belt Environment (RBE) model has been rigorously developed over past 10 years (Fok et al., 2005; Fok et al., 2011; Fok, Moore & Spjeldvik, 2001; Fok et al., 2008; Glocer et al., 2011; 2009; Zheng et al., 2011; 2003). It models both the radiation belts and ring current electrons. The RBE model is similar to the CRCM and is based on the solution of the bounce-averaged advection equation. Additional physics in the model includes the description of processes that do not conserve the first and the second adiabatic invariants via inclusion of diffusion for distribution functions. A radial transport is included implicitly by time-varying magnetic and electric fields, in the same manner as in the CRCM. Fok et al. (2008) consider only pitch-angle diffusion and energy diffusion. Recently, a mixed diffusion term has been added to the RBE model using the Alternating Direction Implicit (ADI) method (Fok et al., 2011; Zheng et al., 2011). The resulting equation is:

$$\frac{\partial \bar{f}_s}{\partial t} + \langle \dot{\lambda} \rangle \frac{\partial \bar{f}_s}{\partial \lambda} + \langle \dot{\phi} \rangle \frac{\partial \bar{f}_s}{\partial \phi} = \frac{1}{G} \frac{\partial}{\alpha_0} \left[G \left(D_{\alpha_0 \alpha_0} \frac{\partial \bar{f}_s}{\partial \alpha_0} + D_{\alpha_0 E} \frac{\partial \bar{f}_s}{\partial E} \right) \right] + \frac{1}{G} \frac{\partial}{\partial E} \left[G \left(D_{EE} \frac{\partial \bar{f}_s}{\partial E} + D_{E \alpha_0} \frac{\partial \bar{f}_s}{\partial \alpha_0} \right) \right] - \left(\frac{\bar{f}_s}{0.5 \tau_b} \right)_{\text{loss cone}}$$

where $G = T(\alpha_0)\sin(2\alpha_0)(E+E_0)\sqrt{E(E+2E_0)}$; E_0 is the rest mass energy; $T(\alpha_0) = 1/2R_0\int_{s_m}^{s_m^*} ds/\cos\alpha$. To perform the diffusion in the (M,K) coordinates, we first map the particle phase space density from the (M,K) to (E,α_0) coordinates, perform diffusion in E,α_0, and then map the updated distribution back to the (M,K) coordinates (Fok et al., 1996).

The bounce-averaged diffusion coefficients are calculated with the Pitch-Angle and Energy Diffusion of Ions and Electrons (PADIE) code (Glauert & Horne, 2005). Corresponding normalized diffusion coefficients D_n is obtained that depend on $f = \omega_{pe}/\Omega_e$, the ratio of plasma frequency to the cyclotron frequency. To obtain cold plasma distribution and f we run Ober et al. (1997) model of plasmasphere. The corresponding diffusion coefficient is calculated by $D = D_n \cdot B_{wave}^2$ where B_{wave} is the wave intensity. Only resonance with lower-band whistler mode chorus is considered. The presence of chorus waves is confined between -15° and 15° magnetic latitude. The intensity of chorus waves is calculated from CRRESS plasma wave data for chorus presented by Meredith et al. (2001; 2003; 2009). The chorus wave amplitudes are based on an averaged study of chorus observations. However, the model has a radial dependence, an MLT dependence, and from a dependence on the level of geomagnetic activity.

A simplified version of the RBE model is currently running in real-time to provide radiation belt now-casting updated every 15min. The geosynchronous fluxes at longitudes of GOES-11 and 13 are extracted from the RBE real-time run and are plotted together with real-time GOES electron (>0.6 MeV) data. The model-data comparison is continually posted at http://mcf.gsfc.nasa.gov/RB_nowcast/.

Fok et al. (2011) study the effect of wave-particle interaction on radiation belt dynamics, including the effect of cross diffusion terms. Fig.4 shows simulated radiation belt fluxes on 23–27 October 2002 (adopted from (Fok et al., 2011, Fig.2)) as a function from L and t. Two energy bins are considered: low energy part of radiation belts (may be considered also as ring current electrons) 20–70 keV and high energy radiation belt electrons 0.6–1.8 MeV. Panels (a) and (b) shows electron fluxes without wave-particle interactions; panels (c) and (d) shows fluxes with pure energy/pitch angle diffusion; panels (e) and (f) shows fluxes with cross diffusion included. The Dst index (black curve) is overlaid on the top panels.

Radiation belt fluxes for different energies respond differently to storm conditions. For lower energies, ExB drift is important. During main phase of the storm, increased convection injects particles from higher L shells to lower L shells at the nightside part of the inner magnetosphere. Without wave-induced losses, freshly injected particles fill the entire outer belt (panel a). For MeV electrons, magnetic drifts dominate over convection. During main phase of strong storm, magnetic field in the inner magnetosphere becomes inflated because of ring current plasma. Inflation of magnetic field causes outward motion, de-acceleration and flux dropout of energetic particles. This is the well known Dst effect (Dessler & Karplus, 1961; Kim & Chan, 1997).

The inclusion of wave effects changes the results dramatically. Panels c and d show the results with pure energy and pitch-angle diffusion from interacting with whistler mode chorus for two energy bins. Panels e and f show the results with inclusion of cross diffusion coefficients.

For 20–70 keV, the inclusion of wave interactions significantly reduces radiation belt fluxes. It means pitch–angle diffusion is dominant. It scatters particles into loss cone causing precipitations. The difference in fluxes between panel (a) and panels (c) is the amount of

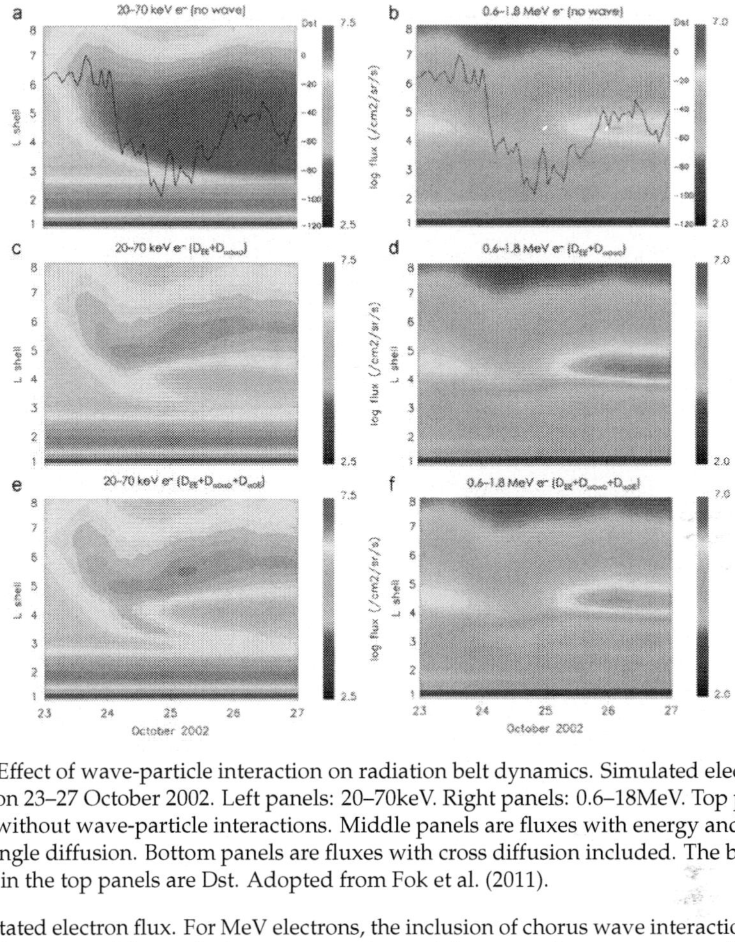

Fig. 4. Effect of wave-particle interaction on radiation belt dynamics. Simulated electron fluxes on 23–27 October 2002. Left panels: 20–70keV. Right panels: 0.6–18MeV. Top panels are fluxes without wave-particle interactions. Middle panels are fluxes with energy and pitch-angle diffusion. Bottom panels are fluxes with cross diffusion included. The black curves in the top panels are Dst. Adopted from Fok et al. (2011).

precipitated electron flux. For MeV electrons, the inclusion of chorus wave interactions causes gradual increase of fluxes during recovery phase of the storm. It means energy diffusion is dominant here.

Cross diffusion tends to moderate the impacts from pure pitch-angle and energy diffusion. For 20–70 keV, cross diffusion weakens pitch-angle diffusion and reduce precipitations. Overall flux calculated with cross diffusion is higher. For MeV electrons, cross-diffusion weakens energy diffusion and flux enhancement. Overall flux calculated with energy diffusion is lower than without it. The magnitude of cross diffusion effect depends from energy and L-shell. For MeV electrons at L-shell ~4, ignoring cross diffusion could cause overestimation of electron flux as much as a factor of 5. At times, cross diffusion may have a comparable effect to pure diffusions (Albert & Young, 2005). The cross-diffusion term can be especially important for ring current electrons with energies ~ 1 keV (Ni et al., 2011) therefore affecting electron precipitations.

6. Coupling the RBE model with MHD BATSRUS/SWMF

In most of previous RBE calculations, we used empirical models of magnetic and electric fields (Fok et al., 2005; Fok et al., 2008; Zheng et al., 2003). Variations in magnetic and electric fields are drivers for radial diffusion in RBE. Radial diffusion plays an important role in transport and energization of electrons in the outer belt. In this mechanism, electrons conserve the first and second adiabatic invariant, but the third invariant is regularly violated by temporal variations of the magnetic and induced electric field (Kellogg, 1959; Schulz & Eviatar, 1969). There is no explicit term for radial diffusion in RBE, rather this is handled implicitly through variations in the magnetic and electric fields. Therefore, a realistic description of these field variations is extremely important. This is the main motivation to couple the RBE model with global MHD model (Fok et al., 2011; Glocer et al., 2011; 2009). It is critical to use magnetic and electric fields from the two-way coupled MHD-ring current model for RBE, because ring current inflates inner magnetosphere magnetic field and changes electric field.

Currently, the RBE is placed inside the Space Weather Modeling Framework (SWMF) (Glocer et al., 2011; 2009). The SWMF consists of $\sim$15 physics-based and several empirical models (modules) describing different processes in Sun-Earth coupled system (Tóth et al., 2005; Tóth et al., 2011). Inside the SWMF, the RBE is coupled with MHD BATSRUS, ring current model RCM and ionospheric solver RIM (for detail see Tóth et al. (2005); Tóth et al. (2011)). Coupling between MHD model and ring current model is done in two-way mode, i.e. with pressure feedback from ring current to MHD. It allows to include inflation of magnetic field during geomagnetic storms as well as realistic variation of magnetic field. For example, Huang, Spence, Hudson & Elkington (2010); Huang, Spence, Singer & Hughes (2010) demonstrate that MHD generates ULF waves associated with variations of magnetic field. These ULF waves are responsible for radial diffusion and result in diffusion rates similar to other estimations.

Recently Fok et al. (2011) and Glocer et al. (2011) use coupled RBE-SWMF to examine 2 events with rapid radiation belt enhancements. Based on careful data analysis from Akebono, TWINS and NOAA satellites, Glocer et al. (2011) conclude that the time scale of these enhancements is about 2 hours and thus it is too short for wave associated energization. Additional analysis of AL index and magnetic field measurements from GOES 11,12 satellites suggests a substorm may play a role in this quick increase of outer belt electrons (Fok et al., 2011; Glocer et al., 2011). The example of flux enhancement is shown on the Fig. 5 (adapted from Fok et al. (2011)). Panel (a) shows the Akebono electron flux of energy $>$2.5 MeV from September 3, 2008, 00 UT to September 4, 12 UT. The Dst index during this time is overlaid in the plot. The Akebono data are averaged over 3 orbit periods ($\sim$ 7.5 h).

Fok et al. (2011) and Glocer et al. (2011) perform RBE–SWMF simulations to understand the cause of the flux enhancement seen in the Akebono electron data. Additionally, the RBE is run in stand-alone mode with empirical models as inputs. Fig. 5(b) shows the RBE electron fluxes calculated with empirical T04 magnetic field model and empirical Weimer electric field model (Tsyganenko et al., 2003; Weimer, 2001). Fig. 5(c) is the RBE flux calculated in the MHD fields simulated from the RBE–SWMF model (De Zeeuw et al., 2004). When comparing the RBE fluxes with Akebono one should remember RBE fluxes are taken from equatorial plane with temporal resolution of 1 h. In contrast, the Akebono measurements are taken along high inclination orbits. Therefore the magnitudes may differ substantially as we are not sampling the same portion of the pitch-angle distribution. However, the temporal variability of high

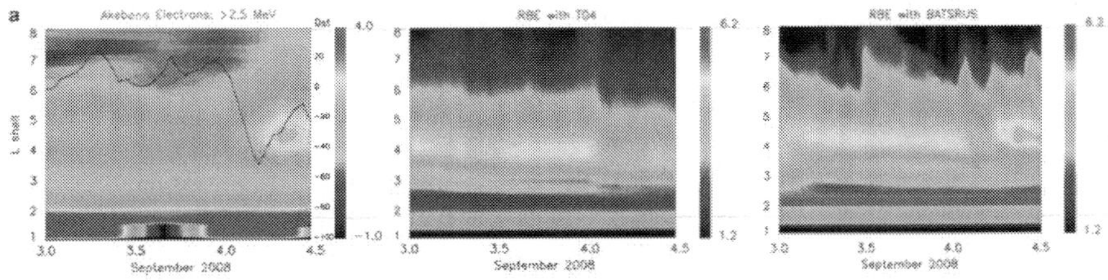

Fig. 5. Radiation belt flux enhancement. Left panel: L–Time plot of Akebono electron flux (>2.5 MeV) on 3–4 September 2008. The black curve is Dst. Middle panel: corresponding RBE simulated flux with T04 magnetic field. Right panel: simulated flux with BATSRUS fields. Adopted from Fok et al. (2011).

latitude fluxes was found to be nearly identical with the equatorial fluxes (Kanekal et al., 2001; 2005). Wave-particle interactions are not included in these RBE calculations. In the quiet period on September 3, the two RBE simulations give similar flux intensity. During the main phase of the storm, both RBE runs produce flux dropout in the outer belt. At 05–06 UT on September 4, RBE-SWMF results show a sudden increase in electron flux consistent with the Akebono data. There is no significant enhancement in the RBE–T04 run during the recovery phase of the storm.

Since wave–particle interactions are not considered in these particular RBE calculations, the enhancement seen in Fig. 5(c) has to be a result of particle transport. Glocer et al. (2011) examine the magnetic configuration during the enhancement. It is found that the MHD model predicts a substorm dipolarization at ∼05 UT on September 4.

Fig. 6 shows the BATSRUS field lines in white and pressure in color on the X–Z plane in a small time window surrounding the flux enhancement. Before the enhancement, the field in the tail gets progressively more stretched. At 05:00 UT the stretching reaches a maximum which coincides with the flux dropout. At 05:15 UT a reconnection site forms close to the Earth and a plasmoid is ejected tailward. Magnetic reconnection which causes a substorm development in this case is controlled by numerical dissipation, but for study of radiation belt response to substorm, it suffices. As the dipolarization proceeds in the magnetic field, field lines convect earthward. Electrons which are gyrating along these collapsing field lines can be accelerated significantly on a time scale of minutes (Fok, Moore & Spjeldvik, 2001; Glocer et al., 2009), much faster than the time scale for energization by whistler mode chorus waves, which is typically of the order of 1-2 days (Summers & Ma, 2000). The T04 model, which is driven by Dst and solar wind parameters, does not contain clear substorm signatures. Empirical magnetic field models of this kind cannot directly simulate substorm reconfiguration unless special tricks are applied (Delcourt et al., 1990; Fok, Moore & Spjeldvik, 2001; Pulkkinen et al., 1991). For moderate storms such as this one on 3–5 September 2008, convection is weak and the dominant energization and transport mechanism is sub-storm reconfiguration and the resulting dipolarization electric field. The rapid enhancements of radiation belt fluxes during modest storms cannot be explained without the consideration of substorm effects.

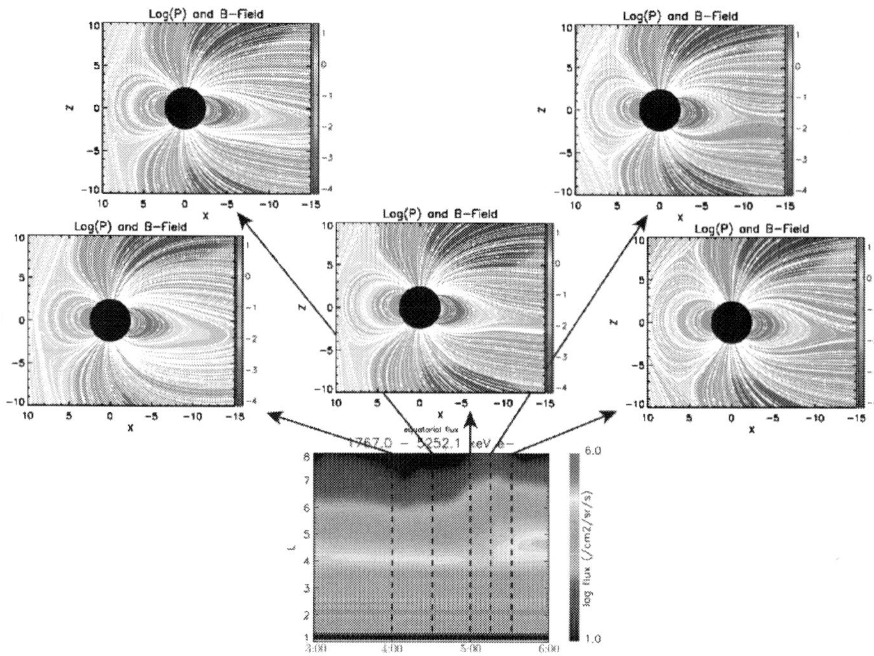

Fig. 6. MHD modeling of radiation belt enhancement and evolution of the magnetosphere. Top and middle panels: MHD plots of magnetic field lines (in white) and the pressure contours (color coded) in Y=0 plane. Bottom panel: The energetic electron flux from RBE from 03:00 to 06:00 UT on 4 September. Adopted from Glocer et al. (2011).

7. Discussion and directions for future work

The near-Earth space environment is described by multiscale physics that reflects a variety of processes and conditions occuring in magnetospheric plasma. Plasma densities vary from 10^6 cm^{-3} in the F layer of dayside ionosphere to less then 1 cm^{-3} in the tail. Ionospheric plasma is highly collisional, and the plasma above ~500 km is essentially collisionless. It has been recently realized that for a successful description of such a plasma system, a complex solution is needed which allows different physics domains to be described using different physical models (De Zeeuw et al., 2004; Fok et al., 2006; Moore et al., 2008; Ridley et al., 2004; Tóth et al., 2005; Tóth et al., 2011; Zhang, Liemohn, De Zeeuw, Borovsky, Ridley, Toth, Sazykin, Thomsen, Kozyra, Gombosi & Wolf, 2007). This is the reason why coupled models are intensively developed and improved during recent years. Besides several cases considered here, the latest examples include: coupling between the RCM (ring current model) and Versatile Electron Radiation Belt (VERB) code (Subbotin et al., 2011); coupling between RAM-SCB code (ring current model) and BATSRUS/SWMF (MHD with the RCM) (Welling et al., 2011; Zaharia et al., 2010); coupling between two ring current models RAM and the RCM (Jordanova et al., 2010); one way coupling between MHD Open Geospace General Circulation Model (Open GGCM) – the RCM (ring current model) (Hu et al., 2010). None of these models however is 'ideal' (including the examples presented in detail earlier) in the sense that all of them have some shortcuts. For example, the RCM assumes an isotropic distribution function which is not consistent with a concept of diffusion in pitch-angle (RCM-VERB code); the RAM-RCM code does not contain a self-consistent magnetic field; the RAM-SCB-BATSRUS/SWMF as well as CRCM-BATSRUS and Open GGCM-RCM work only in one-way mode; coupled RBE-BATSRUS/SWMF code does not have wave interactions so far. To overcome at least some of these shortcuts, we are going to make the following improvements to CRCM/RBE/BATSRUS/SWMF model:

- Create two–way coupling between the CRCM and MHD BATSRUS;
- Combine the CRCM, the RBE and plasmasphere codes into one module of inner magnetosphere electrodynamics;
- Gradually include waves in addition to the chorus wave modes, such as plasmaspheric hiss and electron ion cyclotron waves into the RBE and the CRCM;
- Combine all modules (including two–way coupling between ring current and MHD) and incorporate them into the SWMF.

It should be noted that a coupling of different physical models is always nontrivial and some shortcuts are inevitable. However, each newly coupled model is an incremental improvement to the current state-of-the-art inner magnetosphere models. Each improvement enhances our overall understanding of complex and nonlinear relationships between different domains in the Earth's magnetosphere and the response of magnetosphere to solar wind drivers.

8. Acknowledgments

This research was supported by NASA Science Mission Directorate, Heliophysics Division, Living With a Star Targeted Research and Technology Program, under Work Breakdown Structures: 936723.02.01.06.78 and 936723.02.01.01.27, and Heliophysics Guest Investigators Program, under Work Breakdown Structure 955518. N. Buzulukova thanks E. Mitchell for proofreading of the manuscript.

9. References

Albert, J. M. (1994). Quasi-linear pitch angle diffusion coefficients: Retaining high harmonics, *J. Geophys. Res.* 992: 23741–+.

Albert, J. M. & Young, S. L. (2005). Multidimensional quasi-linear diffusion of radiation belt electrons, *Geophys. Res. Lett.* 321: L14110.

Baker, D. N. (2002). How to cope with space weather, *Science* 297(5586): 1486–1487. URL: *http://www.sciencemag.org/content/297/5586/1486.short*

Borovsky, J. E. & Denton, M. H. (2006). Differences between CME-driven storms and CIR-driven storms, *J. Geophys. Res.* 111: A07S08.

Bourdarie, S., Boscher, D., Beutier, T., Sauvaud, J.-A. & Blanc, M. (1997). Electron and proton radiation belt dynamic simulations during storm periods: A new asymmetric convection-diffusion model, *J. Geophys. Res.* 1021: 17541–17552.

Brandt, P. C., Ohtani, S., Mitchell, D. G., Fok, M.-C., Roelof, E. C. & Demajistre, R. (2002). Global ENA observations of the storm mainphase ring current: Implications for skewed electric fields in the inner magnetosphere, *Geophys. Res. Lett.* 29(20): 1359.

Buzulukova, N., Fok, M.-C., Moore, T. E. & Ober, D. M. (2008). Generation of plasmaspheric undulations, *Geophys. Res. Lett.* 351: L13105.

Buzulukova, N., Fok, M.-C., Pulkkinen, A., Kuznetsova, M., Moore, T. E., Glocer, A., Brandt, P. C., Tóth, G. & Rastätter, L. (2010). Dynamics of ring current and electric fields in the inner magnetosphere during disturbed periods: CRCM-BATS-R-US coupled model, *J. Geophys. Res.* 115: A05210.

De Zeeuw, D. L., Sazykin, S., Wolf, R. A., Gombosi, T. I., Ridley, A. J. & Tóth, G. (2004). Coupling of a global MHD code and an inner magnetospheric model: Initial results, *Journal of Geophysical Research (Space Physics)* 109: A12219.

Delcourt, D. C., Pedersen, A. & Sauvaud, J. A. (1990). Dynamics of single-particle orbits during substorm expansion phase, *J. Geophys. Res.* 952: 20853–20865.

Denton, M. H., Borovsky, J. E., Skoug, R. M., Thomsen, M. F., Lavraud, B., Henderson, M. G., McPherron, R. L., Zhang, J. C. & Liemohn, M. W. (2006). Geomagnetic storms driven by ICME- and CIR-dominated solar wind, *J. Geophys. Res.* 111: A07S07.

Dessler, A. J. & Karplus, R. (1961). Some Effects of Diamagnetic Ring Currents on Van Allen Radiation, *J. Geophys. Res.* 66: 2289–2295.

Ebihara, Y. & Ejiri, M. (2003). Numerical Simulation of the Ring Current: Review, *Space Sci. Rev* 105: 377–452.

Ebihara, Y., Fok, M.-C., Blake, J. B. & Fennell, J. F. (2008). Magnetic coupling of the ring current and the radiation belt, *J. Geophys. Res.* 113: 7221.

Elkington, S. R., Hudson, M. K. & Chan, A. A. (1999). Acceleration of relativistic electrons via drift-resonant interaction with toroidal-mode Pc-5 ULF oscillations, *Geophys. Res. Lett.* 26: 3273–3276.

Fälthammar, C.-G. (1965). Effects of Time-Dependent Electric Fields on Geomagnetically Trapped Radiation, *J. Geophys. Res.* 70: 2503–2516.

Fok, M. C., Ebihara, Y., Moore, T. E., Ober, D. M. & Keller, K. A. (2005). Geospace storm processes coupling the ring current, radiation belt and plasmasphere, *in* J. B. et al. (ed.), *Inner Magnetosphere Interactions: New Perspectives from Imaging*, Vol. 159 of *Geophys. Monogr. Ser.*, AGU, Washington, D.C., pp. 207–220.

Fok, M.-C., Glocer, A., Zheng, Q., Horne, R. B., Meredith, N. P., Albert, J. M. & Nagai, T. (2011). Recent developments in the radiation belt environment model, *J. Atmos. Sol. Terr. Phys.* 73: 1435–1443.

Fok, M.-C. & Moore, T. E. (1997). Ring current modeling in a realistic magnetic field configuration, *Geophys. Res. Lett.* 24: 1775–1778.

Fok, M.-C., Moore, T. E., Brandt, P. C., Delcourt, D. C., Slinker, S. P. & Fedder, J. A. (2006). Impulsive enhancements of oxygen ions during substorms, *J. Geophys. Res.* 111: 10222.

Fok, M. C., Moore, T. E., Kozyra, J. U., Ho, G. C. & Hamilton, D. C. (1995). Three-dimensional ring current decay model, *J. Geophys. Res.* 100: 9619–9632.

Fok, M.-C., Moore, T. E. & Spjeldvik, W. N. (2001). Rapid enhancement of radiation belt electron fluxes due to substorm dipolarization of the geomagnetic field, *J. Geophys. Res.* 106: 3873–3882.

Fok, M.-C., Wolf, R. A., Spiro, R. W. & Moore, T. E. (2001). Comprehensive computational model of Earth's ring current, *J. Geophys. Res.* 106: 8417–8424.

Fok, M., Horne, R. B., Meredith, N. P. & Glauert, S. A. (2008). Radiation Belt Environment model: Application to space weather nowcasting, *J. Geophys. Res.* 113: 3.

Fok, M., Moore, T. E. & Greenspan, M. E. (1996). Ring current development during storm main phase, *J. Geophys. Res.* 101: 15311–15322.

Friedel, R. H. W., Reeves, G. D. & Obara, T. (2002). Relativistic electron dynamics in the inner magnetosphere - a review, *Journal of Atmospheric and Solar-Terrestrial Physics* 64: 265–282.

Galperin, Y. I., Ponomarev, V. N. & Zosimova, A. G. (1974). Plasma convection in polar ionosphere, *Ann. Geophys.* 30(1): 1–7.

Glauert, S. A. & Horne, R. B. (2005). Calculation of pitch angle and energy diffusion coefficients with the PADIE code, *J. Geophys. Res.* 110: A04206.

Glocer, A., Fok, M.-C., Nagai, T., Tóth, G., Guild, T. & Blake, J. (2011). Rapid rebuilding of the outer radiation belt, *J. Geophys. Res.* 116: A09213.

Glocer, A., Toth, G., Fok, M., Gombosi, T. & Liemohn, M. (2009). Integration of the radiation belt environment model into the space weather modeling framework, *J. Atmos. Sol. Terr. Phys.* 71: 1653–1663.

Goldstein, J., Burch, J. L., Sandel, B. R., Mende, S. B., C:son Brandt, P. & Hairston, M. R. (2005). Coupled response of the inner magnetosphere and ionosphere on 17 April 2002, *Journal of Geophysical Research (Space Physics)* 110: A03205.

Harel, M., Wolf, R. A., Reiff, P. H., Spiro, R. W., Burke, W. J., Rich, F. J. & Smiddy, M. (1981). Quantitative simulation of a magnetospheric substorm. I - Model logic and overview, *J. Geophys. Res.* 86: 2217–2241.

Horne, R. B. (2007). Plasma astrophysics: Acceleration of killer electrons, *Nature Physics* 3: 590–591.

Hu, B., Toffoletto, F. R., Wolf, R. A., Sazykin, S., Raeder, J., Larson, D. & Vapirev, A. (2010). One-way coupled OpenGGCM/RCM simulation of the 23 March 2007 substorm event, *J. Geophys. Res.* 115: A12205.

Huang, C.-L., Spence, H. E., Hudson, M. K. & Elkington, S. R. (2010). Modeling radiation belt radial diffusion in ULF wave fields: 2. Estimating rates of radial diffusion using combined MHD and particle codes, *J. Geophys. Res.* 115: A06216.

Huang, C.-L., Spence, H. E., Singer, H. J. & Hughes, W. J. (2010). Modeling radiation belt radial diffusion in ULF wave fields: 1. Quantifying ULF wave power at geosynchronous orbit in observations and in global MHD model, *J. Geophys. Res.* 115: A06215.

Iijima, T. & Shibaji, T. (1987). Global characteristics of northward IMF-associated (NBZ) field-aligned currents, *J. Geophys. Res.* 92: 2408–2424.

Jaggi, R. K. & Wolf, R. A. (1973). Self-consistent calculation of the motion of a sheet of ions in the magnetosphere., *J. Geophys. Res.* 78: 2852–2866.

Jordanova, V. K., Kozyra, J. U., Khazanov, G. V., Nagy, A. F., Rasmussen, C. E. & Fok, M.-C. (1994). A bounce-averaged kinetic model of the ring current ion population, *Geophys. Res. Lett.* 21: 2785–2788.

Jordanova, V. K., Kozyra, J. U., Nagy, A. F. & Khazanov, G. V. (1997). Kinetic model of the ring current-atmosphere interactions, *J. Geophys. Res.* 102: 14279–14292.

Jordanova, V. K., Thorne, R. M., Li, W. & Miyoshi, Y. (2010). Excitation of whistler mode chorus from global ring current simulations, *J. Geophys. Res.* 115: A00F10.

Kanekal, S. G., Baker, D. N. & Blake, J. B. (2001). Multisatellite measurements of relativistic electrons: Global coherence, *J. Geophys. Res.* 106: 29721–29732.

Kanekal, S. G., Friedel, R. H. W., Reeves, G. D., Baker, D. N. & Blake, J. B. (2005). Relativistic electron events in 2002: Studies of pitch angle isotropization, *J. Geophys. Res.* 110: A12224.

Kellogg, P. J. (1959). Van Allen Radiation of Solar Origin, *Nature* 183: 1295–1297.

Khazanov, G. V. (2011). *Kinetic Theory of the Inner Magnetospheric Plasma*, Springer-Verlag.

Kim, H.-J. & Chan, A. A. (1997). Fully adiabatic changes in storm time relativistic electron fluxes, *J. Geophys. Res.* 1022: 22107–22116.

Li, X., Temerin, M., Baker, D. N., Reeves, G. D. & Larson, D. (2001). Quantitative prediction of radiation belt electrons at geostationary orbit based on solar wind measurements, *Geophys. Res. Lett.* 28: 1887–1890.

Liemohn, M. W., Kozyra, J. U., Jordanova, V. K., Khazanov, G. V., Thomsen, M. F. & Cayton, T. E. (1999). Analysis of early phase ring current recovery mechanisms during geomagnetic storms, *Geophys. Res. Lett* 26: 2845–2848.

Lyatsky, W. & Khazanov, G. V. (2008). Effect of geomagnetic disturbances and solar wind density on relativistic electrons at geostationary orbit, *J. Geophys. Res.* 113: A08224.

Lyons, L. R., Thorne, R. M. & Kennel, C. F. (1972). Pitch-angle diffusion of radiation belt electrons within the plasmasphere., *J. Geophys. Res.* 77: 3455–3474.

Meredith, N. P., Horne, R. B. & Anderson, R. R. (2001). Substorm dependence of chorus amplitudes: Implications for the acceleration of electrons to relativistic energies, *J. Geophys. Res.* 106: 13165–13178.

Meredith, N. P., Horne, R. B., Glauert, S. A. & Anderson, R. R. (2007). Slot region electron loss timescales due to plasmaspheric hiss and lightning-generated whistlers, *J. Geophys. Res.* 112: A08214.

Meredith, N. P., Horne, R. B., Thorne, R. M. & Anderson, R. R. (2003). Favored regions for chorus-driven electron acceleration to relativistic energies in the Earth's outer radiation belt, *Geophys. Res. Lett.* 30(16): 160000–1.

Meredith, N. P., Horne, R. B., Thorne, R. M. & Anderson, R. R. (2009). Survey of upper band chorus and ECH waves: Implications for the diffuse aurora, *J. Geophys. Res.* 114: A07218.

Miyoshi, Y. S., Jordanova, V. K., Morioka, A., Thomsen, M. F., Reeves, G. D., Evans, D. S. & Green, J. C. (2006). Observations and modeling of energetic electron dynamics during the October 2001 storm, *J. Geophys. Res.* 111: A11S02.

Moore, T. E., Fok, M.-C., Delcourt, D. C., Slinker, S. P. & Fedder, J. A. (2008). Plasma plume circulation and impact in an MHD substorm, *J. Geophys. Res.* 113: 6219.

Ni, B., Thorne, R. M., Meredith, N. P., Horne, R. B. & Shprits, Y. Y. (2011). Resonant scattering of plasma sheet electrons leading to diffuse auroral precipitation: 2. Evaluation for whistler mode chorus waves, *J. Geophys. Res.* 116: A04219.

Ober, D. M., Horwitz, J. L. & Gallagher, D. L. (1997). Formation of density troughs embedded in the outer plasmasphere by subauroral ion drift events, *J. Geophys. Res.* 102: 14595–14602.

Powell, K. G., Roe, P. L., Linde, T. J., Gombosi, T. I. & de Zeeuw, D. L. (1999). A Solution-Adaptive Upwind Scheme for Ideal Magnetohydrodynamics, *J. Comput. Phys.* 154: 284–309.

Pulkkinen, T. I., Baker, D. N., Fairfield, D. H., Pellinen, R. J., Murphree, J. S., Elphinstone, R. D., McPherron, R. L., Fennell, J. F., Lopez, R. E. & Nagai, T. (1991). Modeling the growth phase of a substorm using the Tsyganenko model and multi-spacecraft observations - CDAW-9, *Geophys. Res. Lett.* 18: 1963–1966.

Rasmussen, C. E. & Schunk, R. W. (1987). Ionospheric convection driven by NBZ currents, *J. Geophys. Res.* 92: 4491–4504.

Reeves, G. D., Henderson, M. C., Skoug, R. M., Thomsen, M. F., Borovsky, J. E., Funsten, H. O., C:Son Brandt, P., Mitchell, D. J., Jahn, J.-M., Pollock, C. J., McComas, D. J. & Mende, S. B. (2003). IMAGE, POLAR, and Geosynchronous Observations of Substorm and Ring Current Ion Injection, *in* A. S. Sharma, Y. Kamide, & G. S. Lakhina (ed.), *Disturbances in Geospace: The Storm-substorm Relationship*, Vol. 142 of *Washington DC American Geophysical Union Geophysical Monograph Series*, pp. 91–+.

Reeves, G. D., Morley, S. K., Friedel, R. H. W., Henderson, M. G., Cayton, T. E., Cunningham, G., Blake, J. B., Christensen, R. A. & Thomsen, D. (2011). On the relationship between relativistic electron flux and solar wind velocity: Paulikas and Blake revisited, *J. Geophys. Res.* 116: A02213.

Ridley, A., Gombosi, T. & Dezeeuw, D. (2004). Ionospheric control of the magnetosphere: conductance, *Ann. Geophys.* 22: 567–584.

Ridley, A. J. & Liemohn, M. W. (2002). A model-derived storm time asymmetric ring current driven electric field description, *J. Geophys. Res.* 107: 1151.

Schulz, M. & Eviatar, A. (1969). Diffusion of equatorial particles in the outer radiation zone., *J. Geophys. Res.* 74: 2182–2192.

Schulz, M. & Lanzerotti, L. J. (1974). *Particle diffusion in the radiation belts*, Springer-Verlag.

Shprits, Y. Y., Chen, L. & Thorne, R. M. (2009). Simulations of pitch angle scattering of relativistic electrons with MLT-dependent diffusion coefficients, *J. Geophys. Res.* 114: A03219.

Shprits, Y. Y., Elkington, S. R., Meredith, N. P. & Subbotin, D. A. (2008). Review of modeling of losses and sources of relativistic electrons in the outer radiation belt I: Radial transport, *Journal of Atmospheric and Solar-Terrestrial Physics* 70: 1679–1693.

Southwood, D. J. & Wolf, R. A. (1978). An assessment of the role of precipitation in magnetospheric convection, *J. Geophys. Res.* 83: 5227–5232.

Spiro, R. W., Heelis, R. A. & Hanson, W. B. (1978). Ion convection and the formation of the mid-latitude F region ionization trough, *J. Geophys. Res.* 83: 4255–4264.

Subbotin, D. A., Shprits, Y. Y., Gkioulidou, M., Lyons, L. R., Ni, B., Merkin, V. G., Toffoletto, F. R., Thorne, R. M., Horne, R. B. & Hudson, M. K. (2011). Simulation of the acceleration of relativistic electrons in the inner magnetosphere using RCM-VERB coupled codes, *J. Geophys. Res.* 116: A08211.

Summers, D. & Ma, C.-y. (2000). A model for generating relativistic electrons in the Earth's inner magnetosphere based on gyroresonant wave-particle interactions, *J. Geophys. Res.* 105: 2625–2640.

Summers, D., Ni, B. & Meredith, N. P. (2007). Timescales for radiation belt electron acceleration and loss due to resonant wave-particle interactions: 2. Evaluation for VLF chorus, ELF hiss, and electromagnetic ion cyclotron waves, *J. Geophys. Res.* 112: A04207.

Toffoletto, F. R., Sazykin, S., Spiro, R. W., Wolf, R. A. & Lyon, J. G. (2004). RCM meets LFM: initial results of one-way coupling, *J. Atmos. Sol. Terr. Phys.* 66: 1361–1370.

Toffoletto, F., Sazykin, S., Spiro, R. & Wolf, R. (2003). Inner magnetospheric modeling with the Rice Convection Model, *Space Sci. Rev.* 107: 175–196.

Tóth, G., Sokolov, I. V., Gombosi, T. I., Chesney, D. R., Clauer, C. R., De Zeeuw, D. L., Hansen, K. C., Kane, K. J., Manchester, W. B., Oehmke, R. C., Powell, K. G., Ridley, A. J., Roussev, I. I., Stout, Q. F., Volberg, O., Wolf, R. A., Sazykin, S., Chan, A., Yu, B. & Kóta, J. (2005). Space Weather Modeling Framework: A new tool for the space science community, *J. Geophys. Res.* 110: A12226.

Tóth, G., van der Holst, B., Sokolov, I. V., Zeeuw, D. L. D., Gombosi, T. I., Fang, F., Manchester, W. . B., Meng, X., Najib, D., Powell, K. G., Stout, Q. F., Glocer, A., Ma, Y.-J. & er, M. O. (2011). Adaptive numerical algorithms in space weather modeling, *J. Comput. Phys.,In Press, Corrected Proof*.

Tsyganenko, N. A. (1995). Modeling the Earth's magnetospheric magnetic field confined within a realistic magnetopause, *J. Geophys. Res.* 100: 5599–5612.

Tsyganenko, N. A. & Mukai, T. (2003). Tail plasma sheet models derived from Geotail particle data, *J. Geophys. Res.* 108: 1136.

Tsyganenko, N. A., Singer, H. J. & Kasper, J. C. (2003). Storm-time distortion of the inner magnetosphere: How severe can it get?, *J. Geophys. Res.* 108: 1209.

Turner, N. E., Cramer, W. D., Earles, S. K. & Emery, B. A. (2009). Geoefficiency and energy partitioning in CIR-driven and CME-driven storms, *J. Geophys. Res.* 71: 1023–1031.

Ukhorskiy, A. Y. & Sitnov, M. I. (2008). Radial transport in the outer radiation belt due to global magnetospheric compressions, *J. Atmos. Sol. Terr. Phys.* 70: 1714–1726.

Varotsou, A., Boscher, D., Bourdarie, S., Horne, R. B., Glauert, S. A. & Meredith, N. P. (2005). Simulation of the outer radiation belt electrons near geosynchronous orbit including both radial diffusion and resonant interaction with Whistler-mode chorus waves, *Geophys. Res. Lett.* 321: L19106.

Weigel, R. S. (2010). Solar wind density influence on geomagnetic storm intensity, *J. Geophys. Res.* 115: A09201.

Weimer, D. R. (2001). An improved model of ionospheric electric potentials including substorm perturbations and application to the Geospace Environment Modeling November 24, 1996, event, *J. Geophys. Res.* 106: 407–416.

Welling, D. T., Jordanova, V. K., Zaharia, S. G., Glocer, A. & Toth, G. (2011). The effects of dynamic ionospheric outflow on the ring current, *J. Geophys. Res.* 116: A00J19.

Wolf, R. A., Spiro, R. W., Sazykin, S. & Toffoletto, F. R. (2007). How the Earth's inner magnetosphere works: An evolving picture, *J. Atmos. Sol. Terr. Phys.* 69: 288–302.

Xing, X. & Wolf, R. A. (2007). Criterion for interchange instability in a plasma connected to a conducting ionosphere, *J. Geophys. Res.* 112: A12209.

Yermolaev, Y. I., Nikolaeva, N. S., Lodkina, I. G. & Yermolaev, M. Y. (2010). Specific interplanetary conditions for CIR-, Sheath-, and ICME-induced geomagnetic storms obtained by double superposed epoch analysis, *Annales Geophysicae* 28: 2177–2186.

Young, D. T., Balsiger, H. & Geiss, J. (1982). Correlations of magnetospheric ion composition with geomagnetic and solar activity, *J. Geophys. Res.* 87: 9077–9096.

Zaharia, S., Jordanova, V. K., Thomsen, M. F. & Reeves, G. D. (2006). Self-consistent modeling of magnetic fields and plasmas in the inner magnetosphere: Application to a geomagnetic storm, *J. Geophys. Res.* 111: 11.

Zaharia, S., Jordanova, V. K., Welling, D. & Tóth, G. (2010). Self-consistent inner magnetosphere simulation driven by a global MHD model, *J. Geophys. Res.* 115: A12228.

Zhang, J., Liemohn, M. W., De Zeeuw, D. L., Borovsky, J. E., Ridley, A. J., Toth, G., Sazykin, S., Thomsen, M. F., Kozyra, J. U., Gombosi, T. I. & Wolf, R. A. (2007). Understanding storm-time ring current development through data-model comparisons of a moderate storm, *J. Geophys. Res.* 112(A11): 4208.

Zhang, J., Richardson, I. G., Webb, D. F., Gopalswamy, N., Huttunen, E., Kasper, J. C., Nitta, N. V., Poomvises, W., Thompson, B. J., Wu, C.-C., Yashiro, S. & Zhukov, A. N. (2007). Solar and interplanetary sources of major geomagnetic storms (Dst < -100 nT) during 1996-2005, *J. Geophys. Res.* 112: A10102.

Zheng, Q., Fok, M.-C., Albert, J., Horne, R. B. & Meredith, N. P. (2011). Effects of energy and pitch angle mixed diffusion on radiation belt electrons, *J. Atmos. Sol. Terr. Phys.* 73: 785–795.

Zheng, Y., Fok, M.-C. & Khazanov, G. V. (2003). A radiation belt-ring current forecasting model, *Space Weather* 1: 1013.

15

Ground-Based Monitoring of the Solar Wind Geoefficiency

Oleg Troshichev
Arctic and Antarctic Research Institute
Russia

1. Introduction

The Earth's magnetosphere is a result of the solar wind impact on the dipole-like geomagnetic field. The form and size of geomagnetosphere are basically determined by the solar wind dynamic pressure, whereas unsteady processes within the magnetosphere affecting the human activity under name of magnetic storms and substorms are due to variations of the magnetic field transported by the solar wind plasma. This field, known as interplanetary magnetic field (IMF), is governed by solar activity and can reverse its direction and vary in amplitude a ten times. As soon as the space-borne measurements of solar wind parameters started, it became clear that southward IMF (opposite in direction to the Earth magnetic field) is one of the most geoeffective solar wind characteristics. The perfect relationship was found between southward IMF (B_{ZS}) and such indicators of magnetic activity as a planetary *Kp* index (Fairfield & Cahill, 1966; Wilcox et al., 1967; Rostoker & Fälthammar, 1967; Baliff et al., 1967), *AE* index of magnetic activity in the auroral zone (Pudovkin et al., 1970; Arnoldy, 1971; Foster et al., 1971; Kokubun, 1972; Meng et al., 1973), magnetic storm *Dst* index (Hirshberg & Colburn, 1969; Kokubun, 1972; Kane, 1974; Russel et al., 1974). Since the correlation of magnetic activity with the solar wind fluctuations distinctly increased when the product of the solar wind speed and southward IMF was taken into account (Rostoker & Fälthammar, 1967; Garrett et al., 1974; Murayama & Hakamada, 1975), the conclusion was made that the interplanetary electric field $\boldsymbol{E}=v\times\boldsymbol{B}_{ZS}$ plays a crucial part in the solar wind-magnetosphere coupling (Rostoker & Fälthammar, 1967).

Various combinations of solar wind parameters (basically, the interplanetary electric field and the solar wind density and speed) were repeatedly examined to establish the best function for description of the solar wind–magnetosphere coupling. The most well-known functions are an parameter $\varepsilon=l_0^2vB^2\sin^4(\theta_c/2)$, firstly presented by Perreault & Akasofu (1978), electric field $E_{KL} = vB_T\sin^2(\theta_c/2)$ introduced in practice by Kan & Lee (1979), and rectifying function $E_Y= vB_{ZS}$ (Reiff & Luhmann, 1986), where v is a velocity of the solar wind, B is IMF intensity, B_Y, B_Z and B_{ZS} are azimuthal, vertical and southward IMF components, B_T is transverse IMF component $B_T= \{(B_Y)^2+(B_Z)^2\}^{1/2}$, and θ_c is an angle between B_T component and the geomagnetic Z-axis. A precise formula for the solar wind-magnetosphere coupling function has not yet been agreed so far, and other functions, in number of more than ten, are also used in practice.

In attempts to derive an universal solar wind–magnetosphere coupling function, Newell et al. (2007, 2008) investigated behavior of 10 variables, characterizing the magnetosphere state, in relation to different coupling functions. The comprehensive investigations did not reveal a unique coupling function applicable for any circumstances and conditions, but it was noted (Newell et al., 2008), that the unique coupling function, if it exists, must involve the solar wind velocity v to the first (or a little higher) power, transverse IMF component B_T to the first (or a little lower) power, and sine of IMF clock angle θ_c to the second (or more) power. It is easy to see that coupling function E_{KL} (Kan & Lee, 1979) is well consistent with these requirements.

Coupling functions are used to characterize the solar wind geoefficiency. As this takes place, the solar wind parameters are measured outside the magnetosphere, at present on board ACE spacecraft positioned in Lagrange point (L1) at a distance of ~ 1.5 millions km from Earth. As a consequence, an actual value of the solar wind parameters at magnetopause can essentially differ from those monitored on board ACE spacecraft, even if they are time-shifted to the magnetosphere. Besides, a very high level of magnetic field turbulence is typical of region between the bow shock and magnetopause with incorporation of nonlinear processes within the boundary magnetosphere (Rossolenko et al., 2009), and it is unlikely to wait that changes in the solar wind parameters are converted in their true shape into the magnetosphere processes, while transmitting a signal through the highly turbulent region. Hence it is very desirable to monitor the solar wind energy that entered into the magnetosphere. It is suggested to use for this purpose a ground-based PC index put forward by Troshichev et al. (1988) as a index of magnetic activity in the polar cap.

2. Physical backgrounds and method for the PC derivation

Distribution and intensity of magnetic activity in the Earth's polar caps is determined by orientation and power of the interplanetary magnetic field, particular types of the polar cap magnetic disturbances being related to the IMF southward B_{ZS} (Nishida, 1968) , azimuthal B_Y (Svalgaard, 1968; Mansurov, 1969) and northward B_{ZN} (Maezawa, 1976; Kuznetsov & Troshichev, 1977) components. Distribution of magnetic disturbances on the ground level is commonly described by systems of equivalent currents being hypothetic currents, providing the observed magnetic effect on the ground surface. Figure 1 demonstrates DP2, DP3 and DP4 current systems derived for polar cap disturbances associated with action of southward, northward and azimuthal IMF components, respectively (Kuznetsov & Troshichev, 1977; Troshichev & Tsyganenko, 1978). In addition, the DP0 disturbances have been separated (Troshichev & Tsyganenko, 1978) which are observed irrespective of the IMF, but well correlate with the solar wind velocity v in the second power (Sergeev & Kuznetsov, 1981) that makes it possible to associate them with the solar wind dynamic pressure. DP2 and DP0 current systems are terminated by the latitudes of Φ=50-60° (Troshichev, 1975) and focuses in these current vortices are located just right where the intense magnetospheric field-aligned currents are regularly observed.

Measurements on board spacecrafts OGO-4 (Zmuda & Armstrong, 1974) and TRIAD (Iijima & Potemra, 1976) showed that field-aligned currents (FAC) are distributed in two regions aligned with the auroral oval. Region 1 FAC system consists of a layer of field-aligned currents on the poleward boundary of the auroral oval, with currents flowing into the ionosphere in the morning sector and flowing out of the ionosphere in the evening sector.

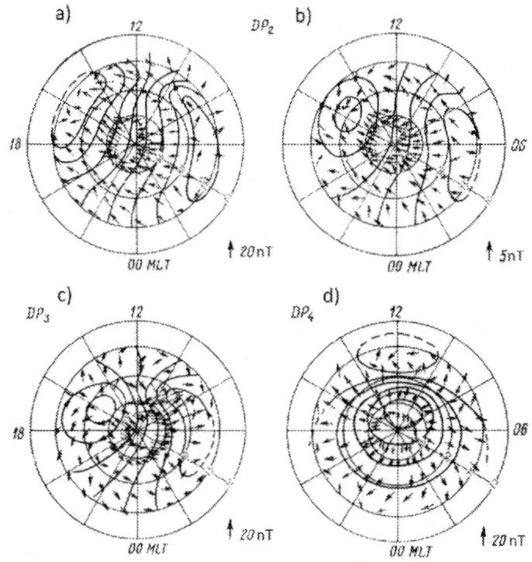

Fig. 1. Current systems of DP2, DP3 and DP4 disturbances generated by variations of IMF components: (a) southward $B_{ZS}=-1nT$, (b) southward $B_{ZS}=-0.25nT$, (c) northward B_{ZN}, (d) azimuthal B_Y (Troshichev and Tsyganenko, 1978).

Region 2 FAC system with the oppositely directed field-aligned currents is positioned on the equatorward boundary of the oval. The currents in Region 1 demonstrate a strong dependency on southward IMF B_{ZS} (Langel, 1975; Mc Diarmid et al., 1977; Iijima & Potemra, 1982) or the interplanetary electric field E_Y (Bythrow & Potemra, 1983). The Region 1 FAC system is observed permanently, even during very quiet conditions, whereas Region 2 system becomes available when magnetic disturbances in the auroral zone are developed (Iijima & Potemra, 1978). Regions 1 and 2 FAC systems are mapped into the equatorial plane of magnetosphere in all magnetosphere models (Antonova et al., 2006). Observations of ion fluxes and magnetic fields made on board DMSP F7 satellite (Iijima et al., 1997), plasma sheet temperature, density, and pressure data inferred from DMSP F8, F9, F10, and F11 satellite measurements at the ionospheric altitudes (Wing and Newell, 2000), measurements of plasma pressure gradients on board the THEMIS satellites (Xing et al., 2009) lead to conclusion that the field-aligned current systems are constantly driven by the pressure-gradient forces generated within the closed magnetosphere while it's coupling with the varying solar wind. The azimuthal pressure gradients required to support the Region 1 field-aligned currents have been derived in studies (Stepanova et al., 2004, 2006; Antonova et al., 2011)

The field-aligned currents patterns typical of low and high magnetic activity have been applied to calculate the systems of electric fields and currents in the polar ionosphere (Nisbet et al., 1978; Gizler et al., 1979; Troshichev et al., 1979). The calculated equivalent current systems turned out to be identical to the experimental systems derived from magnetic disturbances observed in the polar caps. The conclusion was made (Troshichev, 1982) that magnetic activity in polar caps is related to the field-aligned currents responding to changes in the solar wind, and the DP2 disturbances can be taken as an indicator of the geoeffective solar wind impacting on the magnetosphere.

Examination of statistical relationships between the DP2 magnetic disturbances and different interplanetary quantities (Troshichev and Andrezen, 1985) gave the best result for coupling function E_{KL}: the disturbance values δF at the near-pole stations Thule (Greenland) and Vostok (Antarctica) turned out to be linearly linked with E_{KL} value. It means that E_{KL} value can be estimated if the scaling coefficients between quantities E_{KL} and δF is established. Dependence of δF value on ionospheric conductivity is easily taken into account under common conditions, when the polar cap ionosphere is regulated by the solar UV irradiation. These are the physical backgrounds determining a method for the *PC* index derivation (Troshichev et al., 1988, 2006).

The statistically justified regression coefficients a and β, which determine the relationship between the coupling function E_{KL} and vector of DP2 magnetic disturbance δF at stations Thule and Vostok, are derived at first:

$$\delta F = \alpha E_{KL} + \beta \quad (1)$$

These coefficients are calculated for any UT moment of each day of the year since orientation of DP2 disturbances at stations is dependent on local time and slightly changed from summer to winter. To determine the statistically justified orientation of the DP2 disturbances vector (i.e. angle ϕ between the disturbance vector and the dawn-dusk meridian), the correlation between δF and E_{KL} is calculated for all angles ϕ in the range of ±90° from the dawn-dusk orientation on the basis of large set of data (some years). The angle ϕ ensuring the best correlation between values δF and E_{KL} is fixed and the appropriate regression coefficients a and β are calculated. Just these parameters ϕ, a and β are used in further calculations.

The values E_{KL} were calculated from measurements of solar wind parameters in space, shifted to the sub-solar point (12 RE) using the actual solar wind velocity. Then a time delay $\Delta T \sim 20$ min is required for an E_{KL} signal to be transferred from the bow shock position to the polar cap. To take into account the effect of ionospheric conductivity variations and its changes in response to solar activity, the δF values were estimated in reference to quiet daily curve (QDC) with allowance for QDC change from day to day (Janzhura & Troshichev, 2008). Usage of a proper QDC, as a level of reference for δF values, ensures invariance of the parameters α, β and ϕ determining relationship between δF and E_{KL} irrespective of solar activity (Troshichev et al., 2011a). As an evidence, **Figure 2** shows distribution of α, β and ϕ parameter derived for the Vostok station for epochs of solar maximum (Troshichev et al. 2006) and solar minimum (Troshichev et al., 2011a), and for entire cycle of solar activity (Troshichev et al., 2007, Troshichev & Janzhura, 2009). One can see that patterns for ϕ, α and β parameters derived independently for different epochs are totally consistent (if make allowance for some difference in their scales). It means that once derived parameters of α, β and ϕ can be regarded as valid for ever provided the appropriate QDCs are used (additional substantiations can be found in (Troshichev et al., 2011a)).

The parameters a, β and ϕ established for each UT moment of each day of the year are further used for calculation of PC index of any given time

$$PC = \xi\,(\delta F - \beta)/\,\alpha \quad (2)$$

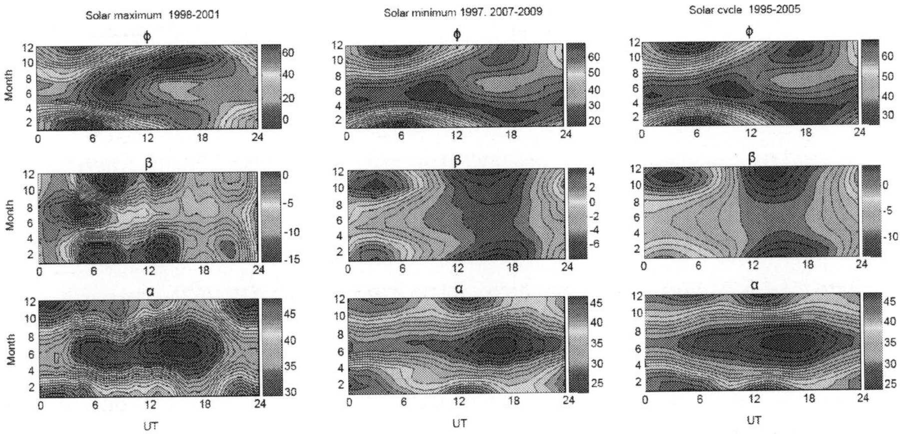

Fig. 2. Parameters ϕ, β and α for the Vostok station, derived independently for epochs of solar maximum (1998–2001), solar minimum (1997, 2007–2008), and the full cycle of solar activity (1995–2005); the axis of abscises being for UT and axis of ordinates being for a month.

The normalization coefficients α and β derived independently for the Thule and Vostok stations eliminate the diurnal and seasonal changes in response of the PC index to changes in E_{KL} field in the summer and winter polar caps. Dimensionality of the scale coefficient ξ is taken equal to 1 for convenience of comparison of PC and E_{KL} values. As an example, **Figure 3** shows the run of the calculated *PCN* (blue) and *PCS* (red) indices in the northern and southern hemispheres in 1998-2001. One can see a remarkable agreement in behavior of the positive *PCS* and *PCN* indices (which are related to DP2 disturbances) irrespective of the season, the largest value of *PC* about +20 mV/m being reached synchronously at both stations. Asymmetry is seen for negative *PC* index which describes the effect of DP3

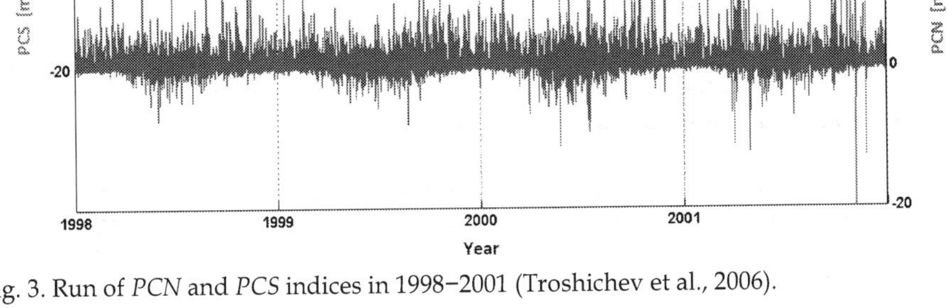

Fig. 3. Run of *PCN* and *PCS* indices in 1998–2001 (Troshichev et al., 2006).

magnetic disturbances responding to the northward IMF influence typical of only the summer polar cap.

3. PC index response to changes in solar wind dynamic pressure and E_{KL}

Statistically justified relationships between *PC* index and changes in the interplanetary electric field E_{KL} and solar wind dynamic pressure P_{SW} were examined in (Troshichev et al., 2007) on the example of the interplanetary shocks, which are commonly accompanied by the clearly defined and significant solar wind dynamic pressure pulses and strong oscillations of the interplanetary magnetic field. To estimate the value E_{KL} and the solar wind dynamic pressure P_{SW} the 5-min averaged data from the ACE satellite for 1998-2002 were used. These parameters were reduced to the magnetopause, the time of the signal passage from the ACE location to the magnetosphere been taken into account with allowance for the real solar wind speed for each particular event. Allowance for additional delay time $\tau_D \sim 20$ minutes typical of the signal passage from the magnetopause to the polar cap and its transformation into magnetic activity was not made in this case.

Only interplanetary shocks with sudden pressure pulses $\Delta P_{SW} > 4$ nPa starting against the background of the steady quiet pressure level lasting no less than 6 hours were examined in the analysis (N=62). The moment of maximum derivative dP/dt was identified as a pressure pulse onset. Just this moment is taken as a key ("zero") date for the epoch superposition method, other characteristics - E_{KL}, *PC*, and real P_{SW} - being related to the key date. *PC* indices were classified as summer and winter ones (*PCsummer* and *PCwinter*), instead of *PCN* and *PCS* indices. To separate the overlapping effects of the electric field and dynamic pressure, the behavior of averaged characteristics P_{SW}, E_{KL} and *PC* was examined under different restrictions imposed in turn on (1) the coupling function E_{KL}, (2) the magnitude of the pressure P_{SW} after the jump, and (3) the rate of the pressure increase (dP_{SW}/dt), the other two quantities being successively kept invariant at the same time.

Analysis of the relationships between the averaged E_{KL} and *PC* quantities under conditions of varying restrictions imposed on the E_{KL} value ($1 > E_{KL} > 0$ mV/m, $3 > E_{KL} > 1$ mV/m, $E_{KL} > 3$mV/m) with the practically arbitrary values of the pressure jump (pressure gradient $\Delta P_{SW} > 2$ nPa and derivative $dP/dt > 0.04$ nPa/min) showed that the *PC* index starts to growth within few minutes after the pressure jump, almost simultaneously with the E_{KL} increase. Nevertheless, the maximal magnetic activity in the polar caps is reached about 15 - 30 minutes after the E_{KL} maximums, the corresponding average E_{KL} and PC values being almost identical. Relationships between averaged P_{SW}, E_{KL} and *PC* under varying restrictions imposed on the pressure magnitude suggest that the dynamic pressure gradient, not the pressure level, affects the *PC* index. Indeed, the *PC* index rose just after the pressure jump and descended about 1 hour later irrespective of the persistent high level of the dynamic pressure.

Figure 4 demonstrates relationship between the averaged P_{SW}, E_{KL} and *PC* values under varying restrictions imposed on the pressure growth rate (dP/dt): (a) $0.2 > (dP/dt) > 0.1$ nPa/min, (b) $0.3 > (dP/dt) > 0.2$ nPa/min and (c) $(dP/dt) > 0.3$ nPa/min, the values of E_{SW} and ΔP_{SW} being arbitrary. One can see that the electric field E_{KL} increases when the dynamic pressure growth rate arises: the sharper is front of the pressure enhancement, the larger are the changes of E_{KL} on the front. Average *PC* indices and E_{KL} start to increase some minutes

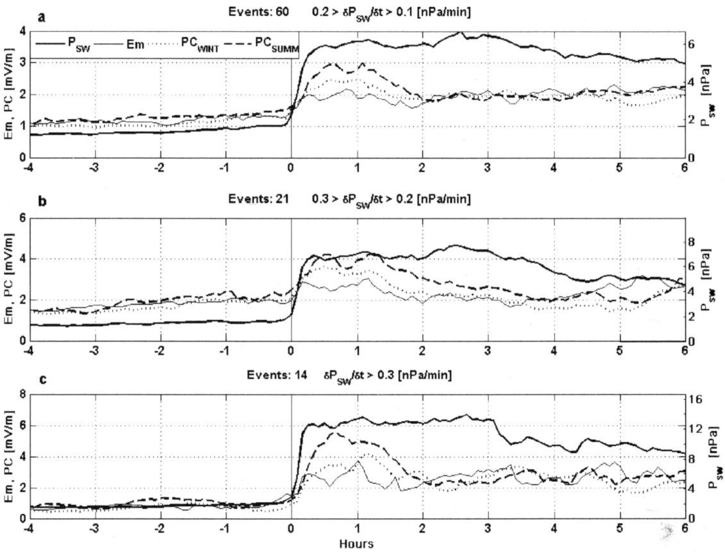

Fig. 4. Relationship between averaged P_{SW}, E_{KL} and PC quantities under varying restrictions imposed on of the pressure growth rate (dP/dt): (a) $0.2 > (dP/dt) > 0.1$ nPa/min, (b) $0.3 > (dP/dt) > 0.2$ nPa/min and (c) $(dP/dt) > 0.3$ nPa/min, the values of E_{KL} and ΔP_{SW} being arbitrary (Troshichev *et al.*, 2007) E_{KL} quantity is denoted here as Em.

after the pressure jump, and the *PC* indices maximums follow the electric field maximums with a common delay time of ~ 15 - 30 minutes. The *PC* index remains increased only 1-1.5 hours after a sharp pressure increase, and then the polar cap magnetic activity quickly decays in agreement with E_{KL} behavior.

The excess of the *PC* index over the appropriate E_{KL} value is typical of the pressure jump conditions: *PCsum* ~ 3 mV/m for E_{KL} ~ 2 mV/m and $0.2 > (dP/dt) > 0.1$ nPa/min (Fig. 4a); *PCsum* ~ 4.2 mV/m for E_{KL} ~ 2.5 - 2.8 mV/m and $0.3 > (dP/dt) > 0.2$ nPa/min (Fig. 4b); *PCsum* ~ 5.6 mV/m for E_{KL} ~ 2.5 - 3 mV/m and $(dP/dt) > 0.3$ nPa/min (Fig. 4c). The discrepancies between the values of E_{KL} and *PC* indices turn out to be proportional to the corresponding values of ΔP_{SW} and can be assigned just to the influence in the solar wind dynamic pressure. It makes it possible to conclude that an effect of pressure gradient ~1 nPa is approximately equivalent to effect of $E_{KL} \approx 0.33$ mV/m.

A decisive argument in favor of the pressure gradients influence on polar cap activity could be provided with events when the pressure jumps are not accompanied by electric field changes. Unfortunately, it is not feasible to find sharp pressure increases, which are inconsistent with no any variations of the interplanetary electric field. However, there is a number of pressure decreases divorced from the electric field. Results of the epoch superposition for 94 events with negative pressure gradients ($-0.1 > dP_{SW}/dt$ nPa/min) are presented in **Figure 5a**, where the moment of pressure sudden decrease is taken as a "zero moment". One can see that both summer and winter *PC* indices started to decrease right after "zero moment" although the mean electric field remained at level of ~ 2.5 mV/m

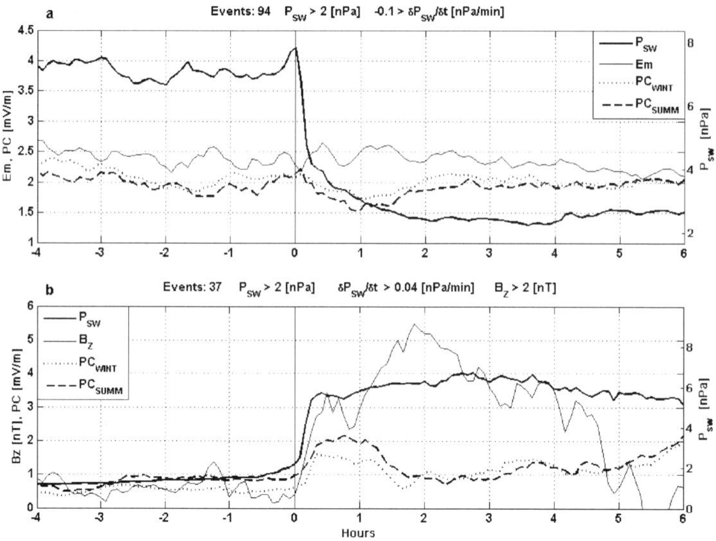

Fig. 5. Relationship between averaged P_{SW}, E_{KL} and PC quantities under conditions of (a) negative pressure gradient ($-0.1 > dP_{SW}/dt$ nPa/min), and (b) northward IMF (Troshichev *et al.*, 2007) E_{KL} quantity is denoted here as Em.

before and after the key date. The average decay of the polar cap magnetic activity lasted about 1.5 hours, and the effect was maximal (~ 0.5 mV/m) in the summer polar cap. A pure effect of the pressure negative jump in these events is estimated as 1 nPa ≈ 0.25 mV/m. The fact that the *PC* index decreases in response to negative pressure gradients suggests that the pressure effect in the PC index is reversible and acts under conditions of the pressure rise and fall.

Figure 5b shows relationship between averaged P_{SW} and E_{KL} under conditions of northward IMF. Since the coupling function E_{KL} reduces to zero when the IMF is northward ($B_Z > 0$), the run of the IMF B_Z component is shown in Figure. One can see that the mean *PC* index increases in response to positive pressure pulses, the ratio 1 nPa ≈ 0.4 mV/m is valid in this case. Thus, the solar wind pressure growth rate (i.e. jump power $\Delta P_{SW}/\Delta t$) proves, after E_{KL}, to be the second most important factor for the *PC* index increase: influence of the dynamic pressure gradient $\Delta P_{SW} = 1$ nPa on the polar cap magnetic activity is roughly equivalent to effect of the coupling function $\Delta E_{KL} = 0.33$ mV/m.

The *PC* index rises in response to the positive dynamic pressure pulses irrespective of IMF polarity (southward or northward). It means that mechanism of the pressure gradients influence on polar magnetic activity is not related to IMF whose effects in polar ionosphere are strongly controlled by the IMF orientation. The *PC* indices in the summer and winter polar caps demonstrate similar response to solar wind dynamic pressure pulses, but the summer *PC* value is persistently higher than the winter *PC* value (up to factor 1.5). The predominant growth of the summer *PC* index is indicative of important role of ionospheric conductance in mechanism of the pressure pulses effect that implies the better conditions for field-aligned currents closure through the well-conducting sunlit ionosphere.

4. Relation of the PC index to magnetospheric substorms

Dynamics of magnetic disturbances in the auroral zone is described by the "auroral indices" *AU* and *AL* that characterize intensity of magnetic disturbances produced, respectively, by eastward and westward electric currents (electrojets) flowing in the morning and evening sectors of auroral zone. Their total, *AE* index, is regarded as a measure of disturbance in the auroral zone. During substorm periods a powerful westward electrojet is developed in the midnight auroral zone as a result of substorm current wedge formation (Birkeland, 1908) in response to strongly enhanced auroral particle precipitation and short-circuiting of neutral sheet currents through the high conductivity auroral ionosphere. That is why a sudden increase of the *AL* index is identified with the magnetospheric substorm onset, the intensity of substorm being evaluated by the *AE* or *AL* indices. The substorm sudden onset is usually preceded by gradual increase of westward and eastward electrojets regarded as a substorm growth phase. Growth phase is related to Region 1 FAC enhancement, which is accompanied by progressive intensification of auroral particle precipitation and formation Region 2 FAC system.

Relationship between the *PC* index and development of the isolated substorms, occurring against the background of magnetic quiescence, was analyzed by Janzhura *et al.* (2007). The following four classes of isolated magnetic disturbances were examined: weak magnetic bays, short magnetic substorms with duration under 3 hours, long substorms lasting more than 3 hours, and extended substorms, which demonstrate, after sudden onset a slow intensity increase, with maximum being retarded for some hours after the sudden onset.

Relationship between the *PC* index and progress of the strongest "sawtooth" substorms, whose intensity periodically increases and decreases, was analyzed by Troshichev & Janzhura (2009). The sawtooth events, developing under influence of a high-speed solar wind with strong fluctuating or a steady southward IMF, are distinguished from usual substorms by a larger local time extent (Lui *et al.*, 2004; Henderson *et al.*, 2006a,b; Clauer *et al.*, 2006). Results (Troshichev et al., 2011b) indicate that a permanently high level of auroral activity is a typical feature of powerful sawtooth substorms. Aurora activity starts long before a magnetic disturbance onset and keeps a high level irrespective of magnetic disturbances. As a result, the close agreement between magnetic disturbance sudden onsets and behavior of aurora and particle injections on the synchronous orbit breaks down in contrast to "classical" substorms which are strongly associated with auroral particle precipitation.

The epoch superposition method, with the time of a substorm sudden onset taken as a key date, was used to reveal regularity in relationship between *PC* and *AE* indices in course of different types of substorm. In case of sawtooth substorms observed in 1998-2001 under conditions of the fluctuating southward IMF (N=43), the relationships between the IMF *Bz* and *By* components, electric field E_{KL}, and *PC* and *AL* indices are also examined. **Figure 6** integrates results obtained in analyses of Janzhura et al. (2007) and Troshichev & Janzhura (2009) for magnetic bays (average *AL* < 200 nT), short (*AL* > 200 nT), extended (*AL* > 200 nT) and sawtooth (*AL* > 500 nT) substorms (notice that the scale at the bottom panel is about

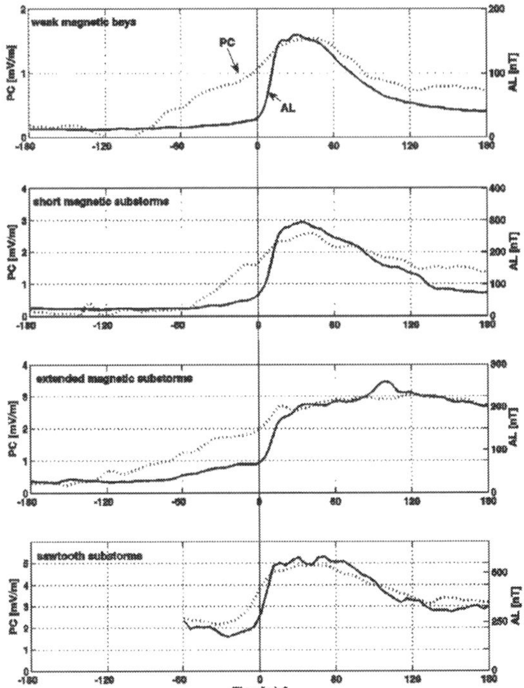

Fig. 6. Relationship between average *PC* and *AL* indices for weak magnetic bays (first panel), isolated short and extended substorms (second and third panels), and for sawtooth substorms (forth panel) (Troshichev and Janzhura, 2009).

twice as large as the scale for isolated substorms). One can see that the substorm sudden onset is preceded by the persistent *PC* index increase irrespective of substorm intensity. Following the generally accepted terminology, we shall name a time interval between a *PC* growth beginning and an explosive escalation of a magnetic disturbance in the auroral zone as a substorm growth phase. The mean duration of the growth phase is about 1 hour for isolated substorms and about 20-30 minutes for sawtooth substorms. As shown in (Troshichev and Janzhura, 2009) the *PC* growth phase is determined by the related E_{KL} function growth. In case of sawtooth substorms both E_{KL} and *PC* quantities stay at a high level after the substorm onset as well. As for magnetic disturbances in the auroral zone, the mean *AL* index keeps at low level before substorm and starts to grow only ~ 10 minutes ahead of the *AL* sudden onset.

These regularities clearly demonstrate that *PC* index variations in course of substorm are controlled by interplanetary electric field E_{KL} and that magnetic disturbances in the auroral zone follow the E_{KL} and *PC* increase. There is a definite *PC* level required for the substorm onset: in case of isolated substorms the threshold is about 1 - 2 mV/m, in case of sawtooth substorms the threshold is above 2 mV/m for. The fall of the *PC* value below 1 mV/m is unconditionally followed by the substorm decay.

Sawtooth disturbances last about 1.5–2 hours like isolated magnetic bays and short substorms, but they principally differ by *PC* behavior after the expansion phase. In case of isolated substorms, the *PC* index returns to a quiet level (<2 mV/m); in case of periodically repetitive substorms *PC* remains on a high level (>3 mV/m). If we consider the *PC* index as a signature of the solar wind energy input into the magnetosphere, an evident conclusion follows that isolated magnetic bays and short substorms are caused by a one-step energy income into the magnetosphere lasting for a short time, whereas isolated long and extended substorms take place if the solar energy is delivered with a different capability for a longer time. Sawtooth substorms are generated when a very powerful energy supply proceeds during a long period.

Thus, the analyses (Janzhura et al., 2007 and Troshichev & Janzhura, 2009) revealed that the *PC* index growth is a precursor of substorms development irrespective of a substorm type (isolated or sawtooth) and intensity. The growth phase duration is determined by the *PC* growth rate: the higher is the rate, the shorter is the growth phase duration. As **Figure 7** shows, the *PC* growth rate is a controlling factor for such important characteristics of magnetic disturbance as an *AL* growth rate before the substorm sudden onset (AL_{GR} = - 0.5 + 55*PC_{GR}, R=0.996) and for maximal intensity of magnetic substorm ($ALmax$ = 114 + 6570*PC_{GR}, R=0.997). According to last relationship, the *ALmax* value will reach ~1600 nT when PC_{GR} is ~ 0.2 mV/m/min, which is equivalent to the *PC* jump by 2 mV/m per 10 minutes, that is observed in the case. Therefore, the average substorm intensity (i.e. magnitude of magnetic disturbance in the auroral zone) is predetermined by the *PC* growth rate during the growth phase.

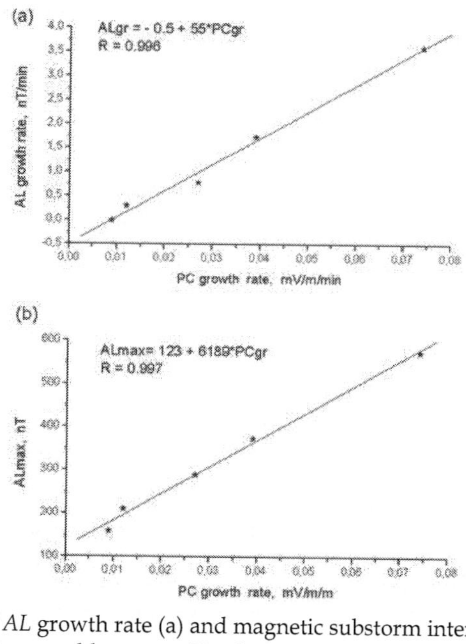

Fig. 7. Dependence of *AL* growth rate (a) and magnetic substorm intensity *ALmax* (b) on *PCgrowth* rate preceding a sudden onset (Troshichev and Janzhura, 2009).

The above presented results suggest that the *PC* index is a signature of solar wind energy that entered into the magnetosphere in course of solar wind–magnetosphere coupling. Indeed, the magnetospheric substorms are always preceded by the *PC* index growth and start as soon as the *PC* index exceeds a definite threshold value. Threshold value appears to be a level when the excess of energy incoming into the magnetosphere is over the ordinary energy dissipating in the magnetosphere; if threshold is not reached, the substorm can not begin, irrespective of how long the solar wind energy was income into the magnetosphere. From this point of view, the substorm growth phase is a period of the enhanced rate of energy pumping into the magnetosphere, not a period of energy storage in the magnetosphere. The large *PC* growth rate indicates that rate of energy pumping into the magnetosphere increases respectively. The greater the energy input, the faster is reached the threshold level; the shorter is the growth phase duration. The higher the energy input rate (i.e. energy input related to the growth phase duration), the larger is excess of incoming energy over the level of ordinary energy dissipation, and the more powerful is substorm. The succeeding substorm development is determined by dynamics of the subsequent input of energy, which is displayed by the *PC* index run. Thus, the solar wind geoefficiency in case of magnetospheric substorms can be monitored with confidence by the *PC* index.

5. Relation of the PC index to magnetic storms

Magnetic storms are the result of a joint action of magnetopause currents (*DCF*), which are proportional to the square root of the solar wind dynamic pressure and ring currents (*DR*) flowing in the inner magnetosphere (Chapman, 1963). The *DR* current ground effect typically far exceeds the *DCF* current effect, that is why the magnetic storm intensity is evaluated by the *Dst* index depicting a longitudinally averaged magnetic field depression at low latitudes (Sugiura, 1976). It is well known (Kamide, 1974; Russel et al., 1974; Burton et al., 1975; Akasofu, 1981) that magnetic storms intensity is dominantly controlled by southward IMF component (B_{ZS}), whereas the solar wind velocity (v) and density (n) are of minor importance. While investigating solar wind–magnetosphere coupling functions, the best result was obtained for functions including the geoeffective interplanetary electric field E_{KL} (Newell et al., 2008; Spencer et al., 2009).

Relationship between the 1-min *PC* index behavior and the storm depression development (*Dst* index) for epoch of solar maximum (1998–2004) has been examined by Troshichev et al. (2011c). Two criteria were used as a basic guideline to choose magnetic storms for the analysis: (1) magnetic storm duration should be longer than 12 hours, (2) magnetic storm depression should be larger than *Dst* = -30nT. On the basis of these criteria, 54 magnetic storms were separated for the period of 1998–2004 with a maximal storm intensity varying in the range from *Dst* = -30nT to *Dst* = -373nT. For convenience of comparison of the $PC(E_{KL})$ behavior with storm development, the main phase of each magnetic depression was divided into two parts: a "growth phase" when magnetic depression increases, and a "damping phase" when magnetic depression decreases, a "recovery phase" being used as before: as a period of a magnetic field slow restoration to the previous undisturbed level.

Since the *Dst* value is determined by a joint action of two, *DCF* and *DR*, current sources, the *Dst* index initial decline can be caused by a *DR* current growth as well as by a solar wind pressure reduction. In addition, during the solar maximum epoch (1998-2004), magnetic storms were usually following one after other, when a new magnetic storm started against the background of the recovery phase of the previous storm. Under these conditions, it would be well to look for another characteristic of storm depression beginnings, independent on the peculiarities of individual magnetic storm development. Referring to results (Janzhura *et al.*, 2007; Troshichev & Janzhura, 2009) and suggesting that a comparable input of the solar wind energy is required for development of magnetospheric substorms and magnetic storms, the value *PC*=2mV/m was considered as a possible threshold level.

Examination of 54 chosen storms showed that all of them occurred under condition *PC*>2 mV/m. To demonstrate that a threshold level *PC*=2 mV/m is a typical feature of relationships between *PC* (or E_{KL}) changes and magnetic storm development, the magnetic storms were separated into six gradations according to their intensity, and the averaged *PC*(E_{KL}) and *Dst* quantities for these gradations were examined. The gradations determined by a minimal *Dst* value are the following: (a) -30>*Dst*>-50nT, (b) -50>*Dst*>-80nT, (c) -80>*Dst*>-100nT, (d) -100>*Dst*>-120nT, (e) -130>*Dst*>-160nT, and (f) -160>*Dst*>-240nT. The method of superposed epochs was used. Correspondingly, the time when the *PC* index

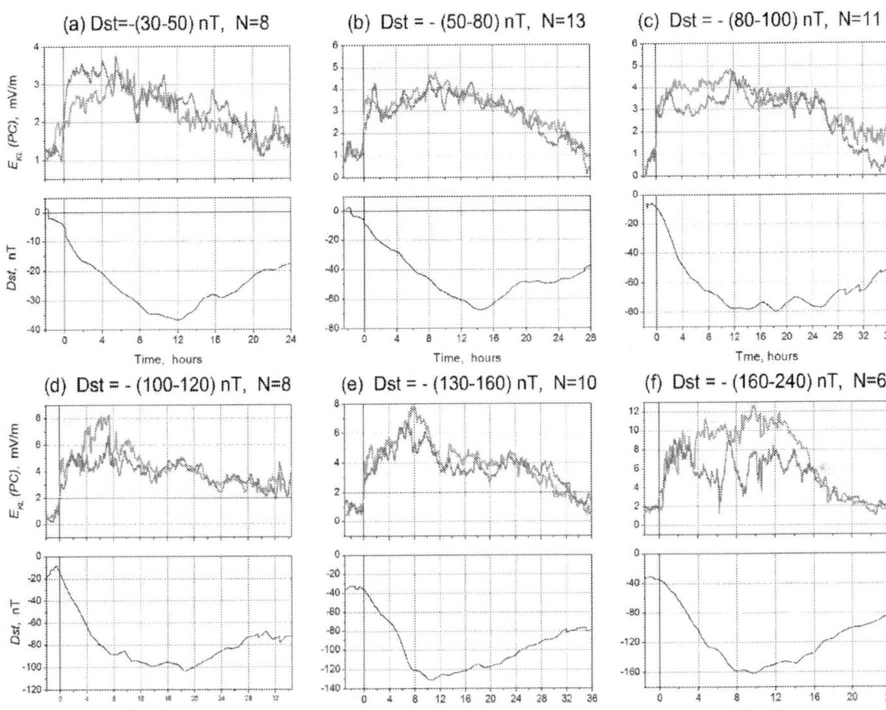

Fig. 8. Relationship between behavior of averaged *PC* index (red) and E_{KL} (green) quantities and development of magnetic storm *Dst* index (black) for 6 gradations of storm intensity: (a) −30 > *Dst* > −50 nT, (b) −50 > *Dst* > −80 nT, (c) −80 > *Dst* > −100 nT, (d) −100 > *Dst* > 120 nT, (e) 130 > *Dst* > 160 nT, (f) 160 > *Dst* > 240 nT. The key date (T=0) is taken as a time of the persistent transition of the E_{KL} value over the level of 2 mV/m (Troshichev et al., 2011c).

persistently rises above the level of 2 mV/m was taken as a time (T=0) of the disturbance beginning, and the time when the *PC* index persistently falls below the level of 2 mV/m was taken as a time of the recovery phase beginning.

Figure 8 shows the behavior of coupling function E_{KL} and the *PC* index, as well as corresponding changes in the *Dst* index for these 6 gradations. Results of case studies and statistical analysis (Figure 8) demonstrate the following regularities: geomagnetic field depression generally starts to develop as soon as *PC* and E_{KL} exceed the threshold of ~2 mV/m; as a rule, *PC* and E_{KL} simultaneously cross the threshold, although sometimes one goes ahead of other; when *PC* (E_{KL}) demonstrates repetitive strong enhancements and decreases, the magnetic storm displays the appropriate multiple depressions with growth and dumping phases; persistent descent of *PC* (E_{KL}) below the threshold level of 2 mV/m is indicative of the end of the storm main phase and transition to the recovery phase; "*PC* saturation effect" is typical of events with E_{KL} values > 6 mV/m.

To derive statistical relationships between *Dst* and mean *PC* (E_{KL}) values, the 1 min quantities *PC* and E_{KL} were averaged over the growth phase duration (the interval from time T=0 to the time of the peak value of *Dst (Dst(peak))*. The averaged values *PCgrowth* and E_{KL}*growth* were compared with value of maximal depression. It turned out (Figure 9) that under conditions of $E_{KL} < 6$ mV/m, the relationship between *Dst(peak)* and *PC* (*Dst* = 24.8-31.8**PC*) is of the same character as between *Dst(peak)* and E_{KL} (*Dst*=24.9-30.9*E_{KL}), although the correlation of *Dst* with E_{KL} (R=-0.74) is much lower than with *PC* (R=-0.87). Under

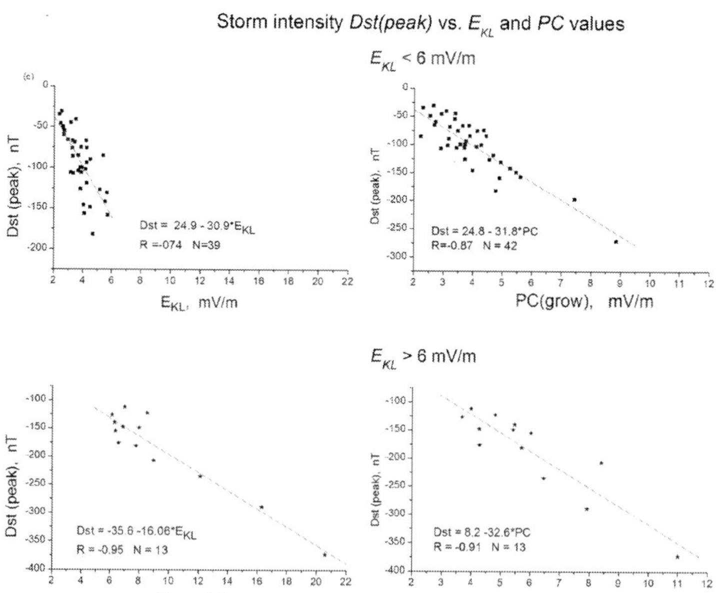

Fig. 9. Relationships between storm intensity *Dst(peak)* and quantities E_{KL}*growth* and *PCgrowth* averaged over the storm growth phase interval, derived for categories of $E_{KL} < 6$ mV/m and $E_{KL} > 6$ mV/m (Troshichev et al., 2011c)

conditions E_{KL}>6 mV/m the efficiency of E_{KL} strongly decreases (*Dst*=-35.6-16.05*E_{KL}), whereas relationship between *Dst(peak)* and *PC* is practically unchanged (*Dst*=8.2-32.6PC). The evident slowing down of depression *Dst(peak)* for conditions of E_{KL} > 5-6 mV/m can be termined as a "*Dst* saturation effect" which is not seen in relations of *Dst* with the *PC* index.

Thus, all examined magnetic storms started as soon as the field E_{KL} and the appropriate *PC* index firmly exceeded the threshold of ~2 mV/m. The storm main phase lasted till the E_{KL} and *PC* steadily stands above this threshold level. The time of the firm descent of the *PC* and E_{KL} quantities below the same level of ~ 2 mV/m is indicative of the storm transition to the recovery phase. The storm intensity *Dst(peak)* demonstrates the steady linear correlation with the *PC* value averaged for the growth phase *PC(growth)* for all storm intensities irrespective of *PC* value. The storm dynamics correlate better with value and changes of the *PC* index than with those of the E_{KL} field that provides a weighty argument in support of the *PC* index as a signature of the solar wind energy that entered into the magnetosphere.

6. Conclusion

The *PC* index has been introduced initially (Troshichev *et al.*, 1988) as a characteristic of the polar cap magnetic activity related to the geoeffective interplanetary electric field E_{KL} which was determined by formula of Kan & Lee (1979). The recent studies (Troshichev *et al.*, 2007; Janzhura *et al.*, 2007, Troshichev & Janzhura, 2009; Troshichev *et al.*, 2011c) show that the *PC* index implication is more significant: the magnetospheric storms and substorms start only if the *PC* index reaches the definite threshold value (~ 2 mV/m for storms, and >1.5mV/m for substorms); the substorm growth phase duration and substorm intensity are determined by the *PC* growth rate and substorms are stopped as soon as *PC* index falls below 1-1.5 mV/m; the storm length is terminated by duration of period, when *PC*>2mV/m, the storm intensity being linearly related to the *PC* index averaged for the storm time interval; development of storms and substorms is better consistent with the *PC* behavior than with the coupling function variations; and so on. In addition, it turned out that the *PC* index adequately responds to sharp changes in the solar wind dynamic pressure. All these experimentally established relationships make it possible to conclude that the *PC* index should be regarded as an adequate and convenient proxy of the solar wind energy that entered into the magnetosphere.

If holding this point of view it becomes obvious why the storm and substorm indicators should correlate better with the *PC* index than with coupling function E_{KL} (since the coupling function E_{KL} characterizes the state of solar wind coupling with the magnetosphere, whereas the *PC* index characterizes the energy that entered into the magnetosphere); why the *PC* index increase precedes storm and substorms (because growth of the entered energy above the level of energy dissipation is followed by realizing the energy excess in form of magnetic disturbances); why the sawtooth substorms demonstrate the distinct periodicity under conditions of steadily high energy supply (because the persistently high entered energy ensures the extreme intensity of field aligned currents discharging the current generator which power is limited by finite plasma pressure gradients in the closed magnetosphere (see Troshichev et al., 2012)).

As far as the *PC* index characterizes the energy that entered into the magnetosphere in course of solar wind - magnetosphere coupling, the index can be used to monitor the solar wind geoefficiency and state of magnetosphere. Since disturbances in magnetosphere are always preceded by energy input, the *PC* index usage makes it possible to realize the space weather nowcasting (including the auroral ionosphere state and even the anomalous processes in polar atmosphere). At present, the *PCN* and *PCS* indices are published at sites:

PCN ftp://ftp.space.dtu.dk/WDC/indices/pcn/

PCS http://www.geophys.aari.ru/

7. References

Akasofu, S.-I. (1981). Energy coupling between the solar wind and the magnetosphere, *Space Sci. Rev.*, vol.28, 121

Antonova, E.; Kirpichev I. & Stepanova, M. (2006). Field-aligned current mapping and the problem of the generation of magnetospheric convection. *Adv. Space Res.*, vol.38, 1637 - 1641, doi:10.1016/j.asr.2005.09.042

Antonova, E,; Kirpichev, I.; Ovchinnikov, I.; Pulinets, M.; Znatkova, S.; Orlova, K. & Stepanova, M. (2011). Topology of High-Latitude Magnetospheric Currents. In: Fujimoto M, Liu W (eds) *The Dynamic Magnetosphere*. IAGA Special Sopron Book Series, V 3, Springer Dordrecht Heidelberg London New York, pp.201-210

Arnoldy, R. (1971). Signature in the interplanetary medium for substorms. *J. Geophys. Res.*, vol.76, 5189-5201

Baliff, J.; Jones, D.; Coleman, P.; Davis, L. & Smith, E. (1967). Transverse fluctuations in the interplanetary magnetic field: a requisite for geomagnetic variability. *J. Geophys. Res.*, vol.72, 4357-4362

Birkeland, K. (1908). *The Norwegian Aurora Polaris Expedition 1902-1903,* vol.1 Christiania

Burton, R.; McPherron, R. & Russell, C. (1975). An empirical relationship between interplanetary conditions and Dst. *J. Geophys. Res.*, vol.80, 4204-4214

Bythrow, P. & Potemra, T. (1983). The relationship of total Birkeland currents to the merging electric field. *Geophys Res. Lett.*, vol.10, 573-576

Chapman, S. (1963). Solar plasma, geomagnetism and aurora, *in Geophysics: The Earth's Environment*. C.DeWitt, (Ed), Gordon and Breach Sci.Pub., New York - London

Clauer, C.; Cai, X.; Welling, D.; DeJong, A. & Henderson, M. (2006). Characterizing the 18 April 2002 storm-time sawtooth events using ground magnetic data. *J. Geophys. Res.*, vol.111, A04S90, doi:10.1029/2005JA011099

Fairfield, D. & Cahill, L. (1966). Transition region magnetic field and polar magnetic disturbances. *J. Geophys. Res.*, vol.71, 6829-6846

Foster, J.; Fairfield, D.; Ogilvie, K. & Rosenberg, T. (1971). Relationship of interplanetary parameters and occurrence of magnetospheric substorms. *J. Geophys. Res.*, vol.76, 6971-6975

Garrett, H.; Dessler, A. & Hill, T. (1974). Influence of solar wind variability on geomagnetic activity. *J. Geophys. Res.*, vol.79, 4603-4610

Gizler, V.; Semenov, V. & Troshichev, O. (1979). The electric fields and currents in the ionosphere generated by field-aligned currents observed by TRIAD. *Planet. Space Sci.*, vol.27, 223-231

Henderson, M.; Reeves, G.; Skoug, R.; Thomsen, M.; Denton, M.; Mende, S.; Immel, T.; Brandt, P. & Singer, H. (2006a) Magnetospheric and auroral activity during the 18 April 2002 sawtooth event. *J. Geophys. Res.*, vol.111, A01S90, doi:10.1029/2005JA011111

Henderson, M.; Skoug, R.; Donovan, E; et al. (2006b). Substorm during the 10 August 2000 sawtooth event. *J. Geophys. Res.*, vol.111, A06206, doi:10.1029/2005JA011366

Hirshberg, J. & Colburn D. (1969). Interplanetary field and geomagnetic variations - a unified view. *Planet. Space Sci.*, vol.17, 1183-1206

Iijima, T. & Potemra, T. (1976). The amplitude distribution of field-aligned currents at northern high latitudes observed by Triad. *J. Geophys. Res.*, vol.81, 2165-2174

Iijima, T. & Potemra, T. (1978). Large-scale characteristics of field-aligned currents associated with substorms. *J. Geophys. Res.*, vol.83, 599-615

Iijima, T. & Potemra, T. (1982). The relationship between interplanetary quantities and Birkeland current densities. *Geophys. Res. Lett.*, vol.4, 442-445

Iijima, T.; Potemra, T. & Zanetti, L. (1997). Contribution of pressure gradients to the generation of dawnside region 1 and region 2 currents. *J. Geophys. Res.*, vol.102: 27069-27081

Janzhura, A. & Troshichev, O. (2008). Determination of the running quiet daily geomagnetic variation, *J. Atmos. Solar-Terr. Phys.*, vol.70, 962–972

Janzhura, A.; Troshichev, O. & Stauning, P. (2007). Unified PC indices: Relation to the isolated magnetic substorms, *J. Geophys. Res.*, vol.112, A09207, doi: 10.1029/2006JA012132

Kamide, Y. (1974). Association of DP and DR fields with the interplanetary magnetic field variations. *J. Geophys. Res.*, vol.79, 49

Kan, J. & Lee, L. (1979). Energy coupling function and solar wind-magnetosphere dynamo. *Geophys. Res. Lett.*, vol.6, 577-580

Kane, R. (1974). Relationship between interplanetary plasma parameters and geomagnetic Dst. *J. Geophys. Res.*, vol. 79, 64-72

Kokubun, S. (1972). Relationship of interplanetary magnetic field structure with development of substorm and storm main phase. *Planet. Space Sci.*, vol.20, 1033-1050

Kuznetsov, B. & Troshichev O. (1977). On the nature of polar cap magnetic activity during undisturbed periods. *Planet. Space Sci.*, vol.25, 15-21

Langel, R. (1975). Relation of variations in total magnetic field at high latitude with parameters of the IMF and with DP2 fluctuations. *J. Geophys. Res.*, vol.80, 1261-1270

Lui, A.; Hori, T.; Ohtani, S.; Zhang, Y.; Zhou, X. Henderson, M.; Mukai, T.; Hayakawa, H. & Mende, S. (2004). Magnetotail behavior during storm time 'sawtooth injections'. *J. Geophys. Res.*, vol.109, A10215, doi:10.1029/2004JA010543

Maezawa, K. (1976). Magnetospheric convection induced by the positive and negative Z components of the interplanetary magnetic field: quantitative analysis using polar cap magnetic records. *J. Geophys. Res.*, vol.81, 2289-2303

Mansurov, S. (1969). A new evidence for relationship between the space and earth magnetic fields. *Geomagn. Aeronomy*, vol.9, 768-770 (in Russian)

McDiarmid, I.; Budzinski, E.; Wilson, M. & Burrows J. (1977). Reverse polarity field-aligned currents at high latitudes. *J. Geophys. Res.*, vol.82, 1513-1518

Meng, C-I.; Tsurutani, B.; Kawasaki, K. & Akasofu, S-I. (1973). Cross-correlation analysis of the AE-index and the interplanetary magnetic field Bz component. *J. Geophys. Res.*, vol.78, 617-629

Murayama, T. & Hakamada, K. (1975). Effects of solar wind parameters on the development of magnetospheric substorms. *Planet. Space Sci.*, vol.23, 75-91

Newell, P.; Sotirelis, T.; Liou, K.; Meng, C-I. & Rich, F. (2007) A nearly universal solar wind magnetosphere coupling function inferred from 10 magnetospheric state variables. *J. Geophys. Res.*, vol.112, A01206, doi:10.1029/2006JA012015

Newell, P.; Sotirelis, T.; Liou, K. & Rich, F. (2008). Pairs of solar wind-magnetosphere coupling functions: combining a merging term with a viscous term works best. *J. Geophys. Res.*, vol.113, A04218, doi:10.1029/2007JA012825

Nisbet, J.; Miller, M. & Carpenter, L. (1978). Currents and electric fields in the ionosphere due to field-aligned auroral currents. *J. Geophys. Res.*, vol.83, 2647

Nishida, A. (1968). Geomagnetic DP2 fluctuations and associated magnetospheric phenomena. *J. Geophys. Res.*, vol.73, 1795-1803

Perreault, P. & Akasofu, S-I. (1978). A study of geomagnetic storms. *Geophys. J. R. Astr. Soc.*, vol.54, 57

Pudovkin, M.; Raspopov, O.; Dmitrieva, L.; Troitskaya, V. & Shepetnov, R. (1970). The interrelation between parameters of the solar wind and the state of the geomagnetic field. *Ann. Geophys.*, vol.26,: 389-392

Reiff, P. & Luhmann, J. (1986). Solar wind control of the polar-cap voltage. In: *Solar Wind – Magnetosphere Coupling*, Kamide, Y. & Slavin, J. (eds), pp 453-476, Terra Sci, Tokyo

Rossolenko, S.; Antonova, E.; Yermolaev, Y.; Verigin, M.; Kirpichev, I. & Borodkova, N. (2008). Turbulent Fluctuations of Plasma and Magnetic Field Parameters in the Magnetosheath and the Low-Latitude Boundary Layer Formation: Multisatellite Observations on March 2, 1996. *Cosmic. Res.*, vol. 46, 373–382 (in Russian).

Rostoker, G. & Falthammar, C. (1967). Relationship between changes in the interplanetary magnetic field and variations in the magnetic field at the Earth's surface. *J. Geophys. Res.*, vol.72, 5853-5863

Russel, C.; McPherron, R. & Burton, R. (1974). On the cause of geomagnetic storms. *J. Geophys. Res.*, vol.79, 1105

Sergeev, V. & Kuznetsov, B. (1981) Quantitative dependence of the polar cap electric field on the IMF BZ component and solar wind velocity. *Planet. Space Sci.*, vol.29, 205-213

Spencer, E.; Rao, A.; Horton, W. & Mays, M. (2009). Evaluation of solar - magnetosphere coupling functions during geomagnetic storms with the WINDMI model. *J. Geophys. Res.*, vol.114, A02206, doi: 10.1029/2008JA013530

Stepanova, M.; Antonova, E., Bosqued, J. & Kovrazhkin, R. (2004). Azimuthal plasma pressure reconstructed by using the Aureol-3 satellite date during quiet geomagnetic conditions. *Adv. Space Res.*, Vol. 33, 737-741

Stepanova, M.; Antonova, E. & Bosqued J. (2006). Study of plasma pressure distribution in the inner magnetosphere using low-altitude satellites and its importance for the large-scale magnetospheric dynamics, *Adv. Space Res.*, Vol. 38, 1631-1636

Sugiura, M. (1976). Hourly values of equtorial Dst for the IGY, *Ann. Int.Geophys.Year*, 35, 1.

Svalgaard, L. (1968). Sector structure of the interplanetary magnetic field and daily variation of the geomagnetic field at high latitudes. *Det Danske meteorologiske institute Charlottenlund*, preprint R-6

Troshichev, O. (1975). Magnetic disturbances in polar caps and parameters of solar wind. In: *Substorms and magnetospheric disturbances*. Nauka, Leningrad, pp 66-83, (in Russian)

Troshichev, O. (1982). Polar magnetic disturbances and field-aligned currents. *Space Sci. Rev.*, vol. 32, 275-360

Troshichev, O. & Tsyganenko, N. (1978). Correlation relationships between variations of IMF and magnetic disturbances in the polar cap. *Geomagn. Res.*, vol.25, 47-59 (in Russian)

Troshichev, O. & Andrezen, V. (1985). The relationship between interplanetary quantities and magnetic activity in the southern polar cap. *Planet. Space Sci.*, vol.33, 415

Troshichev, O. & Janzhura, A. (2009). Relationship between the PC and AL indices during repetitive bay-like magnetic disturbances in the auroral zone. *J. Atmos. Solar-Terr. Phys.*, 71, 1340–1352

Troshichev, O.; Gizler, V.; Ivanova, I. & Merkurieva, A. (1979). Role of field-aligned currents in generation of high latitude magnetic disturbances. *Planet. Space Sci.*, vol.27, 1451-1459

Troshichev, O.; Andrezen, V.; Vennerstrøm, S. & Friis-Christensen, E. (1988). Magnetic activity in the polar cap - A new index. *Planet. Space Sci.*, vol.36, 1095,.

Troshichev, O.; Janzhura, A. & Stauning, P. (2006). Unified PCN and PCS indices: Method of calculation, physical sense and dependence on the IMF azimuthal and northward components. *J. Geophys. Res.*, vol.111, A05208, doi:10.1029/2005JA011402

Troshichev, O.; Janzhura, A. & Stauning, P. (2007). Magnetic activity in the polar caps: Relation to sudden changes in the solar wind dynamc pressure, *J. Geophys. Res.*, vol.112, A11202, doi:10.1029/2007JA012369,

Troshichev, O.; Podorozhkina, N. & Janzhura, A. (2011a). Invariability of relationship between the polar cap magnetic activity and geoeffective interplanetary electric field. *Ann. Geophys.*, vol.29, 1–11, doi:10.5194/angeo-29-1-2011

Troshichev, O.; Stauning, P.; Liou, K. & Reeves, G. (2011b). Saw-tooth substorms: inconsistency of repetitive bay-like magnetic disturbances with behavior of aurora. *Adv. Space Res.*, vol.47, 702–709, doi: 10.1016/j.asr.2010.09.026

Troshichev, O.; Sormakov, D. & Janzhura, A. (2011c). Relation of PC index to the geomagnetic storm Dst variation. *J. Atmos. Solar-Terr. Phys.*, vol.73 611–622, doi:10.1016/j.jastp.2010.12.015

Troshichev, O.; Sormakov, D. & Janzhura, A. (2012). Sawtooth substorms generated under conditions of the steadily high solar wind energy input into the magnetosphere: Relationship between *PC*, *AL* and *ASYM* indices. *Adv. Space Res.*, doi:10.1016/j.asr.2011.12.011

Wilcox, J.; Schatten, K. & Ness, N. (1967). Influence of interplanetary magnetic field and plasma on geomagnetic activity during quiet-sun conditions. *J. Geophys. Res.*, vol.72, 19-26

Wing, S. & Newell, P. (2000). Quiet time plasma sheet ion pressure contribution to Birkeland currents. *J. Geophys. Res.*, vol.105, 7793-7802

Xing, X.; Lyons, R,; Angelopoulos, V.; Larson, D; McFadden, J.; Carlson, C.; Runov, A. & Auster, U. (2009). Azimuthal plasma pressure gradient in quiet time plasma sheet. *Geophys. Res. Lett.*, vol.36, L14105, doi:10.1029/2009GL038881

Zmuda, A. & Armstrong, J. (1974). The diurnal flow pattern of field-aligned currents. *J. Geophys. Res.*, vol.79, 4611-4519

16

The Polar Cap PC Indices: Relations to Solar Wind and Global Disturbances

Peter Stauning
Danish Meteorological Institute
Denmark

1. Introduction

The solar wind plasma flow impinging on the Earths magnetosphere causes a range of geophysical disturbances such as magnetic storms and substorms, energization of the plasma trapped in the Earth's near space, auroral activity, and Joule heating of the upper atmosphere. Auroral activity was characterized already in 1961 by the electrojet indices (AU, AL, AE and AO) based on magnetic observations at auroral latitudes (Davis & Sugiura, 1966). On the contrary, several attempts have been made in the past to scale high-latitude disturbances on basis of polar magnetic variations without reaching a final, agreed and IAGA-adopted Polar Cap index. Fairfield (1968) suggested that the maximum amplitude of the magnetic variations observed from a ring of polar cap observatories could be a better indicator of the overall high-latitude magnetic activity than the auroral electrojet indices. He also noted that the Polar Cap "magnetic activity magnitude" sometimes increased before changes in the AE index.

Kuznetsov & Troshichev (1977) defined a "PC_L" index based on the variability of high-latitude magnetic recordings (much like the K indices), and not equivalent to the present PC "level index". Later, a "PC(Bz)" index based on a composite of the variance and the level of polar magnetic activity was proposed by Troshichev et al. (1979), who used the Polar Cap magnetic activity as a signature of substorm development. The "MAGPC" index suggested by Troshichev & Andrezen (1985) was based on the magnitude (in nT) of 15 min samples of the magnetic variation in the direction of the 03:00-15:00 MLT meridian. The MAGPC index was introduced to provide a measure of the solar wind electric field to be derived from available ground-based magnetic observations in the central Polar Cap. A major problem for these initial "PC" indices was their dependence on the daily and seasonal changes in the ionospheric conductivity with the varying solar illumination. Such variations cause corresponding variations in the "sensitivity" of the disturbance indices in their response to varying solar wind conditions.

The present version of a Polar Cap (PC) index is based on the formulation by Troshichev et al. (1988). The new idea is the scaling on a statistical basis of the magnetic variations to the electric field in the solar wind in order to make the new PC index independent of the local ionospheric properties and their daily and seasonal variations. The PC index concept was further developed by Vennerstrøm (1991) and Vennerstrøm et al. (1991). The development of a PC index was recommended by IAGA in 1999 and the index was later adopted by IAGA as an international standard index on the condition that a unified procedure for the PC index calculations was defined. Unfortunately, this unification has not been accomplished yet, although a close cooperation between the Arctic and Antarctic Research Institute (AARI) responsible for the PCS index, and the Danish Meteorological Institute (DMI) formerly responsible for the PCN index, has solved most of the initial disagreements (Troshichev et al., 2006).

The relative importance of the solar wind dynamical pressure and the interplanetary geo-effective electric field for the various disturbance processes like the build-up of the current systems involved in magnetic storms and substorms is a subject of great controversy. Here, we shall specifically discuss the relations of the transpolar currents characterized by the PC index to the "external" solar wind parameters, E_M and P_{SW}, and we shall also consider the influence of "internal" auroral substorm processes parameterized by the auroral electrojet indices, primarily the AE and AL indices.

Following the introduction in section 1, section 2 briefly outlines the high-latitude response to solar wind forcing while section 3 more specifically discusses the response in the magnetic recordings to IMF variations. Section 4 outlines the calculation procedures for the PC indices, while section 5 discusses in more detail the relations between the PC indices and the merging electric field in terms of timing and amplitude response as well as the effects of the solar wind dynamic pressure on the PC index. Section 6 discusses the relations between the PC index and various further geophysical disturbances such as auroral electrojet activity, Joule and particle heating of the upper atmosphere, and the ring current activity. Finally, there is a concluding summary section.

2. High-latitude response to solar wind-magnetosphere interactions

It was early recognized (e.g., Spreiter et al., 1966) that the solar wind dynamical pressure to a large extent controls the general morphology of the magnetosphere, among other, its size and shape and also affects the internal magnetic field configuration. An important parameters to characterize the impact of the solar wind on the magnetosphere is the dynamical pressure, P_{DYN} :

$$P_{DYN} = D_{SW}\, V_{SW}^2 = M_{SW}\, N_{SW}\, V_{SW}^2 \tag{1}$$

where V_{SW} is the solar wind velocity while the density, D_{SW} , is a combination of solar wind number density, N_{SW}, and average particle mass, M_{SW},. After transition through the bow shock into the magnetosheath the solar wind dynamic pressure (P_{DYN}) is partly converted into thermal pressure (N_{SW} kT). At the magnetopause in Spreiter's model the thermal pressure from the solar wind in the magnetosheath is balanced by the internal "magnetic pressure" (i.e., $N_{SW}\,kT=B_M^2/2\mu_o$). The magnetic field generated by the boundary currents must cancel the geomagnetic field outside the magnetopause and adds to doubling the field strength just inside the boundary.

However, the solar wind electric field is in most cases the dominant factor for the high-latitude magnetospheric electric field structures and related plasma convection processes. A useful parameter is the "merging" (or "reconnection" or "geo-effective") electric field defined by Sonnerup (1974) and Kan and Lee (1979):

$$E_M = V_{SW} B_T \sin^2(\theta/2) \quad (2)$$

This parameter is a combination of the solar wind velocity, V_{SW}, and the transverse component, B_T, ($B_T = (B_Y^2 + B_Z^2)^{1/2}$) of the interplanetary magnetic field (IMF) in the solar wind and includes a strong dependence on the field direction represented by the polar angle θ of the transverse component of the IMF with respect to the direction of the Z-axis in a "Geocentric Solar Magnetospheric" (GSM) coordinate system (i.e., $\tan(\theta) = |B_Y| / B_Z$, $0 \leq \theta \leq \pi$).

The merging electric field is the optimum parameter to characterize the energy transfer from the solar wind to the magnetosphere (Akasofu, 1979). It was shown theoretically by Kan and Lee (1979) that the time dependent power, P(t), delivered by the solar wind dynamo to the Earth's magnetosphere can be expressed in terms of the merging electric field, E_M, (defined in Eq.2) by:

$$P(t) = E_M^2 \ell_o^2 / R \quad (3)$$

where R is the total equivalent resistance connected to the solar wind-magnetosphere dynamo. The resistance includes the dissipation processes in the reconnection region, the energy used in driving the internal magnetospheric convection and building the ring current as well as the energy dissipated in the upper atmosphere by Joule heating and particle precipitation particularly in the auroral regions. The parameter ℓ_o is an interaction length ($\approx$ 7 R_E), which, when projected along geomagnetic field lines to the polar cap ionosphere, defines the width of the open polar cap. Equation (3) indicates the importance of E_M to characterize the energy input although the concept is simplified and the parameters ℓ_o and R may not be quite well-defined or constant.

The interaction between the solar wind and the Earth's magnetosphere is sketched in Fig. 1 (from GEM Source book, 1999). The magnetosphere is delimited by magnetopause currents to provide the transition from the terrestrial magnetic field to the solar wind. In the diagram the direction of the IMF, depicted by the field lines, is assumed to be southward. In consequence, the IMF and the magnetospheric field merge at the front of the magnetosphere, where the terrestrial field is northward. The combined field reach to the ionosphere in the polar cusp region and is then dragged tailward over the polar cap and into the tail region by the plasma motion. In the middle of the tail region the neutral sheet of transverse currents separate the sunward magnetic field in the upper part of the plasma sheet from the antisunward magnetic field in the lower part of the plasma sheet.

The plasma flow just inside the magnetopause and in the northern and southern lobes at high latitudes is antisunward as in the solar wind, while the plasma motion is generally sunward at lower latitudes in the plasma sheet further inside the magnetosphere. The plasma velocity and pressure gradients close to the flanks of the magnetosphere generate the so-called "Region 1" (R1) field-aligned currents (FACs), downward to the ionosphere at the morning side, and upward from the ionosphere at the evening side. Part of the R1 currents that enter or leave the ionosphere closes across the polar cap while another part closes across the auroral oval to the "Region 2" (R2) field-aligned currents. The R2 FAC, upward at the morning side and downward at the evening side, connect to the ring current region at 4-6 Re and close through a partial ring current flowing from the morning to the evening side near equator (Iijima & Potemra, 1976a,b).

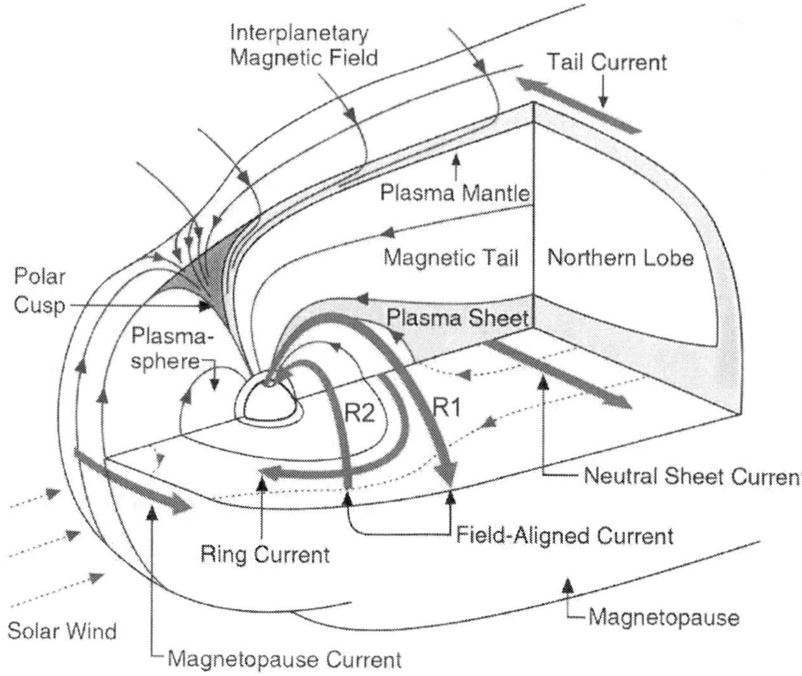

Fig. 1. Simplified sketch of magnetospheric structure (from GEM Source Book, 1999)

The polar ionospheric convection and current systems that could affect the magnetic variations measured from ground within the central polar cap could be divided into distinct categories, DP1, DP2, DP3 (or NBZ), and DP4 (or DPY) characterized by their relation to substorm activity and interplanetary magnetic field conditions (e.g., Wilhjelm et al., 1972; Friis-Christensen & Wilhjelm, 1975; Troshichev et al., 1986; Weimer et al, 2010). Each type of convection pattern comprise field-aligned currents (FAC) and horizontal Hall and Pedersen currents. The magnetic variations at ground can be assigned to a system of "equivalent currents", which in many cases will correspond to the ionospheric Hall currents that dominate the response at ground level, since the effects of the field-aligned currents and the horizontal Pedersen currents tend to cancel each other (Fukishima, 1969). The convective motion and the Hall currents are both transverse to the electric field but in opposite directions. Thus the convection patterns and the equivalent current patterns, although oppositely directed, will bear close resemblance. Both patterns are also close in shape to the equipotential contours.

The DP2 two-cell convection system comprising antisunward convection of plasma over the polar cap at all altitudes from the ionosphere up through the polar magnetosphere to the solar wind is the prevailing system except during intervals of strong northward IMF. However, the DP2 system may be modified to a greater or lesser extent by effects related to the DP1 (substorm) or/and the DP4 (DPY) convection systems. During northward IMF one or both vortices of the DP3 (NBZ) convection system could be formed at very high latitudes inside a weakened DP2 system prevailing at lower latitudes.

3. IMF and solar wind impact on polar geomagnetic variations

The main part, around 97%, of the geomagnetic field is generated by internal sources in the Earth's core and crust. The field from the core, the main field, has relatively slow (secular) variations in magnitude as well as in direction. The crustal remnant magnetization is constant on geological time scales. The remaining field contributions generated by external current systems and their induced counterparts in the ground are generally found by subtracting a set of base line values from the measured field components.

The external contributions can be subdivided into two fractions. One is associated with the effects of the slowly varying solar UV flux and the rather steady flow of solar wind plasma past the Earths magnetosphere during intervals with quiescent conditions. When this part, the so-called "Quiet Day" (QDC) variation, is also subtracted from the observed geomagnetic field, then the remaining field contributions are mainly related to the combination of enhanced solar wind velocities and an appreciable magnetic field in the solar wind, which may strongly influence the interaction of the solar wind flow with the Earths magnetosphere. The interaction manifests itself, among other, through enhanced ionospheric currents and associated magnetic variations in the Polar Regions.

Hourly values of the interplanetary magnetic field at the Earth's position derived from the ACE data base for the year, 2002, are shown in the top field of Fig. 2. These magnetic data are resolved in Geocentric Solar Magnetospheric (GSM) coordinates, B_X (toward the Sun), B_Y (perpendicular to the Earth-Sun line and to the Earth's magnetic dipole axis) and B_Z (completing an orthogonal system) almost in the direction of the (negative) dipole axis. Throughout the diagrams in Fig. 2 the heavy red continuous curves depict smoothed averages.

Typical examples of polar geomagnetic observations are also presented in Fig. 2. The middle field displays the data from Thule, Greenland, located at corrected geomagnetic (CGM) latitude of 85.30°. The bottom field displays data from Vostok, Antarctica, located at corrected geomagnetic latitude of -83.58°. The thin blue traces in the diagram show hourly average values of the field resolved in the H-component (geomagnetic North), E-component (East) and Z-component (down for Thule, up for Vostok) through the year 2002.

In these data several features are noteworthy:

i. There are daily variations in all components during quiet as well as disturbed conditions judging the disturbance level from the IMF data in the top field. The main part of these variations is related to the rotation of the station beneath the more or less stationary polar cap current patterns.

ii. The polar magnetic data, particularly the E-component of the Thule observations, display small but significant steady changes through the year. These changes result from secular variations in the main field.

iii. There are low-frequency modulations on time scales from one to a few weeks seen in the interplanetary B_X and B_Y magnetic field components as well as in the ground-based Thule and Vostok data, most distinctly in the Z-component. These modulations are indicative of the interplanetary sector structure where the preferred field direction is either toward or away from the Sun. The associated changes in the polar magnetic field represent the so-called Svalgaard-Mansurov effect (Svalgaard, 1968; Mansurov, 1969).

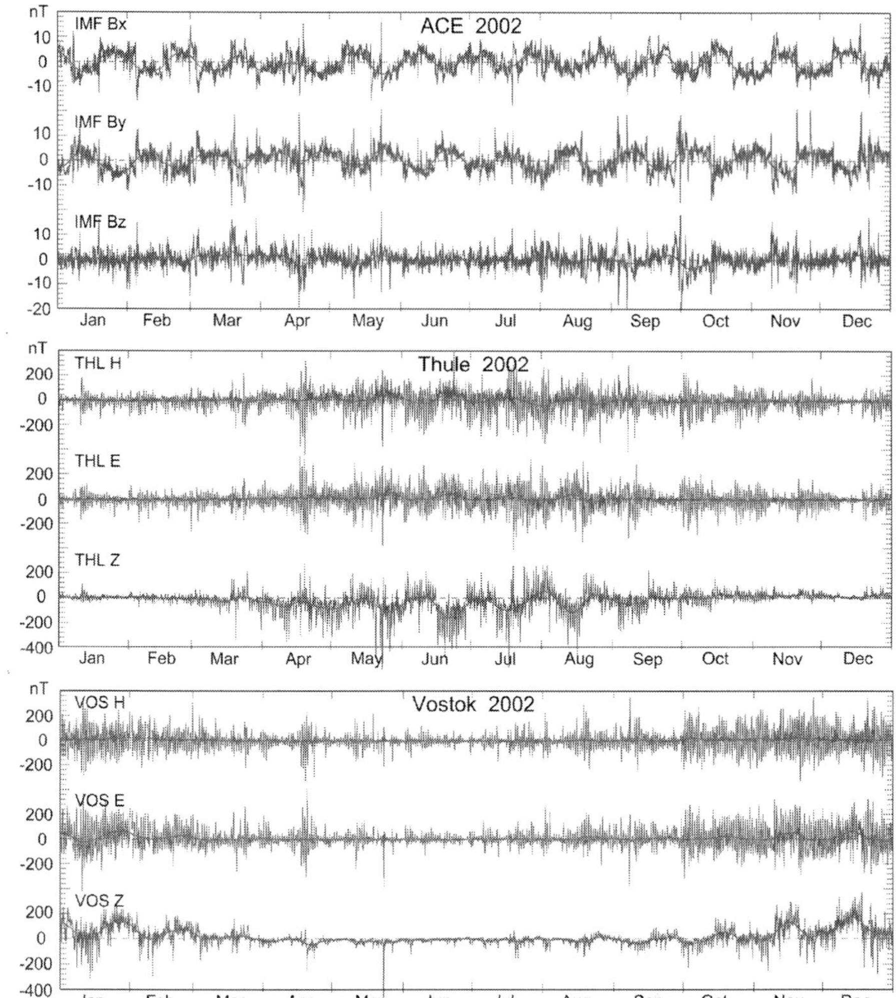

Fig. 2. ACE interplanetary magnetic field data (top) and polar geomagnetic field data from observatories in Thule, Greenland (middle), Vostok, Antarctica (bottom). Hourly values 2001.

iv. Strong high-frequency excursions, here on time scales of one to a few hours, are seen in the interplanetary data as well as in all three components of the ground-based data. In the ground-based data for Thule these excursions are strongest during the northern summer months occurring at the middle of the diagram. In the data from Vostok, correspondingly, the strongest excursions are seen in the southern summer months occurring at either ends of the diagram. The excursions on time scales from minutes to hours observed in the solar wind and at ground in the polar cap are strongly correlated. This correlation is the fundamental basis for defining a Polar Cap (PC) index.

Basically, the Polar Cap (PC) index is defined to be a proxy for the merging electric field by assuming a DP2 (or DP2-like) convection system with antisunward plasma convection and uniform sunward Hall current across the central polar cap. It is further assumed that there is proportionality between the horizontal magnetic field variations, ΔF, in the central polar cap and the merging electric field, E_M, and that the magnetic variations related to E_M have a preferred direction with respect to the Sun-Earth direction. Thus, on a statistical basis from observations of ΔF and E_M:

$$\Delta F_{PROJ} = S \cdot E_M + \Delta F_I \quad (4)$$

where ΔF_{PROJ} is the projection of the magnetic disturbance vector, ΔF, to the "optimum direction", that is, the direction most sensitive to the merging electric field. The parameter S (often named α) is the "Slope", and the residual, ΔF_I, (often named β) is the "Intercept" parameter, named from a visualized graphical display of ΔF_{PROJ} against E_M. Throughout the calculations we express magnetic fields in nT and the electric fields in mV/m.

In order to calibrate magnetic variations with respect to the merging (geo-effective) electric field we now define the PC index by the inverse relation:

$$PC = (\Delta F_{PROJ} - \Delta F_I)/S \quad (\sim E_M \text{ [in mV/m]}) \quad (5)$$

The scaling parameters, that is, the optimum direction angle, the slope and the intercept, are found from Eq. 4 by regression analysis of polar observatory geomagnetic observations against interplanetary satellite data to include daily as well as seasonal variations. This scaling serves to make the resultant PC index independent of the regular daily and seasonal variations in parameters like the ionizing solar UV flux, that are not related to the merging electric field. Furthermore, the index is also now independent of the geographical position as long as the observatory is located within the central polar cap, where the magnetic variations are mostly associated with the rather uniform transpolar electrical currents that, in turn, relate mainly to the merging electric field in the solar wind and not to auroral substorm processes.

4. Calculation of PC indices

The calculation of PC indices according to Eqs. (4) and (5) requires a number of steps. Firstly, the calibration constants, optimum angle, slope and intercept, must be determined from Eq. (4) using lengthy series of correlated values of the merging electric field (Eq. 2) existing in the solar wind but adjusted in timing to apply to the Polar Cap, and geomagnetic variations corrected for secular variations and quiet day variations (QDC). Then the actual magnetic variations, again corrected for secular and quiet day variations, could be used in Eq. (5) with the calibration parameters, selected from tables spanning time of day and day of year, to provide PC index values (Troshichev et al., 2006; Stauning et al., 2006).

4.1 Satellite observations of solar wind and IMF parameters

Data for the merging electric field in the solar wind are needed for the derivation of the scaling parameters. During the time interval from 1975 to present, where high-quality digital magnetic data are available from Thule there are four main sources of solar wind plasma and interplanetary magnetic field data. These sources are IMP 8, Wind, Geotail, and ACE satellite data. Here we concentrate on IMP8 and ACE data for calculations of the scaling parameters, but Geotail and Wind satellite data can be used in special correlation studies since they offer further ranges of satellite positions with respect to the magnetosphere.

The IMP 8 satellite was launched 1973-10-26 into an orbit with inclination varying between 0 deg and 55 deg. Apogee and perigee were around 40 and 25 earth radii, respectively. Data are available up to June 2000. The spacecraft was located in the solar wind for 7 to 8 days of every 12.5 days orbit. Telemetry coverage has been varying between around 60 and 90%.

The ACE satellite was launched 1997-08-25 into a halo orbit about the L1 point. The satellite instruments and telemetry have provided almost 100% recovery of magnetic data and solar wind velocity data since the beginning of 1998. The satellite is still active. The main advantage of using ACE data is their regularity over the years, which are now spanning more than one full solar activity cycle. The main disadvantage of using ACE data is the large distance of around 240 Re (~1.500.000 km) from the Earth to the satellite. This large distance causes a delay of typically around 1 hour for the solar wind with its embedded magnetic field to travel from the satellite position to the encounter with the Earth's magnetosphere.

In most previous satellite-ground correlation studies IMP 8/ACE data have been referred to a fixed position at 12 Re in front of the Earth in order to provide a uniform basis. This modification is accomplished by shifting the data in time with an amount corresponding to the solar wind travel time from the actual satellite position to the GSE (=GSM) X-coordinate of the reference position assuming uniform conditions in planes perpendicular to the Sun-Earth line. For the recent analyses (Weimer & King, 2008), the satellite data have been shifted to the position of the bow shock nose (BSN). A detailed description of the shift is provided at the web site http://omniweb.gsfc.nasa.gov/.

For IMP 8 the upper limit on the X-coordinate is around 40 Re. This leaves a span for the solar wind travel from the satellite to the reference location to deal with distances between -12 Re (at $X_{GSE} = 0$) and +30 Re, which with typical solar wind velocities of 350 km/sec corresponds to delays ranging between around -4 min and +10 min. For ACE the typical delay is around 60 min between solar wind observations at the satellite and the encounter of the same solar wind volume with the Earth's magnetosphere (12 Re or BSN reference position). The above-mentioned delays are just typical values. In actual cases the precise satellite position and the observed solar wind velocity were used to calculate the relevant delays. With the above outlined procedure, all the relevant solar wind field and velocity parameters have been converted into 1-min samples at the reference position. For some of the statistics we proceed to calculate 5-min sorted averages (excluding max and min values, average of rest).

From the reference position (12 Re or BSN) it takes the E_M effects some time to propagate and affect processes in the central polar cap ionosphere. Precisely how the effects propagate to the polar cap ionosphere is still an open question involving magnetospheric convection processes, pressure gradients, field-aligned currents and coupling to the ionospheric potential and current systems. It is shown below that this time interval is around 20 min. Accordingly, the series of merging electric field values referenced to 12 Re or the bow shock nose should be further time-shifted by 20 min before being correlated with polar cap magnetic variations. In the further derivation of *PC* index parameters and other uses of IMF and solar wind data we have selected to base all calculations on IMP 8 and ACE data discarding the Wind and Geotail measurements except for specialized applications.

4.2 Handling of geomagnetic data

It is clear from Fig. 2, particularly in the trace for the geomagnetic E-component measured at Thule, that the magnetic field related to internal sources changes slowly. For Thule these secular changes in the components amount to several tens of nT each year. In Vostok the secular changes are smaller. These changes are unrelated to solar wind and ionospheric effects and must be corrected for before PC index values are derived. Thus the baseline caused by the "internal field" contribution should be derived. During midwinter, at night hours and for very quiet conditions, the "external field" contributions are at minimum, at most a few nT. Consequently, the winter night magnetic recordings are carefully monitored in order to detect the magnetic component values during extremely quiescent conditions. For Thule, these values, considered to represent the baseline related to the "internal field" contributions, are referenced to 1. January and kept since 1973 in a yearly updated "Quiet Winter Night Level", QWNL, table. It is assumed that the baselines for the individual components, with the accuracy needed for PC calculations, vary linearly through the year such that they could be interpolated from the values at day 1 one year to day 1 the following year, or extrapolated beyond 1 January through the actual year.

The "Solar Rotation Weighted" (SRW) calculations of the "Quiet Day Curve" (QDC) at DMI are explained in Stauning (2011). The method builds on superposition of a selection of quiet segments to build a QDC for each component separately for any given day. The samples are weighted according to the variability in the horizontal component. In the superposition of quiet recordings from different days, further weight functions are applied to give preference to intervals close to the QDC day in question and to intervals where the same face of the sun is directed towards the Earth. The solar rotation weight factor takes into account the regular variations in the solar UV radiation, in the solar wind velocity and in the IMF sector structure. The procedure is fully automatic and a quality parameter (the sum of weights) is provided for each hourly QDC value to enable a warning for poorly defined QDC values. Quadratic interpolations are used to provide QDC values with finer than one hour time resolution (e.g., 1-min samples). The baseline values and the QDC values calculated for each day of the data range are now subtracted from the recorded data to provide the series of magnetic variations to be used for PC index calculations. Composite plots of baseline-corrected magnetic recordings from January and July 2002 and the QDC values for these intervals are shown in Fig. 3.

4.3 Optimum direction angle

Searching for a proxy based on polar magnetic disturbances to represent the solar wind merging electric field (E_M = MEF = V_{SW} B_T $\sin^2(\theta/2)$), we may increase the correlation between the horizontal disturbance vector ΔF (corrected for the quiet daily variations) and the E_M by projecting ΔF to a specific direction ("optimum direction"). In principle, this direction is perpendicular to the transpolar DP2 current. The direction is not entirely fixed in space but varies slowly with local time and season. The rotating horizontal magnetic vector is resolved in an X-component (northward in a geographical coordinate system) and a Y-component (eastward). The vertical Z-component is downward in the northern polar cap.

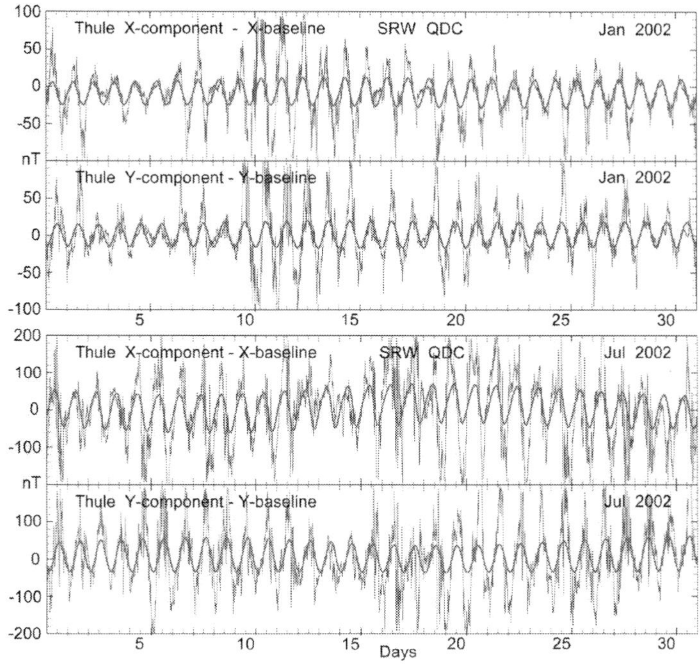

Fig. 3. Thule baseline-corrected magnetic data and QDC for X- and Y-components for January and July 2002.

For the horizontal components we first subtract from the raw data the baseline values and then the QDC values, i.e.:

$$\Delta X = (X_{RAW} - X_{BL}) - X_{QDC} \quad (6a)$$

$$\Delta Y = (Y_{RAW} - Y_{BL}) - Y_{QDC} \quad (6b)$$

Here, X_{BL} is the baseline value for the X-component defined yearly (every 1 January) and now adjusted to the actual day-of-year. X_{QDC} is the reference QDC level provided as tables of hourly values for each day and now adjusted to the proper time. The Y-component is handled correspondingly.

Using the + sign for the southern hemisphere and the – sign for the northern, the projection of the disturbance vector ΔF to the optimum direction is given through:

$$\Delta F_{PROJ} = \Delta X \cdot \sin (V_{PROJ}) \pm \Delta Y \cdot \cos(V_{PROJ}) \quad (7)$$

where the angle V_{PROJ} is defined as a function of UThr (UT time in hours) through:

$$V_{PROJ} = \text{Longitude} + \text{UThr} \cdot 15^{o} + \text{optimum direction angle (ODA)} \quad (8)$$

With this definition, the "Optimum Direction Angle" (ODA) is the angle between the dawn-dusk direction and the optimum direction for the magnetic variation vector. This angle is also the angle between the midnight-noon direction and the transpolar equivalent current direction, which by definition is perpendicular to the related magnetic disturbance vector.

The optimum direction angle values are calculated from statistical analyses to find the maximum correlation between the geomagnetic variations measured in the polar cap and the solar wind merging electric field values derived from interplanetary spacecraft data, which have been time-shifted to apply to polar cap effects. The direction depends on observatory location and varies with day of year, and UT time of day. With this procedure we have calculated and tabulated the optimum direction angle for each UT hour of the day and each month of the year.

4.4 Calculations of slope and intercept

It is the general assumption that there is a linear relation between the projected polar magnetic variations, ΔF_{PROJ}, and the solar wind electric field, E_M, and that this relation can be inverted and used to define a PC index by equivalence (cf. Eqs. 4 and 5). This relation assumes a DP2-type two-cell polar ionospheric "forward" convection, which is the most common convection mode. It applies to solar wind conditions where the interplanetary magnetic field (IMF) is either southward oriented or only weak in magnitude when northward directed. During conditions of strong northward oriented IMF a DP3 reverse convection system may develop in which the transpolar flow is sunward, while the return flow is antisunward.

With an overhead reverse current direction the magnetic deflections at ground are opposite of those of the forward convection mode. Accordingly, the projected disturbance vector may become less than the QDC level or even strongly negative. PC index values calculated during such conditions may turn out to be negative. Since the interplanetary merging electric field (E_M) by definition (Eq. 2) is always non-negative, then the concept of the PC index as a proxy for the E_M breaks down. Hence, reverse convection cases must be excluded from the calculations of slope and intercept.

In a least squares fit to estimate the regression of ΔF_{PROJ} on E_M from a comprehensive and representative data base, the deviations of the magnetic variations from the regression line are minimized whereby the slope, S, and the intercept parameter, F_I , are derived. With this procedure we have calculated and tabulated values of the slope, S, and the intercept parameter, ΔF_I , for each UT hour of the day and each month of the year.

4.5 PCN, PCS and combined PC index, PCC

The PCN index for the northern polar cap is based on data from the Danish geomagnetic observatory in Thule (Qaanaaq) in Greenland while the PCS index for the southern polar cap is based on data from the Russian geomagnetic observatory in Vostok at Antarctica. Parameters for the two observatories are listed in Table 1.

Station name	IAGA code	Geocentric Latitude	Geocentric longitude	Corr. geomag. latitude	Corr. geomag. longitude	UT at MLT noon
Qaanaaq	THL	77.47	290.77	85.30	31.11	15:04
Vostok	VOS	-78.46	106.87	-83.58	54.77	13: 02

Table 1. PC geomagnetic observatories in the northern and southern polar caps. Epoch 2000

Both the PCN and the PCS indices have been calibrated with respect to the merging electric field (MEF=E_M) in a statistical sense. Accordingly, the two index series are also mutually equivalent in a statistical sense. However, differences may arrive as the result both of different conditions in the two polar caps (e.g., different solar illumination) and of different response to forcing from the solar wind as well as different response to substorms. Such differences have the potential for interesting studies (e.g., Lukianova et al., 2002).

Another possible application of the two PC index series uses the possibility to define "summer" or "winter" PC index series based alternating on either PCN or PCS depending on local season. Thereby characteristic seasonal variations may emerge more clearly than otherwise by using one or the other hemispherical index series.

A further possibility is the combination of the two index series into one (Stauning, 2007). In the most common two-cell forward convection cases the projected magnetic variations and the PC index values are positive. However, with overhead reverse convection flow during NBZ conditions the magnetic deflections are opposite to those of the forward convection mode. Accordingly, the projected disturbance vector may become less than the QDC level or even strongly negative. PC index values calculated during such conditions may turn out to be negative. Since the interplanetary merging electric field by definition is always non-negative (cf. Eq. 2) then the concept of the PC index as a proxy for E_M breaks down.

Such cases of reverse convection and large negative PC index values are, by far, most common at daytime in the summer season. In these cases the forward convection mode may still prevail at the local winter season in the opposite polar cap and the PC index values derived there may be small but still positive like E_M. A simple way to accomplish a combined PC index would be to calculate the plain average values of the PCN and PCS data series. However, the problem with large negative PC index values in one hemisphere speaks for constructing a combined PC index (PCC) from non-negative values only. Accordingly:

$$\mathrm{PCC} = [(\mathrm{PCN\ if} > 0 \mathrm{\ or\ else\ zero}) + (\mathrm{PCS\ if} > 0 \mathrm{\ or\ else\ zero})]/2 \tag{9}$$

This option was tested in Stauning (2007) who examined the correlation between E_M and PCN, PCS, and the new PCC values, respectively. From this work, Fig. 4 presents hourly average values of E_M, PCN, PCS and PCC index values through the year 2000. The top field displays E_M values calculated according to Eq.2 from ACE satellite measurements of the interplanetary magnetic field and the solar wind velocity. The parameters measured at the ACE satellite have been shifted to apply to polar cap effects. The second and third fields from the top present PCN and PCS values using identical procedures in the calculation of the two index series. The bottom field presents the PCC index values derived from using Eq. 9.

Fig. 4 indicates a close similarity between the enhancements in E_M and the positive enhancements in PCN and PCS. These enhancements are, of course, also seen in the combined PC index, PCC. Further, the occurrences of negative PCN and PCS index values, preferably during the local summer seasons in the two hemispheres (May-August for PCN, November-February for PCS) should be noted. These events have no proportional (i.e., strongly negative) counterpart in E_M values. The relevant merging electric field values are just small at these times. The new PCC index displays the positive enhancements corresponding to the larger values of E_M and, by its definition, no negative values even during NBZ conditions.

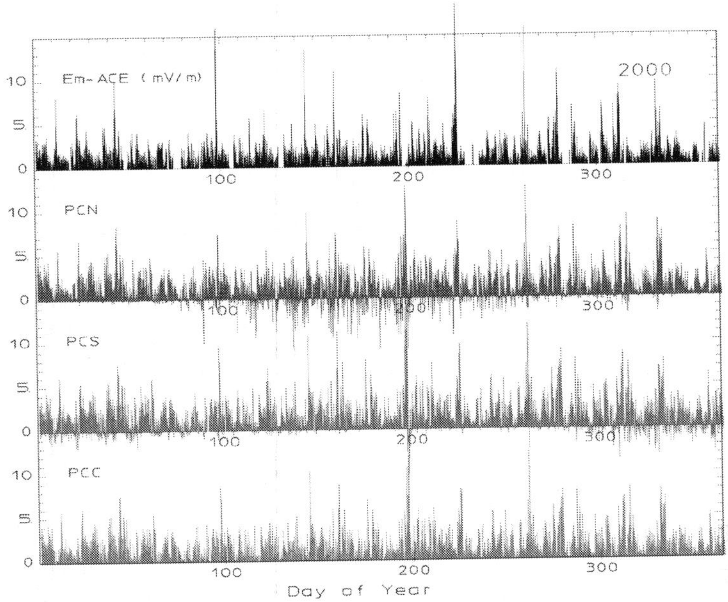

Fig. 4. Hourly averages calculated from 5 minute values of the merging electric field (top), Polar Cap PCN and PCS indices (middle) and new PCC index (bottom) through 2000.

5. Further relations of PC indices to solar wind parameters

The definition and the calculation of polar cap indices are based on an assumed linear dependence (cf. Eqs. 4 and 5) on the merging electric field, E_M (cf. Eq. 2). However, there are a number of additional effects, which could be of importance for various applications of the indices. Such effects, which to some extent violate the premises, include:

i. Delayed PC index response to variations in the merging electric field.
ii. Non-linear effects such as saturation of PC index values at high levels.
iii. Response in the PC index to variations in the IMF Bx component not included in E_M.
iv. PC index response to solar wind dynamic pressure variations.

It is important to be aware of these modifications of the simple straight relations between E_M and the PC indices in order to recognize their signatures or consider whether they could deteriorate analyses of the relations between the PC indices and further terrestrial parameters.

5.1 Timing of PC index variations at changes in the merging electric field

When a change in the merging electric field, E_M, hits the magnetosphere then it takes some time for the polar cap convection pattern, and thus the PC index, to adjust to the new level. The timing issue could be analyzed in two ways. First, one could look for distinct variations, for instance, step-like changes in E_M, which could easily be recognized in the PC indices. Then the relative timing between Em and PC would provide the desired relation. Another approach would be to investigate statistically the correlation between E_M and PC index values with a variable delay in order to disclose the optimum value. We shall proceed both ways.

The top field of Fig. 5a displays a plot of superimposed traces of step-like varying E_M. The data have been shifted to a position of 12 Re in front of the magnetosphere (close to the average position of the bow shock nose). The individual traces have been positioned such that the positive step from a low to a high value occurs at relative timing T=0. The bottom field of Fig. 5a displays the corresponding variations in PCN. The heavy traces in the two plots mark the average variation.

It is easy to see the sudden step in the average E_M trace (top field) from close to zero up to around 2.5 mV/m. The average PCN index values in the bottom field display a gradual change from almost zero and up to 2.5 units, where it levels with the value of E_M. The change in PCN appears to start 9-10 min after the step in E_M. The average PCN index values reach half the final level at around 20 min after the E_M step, while it takes 30-40 min to reach the final level.

The corresponding variations at a negative step in E_M are displayed in Fig. 5b. In the upper field a number of step-like varying Em traces have been superposed with the step placed at relative timing T=0. From the average variations in the corresponding PCN index values it is seen that the change starts around 10 min later than the step in E_M. The average PCN value reaches half the level prevailing before the change at around 20 min after the negative E_M step while it takes around 50 min to reach the final low level.

In summary, following a positive or negative step-like change in the merging electric field impinging on the front of the magnetosphere it takes around 10 min before the change is sensed and the PC index starts to change. At around 20 min after the E_M step the PC index has reached half the final level, while it takes 40-50 min before the change is completed. The positive steps in E_M propagate a little faster to affect the PC index than the negative steps.

The analysis of corresponding features based on a larger and more general selection of data is presented in Fig. 6. For the epoch from 1990 to 1999, PCN data have been plotted against E_M in such diagrams with variable delays from 0 to 30 min in steps of 5 min. The IMP8 satellite data were selected for this study because of the proximity of the satellite to the magnetosphere whereby the uncertainty in the time-shift is minimized (compared, e.g., to ACE or Wind data). At each step the linear correlation coefficient, RX, was calculated along with the parameters S0 (average deviation), S1 (absolute deviation), and S2 (standard deviation). NX is the total no. of samples. The number of samples in each bin of 2 mV/m in E_M is indicated by the size of the large dots and the standard deviation is indicated by the bars. The example shown here uses a delay of 20 min. From the values of RX at different delays, the optimum delay was estimated to be 18 min. At this delay the correlation coefficient was found to be RX=0.748.

A different representation of the correlation between PCN and E_M is displayed in Table 2. Here, the PCN data and IMP8 E_M data through 1990 to 1999 are again correlated. The data have been divided in selections based on the values of the bow shock nose distance, BSN-X, supplied with the satellite data. In addition to confirming that the delay from bow shock nose to polar cap is around 20 min, the correlation study presents further details specified in Table 2.

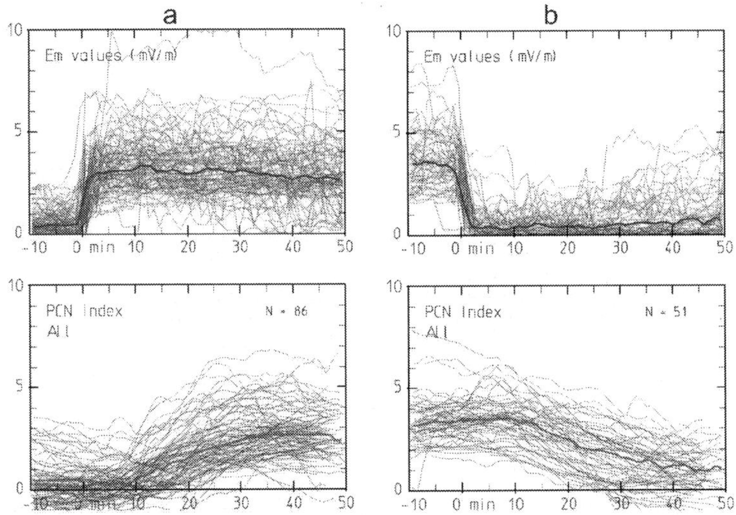

Fig. 5. PCN index variations at steps in the merging electric field at 12 Re.

BSN-X	8	9	10	11	12	13	14	15	16	Re
NX	397	2082	9401	31068	75318	87184	45806	12568	3136	
RXmax	0.45	0.59	0.71	0.74	0.70	0.69	0.69	0.70	0.72	
DTm	12.3	15.5	16.3	17.2	17.4	18.8	20.9	23.2	27.4	min
E_M av	7.7	4.6	2.7	1.7	1.2	0.9	0.7	0.7	0.7	mV/m
PCNav	6.3	4.2	2.4	1.7	1.2	0.8	0.6	0.6	0.6	

Table 2. Results from correlation of PCN with values of E_M at bow shock nose (BSN).

From Table 2 it is clear that for the majority of cases (90%) the BSN distances range between 11 and 15 Re, which is close to the X=12 Re used as a reference position in other analyses. It is also seen that the delay, generally, decreases with decreasing BSN distance. The decreases are larger than the reduction by 17 sec for each 1 Re decrement in distance, which is the time lapse for a typical solar wind speed of 400 km/s. The reduced correlation for the stronger cases (low BSN-X) could indicate that the strongly reduced travel time relies on the mixing of different propagation mechanisms, for instance, effects from accompanying solar wind dynamical pressure enhancements that propagate faster to the Earth than the E_M -associated effects.

Another remarkable feature is the close relation between E_M and PCN for the different ranges of BSN-X. It could be expected that the bow shock moves closer to the Earth when the solar wind is intense, that is, for cases of high solar wind speed and strong, southward IMF in which case the merging electric fields are enhanced. However, it is noteworthy that the average PCN index for each BSN-X interval corresponds closely to the average E_M value except for the highest levels where some saturation of PCN is evident.

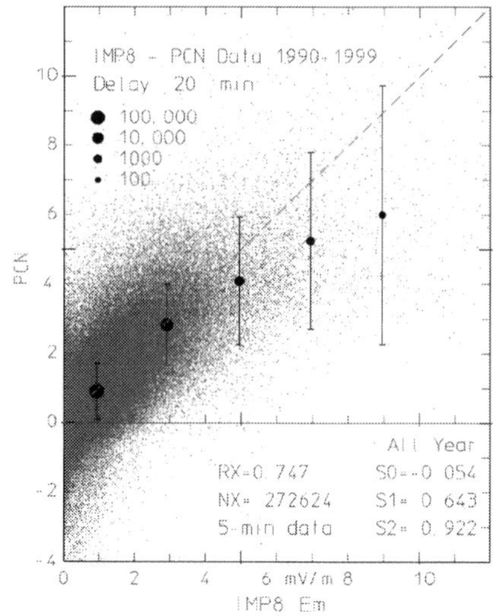

Fig. 6. Correlation scatter plot between E_M at BSN and PCN for 20 min delay.

5.2 PC saturation effects

At high levels of the solar wind forcing, saturation effects are found in many terrestrial parameters, for instance in the polar cap potentials, in the convection velocities, and the cross-polar cap dimension, or in the length of the reconnection line. Borovsky et al. (2009) list 9 different models to explain polar cap potential saturation.

A relevant question is now whether the polar cap indices just reflect the general saturation in the magnetospheric response to strong solar wind forcing or have saturation patterns different from other geophysical disturbance phenomena. Fig, 7 displays in a scatter plot the relations between the combined PCC index and the merging electric field. The data base has been constrained to intervals of 4 days through each of 82 geomagnetic storm intervals (Dst<-100) from the epoch 1995-2005 in order to focus on the strong events with large values of the PC indices. The PCC indices have been preferred over the hemispheric PCN or PCS indices since possible saturation effects are globally extended. From the scatter plot in Fig. 7 the saturation effect is evident. The dotted curve included for reference represents the simple saturation model:

$$PCC = E_M \Big/ \sqrt{\left(1 + (E_M / E_0)^2\right)} \tag{10}$$

The asymptotic electric field value used here is: E_0=10.5 mV/m. The fit to the actual average PCC index values is quite good. Hence, this saturation model could serve to provide a basis for comparison of saturation in the PC index data with saturation effects in further parameters. The model (Eq. 10) might even provide some guidance in a discussion of possible saturation mechanisms.

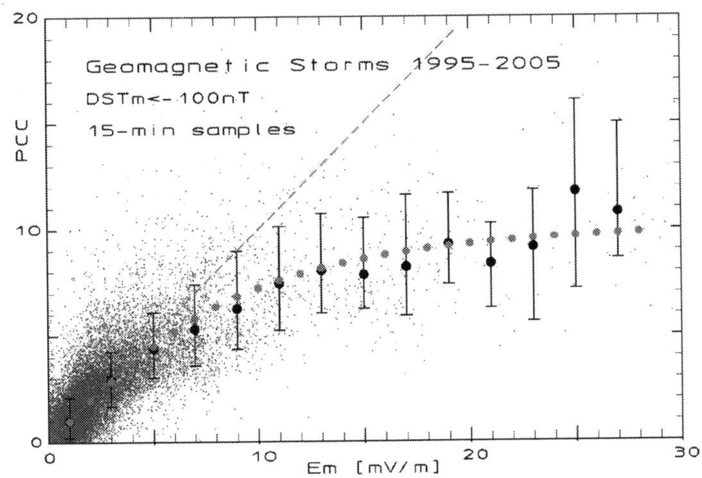

Fig. 7. PCC indices vs. related merging electric field values for all major magnetic storms (Dst<-100 nT) during 1995-2005.

The saturation effect in the PC index may also be seen as a limitation in the response to large values of the IMF Bz parameter. This is illustrated in Fig. 8. The figure displays in a scatter plot the ratio of PCN over E_M as a function of the IMF Bz component. The heavy dots mark the average values within intervals of 2 nT in the Bz component. The Bz component is already included in the expression for E_M (cf. Eq. 2). Ideally, the points should thus be positioned close to and symmetrically around the horizontal dashed line, which indicate a ratio of unity, and should not show any systematic dependence on the IMF Bz (or any other) component.

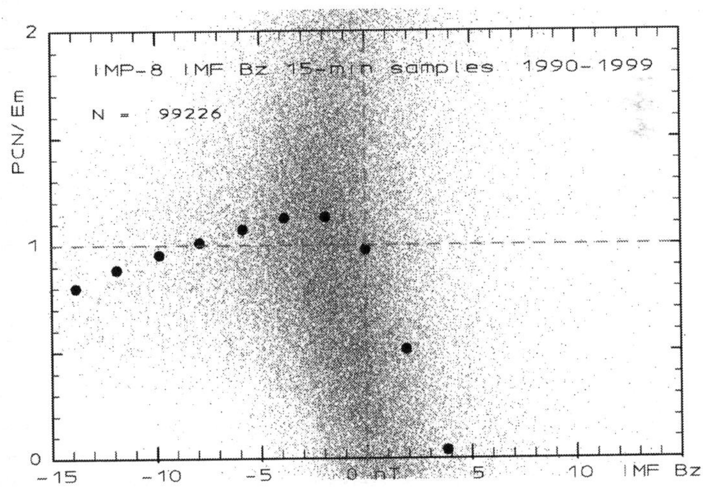

Fig. 8. Scatter plot of the ratio PCN/E_M versus IMF Bz through the epoch 1995-1999.

The left half of the plot displays the ratio PCN/E_M for negative values of the IMF Bz component (southward IMF). The ratio increases from 1.0 at IMF Bz ~ 0 to values between 1.00 and max. 1.13 for IMF Bz values between 0 and -8 nT. Starting from the max. level at an IMF Bz level at around -4 nT, the ratio systematically decreases in an almost linear trend to smaller values ending here at a ratio of 0.8 for IMF Bz ~ -15 nT. This systematic behaviour may - like the above saturation model - provide some guidance to the saturation mechanism.

The right part of Fig. 8 displays the ratio PCN/E_M for positive values of the IMF Bz component (northward IMF). From an average value of this ratio close to unity at IMF Bz ~ 0 the ratio rapidly decreases to zero at IMF Bz ~ +4 nT and thereafter reaches large negative values since the PCN values tend to become negative while the E_M values approach zero in consequence of the IMF direction getting close to due northward in some cases (cf. Eq.2).

5.3 Effects from the IMF Bx component on PC index values

The large-scale systematic variations in the IMF Bx component define the sector structure in the solar wind. The Bx component is positive when the IMF is pointing toward the Sun in "toward" sectors and negative in "away" sectors (Svalgaard, 1968, 1972, 1973; Mansurov, 1969). The IMF Bx component is not included in the expression for the merging electric field (Eq. 2). Hence the scaling of polar magnetic variations to construct a polar cap index is assumed to be independent of this component.

Fig. 9 displays the ratio of PCN over E_M value plotted against the IMF Bx value. The scatter plot has the same format as the plot displayed in Fig. 8. The heavy dots indicate average values of the ratio PCN/ E_M through intervals of 2 nT in the Bx component. These values are all within 0.9 to 1.1, that is, close to unity through most of the displayed IMF Bx range from -15 to + 15 nT . Ideally, the average value of ratio PCN/ E_M should be close to unity regardless of the IMF Bx value. Thus, Fig. 9 provides a confirmation that the IMF Bx component has little influence on the PC values.

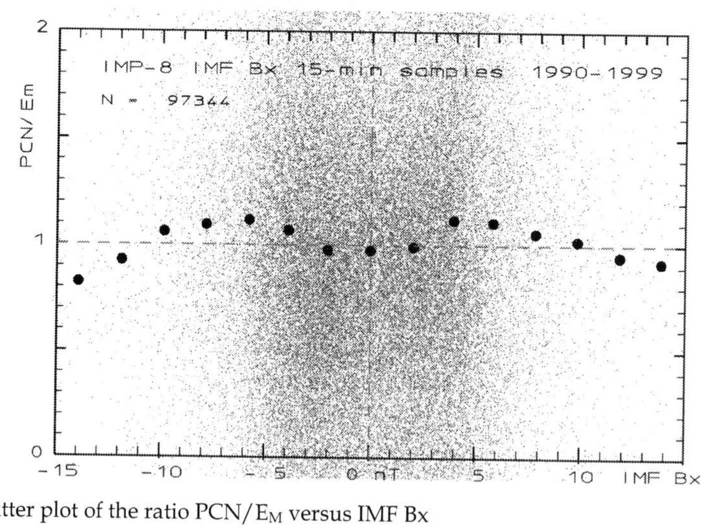

Fig. 9. Scatter plot of the ratio PCN/E_M versus IMF Bx

5.4 Effects from variation in the solar wind dynamic pressure on the PC index

The dynamic pressure in the solar wind (cf. Eq. 1) has profound influence on the general morphology of the magnetosphere. The pressure balance at the magnetopause is an essential factor for determining the dimensions of the magnetosphere (e.g., Spreiter et al., 1966; Shue et al., 1997) and for the location of the bow shock as well (e.g., Farris & Russell, 1994). Furthermore, the solar wind dynamic pressure strongly affects the width of the auroral oval causing auroral activity to expand equatorward during intervals of enhanced pressure (Newell & Meng, 1994). However, the solar wind dynamic pressure and its variations have relatively small effects on the polar cap index compared to the effects of the merging electric field and its variations.

The PC index variations with solar wind dynamic pressure could be divided into 3 types comprising effects of different steady levels, gradual changes, and fast (impulsive) variations, respectively. The sample duration and averaging procedures should be selected accordingly. The variations in the PC index level with solar wind dynamic pressure level are illustrated in Fig. 10. The diagram is based on 15-min average values of the combined PCC index, the solar wind merging electric field (E_M), and the dynamic pressure (P_{SW}) for a series of 82 selected magnetic storm intervals during 1995-2005. Each interval comprises 4 days of which the first includes the storm onset. This selection holds a fair amount of enhanced parameter values. Each of the small points displays the ratio of PCC over the E_M value (vertical axis) against the dynamic pressure (horizontal axis). The horizontal dashed line reflects the "ideal" unity ratio between PCC and E_M. The large "dots" indicate average values of the ratio of PCC over E_M for each interval of width equal to 2 nPa in dynamic pressure. The error bars represent the 68% "standard deviation" range for the distribution of points above and below the average values.

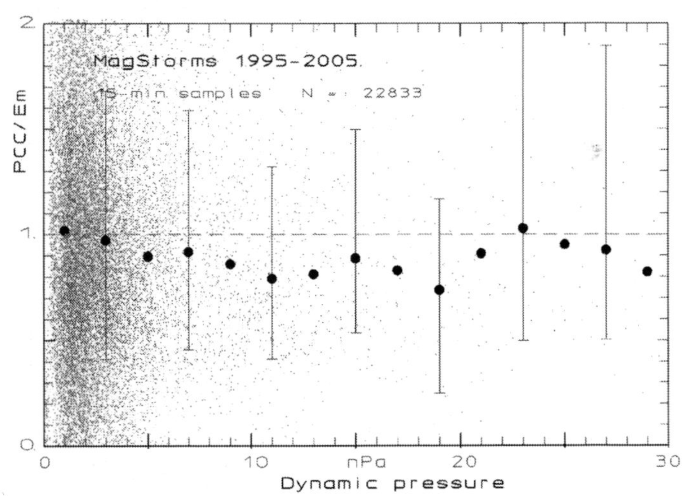

Fig. 10. Effects of varying solar wind dynamic pressure level on the PCC/Em ratio.

Like seen in other scatter plots, the scatter of points in Fig. 10 is quite large. However, it is readily seen that the average ratio of the PCC index over the merging electric field is very close to unity for the bulk of data points representing small and moderate levels of the dynamic pressure up to around 2 nPa. For levels of the dynamic pressure from 2 up to around 10 nPa the PCC index is less than the corresponding E_M value. For still larger values of P_{SW} above 10 nPa the ratio of PCC over E_M is again close to unity but the scatter increases due, among other, to the sparse statistics. Accordingly, through the wide range displayed in Fig. 10 the steady level of the solar wind dynamic pressure has little effect on the polar cap index values.

The relation of the PC index values to gradually changing solar wind dynamic pressure conditions is displayed in Figure 11. The solar wind pressure values used in the figure are 5-min averages of the slope of the pressure variations with time. The solar wind data have been time shifted to apply to the polar cap (cf. section 5.1). The selection of data used to determine the relation of PCC/E_M to the time derivative in the solar wind dynamic pressure is the same set of storm intervals as used previously. The PC index values used in Fig. 11 are 5 min averages of the combined PCC index.

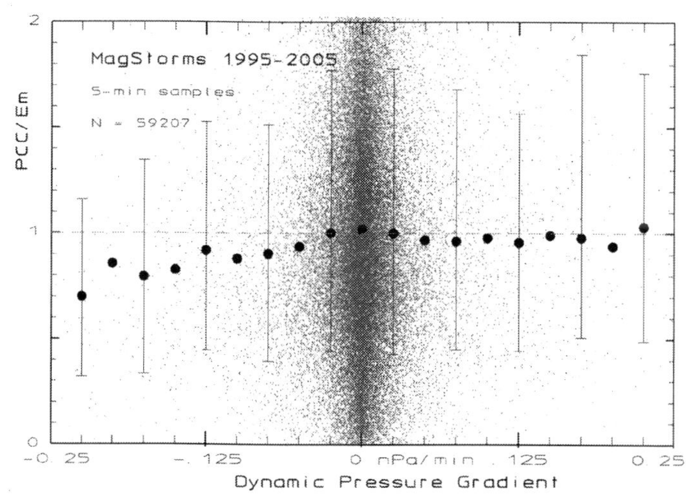

Fig. 11. Effects of dynamic pressure gradient on the ratio of PCC over E_M.

The large dots depict averages of all data samples within each interval of width equal to 0.025 nPa/min. For every other dot error bars have been drawn to indicate the 68% range for values above and below the average value. The average ratio of PCC to E_M (in mV/m) should be close to unity. This is clearly the case for the bulk of data points with time derivative values close to zero or positive. For large negative values of the time derivative the average ratio of PCC over E_M values is less than unity. The effects on the ratio PCC/ E_M and thus on the PC index is fairly small within the range of gradient values displayed in Fig. 11.

The effects on the Polar Cap index of sudden (shock-like) pressure variations are depicted in Fig. 12. The top field displays the dynamical pressure variations referred to the reference position (12 Re). Each of the PCN index traces in the bottom field displays a sequence of an

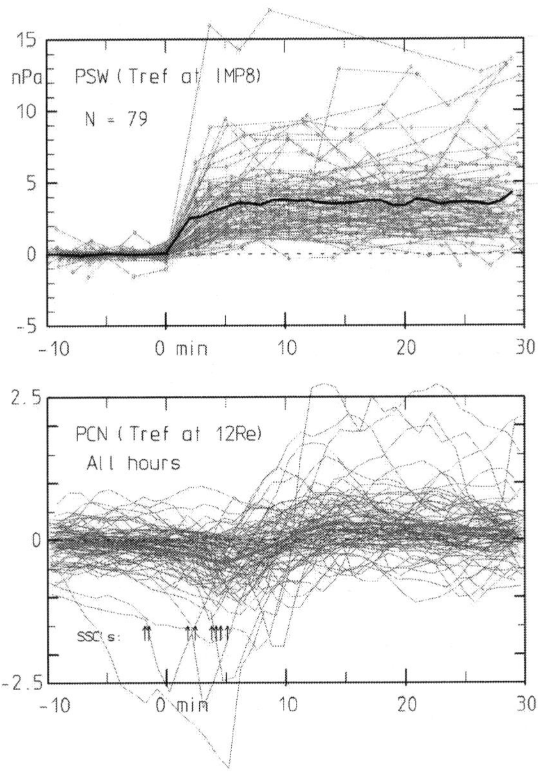

Fig. 12. PCN variations at solar wind PSW steps.

initial negative impulse followed by a positive enhancement in the index value (e.g., Lukianova, 2003; Huang, 2005; Stauning & Troshichev, 2008). The negative peak in the PC index occurs within 2-10 min following the pressure step referred to the reference position at 12 Re in front of the Earth. The width of the negative pulse is typically 3-5 min and the amplitude is around -0.5 to -1.0 units in the PC index. The positive pulse, typically, has a width of 10-20 min and amplitude of 0.5 to 1.0 units. The peak positive amplitude is typically reached ~15 - 20 min after the pressure step impacts at 12 Re. The timing implies that the pressure effects propagate from the reference position (12 Re) to the central polar cap in only a couple of minutes. This is much faster than the propagation of effects related to the E_M and thus suggests different propagations processes (e.g., Araki, 1994; Stauning and Troshichev, 2008).

In summary, the PC index variations with varying level of the solar wind dynamic pressure and with impulsive variations are usually limited to range between - 0.5 and + 1.0 units. The variations at impulsive pressure enhancements could in rare cases amount to a couple of units, which is still fairly modest compared to potential variations with the merging electric field (cf. Fig. 7). The response time for the initial negative pulse in the PC index is a few minutes while the positive trailing impulse develops in 10-15 min suggesting propagation modes different from the usual FAC and convection-based timing of variations in the PC index values with varying merging electric fields.

6. Relations of the polar cap indices to further geophysical disturbance phenomena

The solar wind flow past the magnetosphere generates a cross-magnetosphere electric field along the magnetopause. This field may feed large-scale convection and current systems in the magnetosphere, particularly during southward IMF conditions, and is partly mapped down to create the potential distribution in the polar cap ionosphere, which in turn generate the ionospheric currents responsible for the magnetic variations transformed into the polar cap indices.

Thus, the polar cap indices could be considered to be indicators of the power conveyed from the solar wind to the Earth's magnetosphere and ionosphere. The energy extracted from the solar wind is used to power, among others, the transpolar current and convection systems, the auroral electrojets, the Joule and particle heating of the thermosphere, and the ring currents. Hence, close relations between the polar cap indices and these disturbance phenomena could be expected.

Fig. 13 presents an example case of solar wind and geophysical parameters for a geomagnetic storm event on 12 Sept. 1999. The case has been selected to display data from the three satellites, ACE, Wind and IMP8, which were simultaneously in operation during the event. The ACE and Wind magnetic field and plasma data are displayed in the upper two fields. The traces are plotted in different colours and given different marks. In the next lower fields the merging electric fields and the dynamic pressure based on data from all three satellites have been plotted. The three lower fields display the auroral activity characterized by the AL and AE indices, the ring current intensity characterized by the SYM, ASY, and Dst indices, and the polar cap currents characterized by the PCN, PCS, and PCC indices. The average E_M (MEF) and Psw values have been plotted in the bottom field for reference.

The onset of activity at ground level is indicated by the arrow marked "SSC" (Storm Sudden Commencement) at around 04 UT. During the next ~ 6 hours the Earth was hit by a cloud of enhanced solar wind density causing a strong increase in the dynamical pressure (middle field). This increase, replotted in the bottom field, had no marked influence on the PC indices and also little influence on the auroral electrojet and the ring current indices. At 09 UT the IMF Bz turned southward and the merging electric field increases strongly. This time the PC indices increased markedly and subsequently the PC indices track E_M quite close, and the AE and AL traces indicate strong auroral activity. In the following sections the relations between the polar cap indices and further geophysical disturbances are discussed.

6.1 PC indices and polar cap potentials

The cross polar cap potential difference, Φ_{PC}, is an important parameter to characterize polar cap potentials. This parameter could be investigated by radars, for instance the SuperDarn backscatter radar system (Greenwald et al., 1995), or by direct measurement of the electric field from satellites traversing the polar cap (e.g., Reiff et al., 1981; Rich and Hairston, 1994; Boyle et al., 1997; Hairston et al., 1998).

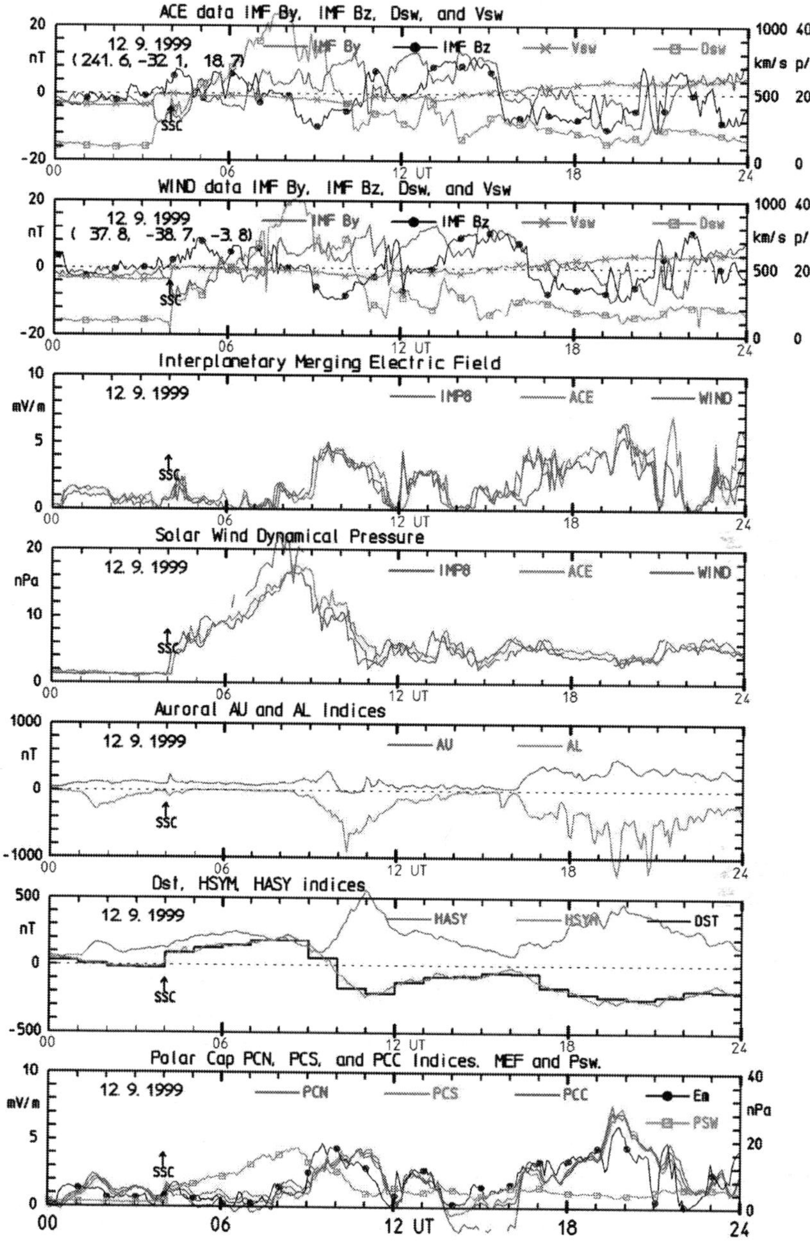

Fig. 13. Example plot of solar wind data from ACE, Wind, and IMP8, and indices for geophysical disturbances, viz. auroral electrojet, ring current, and polar cap currents.

Based on DMSP measurements, comprehensive studies by Boyle et al. (1997) suggest an empirical relation between solar wind parameters, B, V_{SW}, and the IMF clock angle θ, and the cross polar cap potential drop, Φ_{PC}, of the form:

$$\Phi_{PC} = 10^{-4}\, V_{SW}^2 + 11.7\, B \sin^3(\theta/2) \text{ [kV]} \tag{11}$$

Unfortunately, most of the observations relate to quiet or moderately disturbed conditions only. Thus, there were few measurements of Φ_{PC} beyond 100 kV. Hence, the expression does not provide the saturation effects observed during strongly disturbed conditions.

Further developments of empirical relations by Siscoe et al. (2002) provide relations where the height-integrated Pedersen ionospheric conductivities, assumed to be uniform across the polar cap, are included. A problem here is whether to use northern and southern hemisphere conductivities for northern and southern passes, respectively, or to use composite conductivities since the two polar caps are electrically coupled.

Hairston et al. (2005) studied specifically the saturation effects seen during strong magnetic storms. They observed that the cross polar cap potential rarely exceeded 200 kV and only in exceptional cases reached up to 250 kV. The "turn-over" from the non-saturable relation (Eq. 11) from Boyle et al. (1997) and the saturated range would occur at solar wind merging electric field (E_M) values between 4 and 8 mV/m.

This is also the range of E_M values where the polar cap indices start to show substantial saturation effects. Interestingly, the potential data presented in Fig. 10 of Hairston et al. (2005) could well be represented by an expression similar to the above Eq. 10. Their Fig. 10 displays the cross polar cap potential drop in a scatter plot (50 points) against the merging electric field in the solar wind ranging from 0 to 40 mV/m. On basis of these data and the PCC values depicted in Fig. 7 and Eq. 10 here, the relation between the polar cap index, PCC, and the potential would read:

$$\Phi_{PC} \approx 20 \text{ PCC} + 15 \text{ [kV]} \tag{12}$$

The relations between polar cap index and cross polar cap diameter and voltage was investigated by Troshichev et al. (1996). They conclude that the cross polar cap voltage and radius could be expressed by:

$$\Phi_{PC} \approx 19.35 \text{ PC} + 8.78 \text{ [kV]} \tag{13}$$

$$R(\text{morning}) = -.12 \text{ PC}^2 + 2.5 \text{ PC} + 11.0 \text{ [deg]} \tag{14a}$$

$$R(\text{evening}) = -.20 \text{ PC}^2 + 3.0 \text{ PC} + 12.5 \text{ [deg]} \tag{14b}$$

Their expression for Φ_{PC} (Eq. 13) is very close to the above expression (Eq. 12) deduced from the observations reported by Hairston et al. (2005). Their expressions shown here in Eqs. 14a,b for the polar cap radius (angular distance from the pole) show the non-linear enlarging of the open polar cap with increasing values of the PC index.

In the analyses by Troshichev et al. (2000) they derive a non-linear expression for the average ionospheric electric field in polar regions:

$$E_{PC} = 9.29 + 3.76 \text{ PC} - 0.11 \text{ PC}^2 \text{ [mV/m]} \tag{15}$$

For the cross polar cap potential difference, the nonlinear quadratic reduction of the electric field strength is counteracted by the expansion of the polar cap radius. For a moderately strong event the polar cap index could typically have a magnitude PC = 5. In that case the radius $R \approx 21.5°$ and electric field $E \approx 25.3$ mV/m would give a cross polar cap voltage $\Phi_{PC} \approx$ 121 kV, not so different from the results from Eq. 12 ($\Phi_{PC} \approx 115$ kV) or Eq. 13 ($\Phi_{PC} \approx 106$ kV).

Based on SuperDARN radar measurements Fiori et al. (2009) derived a similar relation between Φ_{PC} and the PCN index:

$$\Phi_{PC} \approx 35 + 12\ \mathrm{PCN}\ [\mathrm{kV}] \tag{16}$$

Their values of the cross polar cap voltage are somewhat smaller than those derived from the above expressions, Eqs. 12 and 13. However, the linear (non-saturated) relation between the cross polar cap voltages and the PC indices is maintained.

In summary, the cross polar cap potentials are linearly related to the PC index values, notably to the PCC indices, which reflect the global situation more reliably than the hemispherical indices. The ionospheric electric fields in the polar cap display non-linear saturation effects also in relation to the polar cap index, but for the cross polar cap voltages these effects are to some degree counteracted by the expansion of the polar cap during strong events to result in a linear relation between Φ_{PC} and the PC index.

6.2 PC indices and the auroral electrojets

The Auroral Electrojet indices were introduced by Sugiura and Davis (1966) to measure the over-all electrojet activity in the auroral zone. Index values are based on magnetic data from a number of observatories, usually between 8 and 12, located in the auroral zone. AE index values are derived as the differences over the array of observatories between the upper envelope (AU) and the lower envelope (AL) for superposed variations in the horizontal (H) components. 1-min AU/AL/AE index values scaled in nT are derived and made available by WDC-C2 (Kyoto). AU index values are mostly related to the eastward electrojet in the post-noon sector. The AL index values are most often dominated by contributions from the westward auroral electrojet currents in the morning and midnight sectors. The westward electrojet currents are strongly intensified during substorm activity. Furthermore, substorms are effective in generating intense radiation of energetic auroral particles that, in turn, precipitate to produce increased upper atmospheric electron densities. The resulting increases in ionospheric conductivities in the auroral regions, generally, further enhance the electrojet currents.

Auroral activities (e.g., substorms) are linked to the effects of the solar wind on the magnetosphere. The primary parameter in the solar wind is the merging electric field, E_M, but the dynamic pressure, Psw, also plays a role mainly for the position and width of the electrojet regions. The polar cap magnetic activity characterized by the PC indices could be considered representative of the input power from the solar wind to the magnetospheric processes driving auroral activities. Thus the auroral and polar cap activity levels must be related.

The correlation coefficients between the PCN or PCS indices and the AE index were calculated by Troshichev et al. (2007) for each UT hour through 1998 to 2001. An example of the results is displayed in Fig. 14. The correlation is best at local winter where it would range between 0.80 and 0.85. During local summer the correlation is lower and ranges between 0.7 and 0.75. The results displayed in Fig. 14 also illustrate the solar cycle dependency. The correlation coefficients for PCN and PCS against AE are generally higher by 0.05-0.10 during solar minimum year (1998) than during solar maximum years (1999-2001).

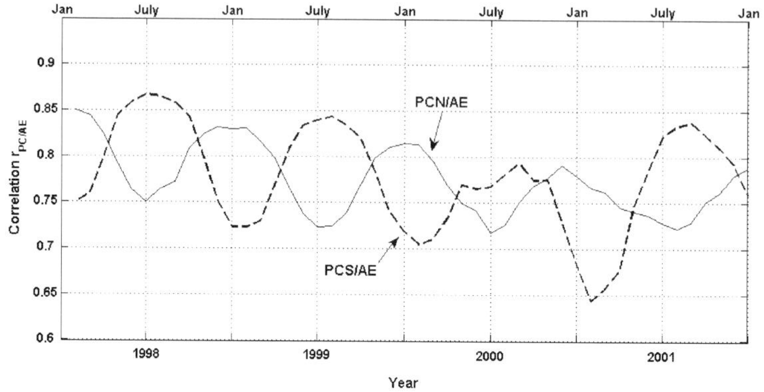

Fig. 14. Seasonal and solar cycle variations of the AE - PC correlations. (from Troshichev et al., 2007)

Another example from a recent correlation study is presented in Fig. 15. Here, values of the auroral electrojet index, AE, are plotted against corresponding values of the combined polar cap index, PCC, using a delay of 5 min. (PCC changes first). Note, that the correlation coefficient, R=0.816, for the full year is considerably higher than the above coefficients based on the individual PCN or PCS indices. Also note that the average AE indices show little indication of saturation (but larger spread) up to PCC levels of ~ 9.

The timing between sudden changes in the solar wind conditions and related changes in the polar cap and, for instance, the onset of auroral substorms as indicated by a sudden negative excursion in the AL index or a sudden increase in the AE index, was analyzed by Janzhura et al. (2007). They investigated isolated substorm events where the growth phase and the onset could be distinctly identified and related to the timing and magnitude of an increase in the PC indices caused by a gradual increase in the solar wind electric field, E_M. They found that when the E_M amplitude (and the PC index) remained below a level of 2 mV/m, then there was no clear indication of directly related substorm onsets. For E_M rising to amplitudes between 2 and 6 mV/m there would be a substorm onset following a growth phase where the PC and AE indices would gradually increase. The average growth phase duration varied systematically with the E_M amplitude from 60 min down to 0. For E_M amplitudes (and related PC levels) above 6 mV/m the substorm activity would immediately follow the E_M increase without any significant delay.

A relevant question is now whether the general activity level of auroral processes as such may influence the polar cap activity level above the direct effects of the solar wind parameters, notably the merging electric field, E_M. Figure 16 presents the relations between

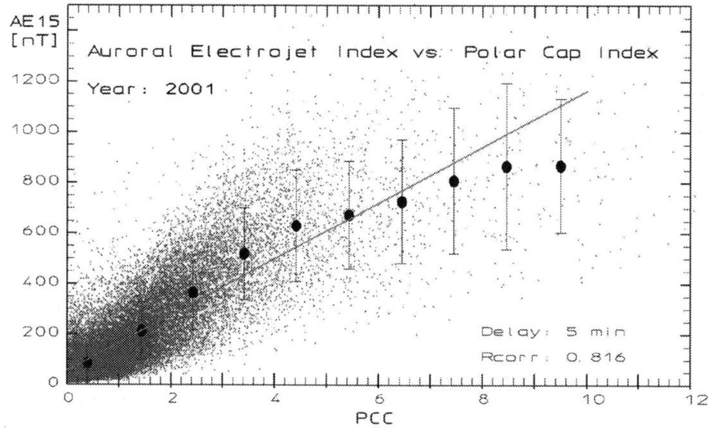

Fig. 15. Scatter plot between values of auroral electrojet indices, AE, and polar cap indices, PCC, through 2001.

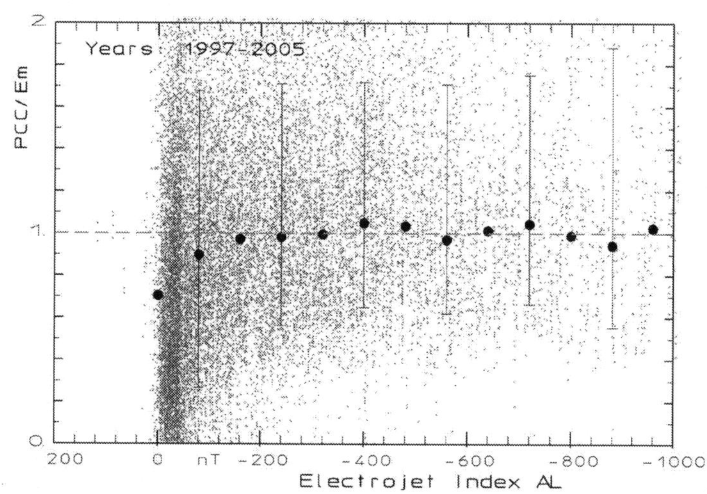

Fig. 16. Scatter plot of the ratio PCC/E_M versus auroral electrojet activity index AL

the ratio of PCC over E_M and the level of the (usually negative) auroral electrojet index, AL. The horizontal dashed line represents the ideal unity ratio between PCC and E_M. A noticeable feature is the variation from a relatively low value of PCC/E_M at around 0.7 for AL values close to zero to almost unity for AL equal to -150 nT. This range comprises quiet and weakly disturbed conditions. For magnetically disturbed conditions with still larger negative values of AL beyond -150 nT, there is hardly any variation in the average ratio of PCC over E_M. There is a substantial scatter in the distribution of values but the average ratio stays close to unity through the range from -150 up to -1000 nT in the auroral electrojet index, AL. Beyond that range the statistics is too sparse for conclusions. The variations in the ratio PCC/E_M at weak to moderate auroral activities (AL>-150 nT), no doubt, relate to the increase in conductivities through ionization by substorm-associated particle precipitation to reach a level adequate to support the ionospheric currents responsible for the PC indices.

A different question is whether the individual steps of the substorm cycle, comprising growth phase, onset, expansion phase, and recovery, affect the polar cap activity variations around the mean level predominantly defined by the solar wind electric field, E_M. In order to investigate this question we have "synchronized" PC index variations to the onset of substorms. The results are displayed in Fig. 17. The data refer to cases with steady solar wind conditions from at least 1 hour before the onset time until 2 hours after. Epoch time zero in the fields is the substorm onset time as defined by a sudden negative excursion in the AL index.

The top field displays the solar wind electric field, E_M, (referred to 12 Re reference position) through an interval spanning from 1 hour before to 3 hours after onset. On top of the individual traces the heavy line marks the average value. The next lower field displays the values of the auroral electrojet indices AU (upper traces) and AL (lower traces). The heavy lines mark the average values within each of the two index data sets. The sharp downward change in AL at epoch time zero is readily identified. The negative peak value in AL is reached at around 10-15 min after onset. The AU indices, notably their average values, hardly indicate any change at all. The next lower field displays the PCC index variation on the same time scale as the AU/AL display. A remarkable feature here is the peak at around 30-40 min after substorm onset, that is, at around 20-25 min after the negative peak in AL (or the corresponding positive peak in AE=AU-AL).

The bottom field repeats the average E_M variations (red line) and average PCC variations (blue line) while the heavy line displays their ratio, PCC/E_M, using the scale to the right. It should be noted that the PCC/E_M ratio has a marked minimum value of approximately 0.7 at around epoch time -30 to -20 min and a maximum value of around 1.3 at epoch time 35-40 min after onset. At 120 min there is another minimum at a value close to 0.8. Thereafter the PCC/ E_M ratio approaches unity.

Using substorm terminology for the PCC/E_M variations, there is clearly a "pre-onset" or "build-up" phase with minimum PCC/E_M ratio prior to substorm onset. There is no steep onset but a gradual "expansion" phase with increasing ratio leading up to a peak value at 35-40 min after substorm onset. Then follows a "recovery" phase with gradual decrease in the PCC/E_M ratio. In the recovery phase the ratio might have reached the low pre-onset level (0.7) at around 120 min after onset had there not been some new events at this time to raise the average PCC/E_M level. These new events are spotted in the PCC plots (second field from bottom). Correspondingly, the pre-onset level is probably influenced to some degree by the increased level experienced during the recovery phase following the preceding event.

In summary of the above analyses the PC index provides a fair indicator of subsequent substorm activity. For quiet conditions with PC<~2 units there are hardly any substorm onsets. The auroral electrojet activity indices AE and AL track possible variations in the PC index. If the PC index increases from quiet to more disturbed conditions then substorm onsets are likely to occur as the PC index increases beyond ~2 units with a delay depending on the actual magnitude of the PC index, typically a few tens of minutes. If the PC index takes values above ~5 units then substorm activity follows immediately. There is some

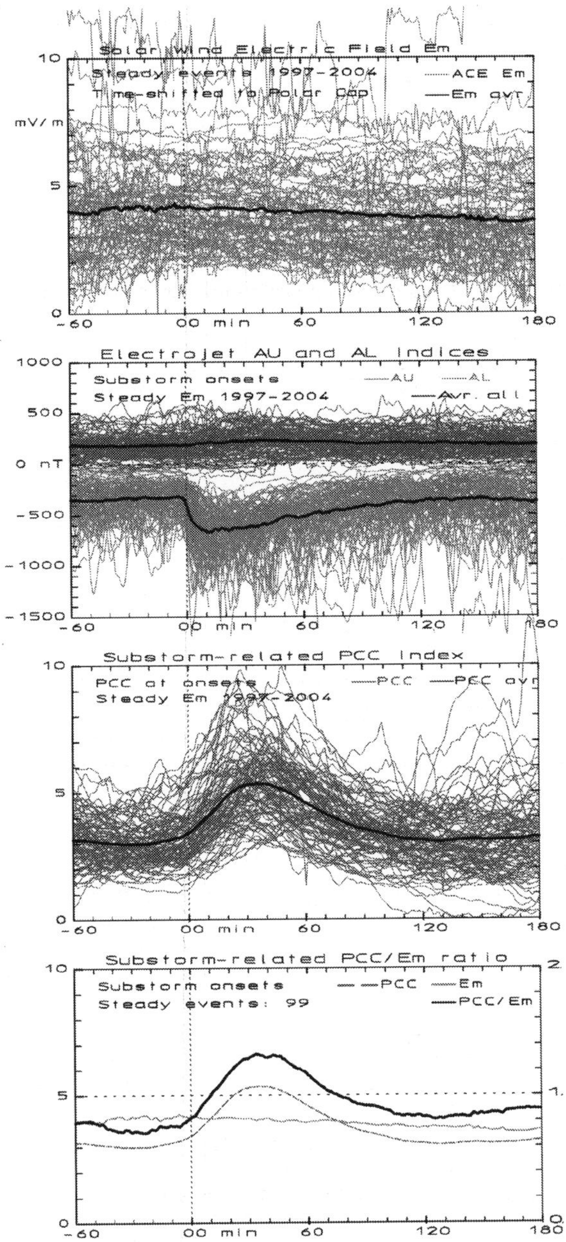

Fig. 17. Solar wind electric fields (top), AL and AU indices, PCC indices, and averages of Em and PCC and their ratio PCC/Em (bottom) for cases with steady interplanetary conditions.

coupling from auroral activity to the amplitude of the polar cap indices during weakly disturbed conditions probably related to the changes in ionospheric conductivities resulting from precipitation of substorm-associated auroral particles. The overall correlation between auroral activity represented by the AE indices and the polar cap indices, PCN or PCS, ranges between 0.60 and 0.85 depending on seasonal and solar cycle conditions. Using the combined PCC index enhances the correlation to reach an average of around 0.82.

6.3 Relations between PC indices and Joule and auroral particle heating

The thermosphere, particularly in the auroral regions, is heated by the amount of energy deposited by Joule heating and by precipitating particles. The heating generates thermospheric vertical motions and composition changes. The associated expansion of the atmosphere could have grave consequences for the tracking of satellites and for other spacecraft operations through the varying drag on space objects.

In various models to calculate drag coefficients, parameters such as the F10.7 cm flux (proxy for solar EUV radiation), the planetary Kp index, or the auroral electrojet index have been used to derive Joule heating values (e.g., Tobiska et al., 2008). Further attempts to calculate Joule heating include estimates of Joule heating in response to IMF input based on large-scale modelling of electrodynamic parameters (e.g., Chun et al., 1999, 2002; Weimer, 2005; Weimer et al., 2011).

Attempts to relate Joule heating to the polar cap index were reported by Chun et al. (1999, 2002). Basically they have derived hemispheric Joule heat production rates from the "Assimilative Mapping of Ionospheric Electrodynamics" (AMIE) procedure through a number of well documented storms. The AMIE procedure uses data from a large number of magnetometers, coherent and incoherent radars, digisondes and satellites with instruments for detection of electron precipitation and electric fields, to produce global patterns of various electrodynamic parameters. AMIE calculates Joule heating rate (jh) as shown in Eq. 17:

$$jh = \Sigma_P E^2 \tag{17}$$

where Σ_P is the ionospheric height-integrated Pedersen conductivity and E the electric field. The Joule heating may be integrated over either hemisphere or globally to produce the total Joule heating power. From the study by Chun et al. (1999), the total Joule heating power for the northern hemisphere (JH) was estimated for each season separately and compared to the corresponding values of the polar cap index, PCN. Their results are shown in Eqs. 18a-d:

$$\text{Winter:} \quad JH = 4.84\ PCN^2 + 16.9\ PCN + 5.6 \quad [GW] \tag{18a}$$

$$\text{Equinox:} \quad JH = 4.14\ PCN^2 + 25\ PCN + 8.9 \quad [GW] \tag{18b}$$

$$\text{Summer:} \quad JH = 14.39\ PCN^2 + 23.7\ PCN + 11.5 \quad [GW] \tag{18c}$$

$$\text{All data:} \quad JH = 4.03\ PCN^2 + 27.3\ PCN + 7.7 \quad [GW] \tag{18d}$$

Chun et al. (1999) noted that the expressions are valid also for negative values of the PCN index, which may occur during reverse convection cases, particularly during the summer season. Further, they compared the results of the above modelling (Eqs. 18a-d) with the simpler models based on either the cross polar cap voltage, Φ_{PC}, or the Auroral electrojet indices, AE, and found them about equally representative for the AMIE-derived Joule heating rates. This result is not so surprising since the cross polar cap potentials as well as the auroral electrojet indices are closely correlated with the polar cap indices as demonstrated in the previous sections.

From Thule PCN data and NOAA SPW Hemispheric Auroral Power (AP) data based on measurements from NOAA POES satellites through the years 1999-2002, and selecting only the northern passes, we have prepared the scatter plot displayed in Fig. 18. The heavy dots mark averages of AP values (in GW) through every unit of the PCN index. From Fig. 21 one may note the minimum in average AP around 10 GW at PCN = 0. The AP values are increasing both for increasing negative and positive PCN values but there is a marked difference between the slopes for positive and negative index values. The slopes for negative index values, particularly for the summer data, are much lower than those for the positive index data. For positive index values the average variations could be described by Eqs. 19a-c:

$$\text{Winter:} \quad AP = 13.5\ PCN + 10 \quad [GW] \tag{19a}$$

$$\text{Equinox:} \quad AP = 13.5\ PCN + 10 \quad [GW] \tag{19b}$$

$$\text{Summer:} \quad AP = 11.5\ PCN + 10 \quad [GW] \tag{19c}$$

These results are not so different from those reported by Chun et al. (2002) taking into account that they did not discriminate against negative PCN index values, which they - looking at their data (their Fig. 9) - might well have done.

Summarizing from the above analyses, there appears to be quite consistent relations between the PC indices and the heating of the thermosphere by Joule heating as well as by

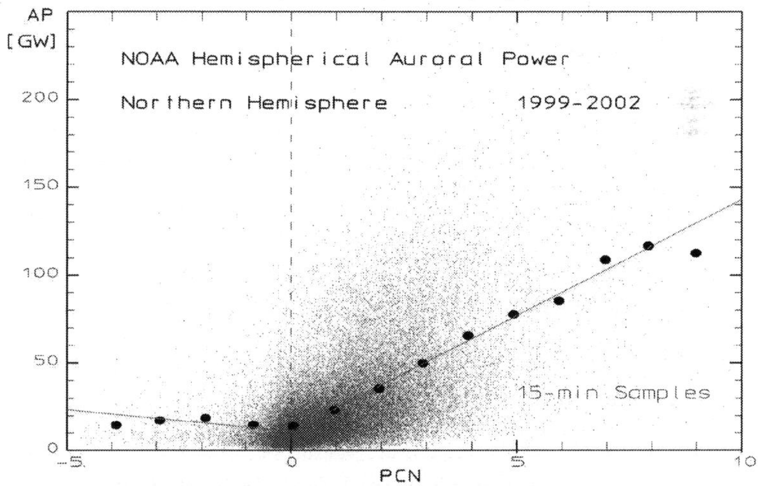

Fig. 18. Scatter plot of Northern Hemispheric Auroral Power data vs. PCN index values.

auroral particle precipitation. There are marked differences between cases with positive (forward convection) and negative (reverse convection) PC index values. Such differences are probably caused by the much reduced size of the active auroral and polar cap regions for the northward IMF Bz cases that causes the major part of the negative PCN values.

6.4 Relations between PC indices and the ring current

The Earth is encircled by currents flowing near equator at distances of typically 4-6 Re. These currents could be divided into the symmetrical part (RCS) that is formed all the way around the Earth, mostly by drifting mirroring energetic electrons and ions, and the partial ring current (RCP). Referring to Fig. 1, field-aligned Region 1 (R1) currents flow from the dawn magnetospheric boundary regions to the ionosphere at the morning side. Part of this current flows equatorward across the auroral region to feed the upward Region 2 (R2) field aligned currents (FACs) emerging from lower latitudes. The R2 FACs flow to the ring current region and add to the ring current at the night side to form a partial ring current. From the evening side of the ring current region, the R2 FACs flow to the ionosphere; part of these currents flows poleward crossing the auroral region and add to the R1 FACs leaving the ionosphere to end at the dusk boundary regions. It is also possible that a substantial part of the partial ring current is built from electrons and ions that simply drift directly between the magnetospheric dawn and dusk boundary regions across the near-tail region.

The ring current intensities are detected from a network of low-latitude magnetometer stations whose data are sent to World Data Centre WDC-C2 in Kyoto and processed to provide indices for the symmetrical as well as the partial ring currents (Sugiura & Kamei, 1981). The hourly symmetrical deflections scaled from the horizontal (H) components give the Dst index. The corresponding symmetrical index scaled from 1-min values of the H components provides the SYM-H index while the corresponding scaling of 1-min values of the D components generates the SYM-D index. Similarly the asymmetrical parts of the 1-min H and D components generate the ASY-H and ASY-D indices.

The asymmetric ring current index, ASY-H, have been provided by WDC-C2 (Iyemori et al., 2000) as 1-min values. For the present statistical study a less detailed time resolution, which reduces the inherent scatter in the data, is considered appropriate. Hence, the ASY-H and the polar cap indices, PCC, have been averaged to form 5-min samples. The two parameter sets have been subjected to linear correlation analysis using a stepwise variable delay between samples of the respective time series assuming that the most appropriate delay gives maximum value of the correlation coefficient. With this delay imposed on all pairs of samples of the time series, a linear relation between two parameter sets is found by least squares regression analysis. The average deviation, S0, the numerical deviation, S1, and the RMS standard deviation, S2, are calculated from the assumed linear relation.

Figure 19 displays ASY-H against PCC. The 15 min delay noted in the figure was found to provide optimum correlation (Rx=0.743). The regression line plotted in the diagram characterizes ASY-H against PCC. For this relation, the scatter parameters, S0, S1, and S2, are listed in the figure. The standard deviation (S2) is ~18 nT. A noteworthy feature in the display is the persistent linear relations of the average ASY-H values on PCC (Eq. 20) up to high levels

$$\text{ASY-H} = 12.1\ \text{PCC} + 11.5\ [\text{nT}] \qquad (20)$$

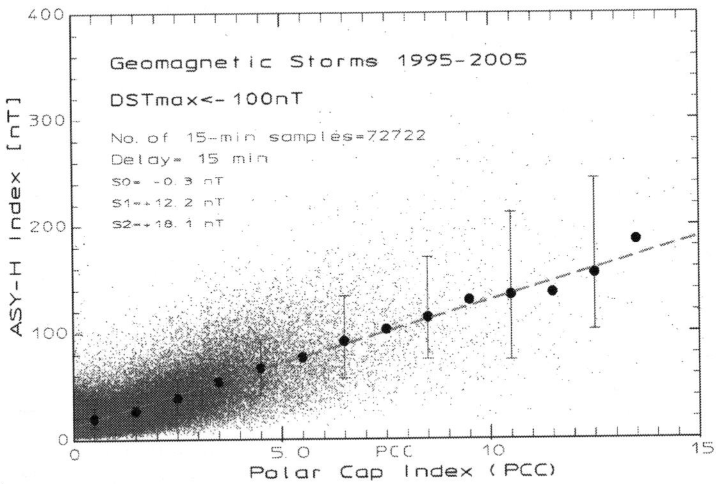

Fig. 19. Scatter plot of ASY-H partial ring current index vs. polar cap PCC index.

The Dst ring current index is closely related to the presence of energetic electrons and ions, notably H^+ and O^+, in the inner magnetosphere. Their different drift motion, positive ions westward and electrons eastward, in the geomagnetic dipole-like field contributes a net westward electric current. A succesful relation between the accumulated kinetic energy of the charged particles and the Dst* index is provided by the Dessler-Parker-Sckopke relation (Dessler & Parker, 1959; Sckopke, 1966) in Eq. 21:

$$Dst^*[nT] = 4.0 \cdot 10^{-30} E_{RC}[keV] \tag{21}$$

where Dst* is the Dst index corrected for the contributions from magnetopause currents (MPC) while E_{RC} is the total kinetic energy of particles trapped in the ring current region.

Thus, the Dst index represents the energy stored in the ring current. Hence the merging electric field, E_M, or, equivalently, the polar cap PC index should be considered to represent a source function for the Dst index rather than being related to its current value. Following Burton et al. (1975) the change in the Dst index with time could be written:

$$dDst^*/dt = Q - Dst^*/\tau \; [nT/hr] \tag{22}$$

where Q in Eq. 22 is the source term while the last term is the ring current loss function controlled by the decay time constant τ here measured in hours. For the small actual MPC corrections, the Dst dependent statistical values provided in Joergensen et al. (2004) are used while the decay function given in Feldstein et al., (1984) is used for the loss term. This function uses two decay time constants, τ = 5.8 hrs for large disturbances where Dst < -55 nT, and τ = 8.2 hrs for small disturbances where Dst > -55 nT. Now, the relation in Eq. 22 has only terms relating to the source function Q and to the observed initial Dst index values.

In Burton et al. (1975) the source term Q is related to the Y_{GSM} component of the solar wind electric field. In the analysis by Stauning (2007), in addition to the dependence on E_M, the relation of Q to the polar cap index PCC was examined for a number of storm event cases during the interval 1995-2002. Here we repeat these analyses using selected large storm events through 1995-2005. In all these cases we first derive the temporal change at time t=T in the hourly Dst* index from the hourly values at t=T-1 and t=T+1 [hrs] by the simple differential term:

$$dDst^*/dt\ (T) = (Dst^*(T+1) - Dst^*(T-1))/2.\ [nT/hr] \tag{23}$$

In order to derive the source function, Q, to be used in Eq. 22, the average slope values defined by Eq. 23 are corrected by adding the decay term defined above using the current Dst* value at t=T. The resulting source function, Q_{obs}, is then related to the potential source parameters with a variable positive or negative delay imposed on the relation. The parameters, the PC indices, are provided at a more detailed time resolution (typically 1-min) than the hourly source function values. By shifting the averaging interval by delays varying on minute scale, hourly averages of the parameters are correlated with the hourly source function values to derive the delay that produces the maximum correlation through the ensemble of storm events. With this delay the best fit linear relation between the source function values and the relevant source parameter is determined by linear regression analysis.

The scatter plots in Fig. 20 present the Dst* source function, Q_{obs} , based on observed hourly Dst values plotted against the combined polar cap index, PCC, as a potential source parameter. The relation between the best fit "equivalent" source function, Q_{eq}, and the source parameter values, PCC, is then expressed in a linear function:

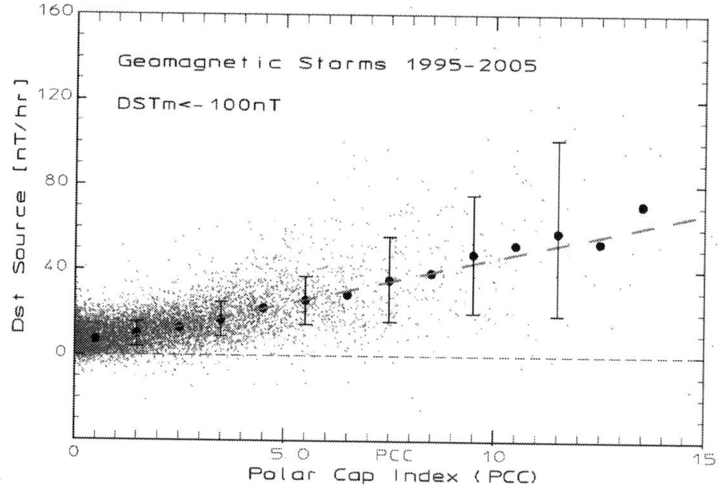

Fig. 20. Scatter plot of Dst* source function vs. polar cap PCC index.

$$Q_{eq} = 4.6\ PCC + 1.2\ [nT/h] \tag{24}$$

with standard deviation equal to 9.4 nT/h.

With continuous time series of source parameter values, specification of the relational constants appearing in Eq. 24, and initial Dst values, it is now possible to integrate Eq. 22 to derive values of an "equivalent" Dst index, Dst_{eq}, through any interval of time. The procedure has been applied to the major geomagnetic storms (Dst<-100 nT) occurring through cycle 23. For each storm interval the calculations have been carried through intervals of 4 days. For each interval the equivalent Dst series was given an initial value equal to the observed Dst at the start of the first day, while all later Dst_{eq} values through the 4 days were derived solely from the integration of Eq. 22 without attachment to observed Dst values.

Figure 21 displays an example for the case of a moderate geomagnetic storm on 7 - 10 April 1995. The peak (negative) Dst value is -150 nT observed at 1800 UT on 7.4.1995. In the figure the observed Dst values are displayed in black line with "dots" and located in the lower part of the diagram. The merging electric field, E_M, is displayed in thin green line in the upper half of the diagram. By its definition E_M is non-negative. Values of the PCC index (also non-negative) are displayed in thin blue line with small squares while values of the asymmetric ring current index, ASY-H, are displayed in red line with crosses mainly in the upper part. Values of the equivalent Dst index calculated from the PCC indices are shown in blue line with squares.

The merging electric field, Em, and the asymmetric index, ASY-H have also been used as source functions similarly to the above derivation of the Dst indices using the PCC indices. Values of the equivalent Dst index calculated from E_m are displayed in the thin green line in

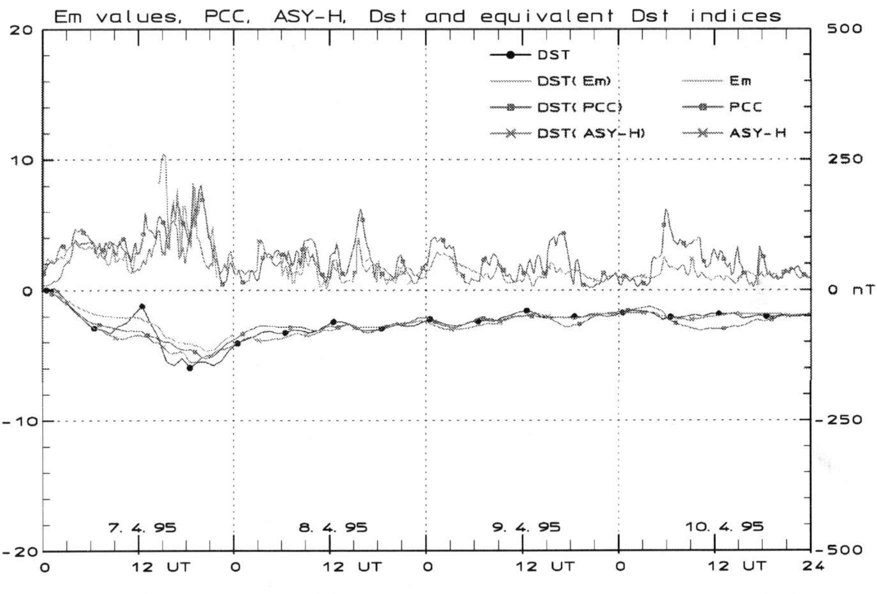

Fig. 21. Display of parameters E_M, PCC, ASY-H (in upper half of fields), and observed ring current index, Dst, and derived ring current indices $Dst_{eq}(E_M)$, Dst_{eq} (PCC), and Dst_{eq} (ASY-H)

the lower part. In this case the series of equivalent Dst values based on E_m terminates after the first day since there is a break in the E_M data at midnight between 7 and 8 April 1995. The equivalent Dst indices using ASY-H in the source function are displayed in red line with crosses through all 4 days. In this case the PCC and the ASY-H index data are continuously available for the integration of Eq. 22 throughout the storm interval.

It is clear from Fig. 21 that there is good agreement between the observed and the equivalent Dst values throughout the 4 days of integration. A similar case is displayed in Fig. 22. The PCS index was not available for the storm. Hence, here the calculations were based solely on the PCN index series. Based on the peak values of the Dst index, this case illustrates one of the strongest geomagnetic storm events of cycle 23. Generally, going from the moderate to the strong storm cases gives less agreement between observed and equivalent Dst values. There is no clear indication in these plots whether the equivalent Dst values based on E_M are better or worse than those based on PCC, PCN, AL or ASY-H.

In order to resolve the question whether using E_M, PCC, PCN, AL or ASY-H in the source function give better agreement the average differences and variances between the observed and the equivalent Dst values were calculated for all cases studied. The results are listed in Table 3. It should be noted that the data sets forming the basis for Dst_{eq} calculations are not completely identical since the E_M data are often interrupted temporarily or non-existing for the storm cases. The reason for the lack of E_M data is often the disabling of the solar wind

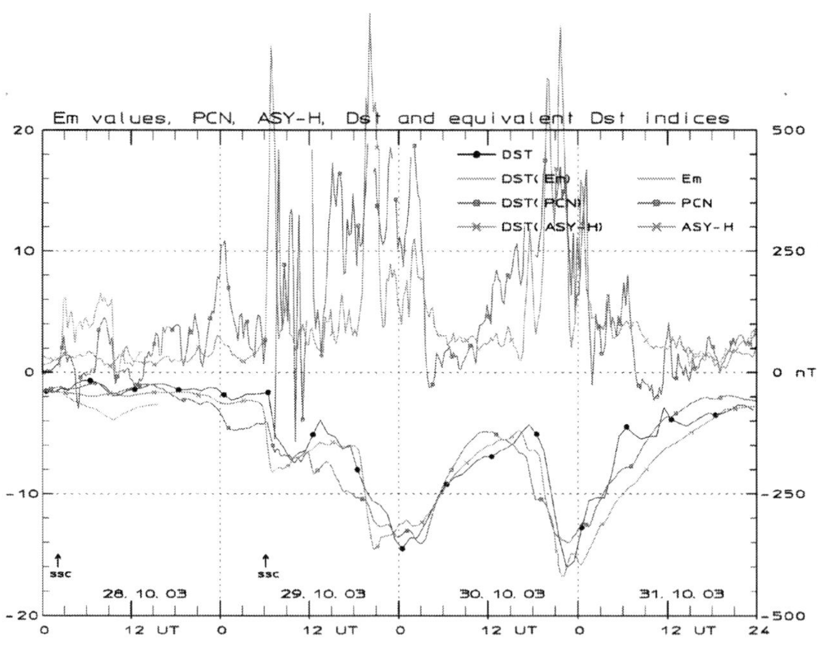

Fig. 22. Observed Dst and equivalent Dst values calculated from PCN, E_M and ASY-H, respectively, through a major magnetic storm of cycle 23.

detectors by strong high-energy solar proton fluxes. Hence the equivalent Dst values based on E_M data are often available during the first part of the storm interval only and the total number of samples is considerably reduced (cf. Table 3). Since the integration extending through 4 days becomes less and less certain with increasing interval length this preference for the first days may give the E_M-based Dst_{eq} values an advantage in the correlation with Dst over the PCC, PCN, AL or ASY-H-based values.

Parameter\Source	E_M field	PCC index	PCN index	AL index	ASY-H index
No. samples	3992	7776	8640	7968	8640
Avr. deviation	-7.6	1.8	2.7	11.5	- 4.9 nT
Std. deviation	27.6	30.6	31.4	34.2	24.7 nT

Table 3. Equivalent Dst versus observed Dst index through major geomagnetic storms 1995-2005

The best source function when comparing the average deviation between the observed Dst and the calculated equivalent values is provided by the PCC index series. Best to reduce the variance (standard deviation) is the ASY-H index followed by the solar wind merging electric field, E_M. The PCC index series gives larger variance than E_M but is continuously available in twice the number of cases. PCN based on Thule data are virtually available continuously. Thus, the PC indices, particularly the PCC index series, may serve to calculate the ring current indices, Dst, SYM-H, and ASY-H, in real time if the magnetic data from Thule and eventually Vostok are available on-line.

7. Conclusions

The polar caps provide terrestrial "windows" to the solar wind and enable monitoring of the merging electric field, E_M, the most important parameter for transfer of energy and momentum from the solar wind into the Earth's environment. The intruding solar wind energy is used to power a range of geophysical processes such as the large-scale electric potential patterns and the associated convection and current systems in the high-latitude magnetospheric and ionospheric regimes. The intensities of the transpolar ionospheric currents are very well suited to monitor the solar wind electric field. The associated magnetic disturbances at ground, when properly scaled to compensate for the regular daily and seasonal variations in conductivities, provide the polar cap index, PC, which can serve as a proxy for the merging electric field.

In addition, the PC indices, in particular when available from both polar caps, may serve to indicate the level of further geophysical disturbances such as polar ionospheric electric fields and plasma motions, auroral activities characterized, for instance, by the auroral electrojet indices (AL, AU, and AE), heating of the upper atmosphere by Joule dissipation (JH) and particle precipitation (AP), mid-latitude magnetic disturbances characterized by the Kp index, and the development of partial and symmetrical ring currents in the equatorial regions characterized by the Dst, the SYM-H, and the ASY-H indices. Such disturbances constitute an essential part of the conditions termed "Space Weather".

The polar cap indices are derived from a single station in each hemisphere, Thule in the northern polar cap and Vostok in the southern, both of which provide the basic magnetic data in high resolution and on-line in real time. These characteristics could be contrasted to those applicable to other indices used to characterize geophysical disturbances such as the auroral electrojet indices requiring data from 8-12 observatories to provide a reliable index, or the ring current indices based on 4-6 observatories. The estimation of Joule or particle heating power need either an extended network of magnetometer stations to run the AMIE procedure or polar satellite passes to derive the transpolar potential structure and the intensities of precipitating particles. Even in comparison with observations from interplanetary satellites, the monitoring of disturbances through the polar cap indices provides the advantage of not being affected by the intense high-energy particle radiation that often accompany the strong solar outburst, where the reliable monitoring of possible geophysical disturbances, that is, adverse space weather conditions, is particularly important.

8. Acknowledgments

The author is indebted to the observatory staffs for their dedicated efforts to provide magnetic data of utmost quality and continuity from the DMI geomagnetic observatory in Qaanaaq (Thule), Greenland, and the AARI geomagnetic observatory in Vostok at Antarctica. The author gratefully acknowledges the access to IMP8, Geotail, WIND and ACE satellite data on solar wind plasma and magnetic field observations provided through the NSSDC and the Caltech Research Laboratory data centres. We are also indepted to the NOAA SWP for supplying the NOAA POES satellite data. Furthermore, we gratefully acknowledge the access to Dst, SYM, ASY, and Auroral Electrojet indices from WDC-C2 at Kyoto University and express our deep appreciation of the efforts invested in the collection and processing of the geomagnetic data.

9. References

Akasofu, S.-I. (1979). Interplanetary energy flux associated with magnetospheric substorms, doi:10.1016/0032-633(79)90119-3, *Planet. Space Sci.*, 27, 425.

Araki, T. (1994). A physical model of geomagnetic sudden commencement, in: *Solar Wind sources of Magnetospheric Ultra-Low-Frequency Waves*, Geophys. Monogr. 81, eds.: M.J. Engebretson et al., pp. 183-200, AGU, Washington, D.C.

Borovsky, J. E., Layraud, B. & Kuznetsova, M. M. (2009). Polar cap potential saturation, dayside reconnection, and changes to the magnetosphere, *J. Geophys. Res.*, 114, A03224, doi:10.1029/2009JA014058.

Boyle, C. B., Reiff, P. H. & Hairston, M. R. (1997). Empirical polar cap potentials, *J. Geophys. Res.*, 102, 111.

Burton, R. K., McPherron, R. L., & Russell C. T. (1975). An empirical relationship between interplanetary conditions and Dst, *J. Geophys. Res.*, *80*, 4204-4214.

Chun, F. K., Knipp,D. J., McHarg, M. G., Lu, G., Emery, B. A., Vennerstrøm, S. & Troshichev, O. A. (1999). Polar cap index as a proxy for hemispheric Joule heating, *Geophys. Res. Lett.*, 26 , 1101-1104.

Chun, F.K., Knipp, D. J., McHarg, M. G., Lacey, J. R., Lu, G., & Emery, B. A. (2002). Joule heating patterns as a function of polar cap index, *J. Geophys. Res.*, *107* (A7), doi:10.1029/2001JA000246.

Davis, T. N. & Sugiura, M. (1966). Auroral electrojet activity index AE and its Universal Time variations, *J. Geophys. Res.* 71, 785.

Dessler, A. J. & Parker, E. N. (1959). Hydromagnetic theory of geomagnetic storms, *J. Geophys. Res.*, *64*, 2239-2259.

Fairfield, D. H. (1968). Polar magnetic disturbances and the interplanetary magnetic field, *COSPAR Space Research VIII*, 107.

Farris, M.H. & Russell, C. T. (1994). Determining the standoff distance of the bow shock: Mach number dependence and use of models, *J. Geophys. Res.*, 99, 17681-17689.

Feldstein, Y. I., Pisarsky, V. Yu, Rudneva, N. M. & Grafe, A. (1984). Ring current simulation in connection with interplanetary space conditions, *Planet. Space Sci*, *32*, 975-984.

Fiori, R. A. D., Koustov, A. V., Boteler, D. & Makarevich, R. A. (2009). PCN magnetic index and average convection velocity in the polar cap inferred from SuperDARN radar measurements, *J. Geophys. Res.* 114, A07225, doi:10.1029/2008JA013964.

Friis-Christensen, E. & Wilhjelm, J. (1975). Polar cap currents for different directions of the interplanetary magnetic field in the Y-Z plane, *J. Geophys. Res.*, 80, 1248-1260.

Fukushima, N. (1976). Generalized theorem for no ground magnetic effect of vertical currents connected with Pedersen currents in the uniform-conductivity ionosphere, *Rep. Ionos. Space Res. Jpn.*, 30, 35.

Greenwald, R. A., Bristow, W. A., Sofko, G. J., Senior, C., Ceriser, J.-C. & Szabo, A. (1995). SuperDual Auroral Radar Network radar imaging of dayside high-latitude convection under northward interplanetary magnetic field: Toward resolving the distorted two-cell versus multicell controversy, *J. Geophys. Res.*, 100, 19.661.

Hairston, M. R., Heelis, R. A., & Rich, F. J. (1998), Analysis of the ionospheric cross polar cap potential drop using DMSP data during the National Space Weather Program study period, *J. Geophys. Res.*, 103, 26,337.

Hairston, M. R., Drake, K. A., and Skoug, R. (2005). Saturation of the ionospheric polar cap potential during the October-November 2003 superstorms, *J. Geophys. Res.*, 110, A09826, doi:10.1029/2004JA010864.

Huang, C.-S. (2005). Variations of polar cap index in response to solar wind changes and magnetospheric substorms, *J. Geophys. Res.*, 110, A01203, doi:10.1029/2004JA10616.

Iijima, T. & Potemra, T. A. (1976a). The amplitude distribution of field-aligned currents at northern high latitudes observed by Triad, *J. Geophys. Res.*, 81, 2165-2174.

Iijima, T. & Potemra, T. A. (1976b). Field-aligned currents in the dayside cusp observed by Triad, *J. Geophys. Res.*, 81, 5971.

Iyemori, T., Araki, T., Kamei, T. & Takeda, M. (2000). Mid-latitude geomagnetic indices "ASY" and "SYM" for 1999, *Geomagnetic indices home page*, ed.: T. Iyemori, (http://swdcwww.kugi.kyoto-u.ac.jp/dstdir/index.html), WDC-C2 for Geomagnetism, Kyoto University.

Janzhura, A, Troshichev, O. A. & Stauning, P. (2007). Unified PC indices: Relation to isolated magnetic substorms, *J. Geophys. Res.*, 112, A09207, doi:10.1029/2006JA012132.

Jorgensen, A. M., Spence, H. E., Hughes, W. J. & Singer, H. J. (2004). A statistical study of the ring current, *J. Geophys. Res.*, *109*, A12204, doi:10.1029/2003JA010090.

Kan, J. R. & Lee, L. C. (1979). Energy coupling function and solar wind-magnetosphere dynamo, *Geophys. Res. Lett.*, *6*, 577.

Kuznetsov, B. M. and Troshichev, O. A. (1977). On the nature of polar cap magnetic activity during undisturbed periods, *Planet. Space Sci.*, 25, 15–21, 1977.

Lukianova, R., Troshichev, O. A. & Lu, G. (2002). The polar cap magnetic activity indices in the southern (PCS) and northern (PCN) polar caps: Consistency and disrepancy, Geophys. Res. Lett., 29(18), 1879, doi:10.1029/2002GL015179.

Lukianova, R. (2003). Magnetospheric response to sudden changes in the solar wind dynamic pressure inferred from polar cap index, *J. Geophys. Res., 108* (A12), 1428, SMP-11. doi:10.1029/2002JA009790.

Mansurov, S. M. (1969). A new evidence of connection between Space and Earth's magnetic fields, *Geomag. and Aeron.*, 9, 768.

Newell, P. T. & Meng, C.-I. (1994). Ionospheric projections of magnetospheric regions under low and high solar wind pressure conditions, *J. Geophys. Res., 99* (A1), 273-286.

Reiff, P. H., Spiro, R. W. & Hill, T. W. (1981). Dependence of polar cap potential drop on interplanetary parameters, *J. Geophys. Res.*, 86, 7639.

Rich, F. J., & Hairston, M. (1994). Large-scale convection patters observed by DMSP, *J. Geophys. Res.*, 99, 3827.

Sckopke, N. (1966). A general relation between the energy of trapped particles and the disturbance field near the Earth, *J. Geophys. Res., 71*, 3125-3130.

Shue, J.-H., Chao, J. K., Fu, H. C., Russell, C. T., Song, P., Khurana, K. K. & Singer, H. J. (1997). A new functional form to study the solar wind control of the magnetopause size and shape, *J. Geophys. Res.*, 102, 9497-9511.

Siscoe, G. L., Crooker, N. U. & Siebert, K. D. (2002). Transpolar potential saturation: Roles of region 1 current system and solar wind ram pressure, *J. Geophys. Res.* 107 (A10), 1321, doi:10.1029/2001JA009176.

Sonnerup, B. (1974). Magnetopause reconnection rate, *J. Geophys. Res.*, 79, 1546-1549.

Spreiter, J. R., Summers, A. L. & Alksne, A. Y. (1966). Hydromagnetic flow around the magnetosphere, *Planet. Space Sci., 14*, 223-253.

Stauning, P., Troshichev, O. A. & Janzhura, A. (2006). Polar Cap (PC) index. Unified PC-N (North) index procedures and quality. *DMI Scientific Report, SR-06-04.* (available at www.dmi.dk/publications/sr06-04.pdf).

Stauning, P., (2007). A new index for the interplanetary merging electric field and geomagnetic activity: Application of the unified polar cap indices, *Space Weather, 5*, S09001, doi:10.1029/2007SW000311.

Stauning, P. & Troshichev, O. A. (2008). Polar cap convection and PC index during sudden changes in solar wind dynamic pressure, *J. Geophys. Res.*, 113, doi:10.1029/2007JA012783.

Stauning, P., Troshichev, O. A. & Janzhura, A. (2008). The Polar Cap (PC) index:. Relations to solar wind parameters and global activity level, *J. Atmos. Solar-Terr. Phys.*, doi:10.1016/j.jastp.2008.09.028.

Stauning, P. (2011), Determination of the quiet daily geomagnetic variations for polar regions, *J. Atmos. Solar-Terr. Phys.*, doi:10.1016/j.jastp.2011.07.004.

Sugiura, M. & Davis, T. N. (1966). Auroral electrojet activity index AE and its universal time variations, *J. Geophys. Res., 71*, 785-801.

Sugiura, M. & Kamei, T. (1981). Description of the hourly Dst index, in: *Geomagnetic indices home page*, ed.: T. Iyemori, (http://swdcwww.kugi.kyoto-u.ac.jp/dstdir/index.html), WDC-C2 for Geomagnetism, Kyoto University.

Svalgaard, L. (1968). Sector structure of the interplanetary magnetic field and daily variation of the geomagnetic field at high-latitudes, *Scientific Report, R-8,* Danish Meteorological Institute.

Svalgaard, L., (1972). Interplanetary magnetic sector structure 1926-1971, *J. Geophys. Res.*, 77, 4027.

Svalgaard, L., (1973). Polar cap magnetic variations and their relationship with the interplanetary magnetic sector structure, *J. Geophys. Res.*, 78, 2064-2078.

Tobiska, W. K., Bouwer, S. D. & Bowman, B. R. (2008). The development of new solar indices for use in thermospheric density modelling, *J. Atmos. Solar-Terr. Phys.*, 70, 803-819.

Troshichev, O. A., Dmitrieva, N. P. & Kuznetsov, B. M. (1979). Polar Cap magnetic activity as a signature of substorm development, *Planet. Space Sci.*, 27, 217.

Troshichev, O. A. & Andrezen, V. G. (1985). The relationship between interplanetary quantities and magnetic activity in the southern polar cap, *Planet. Space Sci.*, 33, 415.

Troshichev, O. A., Kotikov, A. L., Bolotinskaya, B. D., Andrezen, V. G. (1986). Influence of IMF azimuthal component on magnetospheric substorm dynamics, *J. Geomagn. Geoelectr.*, 38, 1075.

Troshichev, O. A., Andrezen, V. G., Vennerstrøm, S. & Friis-Christensen, E. (1988). Magnetic activity in the polar cap - A new index, *Planet. Space Sci., 36,* 1095.

Troshichev, O., Hayakawa, H., Matsuoka, A., Mukai, T. & Tsuruda, K. (1996). Cross polar cap diameter and voltage as a function of PC index and interplanetary quantities, *J. Geophys. Res., 101,* 13,429.

Troshichev, A. O., Lukianova, R. Y., Papitashvili, V. O., Rich, F. J. & Rasmussen, O. (2000). Polar Cap index (PC) as a proxy for ionospheric electric field in the near-pole region, *Geophys. Res. Lett.*, 27, 3809.

Troshichev, O. A., Janzhura, A. & Stauning, P. (2006). Unified PCN and PCS indices: method of calculation, physical sense and dependence on the IMF azimuthal and northward components, *J. Geophys. Res., 111,* A05208, doi:10.1029/2005JA011402.

Troshichev, O. A., Janzhura, A. & Stauning, P. (2007). Magnetic activity in the polar caps: Relation to sudden changes in the solar wind dynamic pressure, *J. Geophys. Res.*, 112, A11202, doi10.1029/2007JA012369.

Vennerstrøm, S. (1991). The geomagnetic activity index PC, PhD Thesis, *Scientific Report 91-3,* Danish Meteorological Institute, 105 pp.

Vennerstrøm, S., Friis-Christensen, E., Troshichev, O. A. & Andrezen, V. G. (1991). Comparison between the polar cap index PC and the auroral electrojet indices AE, AL and AU, *J. Geophys. Res., 96,* 101.

Weimer, D. R. (2005). Improved ionospheric electrodynamics models and application to calculating Joule heating rates, *J. Geophys. Res.*, 110, A05306, doi:10.1029/2004JA010884.

Weimer, D. R. & King, J. H. (2008). Improved calculations of interplanetary magnetic field phase front angles and propagation time delays, *J. Geophys. Res.*, 113, A01105, doi:10.1029/2007JA012452.

Weimer, D. R., Clauer, C. R., Engebretson, M. J., Hansen, T. L., Gleisner, H., Mann, I. & Yumoto, K. (2010). Statistical maps of geomagnetic perturbations as a function of the interplanetary magnetic field, *J. Geophys. Res.*, 115, A10320, doi:10.1029/2010JA015540.

Weimer, D. R., Bowman, B. R., Sutton, E. K. & Tobiska, W. K. (2011). Predicting global average thermospheric temperature changes resulting from auroral heating, *J. Geophys. Res.*, 116, A01312, doi:10.1029/2010JA015685.

Wilhjelm, J., Friis-Christensen, E., Potemra, T. A. (1972). The relationship between ionospheric and field-aligned currents in the dayside cusp, *J. Geophys. Res.* 83, 5586.

17

Sudden Impulses in the Magnetosphere and at Ground

U. Villante and M. Piersanti
Dipartimento di Fisica, Università e Area di Ricerca in Astrogeofisica, L'Aquila Italy

1. Introduction

Sudden Impulses (SI) are rapid variations of the magnetospheric and geomagnetic field which are usually related to the Earth's arrival of sudden increases in the dynamic pressure of the solar wind (SW), generally associated with interplanetary shock waves or discontinuities. Such impulses often precede geomagnetic storms: in this case they are referred as Storm Sudden Commencements (SSC). Incoming SW pressure pulses compress the magnetosphere, increase the magnetopause and tail currents, and possibly other magnetospheric/ionospheric current systems as well; correspondingly, the magnetospheric and ground fields generally increase to a new state over about a two to fifteen minute period. The field variation associated to SI mostly occurs along the north/south component (B_z in the magnetosphere, H in the geomagnetic field). Additional contributions also come from the effects of the tail current (mostly in the nightside sector) and from those of the ring current (during more active magnetospheric conditions). The further development of field aligned currents (FAC) and ionospheric currents typically makes the SI manifestation at ground much more complex than in the magnetosphere: on the other hand, since earliest investigations (Matsushita, 1962; Nishida and Jacobs, 1962), it is well known that different transient waveforms are detected at different ground stations. Given their global simultaneous occurrence and clear onset time, SI provide a good opportunity for understanding the transient response of the magnetosphere and ionosphere to the SW variations.

In the present paper we review some aspects of the SI manifestations, as they are observed in the magnetosphere (basically at geosynchronous orbit, ≈6.6 Re, Re being the Earth radius) and at ground. For previous reviews of the experimental and theoretical aspects of SI events the reader is referred to Matsushita (1962), Nishida and Jacobs (1962), Siscoe et al. (1968), Nishida (1978), Smith et al. (1986), Araki (1994), Tsunomura (1998).

2. The SI manifestation at geosynchronous orbit

2.1 The general aspects and the local time dependence

Figure 1 (after Villante and Piersanti, 2011) shows the aspects of the SI manifestation at different LT (LT being the local time) along the geosynchronous orbit and the relationship between the magnetospheric field change (ΔB, B being the total field) and the increase of the

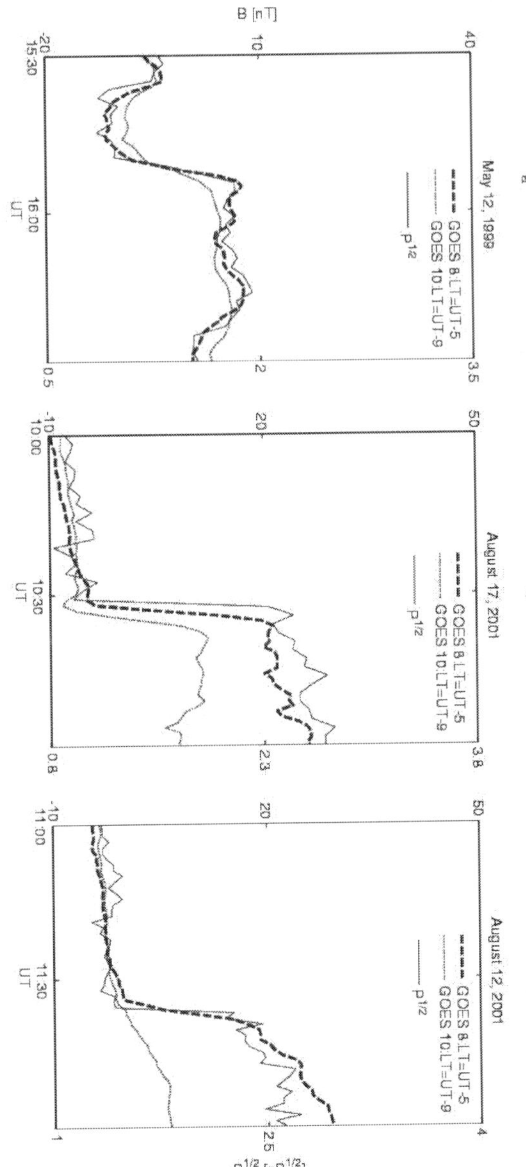

Fig. 1. (after Villante and Piersanti, 2011). Three examples of SI events at geosynchronous orbit. In each panel we show: the square root of the SW dynamic pressure (WIND spacecraft) and the magnetic field amplitude at GOES8 and GOES10. WIND data have been shifted to account for the SW propagation time

square root of the SW pressure $\Delta P^{1/2}$. As shown in panel *a*, the field jump is typically sharp in the entire dayside sector and basically reflects the behaviour of the change of the SW pressure, especially in the subsolar region. The ratio $R=\Delta B/\Delta P^{1/2}$ (hereafter referred as "relative response") at dawn is somewhat smaller (and smoother) than in the noon region ($R\approx 21$ nT/nPa$^{1/2}$ and $R\approx 25$ nT/nPa$^{1/2}$, respectively for the case in panel *a*). Such response attains smaller values in the dark hemisphere ($R\approx 12$ nT/nPa$^{1/2}$ at ≈1:30 LT, $R\approx 17$ nT/nPa$^{1/2}$ at ≈5:30 LT, for the case in panel *b*). Lastly, panel *c* shows that, in some cases, a continuous, small amplitude increase of the magnetospheric field is detected in the night sector, even in presence of a sharp change at dawn.

Although comparisons among different investigations are made ambiguous by the different criteria adopted for the selection of events, for the definition of "magnetospheric response", and for the large variety in its amplitude (and characteristics) in any time sector, an explicit LT modulation of such response has been extensively reported in the scientific literature. On the other hand, since the magnetopause current is mostly enhanced in the dayside sector during the magnetospheric compression, while the enhancement of the tail current produces a negative variation of the magnetospheric field, a strong day/night asymmetry might be expected in the SI manifestation at geosynchronous orbit. Consistently, Patel and Coleman (1970) reported that SI events in the nightside had smaller amplitude than in the dayside. Kokubun (1983) and Kuwashima and Fukunishi (1985) found that the magnetospheric change had highest values at local noon and very small values, or even negative, near midnight. More recently, Lee and Lyons (2004) concluded that SW pressure enhancements generally lead to a magnetospheric compression at all time sectors (with few exceptions in the nightside): it is strongest near noon and decreases toward dawn and dusk. Borodkova et al. (2005, 2006) found that, in general, the changes of the SW pressure (positive and negative) were associated with corresponding variations of the magnetospheric field; they also remarked that all the events without an explicit response were located either before 7:30 LT or after 16:30 LT. Wang et al. (2007) confirmed that the field variation peaked near local noon and decreased toward dawn and dusk. Figure 2a (after Villante and Piersanti, 2011) compares the amplitude of the geostationary response for events simultaneously detected at different LT: it clearly confirms a large data spread of the relative response in any time sector, together with an explicit LT modulation, with greater values at satellite located closer to the noon meridian; as can be seen, negligible and even negative magnetospheric responses are often detected in the dark sector.

2.2 The role of the SW parameters

As for other aspects of the magnetospheric dynamics, particular attention has been dedicated to the possible role of the North/South component of the interplanetary magnetic field ($B_{z,IMF}$). As a matter of fact, Sanny et al. (2002) showed that the variability of the magnetospheric field strength near local noon was independent on the IMF orientation but strongly influenced by changes of the SW pressure. Consistently, Wang et al. (2007) concluded that the IMF orientation does not affect the geosynchronous response significantly (see also Figure 2a). Kuwashima and Fukunishi (1985) and, more recently, Lee and Lyons (2004) suggested that midnight events associated with southward IMF orientations were often characterized by a dipolarization-like change similar to that one occurring during substorms, while the dayside response was mostly compressional; for northward IMF, a compression of the entire magnetosphere was generally observed, with few

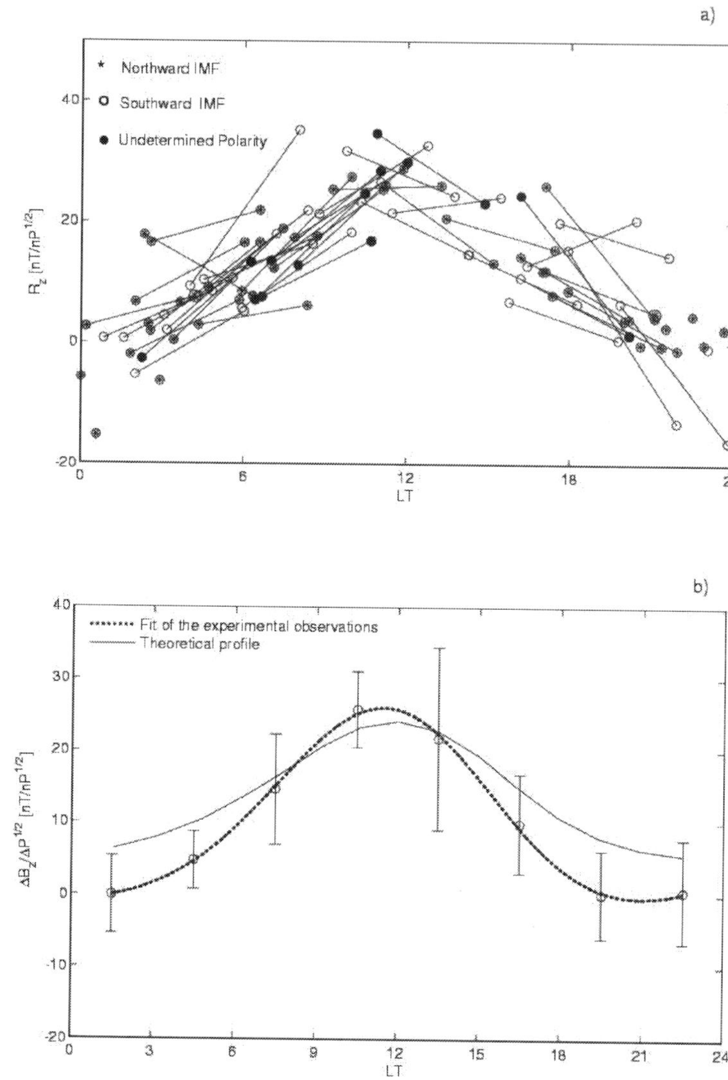

Fig. 2. a) (after Villante and Piersanti 2008) The relative response $R_z = \Delta B_z/\Delta P^{1/2}$ vs. local time (observation of the same event from two spacecraft are connected by a line. Empty circles identify events associated with Southward IMF; stars identify events associated with Northward IMF; black circles identify events associated with undetermined IMF polarity; b) (after Villante and Piersanti 2008) A comparison between average values of $\Delta B_z/\Delta P^{1/2}$ in each 3-h interval and the theoretical profile (solid line), determined considering the magnetic effects of the magnetopause current at geosynchronous orbit. The dotted line represents the fit of experimental measurements.

cases of depression near midnight. Lee and Lyons (2004) suggested that the magnetosphere is very sensitive to small SW pressure enhancements when the IMF is strongly southward for a long period of time. Focusing attention on the midnight sector, Wang et al. (2009) concluded that ≈75% of the negative responses were associated with southward IMF orientations. By contrast, Sun et al. (2011) revealed that the occurrence of positive or negative responses in the midnight sector had no obvious association with the sign of $B_{z,IMF}$, provided that no inversion of $B_{z,IMF}$ exists across the front of the impinging discontinuity.

2.3 The comparison with theoretical models

Kokubun (1983) evaluated the role of the magnetopause and tail currents and concluded that the geosynchronous responses were ≈30% smaller than expected. Figure 2b (after Villante and Piersanti, 2008), compares the 3-hr averages of R_z with the theoretical profiles evaluated assuming that the magnetic field change is determined by the transition between two steady states of the magnetosphere under different SW pressure conditions (i.e. different magnetopause currents, Tsyganenko, 2002a, 2002b). As a matter of facts, the observed average values reveal in the central part of the day a close correspondence with the predicted responses, a feature confirmed by an analysis of individual events (Villante and Piersanti, 2008). It suggests that, in this region, the field jumps are basically determined by the changes of the magnetopause current alone. The occurrence of negligible and negative responses makes the average values smaller than predicted in the dark region. It is worth noting, however, that, even in this region, the positive ΔB_z often show a substantial correspondence with the values predicted for the magnetopause current, especially for higher SW pressure jumps (Villante and Piersanti, 2011): it suggests that the dominant effects of the magnetopause current (i.e. the magnetospheric compression) might extend to a significant portion of the dark magnetosphere. On the other hand, the occurrence of negative variations in the nightside region (unpredictable in terms of the magnetopause current alone, solid line in Figure 2b) reveals, in several cases, a significant role of additional current systems. The interpretation of such events, however, requests a case by case analysis, paying attention to the SW and magnetospheric conditions in the period of interest (Villante and Piersanti, 2011). As a matter of facts, several approaches have been adopted to interpret the characteristics of the SI manifestation in the dark magnetosphere. Interesting results have been recently provided by Sun et al. (2011) who performed a MHD simulation of the nightside response to interplanetary shocks and concluded that when a shock sweeps over the magnetosphere, there exist mainly two regions: a positive response region caused by the compressive effect of the shock and a negative response region which is probably associated with the temporary enhancement of earthward convection in the nightside magnetosphere. In addition, according to their conclusions, a southward IMF would lead to a stronger and larger negative response region, and a higher shock speed would result in stronger negative and positive response region.

3. The SI manifestation at ground

3.1 The general aspects and the local time dependence

As previously underlined, the manifestation of ground SI is, in general, more complex than at geosynchronous orbit. This is because secondary effects like FAC and ionospheric currents contribute significantly in addition to the primary effect of the magnetopause current. On the other hand, it is now clear that FAC may modify the SI field even at middle and low latitudes (Kikuchi et al., 2001; Araki et al., 2006). As a matter of facts, shape and amplitude of the H waveform are strongly dependent on latitude and LT. Basically (Araki, 1994), at auroral latitudes, the waveform consists of two successive pulses with opposite sense (PI and MI, with typical duration of ≈1-2 min and ≈5-10 min, respectively). In the morning a positive pulse precedes and a negative pulse follows. The sense of the pulses is reversed in the afternoon and their amplitude decreases with decreasing latitude (Figure 3). Typically, SI manifestations are more simple at low latitudes, i.e. far from major high latitude and equatorial current systems. Here, the H behaviour becomes more step-like, but a two pulse structure with reduced amplitude is still identified. Figure 4 (after Villante and Piersanti, 2011) compares the aspects of ground events at $\lambda \approx 36°$ (λ being the magnetic latitude) with the SW and geosynchronous observations. It confirms that, independently on LT, the low latitude response of the H component often consists of a simple monotonic increase with amplitude (ΔH) often comparable (although smaller) with the field jump observed at geosynchronous orbit. Interestingly, in any time sector, the H variations are accompanied by explicit negative variations of the D component (ΔD, perpendicular to H in the horizontal plane). In some cases, moreover, even at low latitudes, the main field jump is preceded by transient phenomena which do not appear in the magnetosphere: for example, the event in panel *c* is characterized by the occurrence of a preliminary positive impulse (PPI).

As for geosynchronous orbit, specific attention has been addressed to the LT dependence of the ground response. In a pioneering investigation, Ferraro and Unthank (1951) showed that, at low and middle latitudes, ΔH was larger near midnight than in daytime. More recently, Tsunomura (1998), who analysed events from low to middle latitudes ($\lambda \approx 21°$-$43°$), determined a smaller ΔH during local morning with respect to the rest of the day. Araki et al. (2006, 2009) examined the LT dependence of the average ΔH at $\lambda \approx 35.4°$, separately for summer and winter. They determined, in both seasons, a maximum near midnight, a minimum at 7-8 LT, and a secondary maximum on the dayside; in addition, the amplitude of the LT modulation was much larger in the summer (approximately by a factor ≈3). Russell et al. (1992, 1994a, b) reported that at low and middle latitudes ΔH is maximum around noon during northward IMF conditions; in addition, ΔH was found to decrease in the daytime sector and to enhance significantly in the night time sector in the case of southward IMF. At $\lambda \approx 36°$, Francia et al. (2001) revealed a LT dependence of the relative response, $R_H = \Delta H/\Delta P^{1/2}$, characterized by a depressed value in the morning (≈10 nT/(nPa)$^{1/2}$), a greater amplitude in the evening and night sector (≈14-17 nT/(nPa)$^{1/2}$), and a maximum after the local noon (≈20 nT/(nPa)$^{1/2}$). Similarly, at subauroral latitudes ($\lambda \approx 54°$-$58°$), the geomagnetic response showed strongly depressed values in the morning and enhanced values in the afternoon (≈30 nT/(nPa)$^{1/2}$; Russell and Ginskey, 1995). Recently, Shinbori et al. (2009) conducted a statistical analysis of the relative amplitude from middle ($\lambda \approx 45°$) to equatorial latitudes ($\lambda \approx 15°$) suggesting, between $\lambda \approx 45°$ and $\lambda \approx 36°$, a strong dawn-dusk asymmetry, with minimum and maximum values in the morning (8–9 LT) and afternoon (15– 17 LT), respectively, and some evidence for a new enhancement in the night time sector (20–03 LT). They also showed that the amplitude of the relative response does not vary under specific IMF conditions (northward/southward), suggesting negligible effects from the ring and tail current. As a matter of fact, the emerging overview reveals a LT dependence of the ground response (with amplitude dependent on latitude and season) significantly different than in the magnetosphere, revealing explicit contributions from FAC and ionospheric currents: basically, from low to high latitudes, it is characterized by smaller (even negative, at higher latitudes) values in the morning and greater values in the post-noon and midnight sector, while, at equatorial latitudes, it shows a strong enhancement around 11 LT.

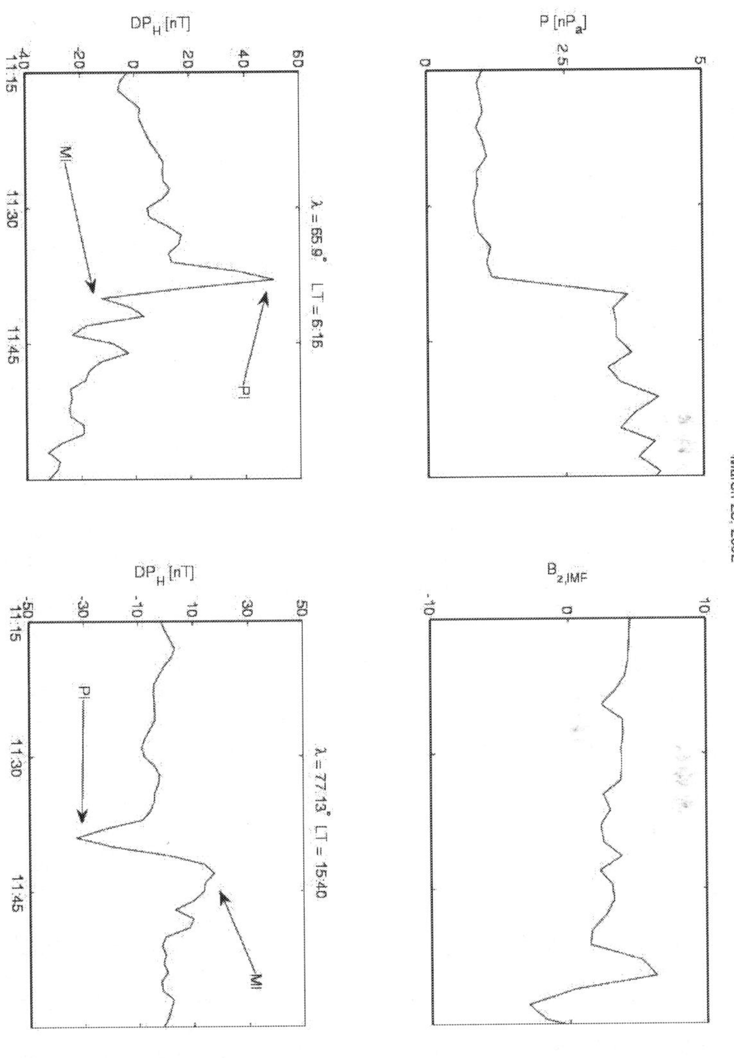

Fig. 3. Two examples of SI events at auroral latitudes in the local morning (right panel) and in local afternoon (left panel).

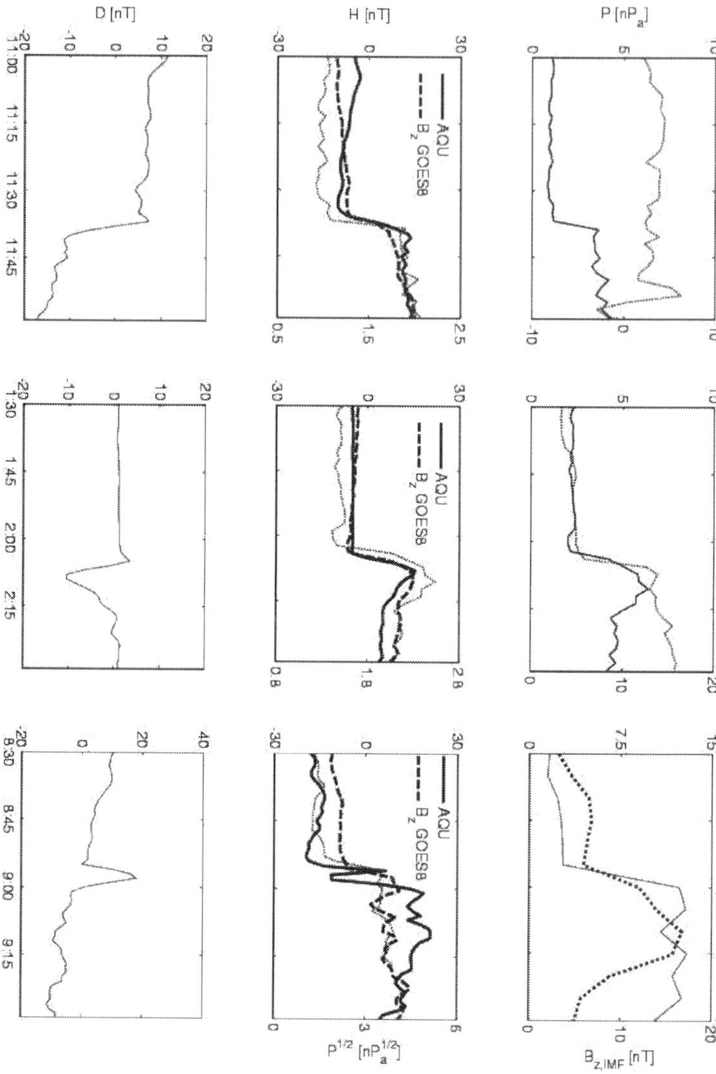

Fig. 4. (after Villante and Piersanti 2008) Three examples of SI events at low latitudes. Top panel: the SW dynamic pressure (solid line); the North/South component of the interplanetary magnetic field (dotted line). Central panel: the geomagnetic field component H at AQU; the magnetic field component B_Z at geosynchronous orbit; the square root of the SW pressure. Bottom panel: the geomagnetic field component D at AQU. WIND data have been shifted to account for the SW propagation time.

3.2 The comparison with theoretical models

The current understanding relates such complex scenario to the combined effects of the magnetospheric and ionospheric current systems. Namely, the total disturbance field (D_{SI}) of the H component is decomposed in different subfields (Araki, 1977, 1994; Araki et al., 1997, 2009):

$$D_{SI}=DL_{MI} + DP_{PI} + DP_{MI}$$

According to models, the direct effect of the increased magnetopause current propagates to low and middle latitudes as a compressive wave and produces a step like increase of the H component (DL_{MI} field, where L stands for low latitudes); its amplitude is largest at the equator and decreases with increasing latitude. A dusk-to-dawn electric field along the compressional wave front induces a twin ionospheric vortex system that produces a preliminary impulse of polar origin (DP_{PI}). The DP_{PI} field manifests as a preliminary reverse impulse (PRI) simultaneously observed at auroral latitudes in the afternoon and near the dip equator on the dayside, as well as a preliminary positive impulse (PPI) observed at auroral latitudes in the morning. On the other hand, if the increased pressure behind the SW discontinuity is kept up, the magnetospheric convection has to adjust itself to the compressed state of the magnetosphere: as a final result, it produces a twin polar vortex system (DP_{MI}) which is opposite to the DP_{PI} field and corresponds to the MI. Such DP_{MI} field is basically driven by an electric field originated in the polar region and transmitted from the outer magnetosphere through FAC which flow into the ionosphere in the morning side and away in the afternoon side. In this scheme, the preliminary impulse is exclusively due to current systems of polar origin, DP_{PI}, whereas the main impulse is due to the combined effect of DL_{MI} and DP_{MI}. On the other hand, Kikuchi et al. (2001) found that preliminary positive pulses tend to appear in the afternoon middle latitudes, and proposed that the generation mechanism is the magnetic effect of the FAC which are accompanied with the dusk-to-dawn electric fields.

The comparison between measurements at ground with those obtained by low altitude satellites above the ionosphere provided important insights on the ionospheric currents. For example, Araki et al. (1984), comparing ground and MAGSAT observations, showed that both H and D components showed variations with opposite sense at satellite and ground, revealing the existence of ionospheric currents associated with SI (at least near dawn and dusk, due to the orbital configuration). Han et al. (2007) examined Oersted and ground data and observed, in the night sector, very similar waveforms above and below the ionosphere; they then concluded that the ionospheric currents do not contribute significantly to nightside SI which, according to their conclusions, were dominantly caused by the enhanced magnetopause currents. By contrast, the waveforms observed by Oersted on the dayside were apparently different from those observed on the ground, reflecting the role of ionospheric currents. Corresponding to the PRI and the MI observed in the H component at the dayside dip equator, Oersted always observed an increase and a clear decrease in the magnetic field, respectively. These observational results suggest that the PRI at the dayside dip equator corresponds to a westward ionospheric current, and an eastward current is excited after the PRI. More recently, a comprehensive analysis of events simultaneously observed at ground and by CHAMP (Luhr et al., 2009) confirmed that night time events at ground are not (or minimally) affected by ionospheric currents. More in general, this analysis also showed that at latitudes smaller than $\lambda \approx 40°$ the amplitude of the field variation was practically the same at ground and satellite; a progressive latitudinal increase of the SI amplitude was determined at higher latitudes both on the ground and at satellite, suggesting the effects of FAC rather than those of currents flowing in the ionosphere.

Figure 5a (after Villante and Piersanti, 2011) shows the 3-hr average values of the ground relative responses $<R_H>$ at $\lambda\approx36°$: as can be seen, the LT modulation is much less pronounced than at geostationary orbit. The average responses in Figure 5a are compared with the theoretical profiles expected for the magnetopause current alone (R_{CF}) and for the global magnetospheric current system (from the magnetopause, tail and ring current, B_T), as evaluated at the winter and summer solstice. Such comparison shows that the observed $<R_H>$ fall, in general, within the limits of the expected profiles from approximately premidnight up to noon (an aspect confirmed by an analysis of single events), suggesting a poor contribution from FAC and ionospheric currents on this component. In the post-noon region, the observed responses overcome the theoretical profiles on average by ≈50% (and occasionally by a factor ≈2-3 in individual cases, Villante and Piersanti, 2011), revealing explicit effects from the additional ionospheric currents in this time sector. Consistently, Shinbori et al. (2009) interpreted such LT (and latitudinal) dependence as the manifestation of superimposed ionospheric currents: in particular, at low and middle latitudes, the observed pattern would be related to currents (producing negative and positive variations of the H component in the dawn and dusk sector, respectively) generated by the enhanced dawn-to-dusk electric field (accompanying FAC) due to the compression of the magnetosphere. Note, in addition, that, the negative responses of the D component ($<R_D>$, figure 5b) are far from the theoretical profiles, suggesting an explicit influence (through the entire day) on this component of the additional current systems.

4. The ULF waves occurrence

SI manifestations are occasionally accompanied by trains of almost monochromatic ULF waves (f≈1-10 mHz, usually referred as "Pc5 pulsations", Villante, 2007) which manifest soon after the main variation and persist for few cycles. According to Zhang et al. (2009), at geosynchronous orbit, the magnitude of ULF waves associated to variations of the SW pressure is larger around noon than at dawn and dusk. Amata et al. (1986) showed that magnetospheric oscillations excited by a SW shock have a complicated structure: a magnetosonic wave propagating from the dayside to the nightside sector suddenly appears and resonant shear Alfvén waves are generated during the propagation of this magnetosonic wave. As regard to ground observations, Ziesolleck and Chamalaun (1993) found wave amplitudes increasing with latitude, while their frequency was not dependent on either latitude or longitude.

A spectacular example of SI-related waves is shown in Figure 6 (after Piersanti et al., 2012) in which, at GOES 8 position (≈04:15 LT), the SI manifestation itself practically consists in the onset of a large amplitude wave mode at f=3.3 mHz; by contrast, GOES 10 (≈00:15 LT) did not observe any similar wave activity in the midnight sector. Waves at the same frequency were detected at all ground stations (Figure 7 and Figure 8; after Piersanti et al., 2012), suggesting an interpretation in terms of a global oscillation mode of the whole magnetospheric cavity. Moreover, the characteristics of the wave amplitude and the behaviour of the polarization pattern suggest that, due to the variable length of the field line through the day, the magnetospheric wave leaked energy to field line resonance at $\lambda\sim66°$ in the morning sector and at $\lambda\sim71°$ in the noon sector.

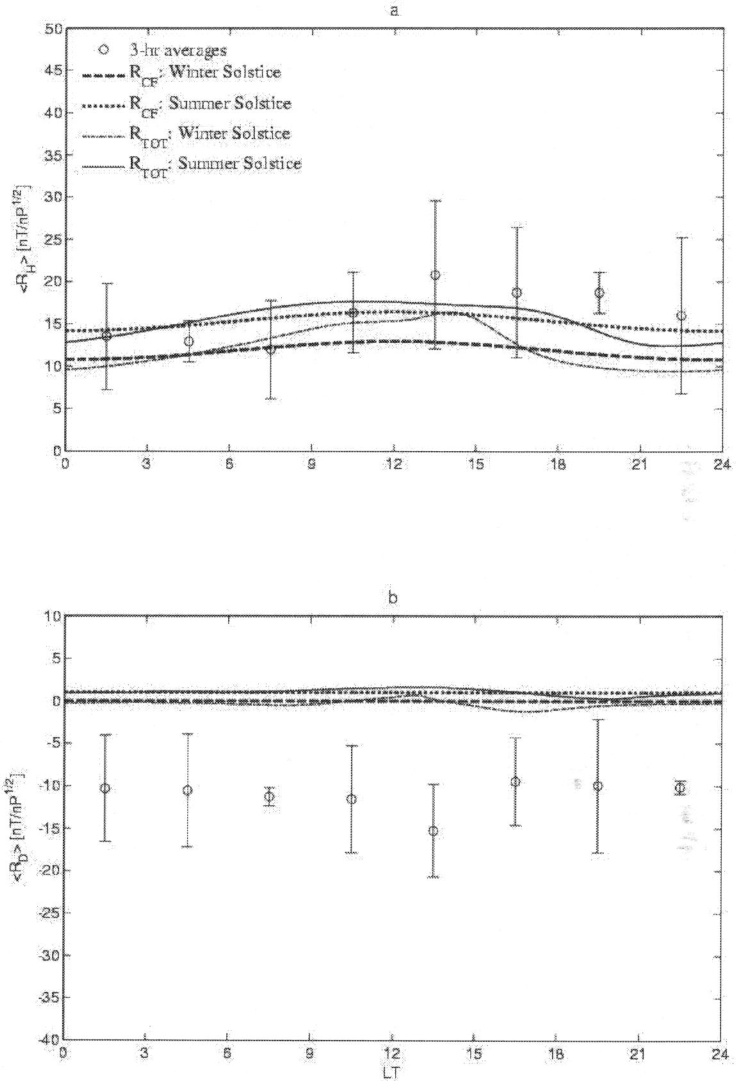

Fig. 5. (after Villante and Piersanti 2008) A comparison between averages values $<R_H>$ in 3-hr intervals and the theoretical profiles determined considering the ground effects of the magnetopause current (B_{CF}-field) and those of the total current system (B_{TOT}-field). Both profiles have been evaluated at winter and summer solstice. b) The same for $<R_D$.

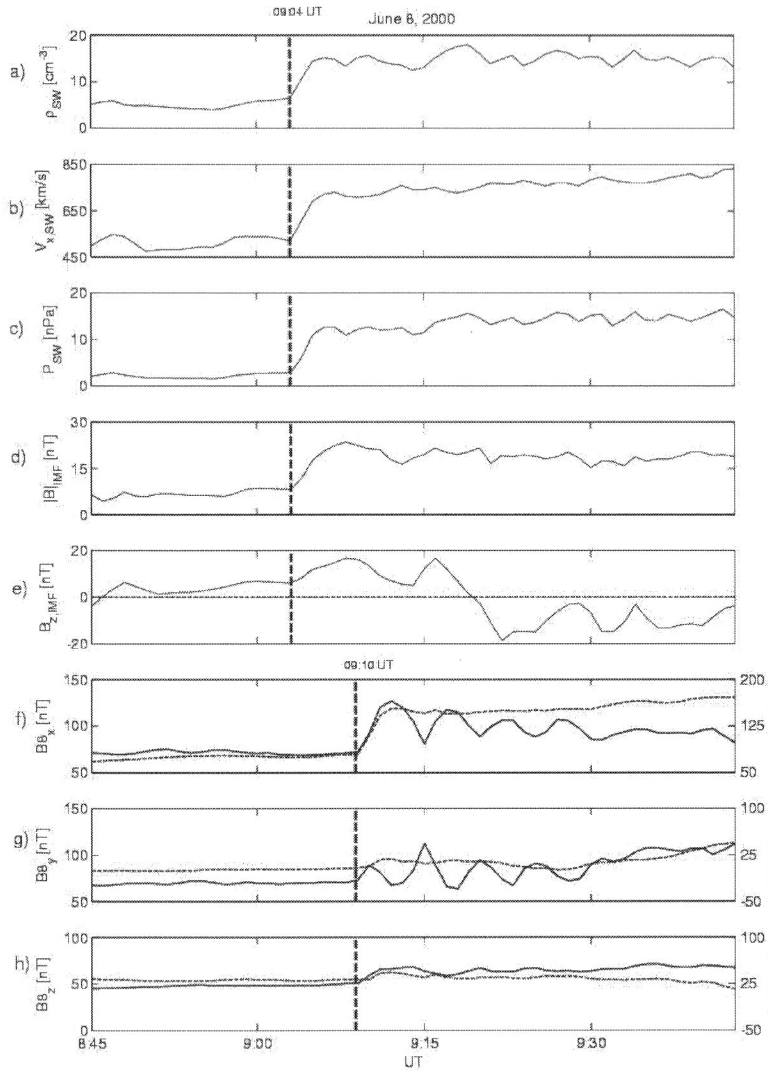

Fig. 6. (after Piersanti et al., 2011) The solar wind (WIND) and the magnetospheric observations (GOES) for the SI event occurred on June 8, 2000: a) the SW density; b) the SW velocity; c) the SW dynamic pressure; d) the IMF | B | ; e) the IMF Bz component; f) The B_X component of the magnetospheric field measured by GOES8 (solid line) and GOES10 (dashed line); g) the B_Y component of the magnetospheric field measured by GOES8 (solid line) and GOES10 (dashed line); The B_Z component of the magnetospheric field measured by GOES8 (solid line) and GOES10 (dashed line). The vertical dashed lines identify the SW pressure jump and the SI occurrence.

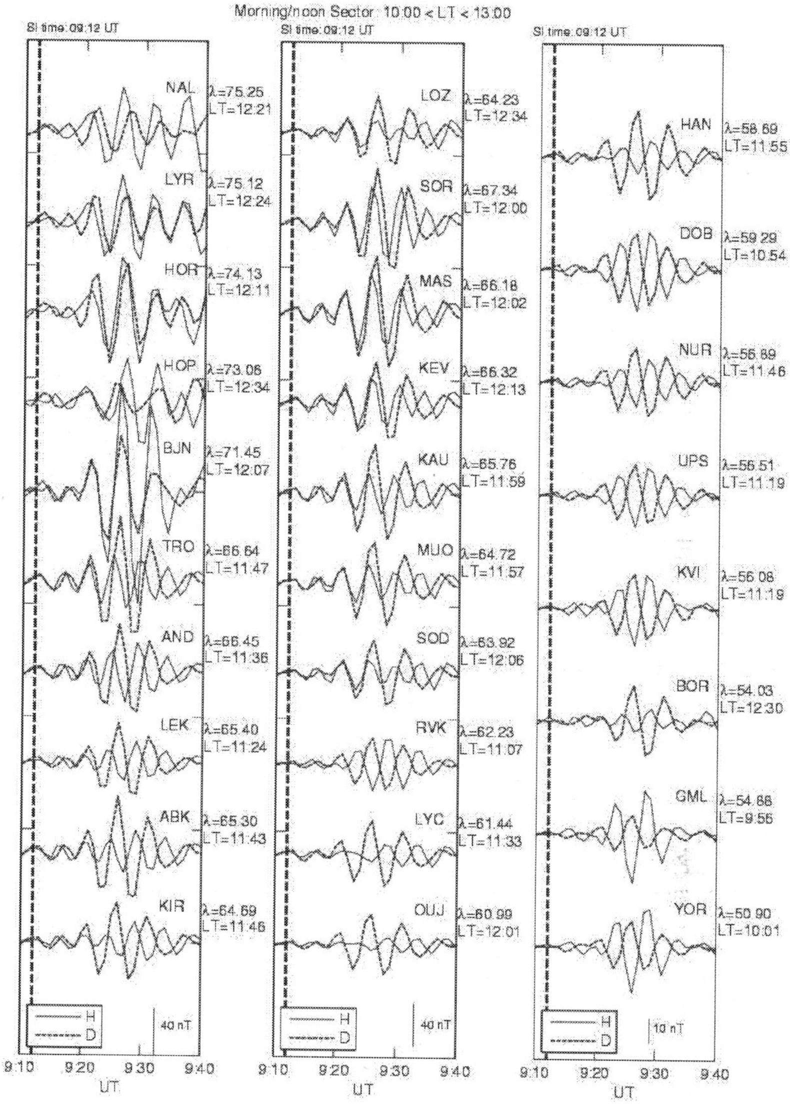

Fig. 7. (after Piersanti et al., 2011) The H (solid) and D (dotted) components of the geomagnetic field at ground stations location in the morning/noon sector (10:00 < LT <13:00) on June 8, 2000 (0910÷0940 UT). The data are filtered at f = 3.3 mHz.

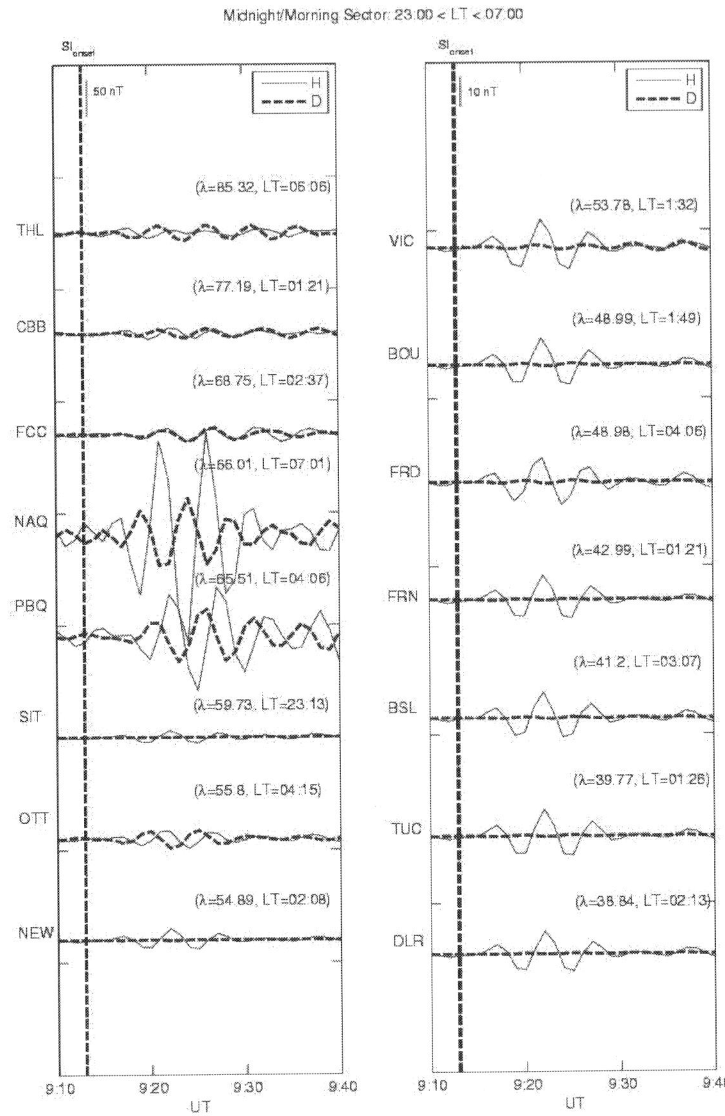

Fig. 8. (after Piersanti et al., 2011) The H (solid) and D (dotted) components of the geomagnetic field at ground stations location in the midnight/morning sector (23:00 < LT < 07:00) on June 8, 2000 (0910÷0940 UT). The data are filtered at $f = 3.3$ mHz.

5. Summary

The impinging on the Earth's magnetosphere of sudden increase of the SW pressure, related to the arrival of shock waves or other discontinuities, typically causes rapid increases of the magnetospheric and geomagnetic field, called sudden impulses (SI). Such pressure pulses, indeed, compress the magnetosphere, increase the magnetopause and tail currents, and possibly other magnetospheric/ionospheric current systems as well.

On the other hand, since the magnetopause current is mostly enhanced in the dayside sector, while the enhancement of the tail current produces a negative variation of the magnetospheric field, a strong day/night asymmetry might be expected in the SI manifestation: consistently, the experimental observations at geosynchronous orbit reveal an explicit LT modulation, with greater responses at satellite located closer to the noon meridian. The comparison between experimental observations and theoretical models reveals that the positive jumps are basically determined by the changes of the magnetopause current alone, in the dayside and in a large portion of the dark magnetosphere. On the other hand, the occurrence of negligible and negative responses in the nightside reveals a significant role of the tail (and ring) current.

At ground, the SI manifestation is more complex than at geosynchronous orbit since secondary effects like FAC and ionospheric currents contribute significantly in addition to the primary effects of the magnetospheric currents. The ground responses show a clear latitudinal dependence and a LT modulation (with smaller values in the morning and higher values in the postnoon region), a behaviour which can be interpreted as the manifestation of the superimposed ionospheric currents.

SI manifestations are occasionally accompanied by trains of almost monochromatic ULF waves which manifest soon after the main variation and persist for few cycles. In some cases they can be interpreted in terms of global oscillation modes of the whole magnetospheric cavity.

6. Acknowledgements

This research activity is supported by Consorzio Area di Ricerca in Astrogeofisica and Università, L'Aquila.

7. References

Amata, E., V. A. Pilipenko, O. A. Pokhotelov, V. A. Troitskaya, and R. V. Shchepetnov, 1986, PSC-5 Pulsations in Geostationary Orbit, *Geomagn. Aeron.*, 26, (2), 283.

Araki, T., 1977, Global structure of geomagnetic sudden commencements, *Planet. Space Sci.*, 25, 372,.

Araki, T., 1994, A physical model of the geomagnetic sudden commencement, in Solar Wind Sources of Magnetospheric Ultra-Low-Frequency Waves, *Geophys. Mono. Series*, 81, 183.

Araki, T., K. Keika, T.

Araki, T., S. Tsunomura and T. Kikuchi, 2009, Local time variation of the amplitude of geomagnetic sudden commencements (SC) and SC-associated polar cap potential, *Earth Planets Space,* 61, e13.

Araki, T., T. Iyemori, and T. Kamei, 1984, Sudden commencements observed by MAGSAT above the ionosphere, *J. Geomag. Geoelectr.*, 36, 507.

Borodkova, N. L., G. N. Zastenker, M. Riazantseva, and J. D. Richardson, 2005, Large and sharp solar wind dynamic pressure variations as a source of geomagnetic field disturbances in the outer magnetosphere (at the geosynchronous orbit), *Planetary and Space Science,* 53, N.1-3, 25.

Borodkova, N. L., J. B. Liu, Z. H. Huang, G. N. Zastenker, C. Wang, and P. E. Eiges, 2006, Effect of change in large and fast solar wind dynamic pressure on geosynchronous magnetic field, *Chinese Physics,* 15, 2458.

Ferraro, V. C. and H.W. Unthank, 1951, Sudden commencements and sudden impulses in geomagnetism: their diurnal variation in amplitude, *Geophys. Pure Appl.*, 20, 2730.

Francia, P., S. Lepidi, U. Villante, P. Di Giuseppe, 2001, Geomagnetic sudden impulses at low latitude during northward interplanetary magnetic field conditions, *J. Geophys. Res.*, 106, 21231.

Han, D.-S., T. Araki, H.-G. Yang, Z.-T. Chen, T. Iyemori, and P. Stauning, 2007, Comparative study of Geomagnetic Sudden Commencement (SC) between Oersted and ground observations at different local times, *J. Geophys. Res,* 112, A05226, doi:10.1029/2006JA011953.

Kamei, H. Yang, and S. Alex, 2006, Nighttime enhancement of the amplitude of geomagnetic sudden commencement and its dependence on IMF-Bz, *Earth Planets Space,* 58, 45.

Kikuchi, T., S. Tsunomura, K. Hashimoto, and K. Nozaki, 2001, Field aligned current effects on mid-latitude geomagnetic sudden commencements, *J. Geophys. Res.*, 106, 15555.

Kokubun, S., 1983, Characteristics of storm sudden commencement at geostationary orbit, *J. Geophys. Res,* 88, 10025.

Kuwashima, M. and H. Fukunishi, 1985, SSC-associated magnetic variations at the geosynchronous altitude, *J. Atmo. Terr. Phys.*, 47, 451.

Lee, D.-Y., and L. R. Lyons, 2004, Geosynchronous magnetic field response to solar wind dynamic pressure pulse, *J. Geophys. Res,* 109, A04201.

Lühr, H.; Schlegel, K.; Araki, T.; Rother, M.; Förster, M., 2009, Night-time sudden commencements observed by CHAMP and ground-based magnetometers and their relationship to solar wind parameters, *Ann. Geophys.*, 27, 5, 1897.

Matsushita, S., 1962, On geomagnetic sudden commencements, sudden impulses, and storm duration, *J. Geophys. Res.*, 67, 3753.

Nishida, A., 1978, Geomagnetic diagnosis of the magnetosphere, *Phys. Chem. Spa., vol. 9,* Springer-Verlag, New York.

Nishida, A., and J. A. Jacobs, 1962, World-wide changes in the geomagnetic field, *J. Geophys. Res.*, 67, 525.

Patel, V. L., P. J. Jr. Coleman, 1970, Sudden impulses in the magnetosphere observed at synchronous orbit, *J. Geophys. Res.*, 75, 7255.

Piersanti, M., U. Villante, C. Waters, I. Coco, 2012, The June 8, 2000 ULF wave activity: a case study, 2012, in press on *J. Geophys. Res.*.

Russell, C. T., Ginskey, M., Petrinec, S. M., Le, G., 1992, The effect of solar wind dynamic pressure changes on low and mid-latitude magnetic records, *Geophys. Res. Lett.*, 19, 1227.

Russell, C. T., M. Ginskey, 1995, Sudden impulses at subauroral latitudes: response for Northward interplanetary magnetic field, *J. Geophys. Res.*, 100, 23695.

Russell, C. T., M. Ginskey, S. Petrinec, 1994a, Sudden impulses at low-latitude stations: steady state response for Northward interplanetary magnetic fields, *J. Geophys. Res.*, 99, 253.

Russell, C. T., M. Ginskey, S. Petrinec, 1994b, Sudden impulses at low-latitude stations: steady state response for Southward interplanetary magnetic fields, *J. Geophys. Res.*, 99, 13403.

Sanny, J., J. A. Tapia, D. G. Sibeck, and M. B. Moldwin, 2002, Quiet time variability of the geosynchronous magnetic field and its response to the solar wind, *J. Geophys. Res.*, 107(A12), 1443, doi:10.1029/2002JA009448.

Shinbori, A., Y. Tsuji, T. Kikuchi, T. Araki, and S. Watari, 2009a, Magnetic latitude and local time dependence of the amplitude of geomagnetic sudden commencements, *J. Geophys. Res.*, 114, A04217.

Shinbori, A., Y. Tsuji, T. Kikuchi, T. Araki, and S. Watari, 2009b, Magnetic latitude and local time dependence of the amplitude of geomagnetic sudden commencements, *11th IAGA Scientific Assembly*, Sopron.

Siscoe, G. L., V. Formisano, and A. J. Lazarus, 1968, Relation between geomagnetic sudden impulses and solar wind pressure changes: An experimental investigation, *J. Geophys. Res.*, 73, 4869.

Smith, E. J., J. A. Slavin, R. D. Zwickl and S. J. Bame, 1986, Solar Wind-Magnetosphere Coupling and the Distant Magnetotail, *Solar Wind-Magnetosphere Coupling*, 717, eds. Y. Kamide and J. A. Slavin, Terra-Reidel, Tokyo.

Sun, T. R., C. Wang, H. Li, and X. C. Guo, 2011, Nightside geosynchronous magnetic field response to interplanetary shocks: Model results, *J. Geophys. Res.*, 116, A04216, doi:10.1029/2010JA016074.

Tsunomura, S., 1998, Characteristics of geomagnetic sudden commencement observed in middle and low latitudes, *Earth Plan. Spa Phys., 50*, 755.

Tsyganenko, N. A., 2002a, A model of the near magnetosphere with a dawn-dusk asymmetry. 1. Mathematical structure , *J. Geophys. Res.*, 107, SMP 10–1.

Tsyganenko, N. A., 2002b, A model of the near magnetosphere with a dawn-dusk asymmetry. 2. Parameterization and fitting to observations, *J. Geophys. Res.*, 107, SMP 12–1.

Villante, U. and M. Piersanti, 2008, An analysis of sudden impulses at geosynchronous orbit, *J. Geophys. Res.*, 113, A08213.

Villante, U. and M. Piersanti, 2011, Sudden Impulses at geosynchronous orbit and at ground, *J. Atm. Sol-Terr. Phys.*, 73, 1, 61.

Villante, U., 2007, Ultra Low Frequency Waves in the Magnetosphere, , in: Y. Kamide/A. Chian, *Handbook of the Solar-Terrestrial Environment.* pp. 397, doi: 10.1007/11367758_16, Springer-Verlag Berlin Heidelberg.

Wang, C., et al., 2007, Response of the magnetic field in the geosynchronous orbit to solar dynamic pressure pulses, *J. Geophys. Res.*, 112, A12210.

Wang, C., J. B. Liu, H. Li, Z. H. Huang, J. D. Richardson, and J. R. Kan, 2009, Geospace magnetic field responses to interplanetary shocks, *J. Geophys. Res.*, 114, A05211, doi:10.1029/2008JA013794.

Zhang X. Y., Q. G. Zong, Y. F. Wang, H. Zhang, L. Xie, S. Y. Fu, C. J. Yuan, C. Yue, B. Yang and Z. Y. Pu, 2010, ULF waves excited by negative/positive solar wind dynamic pressure impulses at geosynchronous orbit, *J. Geophys. Res.*, doi:10.1029/2009JA015016.

Ziesolleck, C. W. S., and F. H. Chamalaun, 1993, A Two-Dimensional Array Study of Low-Latitude PC 5 Geomagnetic Pulsations, *J. Geophys. Res.*, 98(A8), 13,703, doi:10.1029/93JA00637.

18

Turbulence in the Magnetosheath and the Problem of Plasma Penetration Inside the Magnetosphere

Elizaveta E. Antonova[1,2], Maria S. Pulinets[1], Maria O. Riazantseva[1,2], Svetlana S. Znatkova[1], Igor P. Kirpichev[1,2] and Marina V. Stepanova[3]

[1]*Skobeltsyn Institute of Nuclear Physics, Moscow State University, Moscow*

[2]*Space Research Institute RAS, Moscow*

[3]*Physics Department, Universidad de Santiago de Chile*

[1,2]*Russia*

[3]*Chile*

1. Introduction

Chapman & Ferraro (1931) introduced the concept of confinement of the Earth's magnetic field in a cavity carved in the solar plasma flow. The balance between the Earth's magnetic field (more accurately between the magnetic pressure at the boundary of the cavity) and the solar wind dynamic pressure was considered as the condition of the formation of the boundary of the cavity. Chapman-Ferraro model is called a closed magnetosphere. Low energy particles can not penetrate through the boundary of the cavity. Dungey (1961) made the most drastic revision of Chapman-Ferraro's original theory. Dungey envisaged that the connection process, called reconnection, takes place on the dayside magnetopause and that the connected field lines are then transported in the antisolar direction by the solar wind, resulting in the magnetotail. Subsequently, the field lines are reconnected there and then transported back to the dayside magnetosphere. Such process takes place when interplanetary magnetic field (IMF) has the southward direction. The large scale reconnection takes place at high latitudes when IMF has the northward direction. The scheme shown on Fig. 1 demonstrates Dungey's concept of reconnection at the dayside magnetopause when IMF has southword (a) and northward (b) directions. The model of Dungey qualitatively accounts for such phenomena as the inward motion of the dayside magnetopause, equatorward motion of the cusp, expansion of the auroral oval, increase in magnetotail magnetic field strength, and expansion of the magnetotail radius which occur when the IMF turns southward. It can also easily explain the penetration of the plasma of solar wind origin inside the magnetosphere. That is why this concept for a long period was the dominant concept in the physics of the magnetosphere and was widely used for the description of different phenomena including the formation of boundary layers (see, for example, the review Lavraud et al. (2011)). However step by step a number of observations ant theoretical arguments have appeared which give the possibility to throw doubts on the applicability of the scheme shown on Fig. 1 for the real situation.

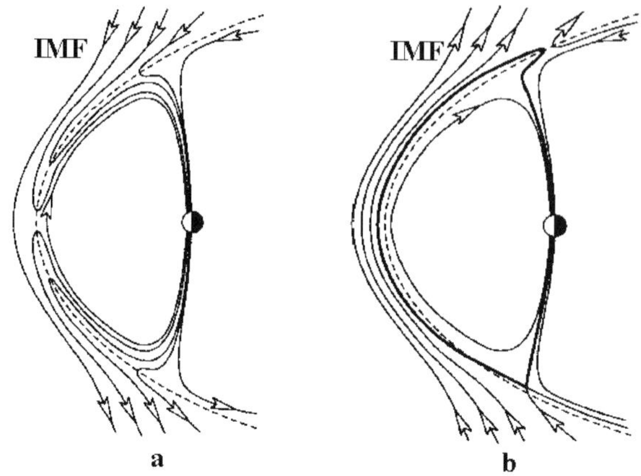

Fig. 1. Sketch illustrating the process of reconnection at the dayside magnetopause when IMF has southward (a) and northward (b) directions

In this paper, we try to summarize arguments demonstrating principal differences of the scheme shown on Fig.1 and the picture, which corresponds to the results of experimental observations. We discuss the process of particle penetration through the magnetopause and try to select arguments demonstrating the formation of the changes of the form of the magnetopause and particle penetration inside the magnetosphere as the results of the change of conditions of pressure balance at the magnetopause. THEMIS mission multisatellite observations available at (http://www.nasa.gov/ mission_pages/themis/) were used for illustration of the main features of magnetic field and plasma observations near the magnetopause. The paper is organized as follows. Section 2 contains the analysis of the properties of turbulence in the magnetosheath. Section 3 is dedicated to the condition of pressure balance at the magnetopause. We discuss the applicability of the frozen in condition for the description of plasma flow in the magnetosheath in Section 4. Section 5 contains conclusions and discussions.

2. Turbulent magnetosheath and magnetic field near the subsolar magnetopause

The magnetopause is formed not as the boundary between the solar wind and the geomagnetic cavity. Magnetosheath plasma and magnetic field come into a contact with magnetopause. The magnetosheath is a region through which mass, energy and momentum are transported from the solar wind into the Earth's magnetosphere. There is a significant number of experimental results showing the high level of plasma turbulence in the magnetosheath (see Luhmann et al. (1986), Sibeck et al. (2000), Zastenker et al. (1999, 2002), Lucek et al. (2001), Němeček et al. (2000a,b; 2002a,b), Shevyrev & Zastenker (2005), Shevyrev et al. (2007), Gutynska et al. (2008), Savin et al. (2008), Rossolenko et al. (2008), Ŝafránková et al. (2009), Znatkova et al. (2011) and references therein). Fig. 2 shows an example of plasma and magnetic field fluctuations in the magnetosheath measured by Geotail satellite March 2, 1996. It is possible to see that the amplitude of magnetic field fluctuations is much larger than the averaged field and constitutes ~10-20 nTl.

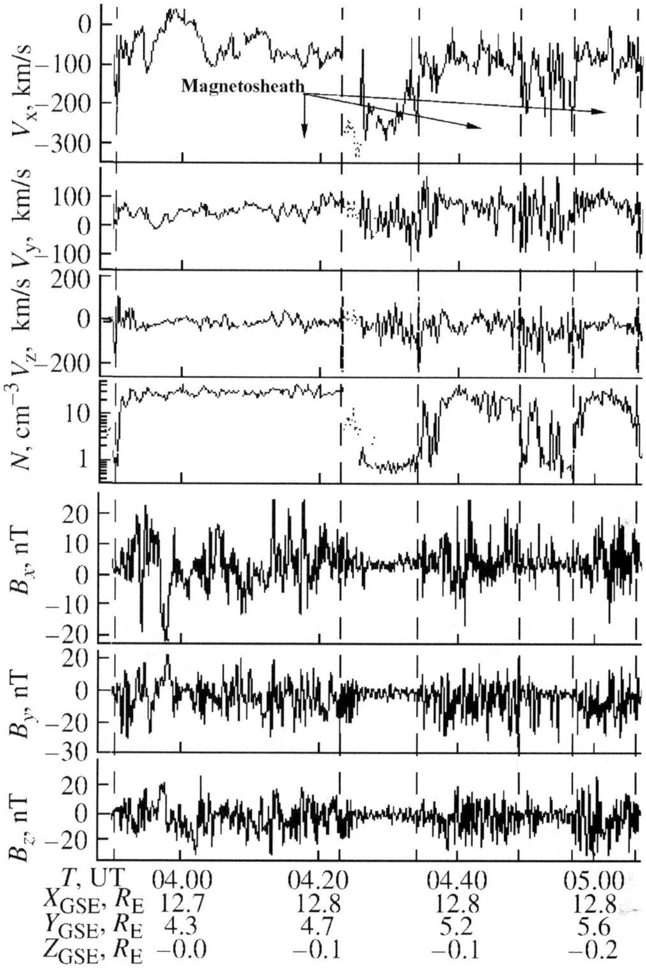

Fig. 2. Results of measurements on the Geotail satellite on March 2, 1996

Zastenker et al. (2002) discussed the origin of magnetosheath variations and showed that a part of these variations is from propagation and/or amplification of solar wind or interplanetary magnetic field (IMF) disturbances, which pass through the bow shock, and a part of these variations originates inside the magnetosheath. Gutynska et al. (2008) present the results of a statistical survey of the magnetosheath magnetic field fluctuations and other parameters using two years of Cluster observations. They have found that the correlation length of the turbulence in the magnetosheath is approximately ~$1R_E$ in the frequency range 0.001- 0.125 Hz and does not depend significantly on the magnetic field or plasma flow direction. When the plasma flow velocity in the magnetosheath is about ~ 200 km/s the distance ~$1R_E$ is traversed by plasma during approximately ~ 30 seconds.

The existence of high level of turbulence in the magnetosheath suggests that the direction of magnetic field near magnetpause can not coincide with the direction of IMF. Ŝafránková et al. (2009) determine a probability of simultaneous observations of the same sign of the magnetic field B_Z component in the solar wind and magnetosheath. They conclude that the probability of observations of the same B_Z sign in the solar wind and in the magnetosheath is surprisingly very low from a general point of view. It was shown that regardless of the solar cycle phase, the probability to observe the same B_Z sign in the solar wind and in the magnetosheath is close to 0.5 (random coincidence) for IMF $|B_z| < 1$ nT, and it is a rising function of the B_Z value.

Solar wind is the turbulent medium (see Riazantseva et al. (2005, 2007) and references in these works). Therefore, solar wind parameters may change during the propagation to the Earth's orbit from the position of such satellite in the solar wind as ACE and Wind till the orbit of the Earth. That is why to assess the effect of magnetosheath turbulence on the magnetic field parameters changing during the propagation through the magnetosheath to the magnetopause these parameters should be compared directly in front of the shock wave and near the magnetopause. At the same time measuring of the solar wind should be carried out upstream the foreshock region which makes a strong disturbance in the solar wind before the shock front. The opportunity of such a comparison has appeared only with the start of the five-satellite THEMIS mission (Angelopoulos, 2008; Sibeck & Angelopoulos, 2008). One of THEMIS satellites during summer in the north hemisphere performed measurements in the solar wind, while the other occasionally crossed the magnetopause on the dayside.

To obtain the dependences of the component of magnetic field before the magnetopause with magnetic field before the bow shock we used results of THEMIS mission (http://cdaweb.gsfc.nasa.gov/) for the period from June 25 to October 10, 2008. During this period the orbits of spacecrafts deployed by the precession in such a way that their apogees were located close to the Earth-Sun line, i.e. the configuration convenient for studying the interactions on the dayside of the Earth's magnetosphere takes place. The intervals when one of the spacecrafts was localized in the solar wind, and another crossed the magnetopause near the subsolar point were picked out. The events were selected when the deviation of the probe from the x-axis did not exceed 7 R_E. The moment of crossing of the magnetopause was fixed by the distinctive changes in plasma parameters and magnetic field, determined according to the Electrostatic Analyzer ESA (McFadden et al. (2008)) and the Flux Gate Magnetometer FGM (Auster et al. (2008)) on the probe. Parameters of the interplanetary magnetic field (IMF) were determined by FGM. The events in which the solar wind did not suffer significant variations were chosen. The value of the standard deviation of the absolute value of the magnetic field from the average for the selected periods does not exceed 2 nT, the flow velocity was less than 650 km/sec.

The parameters of the magnetic field, measured by one of the spacecraft after crossing the magnetopause, were compared with the IMF parameters, observed by another spacecraft. The following quantities were used as analyzed parameters: the magnitude and the three components of the magnetic field. Mean value and dispersion were calculated for each variable.

The magnetic field parameters near the magnetopause were averaged over periods of 30 and 90 seconds after crossing the magnetopause (what was fixed simultaneously by changes in the parameters of plasma and magnetic field). Values of the magnetic field, averaged over the spin resolution of the probe, equal to 3 s, i.e. field directly close to magnetopause was also analyzed. The solar wind parameters were averaged over a maximum period of 90 s taking into account the time shift of solar wind propagation from the spacecraft performing measurements in the solar wind to the magnetopause. The shift was calculated as the time of the solar wind passing the difference between x-coordinates of the spacecrafts in the approximation of the radial propagation of the solar wind. Solar wind velocity was determined from the data of THEMIS probe located in the solar wind. The solar wind velocity in the magnetosheath was considered as reduced by about two times as a result of thermalization. The magnetosheath thickness was supposed to be approximately ~2 R_E. For each case, the time shift was calculated individually for the specific spacecraft coordinates. Since the errors of the order of ten seconds are possible when calculating the time shift, the averaging of values in the solar wind was made for a maximum period of 90 seconds to minimize them. 26 events were analyzed.

Fig. 3–6 show the dependences of the magnetic field parameters near the magnetopause on the solar wind parameters. A set of three curves is given for each parameter. The first distribution is plotted for the instantaneous values (three second averaging) after crossing the magnetopause (panels a), the second – for the averaged over a 30-seconds interval after crossing (panels b), the third – for the averaged over a 90-seconds interval (panels c). The dependencies on the corresponding averaged solar wind parameters are shown. Averaging in the solar wind is realized for a maximum period of 90 seconds (taking into account the time shift of the solar wind propagation to the magnetopause) in order to minimize errors due to deviation of the estimated solar wind delay from the real. For each point, an error calculated as the standard deviation over the averaging periods is also shown. On the charts for the instantaneous values, the errors are shown only for the averaging in the solar wind, because averaging near the magnetopause was not carried out.

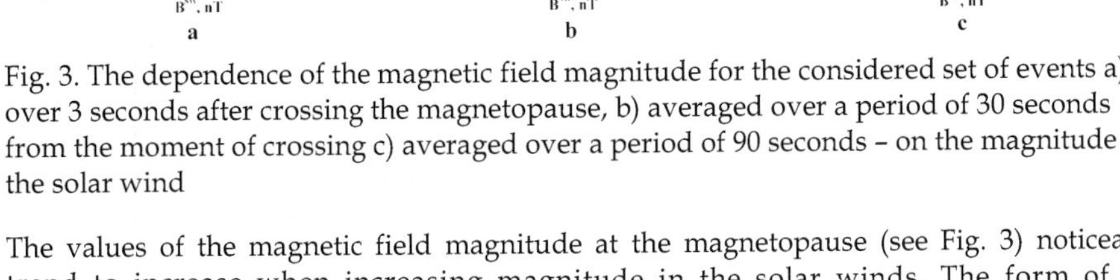

Fig. 3. The dependence of the magnetic field magnitude for the considered set of events a) over 3 seconds after crossing the magnetopause, b) averaged over a period of 30 seconds from the moment of crossing c) averaged over a period of 90 seconds – on the magnitude in the solar wind

The values of the magnetic field magnitude at the magnetopause (see Fig. 3) noticeably trend to increase when increasing magnitude in the solar winds. The form of the distribution remains essentially unchanged when the period of averaging is increased. In accordance with Fig. 4, the X-component of the magnetic field at the magnetopause does not depend on the corresponding value in the solar wind and fluctuates around zero, which is in accordance with the assumption of magnetopause as a tangential discontinuity. As well as for the field magnitude, the increase in averaging interval does not change the form of the

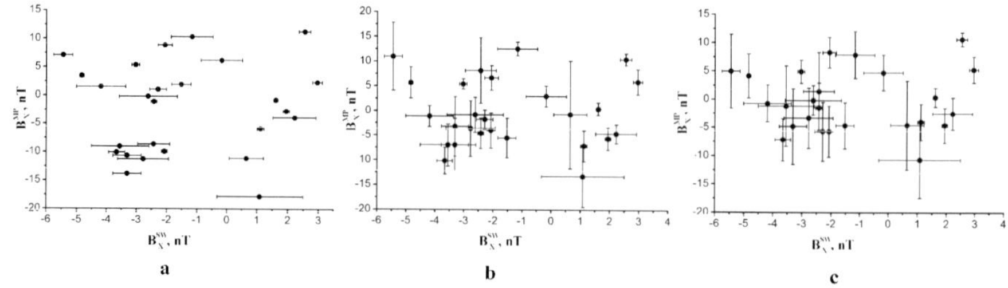

Fig. 4. The dependence of the x-component of the magnetic field for the considered set of events a) over 3 seconds after crossing the magnetopause, b) averaged over a period of 30 seconds from the moment of crossing c) averaged over a period of 90 seconds - on the B_x in the solar wind

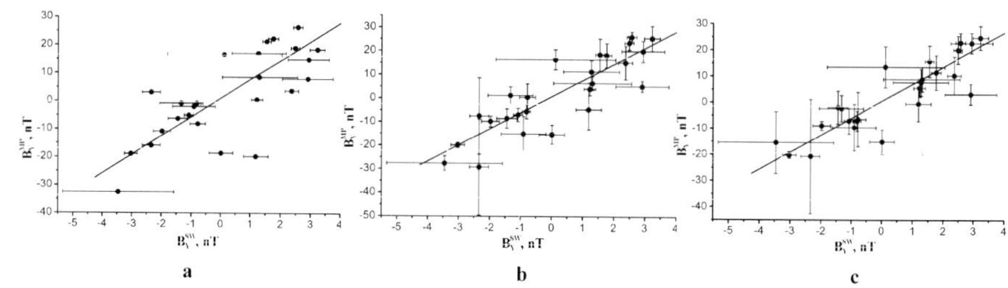

Fig. 5. The dependence of the y-component of the magnetic field for the considered set of events a) over 3 seconds after crossing the magnetopause, b) averaged over a period of 30 seconds from the moment of crossing c) averaged over a period of 90 seconds - on the B_y in the solar wind

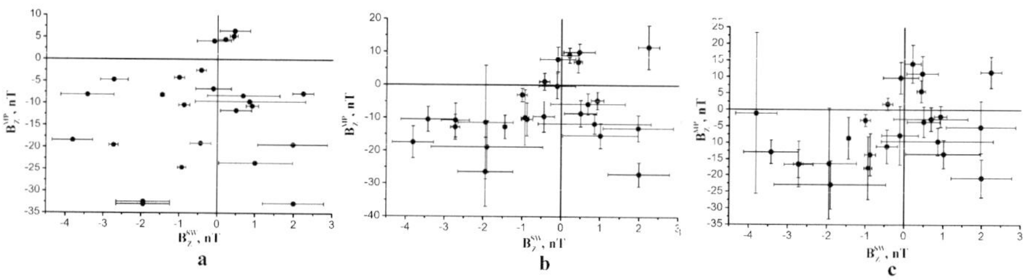

Fig. 6. The dependence of the z-component of the magnetic field for the considered set of events a) over 3 seconds after crossing the magnetopause, b) averaged over a period of 30 seconds from the moment of crossing c) averaged over a period of 90 seconds - on the B_z in the solar wind

distribution of points on the graph. A good linear dependence of the Y-component of the magnetic field at the magnetopause (see Fig. 5) on that in the solar wind is obtained. Therefore B_Y-component at the magnetopause, as it was shown at a stage of the first studies (see Fairfield (1967)), is comparatively well correlated with IMF B_Y. The correlation coefficient increases when increasing averaging period and the errors of the parameters of the approximation and standard deviation is decreased. The results obtained for the Z-component of the magnetic field (see Fig. 6) are of great interest. There is a vague tendency towards an increase in the value of this component with the increasing of corresponding value in the solar wind. However, in at least a quarter of cases (8 out of 26 for the instantaneous values, and 7 out of 26 for the values averaged over a period of 30 seconds and 90 seconds) the sign of the Z-component at the magnetopause changes compared with the sign of B_Z in the solar wind from positive value (in the solar wind) to negative (at the magnetopause), and in a few cases (1 for the instantaneous and 2 averaged over 30 and 90 seconds values) from negative to positive value. Therefore, the correlation of B_Z component near the magnetopause and IMF B_Z is practically absent.

3. Pressure balance at the magnetopause

The problem of pressure balance at the magnetopause continues to be one of the most actual problems of the physics of the magnetosphere as the solution of this problem is deeply connected with the solution of the problem of particle, momentum and energy penetration inside the magnetosphere and the formation of boundary layers. It is necessary to mention that the condition of pressure balance at the magnetopause was not analyzed in connection with Dungey's picture. Sibeck et al. (1991) named Dungey's model the "onion peel" model of magnetic merging. They stressed that "onion peel" model violates pressure balance at the magnetopause and therefore does not lead to a quantitative prediction of the magnetopause location as a function of IMF orientation. Sibeck et al. (1991), Sibeck (1994) verify the pressure balance relationship between the solar wind dynamic pressure and the location of the subsolar magnetopause. It was shown, that the pressure balance between the incident solar wind and the magnetospheric magnetic field determines the location of the dayside magnetopause.

The analysis of the validity of pressure balance at the magnetopause was produced using data of AMPTE/IRM by Phan et al. (1994), Phan & Paschman (1996). Plasma moments are obtained every 4.35 s. The total pressures P_{tot} are obtained by Phan et al. (1994), Phan & Paschman (1996) for cases of small and large magnetic shear across the magnetopause. The perpendicular thermal pressure $P_{\perp}$ and the magnetic pressure are used to calculate the total pressure. In the low-shear case, both perpendicular plasma pressure and magnetic pressure change significantly across the magnetopause and magnetosheath regions but their sum P_{tot} remains rather constant throughout these regions. In the high-shear case, the magnetosheath magnetic and plasma pressures both remain rather uniform in the entire region within 20 min preceding the magnetopause crossing, so that P_{tot} is also constant. Across the magnetopause, the plasma and magnetic pressures vary significantly, and their sum generally has a small jump across this boundary: a deficiency of P_{tot} on the magnetosheath side and an excess on the low latitude boundary layer (LLBL) side of the magnetopause are often observed. On average, the change of P_{tot} is less than 10%.

Panov et al. (1998) study the pressure balance at high latitude magnetopause using data of CLUSTER mission. Suggestion about thermal pressure isotropy was used during calculation of P_{tot}. It was shown that for most of the analyzed 154 magnetosheath-magnetosphere transitions the pressure balance between the two sides of the transitions was fulfilled within the error bars, the magnetosheath-to-magnetosphere pressure ratio was close to unity, within the range of 0.5 to 2.

THEMIS mission multisatellite observations give the opportunity to clarify the conditions of pressure balance near the subsolar point at the magnetopause having simultaneous observations inside and outside the magnetopause by two satellites quite near to its surface with 3 s resolution. THEMIS Flux Gate Magnetometer (FGM) and the Ion and Electron Electrostatic Analyzer (ESA) data from the THEMIS satellite mission can be used to determine the total plasma pressure with 3 s resolution (McFadden et al., 2008). The magnetometer measures the background magnetic field and its low frequency fluctuations (up to 64 Hz) with amplitudes of 0.01 nT in a range extending over six orders of magnitude (Auster et al., 2008). The electrostatic analyzers measure plasma over the energy range from a few eV up to 30 keV for electrons and 25 keV for ions (McFadden et al., 2008).

An analysis of the pressure balance on the magnetopause near the subsolar point was made for 18 crossings of the magnetopause by the THEMIS project satellites under magneto-quiet conditions by Znatkova et al. (2011). Dynamic and static pressures of plasma are determined, as well as magnetic pressure in the magnetosheath, and magnetic and plasma static pressure inside the magnetosphere. Variations of the total pressure have been studied in the case when one satellite is located inside the magnetosphere and another one stays in the magnetosheath near the magnetopause. It is demonstrated, that for 18 investigated events the condition of pressure balance at the subsolar point is valid on average with an accuracy of 7%, within measurement errors and under applicability of the approximation of anisotropic magnetic hydrodynamics to collisionless plasma of the magnetosheath and magnetosphere.

For this study the event July 22, 2007 of the magnetopause crossing have been selected, using the following criteria: the interplanetary magnetic field has a stable northward orientation and magnetosphere was very quite to exclude the contribution of changes of magnetic fields inside the magnetosphere as a source of magnetopause stress balance destruction. Fig. 7 shows the positions of satellite orbits in GSM coordinate system. Fig. 8a shows interplanetary magnetic field and solar wind dynamic pressure for analyzed event in accordance with Wind data shifted on the time delay of solar wind flow from Wind to the magnetopause (http://cdaweb.gsfc.nasa.gov/). Wind data were used as all THEMIS satellites were near to the magnetopause for the analyzed event. Arrow shows the selected moments of magnetopause crossings. It is possible to see analyzing Fig. 8a that IMF has stably northward orientation. Variations of geomagnetic parameters

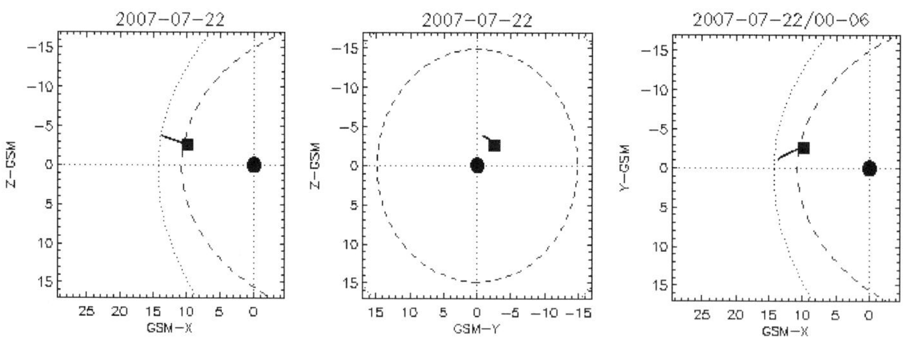

Fig. 7. Positions of THEMIS-C, -D,-E orbits for the event July 22, 2007

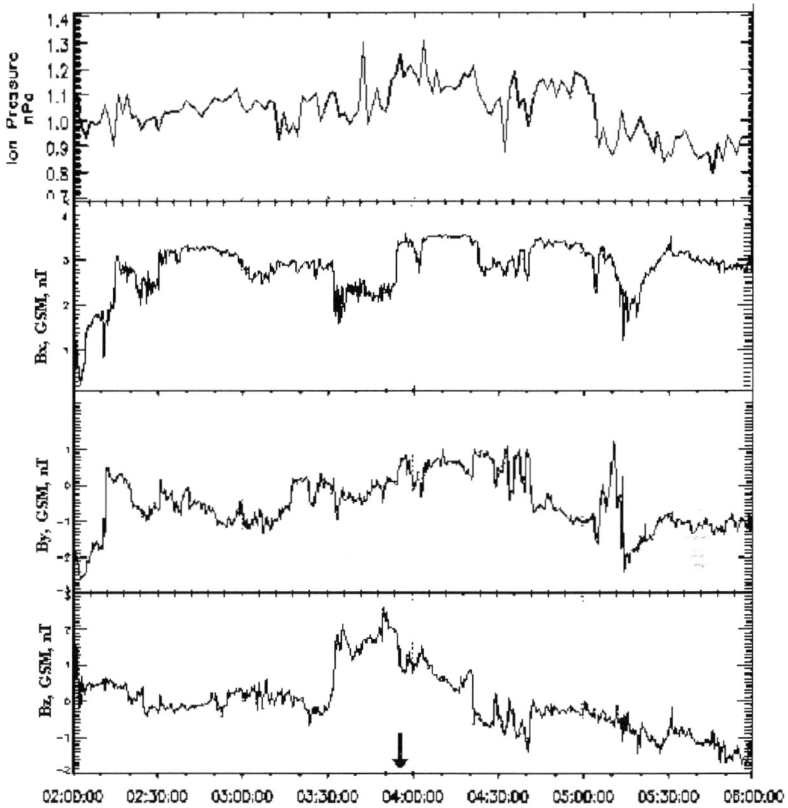

Fig. 8a. Solar wind dynamic pressure and IMF parameters for the event July 22, 2007

(http://swdcdb.kugi.kyoto-u/ac/jp) for the event July 22, 2007 are shown in Fig. 8b. Arrow shows the time for the discussed magnetopause crossings. Dst=-15 hT for the event July 22, 2007. It is possible to see analyzing Fig. 8b that magnetosphere was rather quite having the AE index below 100 nT.

Fig. 9 shows the results of the calculation of dayside magnetic field lines using Tsyganenko-2001 magnetic field model (Tsyganenko, 2002a,b). Stars show the positions of the regions at the magnetic field line where the magnetic field has the minimal value. The values of magnetic field in these regions are also shown. Magnetic field has minimal values at the equator in the inner magnetosphere. These minima are shifted from the equator near the magnetopause. Squares show the position of the equator. Figures near squares show the values of the magnetic field at the equator. It is necessary to mention that Tsyganenko model predicts the existence of closed field lines (dashed lines on Fig. 9) which correspond to the values of the magnetic field at the subsolar point, which can not produce the necessary magnetic pressure for the compensation of the pressure in the magnetosheath.

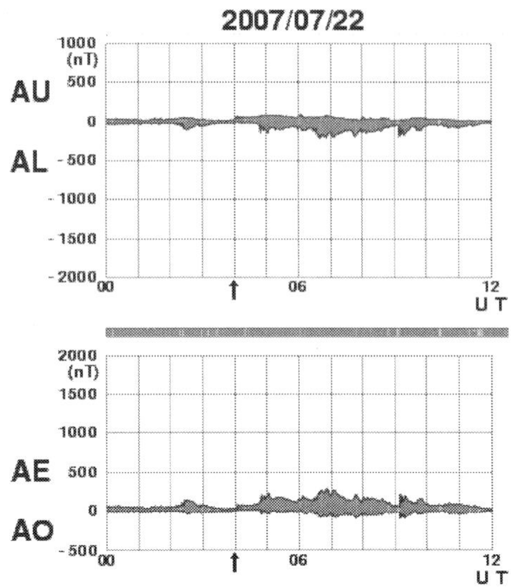

Fig. 8b. AU, AL, AE, AO indexes for the selected time interval in accordance with (http://swdcdb.kugi.kyoto-u/ac/jp)

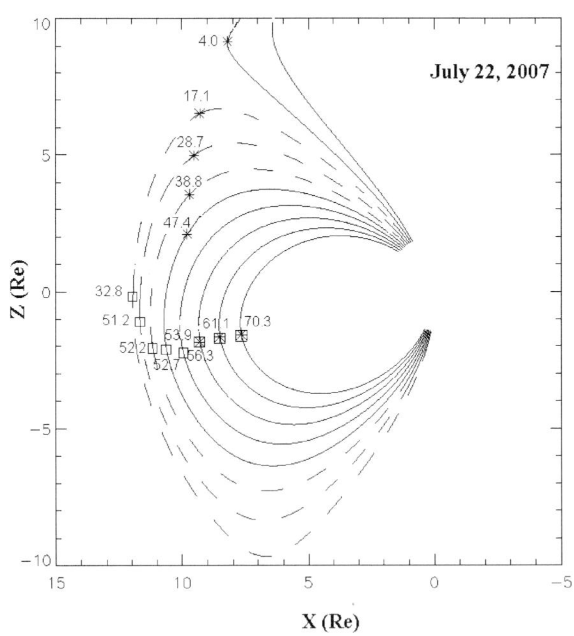

Fig. 9. Forms of the daytime magnetic field line at the Y=0 plane calculated using Tsyganenko-2001 model for solar wind parameters of the event July 22, 2007

Fig. 10 shows the example of magnetic field variations at the moment of THEMIS-D satellite crossing of the magnetopause July 22, 2007. It is possible to see great variations in the amplitude and direction of the magnetic field after the magnetopause crossing, typical for the magnetosheath. Fluctuations of magnetic field ~20 nT are observed. It is necessary to stress that magnetic field has the southward direction after magnetopause crossing in spite of the northward IMF orientating. The solar wind parameters are comparatively stable. Qualitatively the same pictures were observed on other THEMIS satellites. Comparison of the values of the model geomagnetic field (Fig. 9) and the measured geomagnetic field (Fig. 10) shows that amplitudes of magnetic fluctuations in the magnetosheath reach approximately 30 % of the value of magnetic field inside the magnetosphere at the subsolar point. Therefore, fluctuations of magnetic pressure in the magnetosheath can produce ~10% of fluctuations of the total pressure in the magnetosheath. Nevertheless, the same fluctuations at high latitudes, where the magnetic field at the magnetic field line is much smaller than the magnetic field at the subsolar point, the contribution of fluctuations of magnetic pressure is comparable with the magnetic pressure inside the magnetosphere.

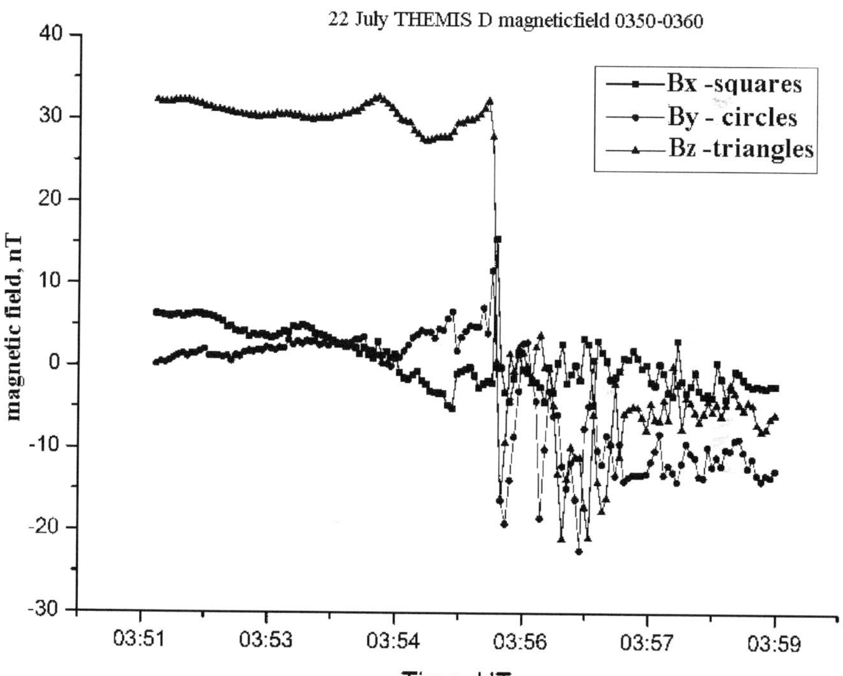

Fig. 10. Variations of THEMIS-D magnetic field July 22, 2007 on the interval 03:51-03:59 UT

The total pressure inside the magnetopause is calculated as a sum of static and magnetic pressure in accordance with the relation $P_{int}=P_{\perp}+(B^2/2\mu_0)$, where μ_0 is the magnetic permeability of vacuum; the integral pressure in the magnetosheath as a sum of dynamic, static and magnetic pressure $P_{int}= n_p\ m_p\ (v_p)^2+\ P_{\perp}+(B^2/2\mu_0)$ (see the results of Phan et al.

(1994) and discussion in (Znatkova et al., 2011)), where n_p and v_p are density and velocity (mainly X-component) of protons, m_i is the mass of proton, $P_\perp$ is the component of plasma pressure perpendicular to the magnetic field line. Fig. 11 a,b,c show the total plasma pressure for pairs of satellites for the periods when one satellite was inside the magnetosphere and the other one was inside the magnetosheath. Dashed line shows the results of calculations inside the magnetopause where the MHD analysis of pressure balance cannot be used due to dominance of kinetic effects. As it can be seen in Fig. 11 the total plasma pressure inside and outside magnetosphere is well balanced for all analyzed crossings of the magnetopause just as in the cases discussed by Znatkova et al. (2011). The main difference of the events July 22, 2007 and July 18, 2007 is connected to the total pressure dynamics. The total pressure is nearly constant for the event July 18, 2007 discussed by Znatkova et al. (2011). The total pressure growth is observed for the event July 22, 2007.

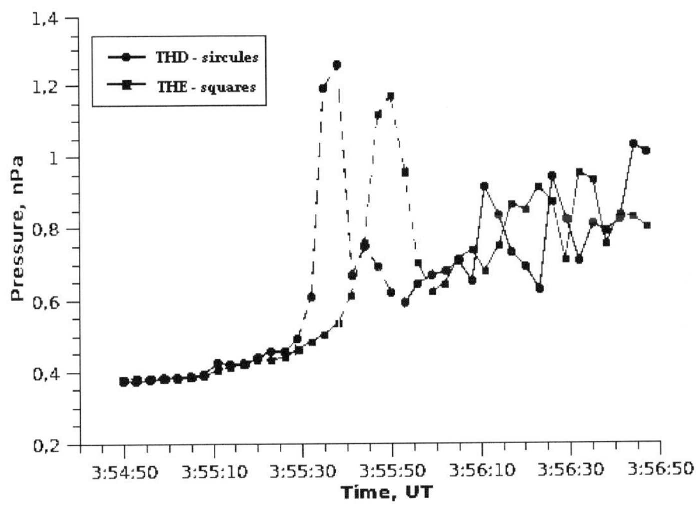

Fig. 11a. Simultaneous measurements of total pressure by the Themis-D and -E satellites for the event July 22, 2007

4. On the applicability of the frozen in condition to the processes in the magnetosheath

The most popular model of plasma flow in the magnetosheath is the gasdynamic model of Spreiter (see Spreiter et al. (1966), Spreiter & Alksne (1969), Spreiter & Stahara (1980)). The validity of this model was demonstrated by Němeček et al. (2000b), Zastenker et al. (2002). It was shown, that Spreiter model describes the parameters of averaged flows of the magnetosheath plasma rather well. MHD models also give the possibility to describe plasma flow parameters. The distribution of magnetic field in the magnetosheath is analyzed in such models using the frozen-in magnetic field approximation. However, it is

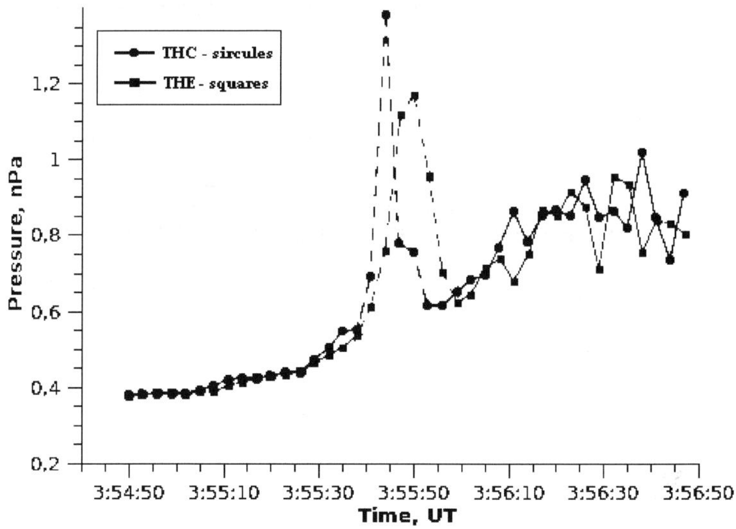

Fig. 11b. Simultaneous measurements of total pressure by the Themis-C and -E satellites for the event July 22, 2007

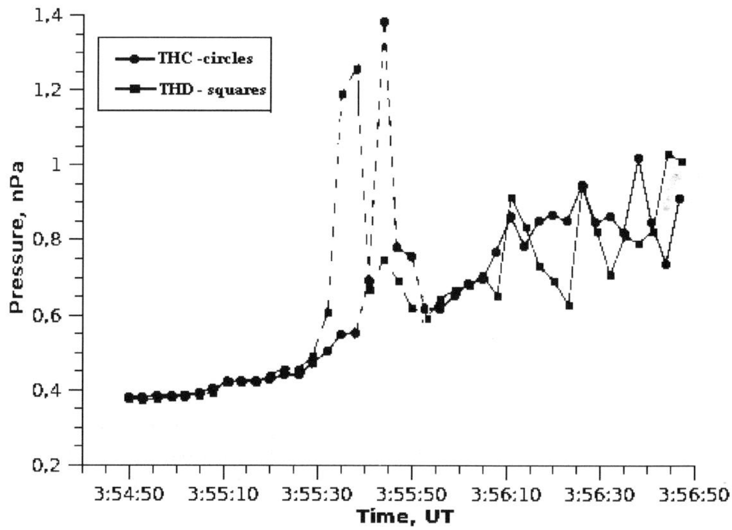

Fig. 11c. Simultaneous measurements of total pressure by the Themis-C and -D satellites for the event July 22, 2007

necessary to stress that frozen-in field condition is the result of the reduction of generalized Ohm's low which have the form in the case of isotropic plasma pressure (see Bittencourt (2004))

$$\mathbf{j}+\tau_{ei}\partial\mathbf{j}/\partial t=\sigma(\mathbf{E}+[\mathbf{VB}]+(\nabla p_e-[\mathbf{jB}])/en)$$

where **j** is the current density, τ_{ei} is the electron to ion collision time, **V** is the plasma velocity, **E** is the electric field, **B** is the magnetic field, σ is the conductivity, n is the plasma density and p_e is the electron pressure. $\mathbf{E}+[\mathbf{VB}]=0$ only if $\sigma\rightarrow\infty$ and it is possible to neglect by electron pressure gradient and Hall term. The Reynolds number in the case of Coulomb collisions in the magnetosheath is extremely high (2.5×10^{10} in accordance with (Borovsky & Gary, 2009)). However the existence of high level of turbulence requires the generalized definition of Reynolds number as a ratio of a dissipation time scale τ_{diss} to a convection time scale τ_{conv} for a flow structure ($R=\tau_{diss}\ /\ \tau_{conv}$) i.e. the introduction of an effective Reynolds number. Borovsky & Gary (2009) analyzing Landau damping and Bohm diffusion show that the effective Reynolds number is considerably reduced in the magnetosheath in comparison with the Coulomb collision Reynolds number. However, the evaluation of Reynolds number in the magnetosheath gives values >>1. That is why $\sigma\rightarrow\infty$ approximation can be used and $\mathbf{E}+[\mathbf{VB}]=([\mathbf{jB}]-\nabla p_e)/en$. Plasma velocity in the magnetosheath becomes low in comparison with Alfven and sound velocities near the subsolar magnetopause. This means that the equation of motion is reduced to the condition of magnetostatic equilibrium $[\mathbf{jB}]=\nabla(p_e+p_i)$. This means that $\mathbf{E}+[\mathbf{VB}]=\nabla p_i/en$ and frozen-in field condition can not describe the magnetic field in the magnetosheath. The observed plasma pressure anisotropy in the magnetosheath does not lead to principal changes of this conclusion. Results of Sections 2 and 3 also support this conclusion.

The violation of the frozen-in field condition is typical for the plasma sheet of the Earth (see, for example, Borovsky & Bonnell (2001), Troshichev et al. (2002), Stepanova et al. (2009, 2011)). Phan et al. (1994) considered the obtained using AMPTE/IRM results as an evidence for violation of the frozen-in field condition in the magnetosheath as well. Reconnection concept is based on the suggestion of the validity of frozen-in condition, which can be destroyed in a number of points or lines at the magnetopause. However, the results of observations and theoretical analysis show that frozen-in field condition is not applicable for the real magnetosheath and it is necessary to explain the observed penetration of plasma inside the magnetosphere using other suggestions.

5. Conclusions and discussion

Produced analysis demonstrates the real changes of the orientation of magnetic field in the magnetosheath in comparison with the magnetic field before the bow shock including the change of its sign. Poor correlation between the magnetic field in the magnetosheath with the IMF had been noted earlier (see Coleman (2005), Šafránková et al. (2009) and references in these papers). The presented results imply that the poor correlation, even at a relatively long averaging interval of 90 s, comparable with the time of solar wind plasma propagation through magnetosheath, is connected with the magnetosheath turbulization. In this study, due to the limited statistics (limited number of crossings the magnetopause by one of the spacecrafts, when the other was performing measurements upstream the foreshock region) we does not distinguish events with quasiperpendicular and quasiparallel shock waves. In accordance with the results of Shevyrev and Zastenker (2005) one can expect that the average level of fluctuations behind quasiperpendicular and quasiparallel shock waves will differ by about a factor of 2.

Presented results demonstrate the validity of the condition of pressure balance at the magnetopause with comparatively high accuracy and support the results obtained by Phan et al. (1994) using AMPTE/IRM observations. The main difference with Phan et al. (1994) results obtained in this work is connected to a possibility to make simultaneous measurements inside and outside the magnetopause at comparatively small distances using the particle and magnetic field data of the THEMIS satellite mission. Current analysis show that the total pressure is nearly constant during the all satellite crossing of the magnetosheath despite high level of turbulence constantly observed in the magnetosheath. These results show that the total pressure balance is the main condition determining the magnetosheath dynamics and the position of the magnetopause.

Significant number of observations in the magnetosheath, including the THEMIS satellite observations shows the great level of plasma turbulence. Fluctuations of the value and the direction of the magnetic field are constantly present even during periods of comparatively stable orientation of the interplanetary magnetic field. The amplitudes of these fluctuations are comparable with the minimal values of the magnetic field at the dayside magnetic field line (see Rossolenko et al. (2008) and Fig. 9 of this work). These fluctuations create an obstacle for the widely accepted picture shown on Fig. 1 of magnetic reconnection at the magnetopause. Real magnetic field near the magnetopause has different values and orientations at different points of the magnetopause. Which means that ordinarily discussed reconnection picture is nonapplicable for description of the processes near magnetopause. Fig. 12 shows more realistic then the scheme shown on Fig. 1 scheme of magnetic field in the magnetosheath when IMF has northward orientation. Large fluctuations of magnetic field at high latitudes where the magnetic field on the magnetospheric field lines is small (see Fig. 9) create the favorable conditions for field line interconnection and magnetosheath plasma penetration inside the magnetosphere. The events having large amplitudes will be selected as flux transfer events.

IMF, especially its B_z-component, is the major factor controlling the geomagnetic activity. It is usually assumed that this control is performed due to the processes of reconnection of the IMF and the magnetic field on the magnetopause and inside the magnetosphere. Numerous studies of turbulence in the magnetosheath, including the above analysis, give reason to reconsider such suggestion. The high level of magnetic field fluctuations in the magnetosheath, even for the relatively large averaging intervals, indicates that at different points of the magnetopause the magnetic field has different orientations, poorly correlated with the orientation of the IMF. Correspondingly [**VB**] also has different orientation at different points of the magnetosheath and near the magnetopause. That is why very popular idea on the solar wind electric field penetration inside the magnetosphere has the real obstacles. It is necessary to mention also that the suggestion about large-scale magnetospheric convection generation by boundary layer processes also meets with some obstacles (for example, boundary layers are mapped at the ionosphere near noon in accordance with Newell and Meng (1992)). However, this subject requires analysis that is more careful.

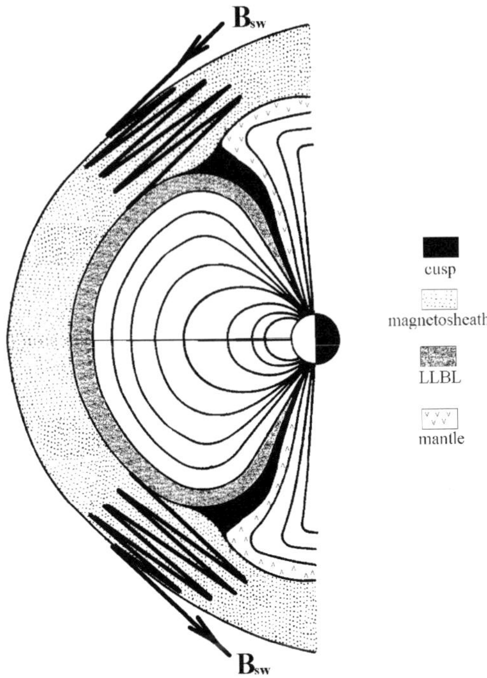

Fig. 12. Sketch illustrating the distribution of magnetic field in the magnetosheath when IMF has northward direction

The ideas about the role of large-scale reconnection processes at the magnetopause and formation of large-scale neutral lines were involved for explaining a relatively good correlation of IMF and large-scale magnetospheric convection. However, it is possible to explain such correlation without postulating the dominant role of reconnection processes at the magnetopause. Let to remind that Sibeck et al. (1991) made a conclusion that "onion peel" model cannot explain the change of magnetopause position under the influence of the interplanetary magnetic field. Sibeck et al. (1991), Tsyganenko & Sibeck (1994), Sibeck (1994) focused on the changes of values of sources of magnetic field inside the magnetosphere under the influence of the IMF and show that such changes can explain the change of magnetopause position under the influence of IMF. Results obtained by Sibeck et al. (1991), Tsyganenko & Sibeck (1994), Sibeck (1994) select the pressure balance at the magnetopause as the main condition determining the dynamics of the magnetopause. It was also shown that the changes of the dayside part of Region 1 currents of Iijima and Potemra under the influence of IMF could produce comparatively large changes in the magnetopause position. Developed by Sibeck et al. (1991) approach is not based on the suggestion of the validity of frozen-in condition, which can be destroyed in a number of points or lines at the magnetopause. The only think, which requires the explanation, is the well-known dependence of the magnetic field inside the magnetosphere on the IMF value and orientation.

Observed distribution of the plasma pressure inside the magnetosphere and of the Region 1 currents of Iijima and Potemra show that Region 1 current generation is a consequence of the existence of azimuthal plasma pressure gradients inside the magnetosphere (see (Iijima et al., 1997; Wing & Newell, 2000; Stepanova et al. 2004; Xing et al., 2009)). These results support the scenario proposed by Antonova & Ganushkina (1997) based on the analysis of the geometry of the high latitude magnetosphere as **j**=rot**B**. In this model, the field-aligned currents appear due to divergence of transverse magnetospheric currents. Therefore, the modulation of currents inside the magnetosphere by large-scale IMF can explain the change of magnetopause position under the influence of IMF irrespective to the orientation of the magnetosheath magnetic field near the magnetopause. It was stressed that the external source of the magnetic field in the condition of magnetostatic equilibrium when the gradient of plasma pressure is equal to the Ampere's force (i.e.$\nabla p=[\mathbf{jB}]$)) produces the increase of current in the case of the decrease of **B** (the addition of field with southward orientation) and the decrease of current in the case of the increase of **B** (the addition of field with northward orientation) when the plasma pressure gradients change slowly. Therefore, penetration of IMF inside the magnetosphere irrespective of magnetic field fluctuations in the magnetosheath can produce the necessary current modulation. The characteristic time of such modulation is determined by the Alfven travel time (i.e. the time of MHD wave flow from the magnetopause to the distances ~$10R_E$, where plasma pressure gradients which support the Region 1 currents are mainly concentrated according to Xing et al. (2009)). This time is ~2-3 min. Correspondingly, the characteristic time of the change of Region 1 currents and dawn-dusk electric field is of the same order of magnitude. Such estimation is in agreement with the results of radar observations of Ruohoniemi and Greenwald (1998), Ruohoniemi and Baker (1998), Ruohoniemi et al. (2001, 2002) who obtained a small time delay in the response of the high-latitude ionospheric convection to the IMF variations.

Change of the magnetic configuration under the influence of external magnetic field takes place even in the case of vacuum configuration. MHD models of magnetosphere, which do not suggest the validity of the frozen-in condition, clearly demonstrate such influence. Therefore, the solution of the problem does not require the action of reconnection processes at the magnetopause as a reason of IMF influence to the magnetosperic processes. Inversely, such reconnection processes can be a consequence of stress disbalance at the local regions of the magnetopause between the total pressure at the magnetosheth and the mainly magnetic pressure inside the magnetopause. Change of field line topology in such a case has a character of topological reconnection. Large-scale change of magnetic configuration leads to the magnetosheath plasma capture inside the magnetosphere. Such capture was observed, for example, 3 June 2007 by THEMIS satellites and interpreted as an action of double reconnection by Lee at al. (2008). It is interesting to mention that topological reconnection does not require local dissipation (and corresponding plasma heating). High level of magnetosheath turbulence (different values and orientation of magnetic field at different points near the magnetopause) suggests that the discussed reconnection have patchy character. The existence of topological reconnection does not exclude the action of different scenario of classical local reconnection with local destruction of magnetopause current due to development of local instabilities and current dissipation. It only gives the possibility to overcome difficulties related to the presence of high level of turbulence in the magnetosheath.

Suggested scenario helps to overcome difficulties connected with the discussed problem of particle penetration inside the magnetosphere in the conditions of high level of magnetic fluctuations in the magnetosheath. However, it does not exclude the traditionally discussed mechanisms of local destructions of magnetopause current sheet and flux transfer events formation, particle diffusion and development of Kelvin-Helmholts instability at the magnetopause. Future studies will give the possibility to evaluate the importance of each of such mechanisms.

6. Acknowledgments

The authors thank the group of developers of THEMIS mission and the support group of spacecraft data website http://www.nasa.gov/mission_pages/themis/.

We are grateful to V. Angelopoulos and participants in NASA-grant NAS5-02099 for the THEMIS mission realization. In particular: C. W. Carlson and J. P. McFadden for the use of ESA data, K.-H. Glassmeier, U. Auster, and W. Baumjohann for the FGM data, developed under the guidance of Technical University of Braunschweig and with financial support from the German Ministry of Economics and Technology, as well as the German Air and Space Center (DLR) under contract 50 OC 0302.

The work was supported by the grants of President of Russian federation MK-1579.2010.2, Russian Foundation for Basic Research 10-05-00247-a, FONDECYT grant 1110729.

7. References

Angelopoulos, V. (2008). The THEMIS Mission. *Space Sci. Rev.*, Vol. 141, pp. 5–34. doi: 10.1007/s11214-008-9336-1.

Antonova, E. E.; & Ganushkina, N. Yu. (1997). Azimuthal hot plasma pressure gradients and dawn-dusk electric field formation. *J. Atmos. Solar-Terr. Phys.* Vol. 59, No.11, pp.1343-1354.

Auster, H.U.; Glassmeier, K.H., Magnes, W. , Aydogar, O. , Baumjohann, W. et al. (2008). The THEMIS Fluxgate Magnetometer. *Space Sci. Rev.* Vol. 141, pp. 235–264.

Bittencourt, J.A. Fundamentals of Plasma Physics, Springer, 2004, 580 p.

Borovsky, J.E.; & Bonnell, J. (2001). The dc electrical coupling of flow vortices and flow channels in the magnetosphere to the resistive ionosphere. *J. Geophys. Res.* Vol. 106, No A12, pp. 28967-28994.

Borovsky, J. E.; & Gary, S. P. (2009). On shear viscosity and the Reynolds number of magnetohydrodynamic turbulence in collisionless magnetized plasmas: Coulomb collisions, Landau damping, and Bohm diffusion. *Physics of Plasmas.* Vol. 16, 082307. Available from http://pop.aip.org/pop/copyright.jsp

Chapman, S.; & Ferraro, V.C.A. (1931). *Terr. Magn.* Vol. 36, pp. 77.

Coleman, I. J. (2005). A multi-spacecraft survey of magnetic field line draping in the dayside magnetosheath. *Annales Geophysicae.* Vol. 23, pp. 885–900.

Dungey, J.W. (1961). Interplanetary magnetic field and auroral zone. *Phys. Rev. Lett..* Vol. 6, No 1, pp. 47-49.

Fairfield, D. H. (1967). The ordered magnetic field of the magnetosheath. *J. Geophys. Res.* Vol. 72, No 23, pp. 5865-5877.

Gutynska, O.; Ŝafránková, J. & Němeček, Z. (2008). Correlation length of magnetosheath fluctuations: Cluster statistics. *Annales Geophysicae.* Vol. 26, pp. 2503–2513.

Iijima, T.; Potemra, T.A. & Zanetti, L.J. (1997). Contribution of pressure gradients to the generation of dawnside region 1 and region 2 currents. *J. Geophys. Res.* Vol. 102, No A12, pp. 27069-27081.

Lavraud, B.; Foulton, C., Farrugia, C.J. & Eastwood, J.P. (2011). The magnetopause, its boundary layers and pathways to the magnetotail. *The Dynamic Magnetosphere,* ed. W. Liu and M. Fujimoto, IAGA Special Sopron Book Series, Springer, pp. 3-28.

Li, W.; Raeder, J., Øieroset, M., & Phan, T.D. (2009). Cold dense magnetopause boundary layer under northward IMF: Results from THEMIS and MHD simulations. *J. Geophys. Res.* Vol. 114, A00C15, doi:10.1029/2008JA013497.

Luhmann, J. G.; Russell, C. T., & Elphic, R. C. (1986). Spatial distributions of magnetic field fluctuations in the dayside magnetosheath. *J. Geophys. Res.* Vol. 91, pp. 1711–1715.

Lucek, E. A.; Dunlop, M.W., Horbury, T. S., Balogh, A., Brown, P., Cargill, P., Carr, C., Fornacon, K.-H., Georgescu, E., & Oddy, T. (2001). Cluster magnetic field observations in the magnetosheath: four-point measurements of mirror structures. *Ann. Geophys.* Vol. 19. pp. 1421–1428.

McFadden, J.P.; Carlson, C.W. , Larson, D., Ludlam, M. , Abiad, R. Elliott, B. Turin, P. Marckwordt, M. , & Angelopoulos, V. (2008). The THEMIS ESA plasma instrument and in-flight calibration. *Space Sci. Rev.* Vol. 141, pp. 277–302.

Němeček, Z.; Ŝafránková, J., Přech, L., Zastenker, G. N., Paularena, K. I., & Kokubun, S. (2000a). Magnetosheath study: INTERBALL observations. *Adv. Space Res.* Vol. 25, pp. 1511–1516.

Němeček, Z.; Ŝafránková, J., Zastenker, G. N., Pišoft, P., Paularena, K. I., & Richardson, J. D. (2000b). Observations of the radial magnetosheath profile and a comparison with gasdynamic model predictions. *Geophys. Res. Lett.* Vol. 27, pp. 2801–2804.

Němeček, Z.; Ŝafránková, J., Zastenker, G., Pišoft, P., & Jelínek, K. 2002a. Low-frequency variations of the ion flux in the magnetosheath. *Planet. Space Sci.* Vol. 50, pp. 567–575.

Němeček, Z.; Ŝafránková, J., Zastenker, G. N., Pišoft, P., & Paularena, K. I. (2002b). Spatial distribution of the magnetosheath ion flux, *Adv. Space Res.* 30(12), 2751–2756.

Newell, P.T. & Meng, C.-I. (1992). Mapping the dayside ionosphere to the magnetosphere according to particle precipitation characteristics. *Geophys. Res. Lett.* Vol. 19, No 6, pp. 609-612.

Panov, E. V.; Büchner, J., Fränz, M., Korth, A., Savin, S. P., Rème, H., & Fornaçon, K.-H. (2008). High-latitude Earth's magnetopause outside the cusp: Cluster observations. *J. Geophys. Res.* Vol. 113, A01220, doi:10.1029/2006JA012123.

Phan, T.-D.; Paschman G., Baumjohanann W., Sckopke N., & Lühr H. (1994). The magnetosheath region adjacent to the dayside magnetopause: AMPTE/IRM observations. *J. Geophys. Res.* Vol. 99, No A1, pp. 121-141.

Phan, T.-D & Paschman G. (1996). Low-latitude dayside magnetopause and boundary layer for high magnetic shear: Structure and motion. *J. Geophys. Res.* Vol. 101, No A4, pp. 7801-7815.

Riazantseva, M.O.; Zastenker G.N., Richardson J.D., & Eiges P.E. (2005). Sharp boundaries of small- and middle-scale solar wind structures. *J. Geophys. Res.* Vol. 110, A12110, doi:10.1029/2005JA011307.

Riazantseva, M.O.; Khabarova O.V., Zastenker G.N., & Richardson J.D. (2007). Sharp boundaries of solar wind plasma structures and their relationship to solar wind turbulence. *Adv. Space Res.* Vol. 40, pp. 1802–1806.

Rossolenko, S. S.; Antonova, E. E. , Yermolaev, Yu. I., Verigin, M. I., Kirpichev, I. P., & Borodkova, N. L. (2008). Turbulent fluctuations of plasma and magnetic field parameters in the magnetosheath and the low-latitude boundary layer formation: Multisatellite observations on March 2, 1996. *Cosmic Research.* Vol. 46, No 5, pp. 373–382.

Ruohoniemi, J. M.; & Greenwald, R. A. (1998). The response of high latitude convection to a sudden southward IMF turning. *Geophys. Res. Lett.* Vol. 25, No 15, pp. 2913-2916.

Ruohoniemi, J.M.; & Baker, K.B. (1998) Large-scale imaging of high-latitude convection with Super Dual Auroral Radar Network HF radar observations, *J. Geophys. Res.* 103(A9), 20797-20811.

Ruohoniemi, J.M.; Barnes, J.M., Greenwald, R.A., Shepherd, S.G., & Bristow, W.A. (2001). The response of high latitude ionosphere to the CME event of April 6, 2000: a practical demonstration of space weather now casting with the SuperDARN HF radars. *J. Geophys. Res.* Vol. 106, No A12, pp. 30085-30097.

Ruohoniemi, J. M.; Shepherd, S.G., & Greenwald, R.A. (2002). The response of the high-latitude ionosphere to IMF variations. *J. Atmosph. and Solar-Terr. Physics.* Vol. 64, No. 2, pp. 159-171.

Ŝafránková, J.; Hayosh, M., Gutynska, O., Němeček, Z., Přech, L. (2009). Reliability of prediction of the magnetosheath B_Z component from interplanetary magnetic field observations. *J. Geophys. Res.* Vol. 114, A12213, doi:10.1029/2009JA014552.

Savin, S.; Amata, E., Zelenyi, L., Budaev, V. et al. (2008). High energy jets in the Earth's magnetosheath: Implications for plasma dynamics and anomalous transport. *JETP Letters.* Vol. 87, No 11, pp. 593–599.

Shevyrev, N. N.; & Zastenker, G. N. (2005). Some features of the plasma flow in the magnetosheath behind quasi-parallel and quasi-perpendicular bow shocks. *Planet. Space Sci.* Vol. 53, pp. 95–102.

Shevyrev, N. N., Zastenker, G. N., & Du, J. (2007). Statistics of low-frequency variations in solar wind, foreshock and magnetosheath: INTIERBALL-1 and CLUSTER data. *Planet. Space Sci.* Vol. 55, No 15, pp. 2330–2335.

Sibeck, D.G.; Lopez, R.E., & Roelof, E.C. (1991). Solar wind control of the magnetopause shape, location and motion. *J. Geophys. Res.* Vol. 96, No A4, pp. 5489-5495.

Sibeck, D.G. (1994). Signatures of flux erosion from the dayside magnetosphere. *J. Geophys. Res.* Vol. 99, No A5, pp. 8513-8529.

Sibeck, D.G.; & Angelopoulos, V. (2008). THEMIS science objectives and mission phases,. *Space Sci. Rev.*. Vol. 141, pp. 35–59, doi: 10.1007/s11214-008-9393-5.

Sibeck, D.G.; Phan T.-D., Lin R.P. et al. (2000). A survey of MHD waves in the magnetosheath: International Solar Terrestial Program observations. *J. Geophys. Res.* Vol. 105, No A1, pp. 129–138.

Spreiter, J.R.; Summers, A.L., & Alksne, A.Y. (1966). Hydromagnetic flow around the magnetosphere. *Planet. Space. Sci.*. Vol. 14, No 3, pp. 223-253.

Spreiter, J.R.; & Alksne, A. Y. (1969). Plasma flow around the magnetosphere. *Rev. Geophys.*. Vol. 7, pp. 11-50.

Spreiter, J.R.; & Stahara, S.S. (1980). A new predictive model for determining solar wind-terrestrial planet interaction,. *J. Geophys. Res.* Vol. 85, No 12, pp. 6769- 6777.

Stepanova, M.V.; Antonova, E.E., Bosqued, J.M., Kovrazhkin, R. (2004). Azimuthal plasma pressure reconstructed by using the Aureol-3 satellite date during quiet geomagnetic conditions. *Adv. Space Res.* Vol. 33, No 5, pp. 737–741, doi:10.1016/S0273-1177(03)00641-0.

Stepanova M.; Antonova, E. E., Paredes-Davis, D., Ovchinnikov, I. L. & Yermolaev, Y. I. (2009). Spatial variation of eddy-diffusion coefficients in the turbulent plasma sheet during substorms. *Ann. Geophys.* Vol. 27, pp. 1407–1411.

Stepanova, M.; V. Pinto, J. A. Valdivia, & E. E. Antonova (2011) Spatial distribution of the eddy diffusion coefficients in the plasma sheet during quiet time and substorms from THEMIS satellite data. *J. Geophys. Res.,* Vol. 116, A00I24, doi:10.1029/2010JA015887.

Troshichev, O.A.; Antonova, E.E., & Kamide, Y. (2002). Inconsistence of magnetic field and plasma velocity variations in the distant plasma sheet: violation of the "frozen-in" criterion? *Adv. Space Res.* Vol. 30, No. 12, pp. 2683-2687.

Tsyganenko, N.A.; & Sibeck, D.G. (1994). Concerning flux erosion from the dayside magnetosphere. *J. Geophys. Res.* Vol. 99, No A7, pp. 13425-13436.

Tsyganenko, N.A. (2002a). A model of the near magnetosphere with a dawn-dusk asymmetry: 1. Mathematical structure. *J. Geophys. Res.* Vol. 107, No A8, doi: 10.1029/2001JA000219.

Tsyganenko, N.A. (2002b). A model of the near magnetosphere with a dawn-dusk asymmetry: 2. Parameterization and fitting to observations. *J. Geophys. Res.* Vol. 107, No A8, doi: 10.1029/2001JA000220.

Zastenker, G. N.; Nozdrachev, M. N., Němeček, Z., Ŝafránková, J., Přech, L., Paularena, K. I., Lazarus, A. J., Lepping, R. P., & Mukai, T. (1999). Plasma and magnetic field variations in the magnetosheath: Interball-1 and ISTP spacecraft observations, in: Interball in the ISTP Program, edited by: Sibeck, D. G. and Kudela, K., *NATO Science Series.* Vol. 537, pp. 277–294.

Zastenker, G. N.; Nozdrachev, M. N., Němeček, Z., Ŝafránková, J., Paularena, K. L., Richardson, J. D., Lepping, R. P., & Mukai, T. (2002). Multispacecraft measurements of plasma and magnetic field variations in the magnetosheath: Comparison with Spreiter models and motion of the structures. *Planet. Space Sci.* Vol. 50, pp. 601–612.

Znatkova, S. S.; Antonova, E. E., Zastenker, G. N. &Kirpichev, I. P. (2011). Pressure balance on the magnetopause near the subsolar point according to observational data of the THEMIS project satellites. *Cosmic Research.* Vol. 49, No. 1, pp. 3–20.

Wing, S.; & Newell, P.T. (2000). Quite time plasma sheet ion pressure contribution to Birkeland currents. *J. Geophys. Res.* Vol. 105, No A4, pp. 7793-7802.

Xing, X.; Lyons, L.R, Angelopoulos, V., Larson, D., McFadden, J., Carlson, C., Runov, A., & Auster, U. (2009). Azimuthal plasma pressure gradient in quiet time plasma sheet. *Geophys. Res. Lett.* Vol. 36, No 14, L14105. doi:10.1029/2009GL038881.

19

Solar Wind Sails

Ikkoh Funaki[1] and Hiroshi Yamakawa[2]
[1]*Japan Aerospace Exploration Agency*
[2]*Kyoto University*
Japan

1. Introduction

Magnetic sail (MagSail) is a unique but never-realized interplanetary propulsion system. The original idea of MagSail proposed by Zubrin is depicted in Fig.1 (Andrews & Zubrin, 1990). A MagSail spacecraft has a hoop coil, and it produces an artificial magnetic field to reflect the solar wind particles approaching the coil, and the corresponding repulsive force exerts on the coil to accelerate the spacecraft in the solar wind direction. MagSail and its derivatives are usually called as solar wind sail in contrast to solar light sail, which is propelled by the solar light pressure. Before realizing the ideas of solar wind sail, however, fundamental studies are required from both physical and engineering points of view, in particular, experimental validation of these ideas are very important before applying these concepts to realistic spacecraft design.

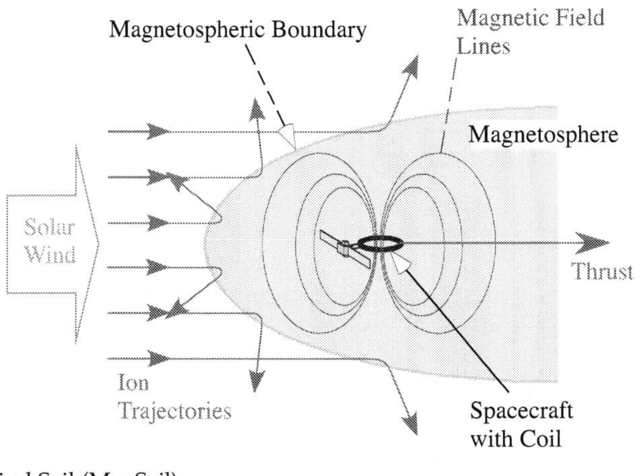

Fig. 1. Solar Wind Sail (MagSail).

To simulate a solar wind sail in laboratory, we fabricated a plasma wind tunnel facility (Funaki, 2007). Our special effort in laboratory experiment is directed to satisfy the similarity law associated with plasma flows of MagSail and its derivatives. Also, the laboratory experiment is intended for direct thrust measurement, which was never tried in the past. In this chapter, after introducing the basics of sail propulsion using the solar wind, details of experimental facility and experimental results of MagSail in laboratory are described.

2. Principle of MagSail

In this section, the idea and feature of MagSail are introduced and its thrust formula is provided.

2.1 Original MagSail by Zubrin

In 1991, Zubrin released the idea of interplanetary MagSail, which deploys superconducting cable after launched into space (Zubrin & Andrew, 1991). Before the launch of a MagSail spacecraft, a loop of superconducting cable, which is millimeter in diameter is attached on a drum onboard the spacecraft. After launch, the cable is released from the spacecraft to form a loop of tens of kilometers in diameter, then, a current is initiated in the loop. Once the current is initiated, it will be maintained in the superconductor without operating power supply. The magnetic field created by the current will impart a hoop stress to the loop for aiding the development and eventually forcing it to a rigid circular shape.

When MagSail is operated in interplanetary space, charged particles approaching the current loop are decelerated/deflected according to the B-field they experience. If the interacting scale length between the plasma flow and the magnetic field is large as in the cases of some magnetized planets like Earth and Jupiter, a magnetosphere (a magnetic cavity or a magnetic bubble) is formed around the current loop (Fig. 2). The solar wind plasma flow and the magnetic field are divided by a magnetopause, at which ions entering the magnetic field are reflected except near the polar cusp region where the ions can enter deep into the magnetic bubble. Due to the presence of the magnetosphere, the solar wind flow is blocked, creating a drag force exerting on the loop; thus a MagSail spacecraft is accelerated in the direction of the solar wind. The solar wind in the vicinity of the Earth is a flux of 10^6 protons and electrons per cubic meter at a velocity of 400 to 600 km/s. The maximum speed available for MagSail would be that of the solar wind itself.

2.2 Thrust of MagSail

One can easily imagine that a force on the current loop depends on the area blocking the solar wind. By increasing the blocking area, a large thrust force is expected. The force on the coil of a MagSail is therefore formulated as,

$$F = C_d \frac{1}{2} \rho_{sw} u_{sw}^2 S \tag{1}$$

where C_d is thrust coefficient, $1/2\rho_{sw}u_{sw}^2$ is the dynamic pressure of the solar wind, and S is the blocking area. In Eq.(1), $\rho_{sw}=m_i n$ is the density of the solar wind, m_i is the mass of an ion, n is the number density, and u_{sw} is the velocity of the solar wind.

To calculate F from Eq.(1), the blocking area S should be specified, but S is not priori given. It is therefore customary to employ the characteristic length, L, and to approximate the blocking area as $S=\pi L^2$. In this paper, we choose the standoff distance as the characteristic

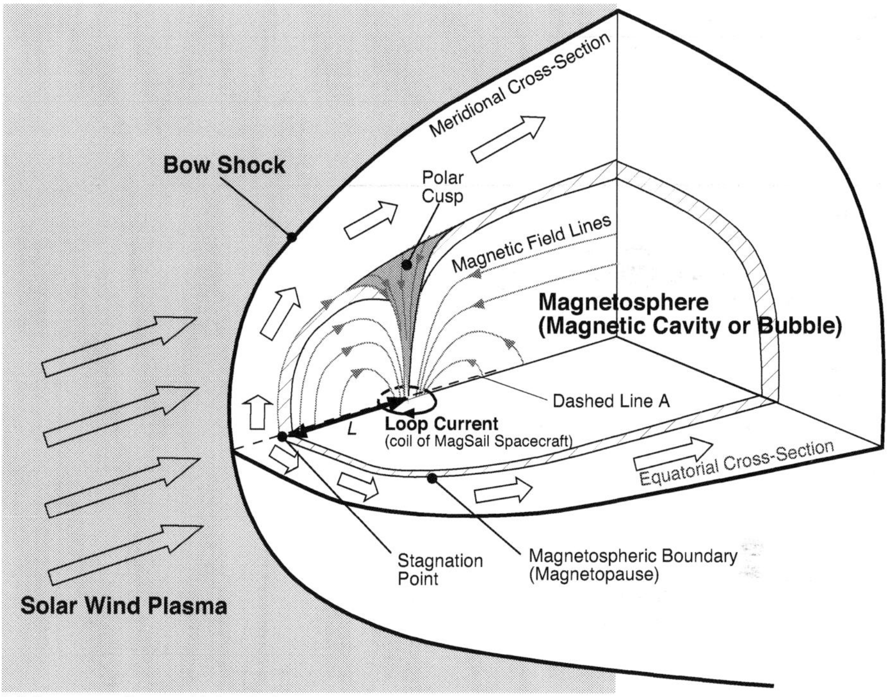

a) Magnetosphere of MagSail

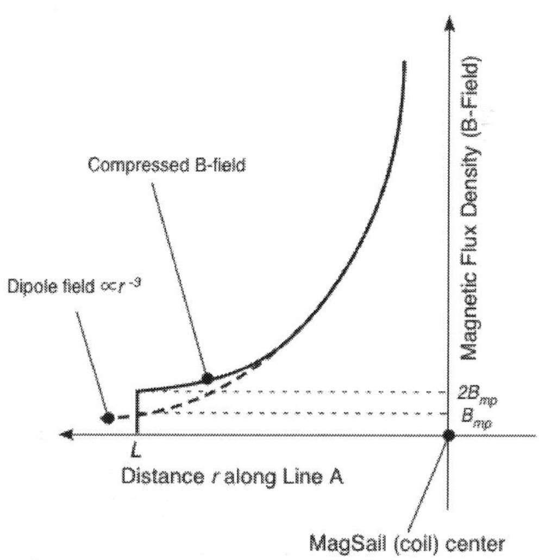

b) B-field distribution along stagnation line (dashed line A in a))

Fig. 2. Expected plasma and magnetic fields of MagSail.

length, which is the distance between the stagnation point and the center of the coil. L is derived from a pressure balance at the stagnation point (Obayashi, 1970) as

$$n m_i u_{sw}^2 = \frac{\left(2B_{mp}\right)^2}{2\mu_0}, \tag{2}$$

where B_{mp} is expressed using the magnetic moment, M_m, as

$$B_{mp} = \frac{\mu_0 M_m}{4\pi L^3}. \tag{3}$$

From these two equations, the standoff distance, L, is directly obtained as,

$$L = \left(\frac{\mu_0 M_m^2}{8\pi^2 n m_i u_{sw}^2}\right)^{1/6}. \tag{4}$$

Because L is uniquely determined from solar wind properties as well as the magnetic moment of MagSail, L is a reasonable choice for the characteristic length of the flow around MagSail.

Using eq.(1) and $S=\pi L^2$, the correlation between L and F is derived as

$$F = C_d \frac{1}{2} \rho u_{sw}^2 \pi L^2 \tag{5}$$

in which C_d in Eq.(5) is a fitted curve to numerical simulation results (Fujita, 2004):

$$C_d = \begin{cases} 3.6\exp\left(-0.28\left(\dfrac{r_{Li}}{L}\right)^2\right) & for \quad \dfrac{r_{Li}}{L} < 1 \\ \dfrac{3.4}{\left(r_{Li}/L\right)}\exp\left(-0.22\left(\dfrac{L}{r_{Li}}\right)^2\right) & for \quad \dfrac{r_{Li}}{L} \geq 1 \end{cases} \tag{6}$$

where r_{Li} is ion's Larmor radius defined afterwards. Thrust by Eqs.(5) and (6) is plotted in Fig.3 The MagSail by Zubrin (Zubrin & Andrews, 1991) required a spacecraft with a large

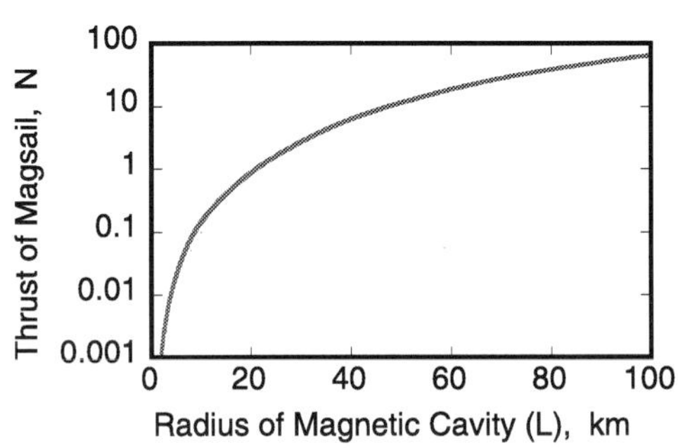

Fig. 3. Theoretical prediction of thrust by MagSail.

hoop coil of 64 km in radius to form 100-km-radius blocking area (which corresponds to approximately 30-N-class thrust). However, the dimension of the hoop coil was too large to realize. We therefore try to design a smaller MagSail than Zubrin proposed. From Fig. 3, one may notice that for L=20 km, 1 N is obtained. We selected the target of MagSail as 1 N, which corresponds to L=20 km; this is an adequate thrust level for medium-size (1,000-kg-class) deep space explorers.

2.3 Scaling parameters of MagSails

Empirical equation (6) indicates that thrust by MagSail is significantly influenced by the non-dimensional parameter, r_{Li}/L. In this section, similarity law for MagSail is explained after introducing two important scaling lengths, ion Lamor radius and skin depth.

2.3.1 Ion Larmor radius and skin depth

The boundary between a solar wind plasma and a magnetosphere is called magnetospheric boundary or magnetopause, where induced currents divide the plasma and the magnetic field regions. Magnetospheric boundary is schematically plotted in Fig.4. As in Fig.4, when charged particles impinge on the magnetospheric boundary from outside the magnetospheric boundary, ions can enter deep into the magnetosphere, whereas electrons are trapped and they are reflected back to the sun direction at the surface of the boundary. It is expected that the penetration depth of ions is comparable to ion Larmor radius:

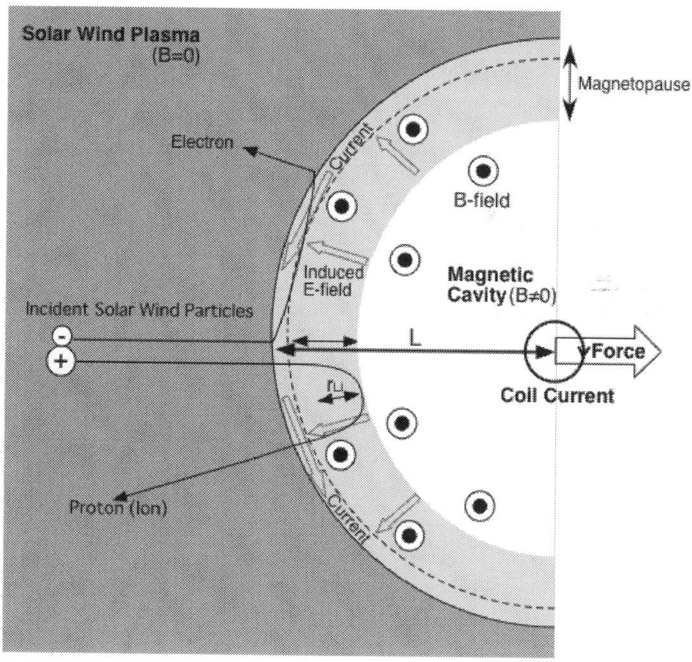

Fig. 4. Charged particle behaviour near magnetopause.

$$r_{Li} = \frac{m_i u_{sw}}{e \cdot 2B_{mp}} \tag{7}$$

from which r_{Li}=72 km if we put m_i=1.67x10^{-27} kg, u_{sw}=400 km/s, e=1.6x10^{-19} C, and B_{mp}=29 nT.

Disparity between ion's and electron's penetration depth at this boundary will also cause charge separation; this sheath region will induce an outward electric field. By this electric field, an electron is accelerated to a high velocity, resulting in a motion with large Larmor radius in the magnetosphere. Such electron motion dominates magnetospheric current at the outer edge of the magnetopause. In contrast to the electron motion at the boundary, ions are decelerated by the same electric field. The theoretical length of charge separation is the skin depth:

$$\delta = c/\omega_p \tag{8}$$

which is the decaying length of an electromagnetic wave incident on a plasma (Bachynski & Osborne, 1965); in eq.(8), c is the light speed and ω_p is the plasma frequency. For the solar wind plasma, δ ~1 km. Realistic penetration depth takes a value between δ and r_{Li} when background neutralizing electrons reduce the magnitude of the outward electric field (Nishida, 1982).

2.3.2 Non-dimensional scaling parameters of MagSail

In the following, four non-dimensional parameters and similarity law for MagSail in space are obtained in analogy to the similarity law of geo-magnetophysics (Bachynski & Osborne, 1965). The solar wind is a super sonic plasma flow, and it consists of collisionless particles: typical mean free path of the solar wind is about 1 AU. These features are described by high Mach number, M > 1 as well as very high magnetic Reynolds number, Rm >>1. M and Rm are defined as follows:

$$M = \frac{u_{sw}}{\sqrt{\gamma R T_{sw}}} \tag{9}$$

$$Rm = \sigma \mu_0 u_{sw} L \tag{10}$$

where σ is electric conductivity, R is gas constant, γ is specific heat ratio, and μ_0 is permeability in vacuum. Putting typical plasma velocity and temperature of the solar wind (u_{sw}=400 km/s and T_i~10 eV) into the above equation, M~8 is obtained. Also, the assumption of Coulomb collision gives σ=2x10^4/Ωm and Rm~10^8 for L=10 km. In addition to these two non-dimensional parameters, we defined r_{Li}/L, and δ/L, hence four non-dimensional parameters in total are introduced and they are listed in Table.1. Among them,

Parameters	MagSail in space	MagSail in laboratory
Mach number	~ 8	> 1
Ratio of ion larmor radius to L (r_{Li}/L)	~ 1	~ 1
Ratio of skin depth to L (δ/L)	< 1	< 1
Magnetic Reynolds' number (Rm)	~10^8	~ 10

Table 1. Non-dimensional parameters of MagSail.

the parameters Rm, r_{Li}/L, and δ/L, are functions of the size of the magnetosphere, in which L was selected as 10 km<L< 100 km in our study. Corresponding non-dimensional parameters are $0.72 < r_{Li}/L < 7.2$ (the ion gyration radius is comparable to or larger than L), which is in contrast to the MHD scale requiring $r_{Li}/L<<1$), and lastly, $\delta/L < 0.03$ (the skin depth is much smaller than L).

Since $\delta << r_{Li}$, the effective thickness of magnetopause is considered to be r_{Li}. As shown in Fig.5, if the thickness of the magnetopause is small enough in comparison to L, almost all of the incident ions are reflected at the magnetospheric boundary, hence large thrust on the coil of the MagSail is expected. Vice versa, if the thickness of the magnetopause is much larger than L, no interaction between the plasma flow and the magnetic field is anticipated. We treat a transitional region from the MHD scale (thin magnetopause mode) to the ion kinetic scale (thick magnetopause mode) in this experiment.

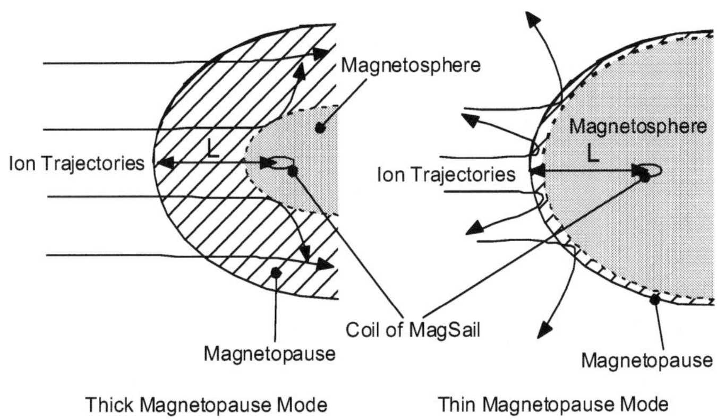

Fig. 5. Thick and thin magnetopause of MagSail.

3. Experimental setup for MagSail

Based on the above scaling consideration, experimental setup for MagSail is developed (Funaki, et al., 2006). The setup consists of a high-power solar wind simulator and a coil simulating MagSail spacecraft, both of which are operated in a quasi-steady mode of about 0.8-ms duration. In our experiment of MagSail (Fig.6), a coil simulating MagSail spacecraft (20-75 mm in diameter) was located at a downstream position of the solar wind simulator, and a plasma jet was introduced to a magnetic field produced by the coil. Typical snapshot is shown in Fig.7, in which plume jet as well as the coil simulating MagSail can be found.

A magnetoplasmadynamic (MPD) arcjet in Fig.8 was used as the solar wind simulator (SWS). The discharge chamber of the MPD arcjet is 50 mm in inner diameter and 100 mm in length. The MPD arcjet consists of eight molybdenum anode rods (8 mm in diameter) azimuthally located, and a short thoriated tungsten cathode rod 20 mm in diameter; they are surrounded by an annular floating body and insulators. These electrodes enable stable operate from a low current discharge range to an erosive high current discharge range. The SWS is attached on the space chamber inner wall as shown in Fig. 6.

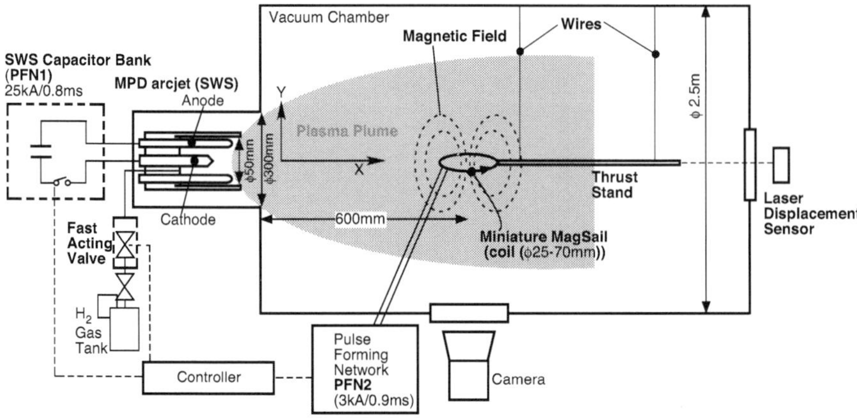

Fig. 6. Schematics of MagSail experimental facility.

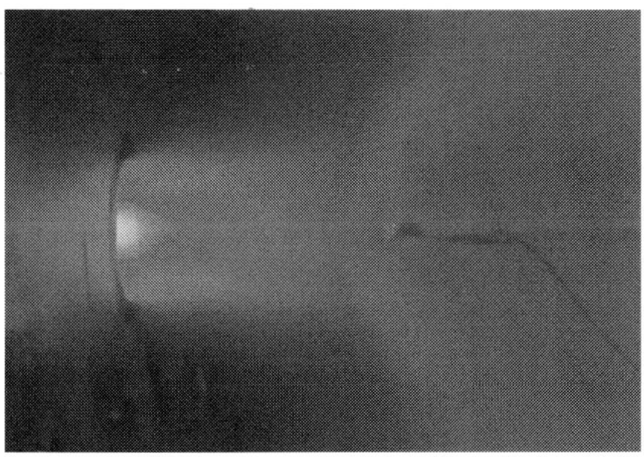

Fig. 7. Experiment of MagSail; typical snap shot.

Fig. 8. Magnetoplasmadynamic(MPD) arcjet as solar wind simulator (SWS).

A fast-acting valve (FAV) allowed us to feed gaseous propellants to the MPD arcjet. The hydrogen mass flow rate was controlled by adjusting the reservoir pressures. Timing of experiment is explained in Fig.9. After a gas pulse reaches its quasi-steady state, a pulse-forming network (PFN) is triggered. The PFN for SWS supplies the discharge current up to 20 kA with a flat-topped waveform in quasi-steady mode as shown in Fig.9. PFN for the coil (PFN2) supplies rather small current below 3 kA, and 20- to 75-mm-diameter 20-turn coil is required to produce up 2.0-T B-field at the center of the coil. After coil current (MagSail operation) reaches its steady state, discharge of the SWS (MPD arcjet) is initiated.

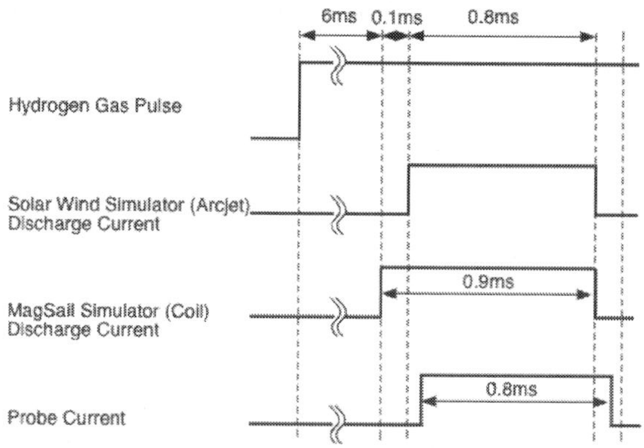

Fig. 9. Timing of MagSail experiment.

4. Experimental results of MagSail

In this section, after operational characteristics of MagSail in laboratory are described, thrust measurement results are provided. Then, the similarity law on a plasma flow of MagSail is discussed.

4.1 Operation of MagSail in lab

Discharge current profiles of the solar wind simulator (SWS) MPD arcjet and the coil are plotted in Fig.10. Flat-topped quasi-steady discharge continues about 0.8 ms in the case of SWS, and 1.0 ms in the case of the coil simulating MagSail spacecraft. High speed photos of MagSail experiment are shown in Fig.11. After the coil current (MagSail operation) reach its steady state, discharge of the MPD arcjet is initiated (Fig.11(a)). In Figs.11(b) and (c), it is observed that a wave front expands radially. This shock wave is followed by a quasi-steady plasma flow, that interacts with the magnetic field created by the solenoidal coil (Fig.11(d)).

During the quasi-steady interaction (0.25-1.0 ms in Fig.10), it was found that the plasma flow at the coil position was fluctuating. Averaged plasma parameters during the quasi-steady interaction were evaluated by Langmuir probe diagnostics as well as time-of-flight velocity measurement at the coil position. Results are listed in Table.2. The size of SWS's plume is 0.7 m in diameter (FWHM) at the coil position, and it is adequate for the laboratory experiment of Magnetic sail, which has typically 10-cm-size magnetic cavity. Scaling parameters can be calculated from the measurement in Table.2, but they are discussed afterwards in this chapter.

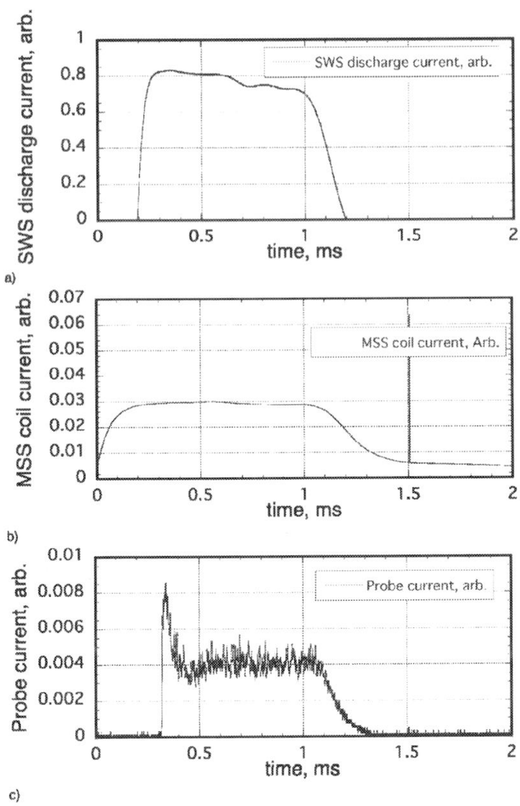

Fig. 10. Operation of solar wind simulator; a) discharge current profile of SWS, b) coil current profile, and c) plasma plume probe current profile (ion saturation current) at the coil position for H_2 0.4 g/s, charging voltage of PFN for SWS is 4 kV, and charging voltage of PFN for coil is 1.5 kV.

Plasma stream form hydrogen MPD solar wind simulator	
Velocity	20–45 km/s
Plasma density	10^{18}–10^{19} /m^3
Electron temperature	1 eV
Radius of plasma stream at the coil position	0.2–0.35 m
Plasma duration	0.8 ms
Coil current simulating MagSail in operation	
Radius of coil	9-37.5 mm
B-field at the center of coil	0–2.0 T
Duration of exciting current	0.9 ms

Table 2. Operating conditions and plasma parameters of solar wind simulator and coil simulating MagSail spacecraft.

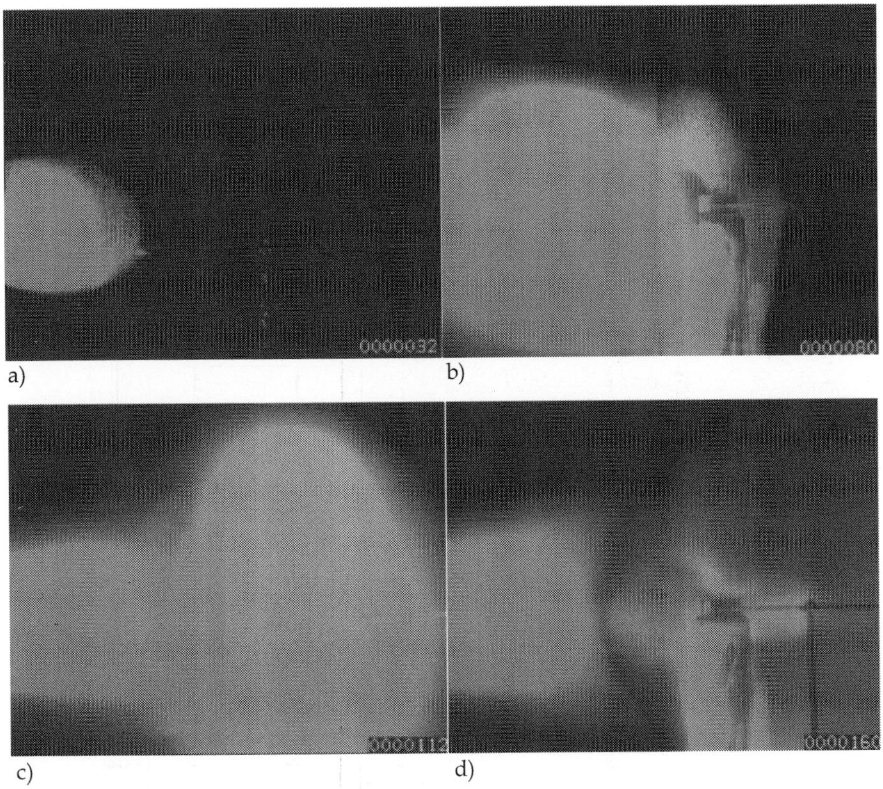

Fig. 11. High speed photos of pure MagSail experiment; a) just after initiating SWS (t=32 μs has passed after the discharge is initiated), b) t=80 μs, c) t=112 μs, and d) t=160 μs; d) corresponds to a quasi-steady state operation) in the case of 45 km/s, $2\text{x}10^{19}$ m^{-3} plasma flow from SWS, B-field at the center of coil is 1.8 T.

Another view of MagSail experiment is shown in Fig.12, in which a shutter camera was used to capture MagSail and its flowfield during a quasi-steady interaction. The most important feature of the interaction is the magnetospheric boundary between the SWS plasma flow and the low-energy plasma in the magnetic cavity. At the location of magnetospheric boundary, significant change of the magnetic field strength was found by a magnetic field measurement using a magnetic probe (Ueno, et al, 2009).

As far as we see Fig.12, the interaction seems very stable during quasi-steady operation of MagSail. However, if we see them with a high-speed camera, oscillatory magnetic cavity was observed (Oshio, et al., 2007; Oshio, et al., 2011). Thrust is hence produced under turbulent environment. Figure 13 shows high-speed photos when the discharge current of the MPD arcjet (J_{SWS}) is 11.6 kA. In Fig. 13, t = 0 corresponds to the time initiating the discharge of the SWS, and the time difference between each frame photograph is 2 μs, and shutter time is 0.5 μs. The interaction has already reached quasi-steady state in Fig. 13, in

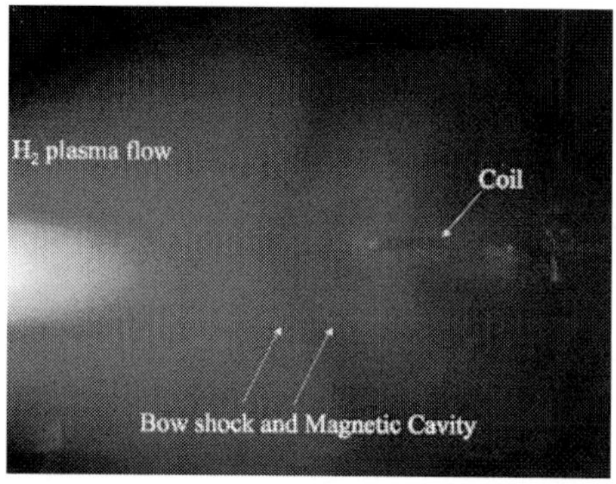

Fig. 12. Typical flow around MagSail during quasi-steady operation.

spite, an oscillating magnetosphere and plasma plume was observed. For example, when comparing the photos at 506 μs and at 516 μs in Fig. 13, the magnetospheric boundary is different. Also, when we see them in a longer time scale, it is found that the magnetospheric size shrinks and expands by about 20% repeatedly, and the averaged magnetospheric size is 0.15 m in the case of Fig. 13.

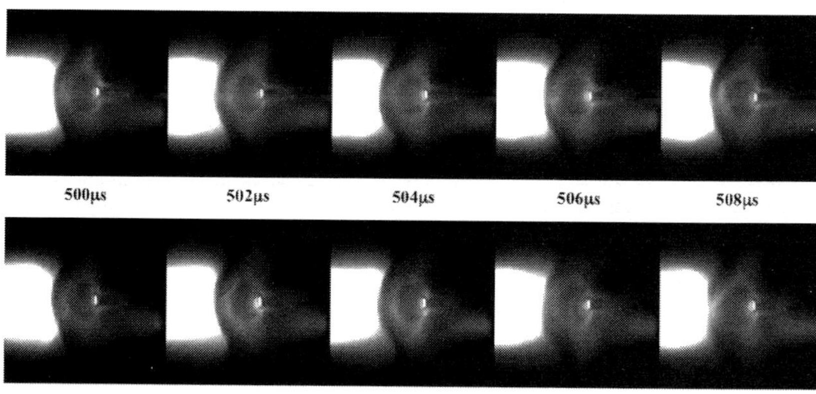

Fig. 13. High-speed photos of Magsail's magnetosphere during its quasi-steady operation; time corresponds to elapsed time after discharge of SWS is initiated. (J_{sws}=11.6 kA, J_{coil}=2 kA).

This magnetospheric fluctuation is characterized by high-speed photography based on the fact that the location of magnetopause corresponds to the dark regions in photos. Figure 14 shows the fluctuation of magnetospheric size for two plasma parameters (the discharge currents of SWS are 7.1 kA and 11.6 kA). The magnetospheric size is defined as a distance from coil center to the dark region. The averaged magnetospheric size is 145 mm in Fig.14a), and 80 mm in Fig. 14b). The magnetospheric size for J_{sws}=11.6 kA is larger than the case for J_{sws} =7.1 kA by about 10 mm, because the dynamic pressure of the simulated solar wind is smaller. The amplitudes of these fluctuations are about 10 mm, which corresponds to the fluctuation of thrust of about 25%. The power spectrum densities show that the dominant

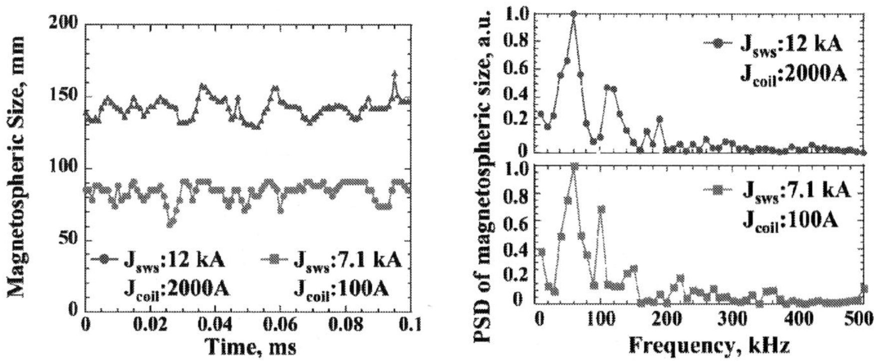

a) Time history of *L* b) Power Spectrum Density of *L*

Fig. 14. Magnetospheric size (L) fluctuation of Magsail; obtained from image analysis.

frequency was about 60 kHz. This frequency corresponds to a natural mode of bouncing magnetosphere. Such a bouncing magnetosphere is also expected in space in a frequency range of 1-10 Hz in the case of moderately sized MagSail (*L*~100 km).

4.2 Thrust measurement of MagSail

The magnetic cavity is blocking the plasma flow emitted from the MPD arcjet to produce a force exerting on a miniature MagSail spacecraft (coil). To evaluate this solar wind momentum to thrust conversion process, impulse measurements were carried out by the parallelogram-pendulum method (Ueno, et al., 2009; Ueno, et al., 2011). For the measurement, a coil simulating MagSail was mounted on a thrust stand suspended with four steel wires as shown in Fig.6.

The impulse of a Magnetic Sail is given by the following equation:

$$(F\Delta t)_{MagSail} = (F\Delta t)_{Total} - (F\Delta t)_{SWS} \tag{11}$$

When only the solar wind simulator is operated, the pressure on the coil surface produces impulse; this impulse corresponds to $(F\Delta t)_{SWS}$ in Eq. (5). If the coil current is initiated during the solar wind operation, the impulse, $(F\Delta t)_{Total}$, becomes larger than $(F\Delta t)_{SWS}$. Thrust by a Magnetic Sail is defined as the difference between the two impulses divided by the SWS operation duration (Δt = 0.8 ms):

$$F_{MagSail} = \frac{(F\Delta t)_{MagSail}}{\Delta t} \tag{12}$$

In the experiment, the displacement of the pendulum was measured with a laser position sensor. For the calibration of the pendulum and position sensor combination, impulses of known magnitude were applied to the coil simulating MagSail.

Displacement waveforms of the thrust stand were observed as in Fig.15 when a coil was immersed into the plasma flow. We can see that the maximum displacement for 1.1-kA coil

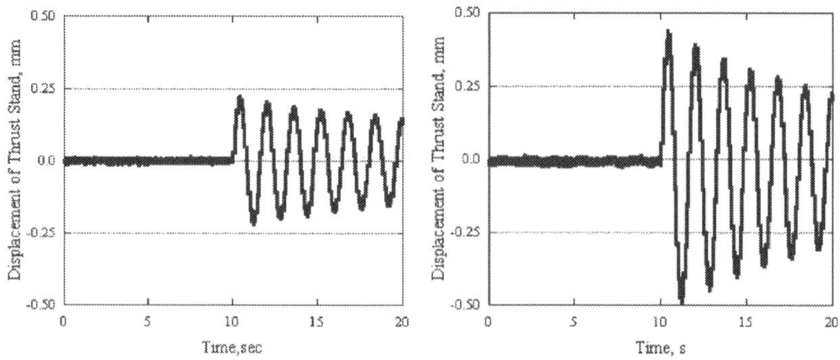

Fig. 15. Thrust stand's swing when 25-mm-diameter coil was immersed into hydrogen plasma flow (u_{sw}=47 km/s, n=1.8x10^{19} m^{-3}); a) without coil current, and b) 1.1 kA coil current (Simulator was initiated at 10s).

current in Fig.15b) is about two times larger than that without a coil current in Fig.15a). Figure 16 shows thrust data of MagSail for various magnetic moments where the magnetic moments were derived from the coil geometry and the coil current. It was confirmed that the thrust level is increased when increasing the magnetic moment of coils and thrust is proportional to $(M_m)^{2/3}$, which is consistent with eq.(5) because $L \propto (M_m)^{1/3}$.

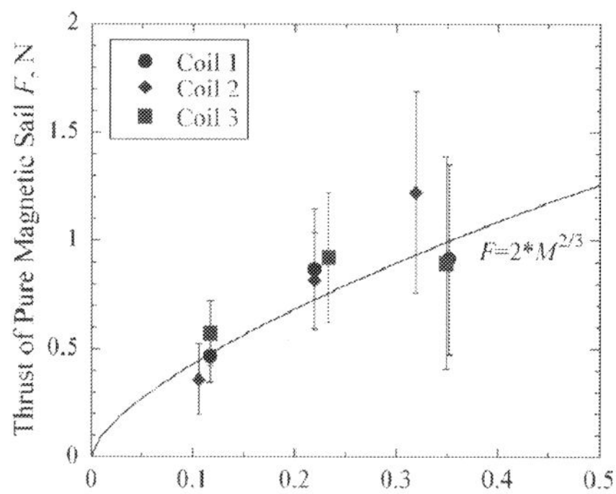

Fig. 16. Thrust vs. magnetic moment in the case of MagSail; SWS was operated for hydrogen plasma flow (u_{sw}=47 km/s, n=1.8x10^{19} m^{-3}), and three types of coils (radius=25 mm (coil 1,2) or 35 mm (coil 3)) are positioned at X=0.6 m.

If ions in a solar wind plasma penetrate into the magnetic cavity, the ions experience Larmor gyration. As was shown in Figs.4 and 5, if $r_{Li}<L$ at magnetospheric boundary, ions are reflected, but if $r_{Li}>L$, ions will not be reflected at the magnetospheric boundary but will penetrate deep into the magnetic cavity without producing thrust by MagSail. As a result, small $r_{Li}<L$ is required to obtain a significant thrust level. Such trend can be summarized in Fig.17, in which C_d (non-dimensional thrust) becomes smaller at a higher r_{Li}/L value.

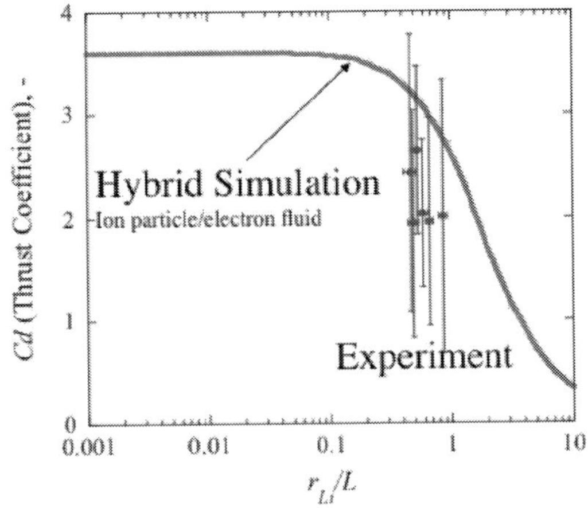

Fig. 17. Non-dimensional thrust characteristics of MagSail; C_d for various r_{Li}/L.

Through the experiment of MagSail, MHD to ion scale MagSail (r_{Li}/L=0.1~1.0) was demonstrated in laboratory.

Seeing the detail of Fig.17, however, one can see that laboratory experiment shows rather small thrust level than that by numerial simulation. This discrepancy is caused as a result of the collisional effect that is inevitable in laboratory [Kajimura et al., 2010], and it means that the laboratory experiment cannot predict the entire feature of collisionless plasmas in space. In the laboratory experiment, the momentum of a plasma flow in a vacuum chamber decreases as a result of collisions with neutral particles, leading to a decrease in thrust level.

4.3 Checking the similarity law

As described before, laboratory experiment is prepared so that the plasma flow follows the similarity law of MagSail/MPS in space. In the following, similarity law including the collisional effect in laboratory is checked based on measured plasma parameters.

When the hydrogen propellant is ionized, protons without a bounded electron are produced. Most of the plume plasma jet released from the SWS-MPD consists of atoms and protons, but molecules are minor species. By estimating the degree of ionization, the scaling parameters in our scale model are to be discussed.

From the data in Table.1, the radius of the plume plasma was 0.35 m at the coil position (X=0.7 m), which indicates that the plume divergence angle was 45 degree. The mass flow rate of a plasma jet from the MPD arcjet is the sum of a neutral flow rate and an ion flow rate,

$$\dot{m} = \left(m_n n_n u_n + m_i n_i u_i\right) A \tag{13}$$

From eq.(13), n_n=1.7x10^{19} m^{-3} is obtained for $\dot{m}$ =0.4 g/s, u_n=u_i=45 km/s, and the cross-sectional area of SWS's plume $A=\pi$ (0.35)^2 m^2. In a realistic MPD arcjet flow, however, the velocity slip between the ions and the atoms are usually found. Including this effect, n_n=3.4x10^{19} m^{-3} for u_i=45 km/s, and u_n= 22.5 km/s. This estimation gives ionization ratio

α=0.2. Using the obtained species concentrations, the mean free paths can be calculated, and an example is shown in Table 3 for typical operational parameters of SWS.

Variable		Expression or assumption	Value
Electron density	n_e	measured typical value	2.0×10^{18} m^{-3}
Electron temperature	T_e	measured typical value	11640 K
Ion/atom temperature	T_i	$T_i \approx T_e$ is assumed	11640 K
Ion velocity	u_i	measured typical value	45 km/s
Thermal velocity	$u_{th,e}$	$\sqrt{8kT_e/m_e\pi}$	6.7×10^5 m/s
	$u_{th,i}\left(\approx u_{th,n}\right)$	$\sqrt{8kT_i/m_i\pi}$	1.6×10^4 m/s
Size of magnetic cavity	L	design value	0.1 m
Mean free path	λ_{ei}	$u_{th,e}/\nu_{ei}$	0.012 m
	λ_{en}	$1/n_n\sigma_{en}$	0.074 m
	λ_{nn}	$1/\sqrt{2}n_n\pi d^2$	0.60 m
Collision frequency	ν_{ei}	$3.64\times10^{-6}n_e\ln\Lambda/T_e^{3/2}$	5.4×10^7 /s
	ν_{en}	$u_{th,e}/\lambda_{en}$	9.1×10^6 /s
	ν_{nn}	$u_{th,n}/\lambda_{nn}$	2.6×10^4 /s
Electric conductivity	σ	$n_e e^2/m_e\left(0.51\nu_{ei}+\nu_{en}\right)$	2040
Non-dimensional variables		**Expression**	**Value**
Knudsen Number	(electron-ion)	λ_{ei}/L	0.12
	(electron-atom)	λ_{en}/L	0.74
	(atom-atom)	λ_{nn}/L	6
Magnetic Reynolds Number	Rm	$\sigma\mu_0 u_i L$	12

Table 3. Operational, measured, and scaling parameters of MagSail in laboratory.

In this experiment, coil diameter (18 mm) and magnetospheric size (L~100 mm) are smaller than the mean free path of hydrogen atoms, hence neutral flow behaves as collisionless particles. Proton and hydrogen atoms also behave as collisionless particles, but electron-heavy particle collision is significant and hence the mean free path of electrons is smaller than L. Let's consider electron motions in laboratory experiment. An electron coming into the magnetosphere is trapped in the magnetosphere, and then it makes a mirror movement in a magnetic flux tube. During such a mirror motion, an electron moves to another magnetic flux tube as a result of a collision with a heavy particle, or completely released from the magnetosphere. Such diffused electron motions will change the skin depth of a magnetopause from δ to δ_D (Bachynski & Osborne, 1965)

$$\delta_D = \frac{c^2\left(0.51\nu_{ei}+\nu_{en}\right)}{\omega_p\,u_{sw}} \tag{14}$$

It is derived that the ratio of δ_D to L is equal to the reciprocal of Rm as

$$\frac{\delta_D}{L} = \frac{1}{Rm} \tag{15}$$

Also, converting the mean free path data to collision frequencies, electric conductivity is evaluated as

$$\sigma = \frac{e^2 n_e}{m_e (0.51 \nu_{ei} + \nu_{en})} \tag{16}$$

where ν_{ei} is electron-ion collision frequency and ν_{en} is electron-neutral collision frequency. Using the values in Table.3, σ is calculated as 2040/ Ωm; then Rm=12 is obtained for u_i=45 km/s and L=0.1m. In this case, the conditions for Rm (Rm>>1) and δ_D/L <<1 are satisfied.

By further tuning SWS (by increasing Rm) more suitable plasma flow for MagSail in laboratory might be provided. To further enlarge Rm, n_n should be decreased and T_e should be increased (see the equations in Table.3).

5. Extension of MagSail: M2P2 and magnetoplasma sail

5.1 The principle of M2P2 and magnetoplasma sail

For a MagSail in space, we must create a large magnetosphere to produce a significant thrust level because the dynamic pressure of the solar wind is very small even around the Earth. Using eqs.(4) and (5), thrust level of Magsail F is rewritten as

$$F = C_d \frac{1}{4} \left(\rho u_{sw}^2 \right)^{5/3} \left(\pi \mu_0 \right)^{1/3} \cdot M_m^{\ 2/3} \tag{17}$$

This equation shows that F is proportional to $(M_m)^{2/3}$, where $M_m = \mu_0 n J \pi r_c^2$ is the magnetic moment. Hence, a coil with a large diameter (r_c), a high current (J), or many turns (n) are required to obtain a high thrust level. In the case of MagSail, a large magnetosphere might be formed using a coil of several tens of kilometer in diameter; for example, Zubrin proposed 10-N-class MagSail spacecraft equipped with a 64-km-diameter coil. This design is, however, impractical at the current technology level.

In order to overcome this issue, the idea to make a large magnetosphere by a compact coil diameter (~several meters) with a plasma jet was proposed instead of employing a large-scale coil. This propulsion system, illustrated in Fig.18, is called mini-magnetospheric plasma propulsion (M2P2) or MagnetoPlasma Sail (MPS). This idea gained great interest because the study by Winglee et al. (Winglee, et al., 2001) suggested the possibility of dramatically large acceleration by M2P2 spacecraft in deep space.

The M2P2/MPS sail uses an artificially generated magnetic field, which is inflated by the injection of high- or low-energy plasma. As shown in Fig.19, based on either magnetic field inflation concept either by high-velocity plasma jet or by equatorial ring-current, either of which initiates additional plasma current inside a magnetosphere to increase the magnetic moment of the system. Magnetic field inflation by plasma jet is carried out when a collisionless plasma jet is introduced in a magnetic field, and magnetic field is frozen into the plasma flow so that the plasma flow expands the magnetic field. In contrast to the method

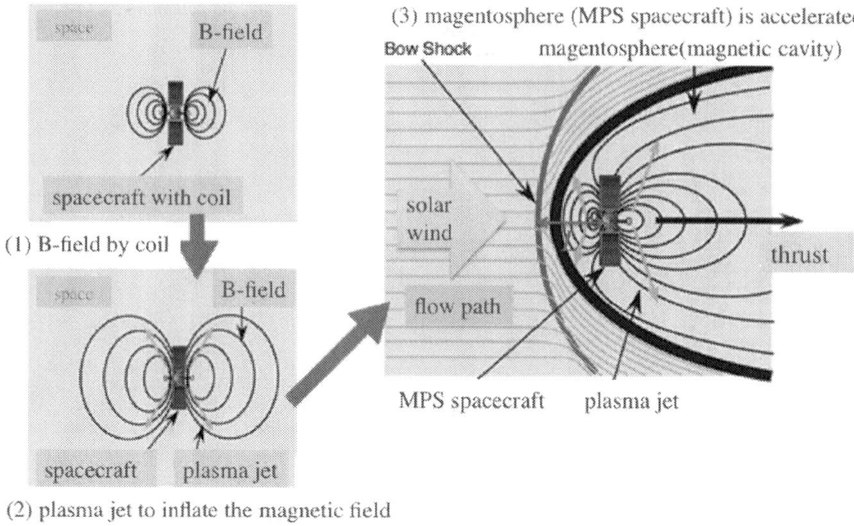

Fig. 18. M2P2/MPS concept.

using a radial flow continuously flowing out, when a low-energy plasma is trapped in the magnetosphere, equatorial ring-current is initiated due to drift motions of both ions and electrons. These plasma currents may allow the deployment of the magnetic field in space over large distances (comparable to those of the MagSail) with the B-field strengths that can be achieved with existing technology (i.e., conventional electromagnets or superconducting magnets). Additionally, one potential significant benefit of the M2P2/MPS would be its small size of the hardware (even though the magnetic bubble is very large); this would eliminate the need for the deployment of large mechanical structures that are presently envisaged for MagSail or solar light sails (Frisbee, 2003, Yamakawa, et al., 2006).

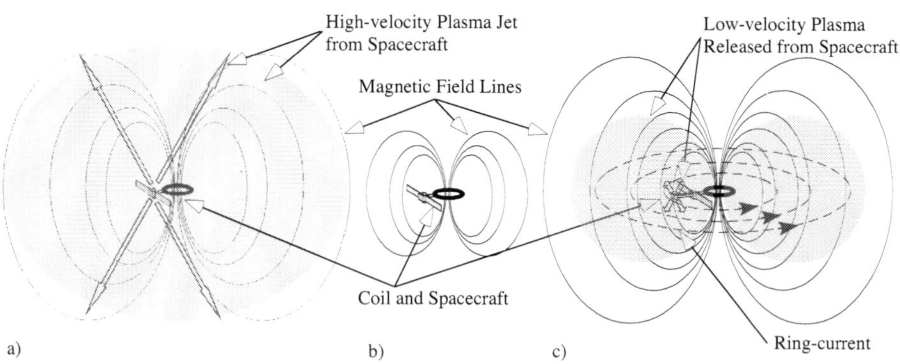

Fig. 19. Magnetic field inflation concept by high-velocity plasma jet or by equatorial ring-current.

However, feasibility of the M2P2/MPS sail concept is not established yet (Khazanov, et al., 2005; Omidi & Karimabadi, 2003}. From the physical point of view, we have questions on the principle of thrust production: under what operational condition is the M2P2/MPS concept valid?; is efficient thrust production really possible for M2P2/MPS? These are still open questions. Also, from the spacecraft design point of view, it is not known that if M2P2/MPS obtains much higher acceleration than that of existing technology such as electric propulsion or solar sail. We are continuing our efforts on theoretical and experiment studies to reveal these issues.

5.2 Preliminary experiment of M2P2/MPS

In contrast that a lot of experiments revealed the physics and thrust characteristics of MagSail, only a few data are available in the case of M2P2/MPS. The first experiment on the magnetic field inflation concept was tried by Winglee and his group (Winglee, et al., 2001; Ziemba, et al., 2001; Slough, et al., 2001; Ziemba, et al., 2003; Giersch, et al., 2003). A high-density helicon or an arc plasma source is located in a drum-like solenoid to observe magnetic field inflation. The demonstration was conducted in a space chamber of 1 m in diameter. He also tried full demonstration of M2P2. However, due to facility limitaion, they cannot demonstrate some important features of M2P2; for example, 1) the entire interaction of M2P2 is not tested due to limited diameter of the solar wind simulator, and 2) what amount of momentum transfers from the solar wind to the spacecraft (coil) are not validated.

Scaled down model approach by our group with a very small plasma source from the center of the coil seems reasonable to simulate the whole system of a M2P2/MPS sail. In Fig.20, an experimental system for MPS is depicted, which consists of a high-power magnetoplasmadynamic (MPD) solar wind simulator (MPD_SWS) and a miniature MPS spacecraft (Funaki, et al., 2007, Ueno, et al., 2009). Close-up views around the miniature MPS spacecraft (scale model) are shown in Fig.21. The miniature MPS spacecraft has a solenoidal coil and MPD arcjets (MPD_Inf, 20-mm-diameter small MPD arcjets) for plasma jet injection. All of these devices (the MPD_SWS, the MPD_Inf, and the coil) are operated again in a quasi-steady mode of about 1 ms duration. To describe the physics of the magnetic field inflation process, the most important parameter is r_{Li}/L_{inf}. In this parameter, L_{inf} is the frozen-in point where local β value is unity, so the B-field inflation occurs. Strictly speaking, L_{inf} is defined as a distance from the coil center to a typical frozen-in point. Also, the B-field inflation process requires small ion gyration radius when the B-field inflation is possible in the MHD approximation. Hence, two inequalities, $r_{Li}/L_{inf} <1$ and $L_{inf} <L$, are added to the inequalities in Table.1. In our preliminary experiment, the miniature MPS spacecraft with a coil of 50-75 mm in diameter was located in a downstream position of the MPD arcjet (SWS) to produce up to 2-T magnetic field at the center of the coil. Into this magnetic field produced by the miniature MPS spacecraft, a plasma jet from the MPD_SWS was introduced to see possible interaction.

Typical snapshot of MPS experiment is shown in Figs.22 and 23. It is obvious that the incoming solar wind plasma flow is blocked by the magnetic cavity in both the MagSail mode as shown in Fig.23(a) and in the miniature MPS mode in Fig.23(b). Comparing these photos, it is observed that the magnetosphere is inflated from an arched shape in Fig.23(a) to a rather flat shape in Fig.23(b) by the plasma injections from inside the solenoidal coil. The

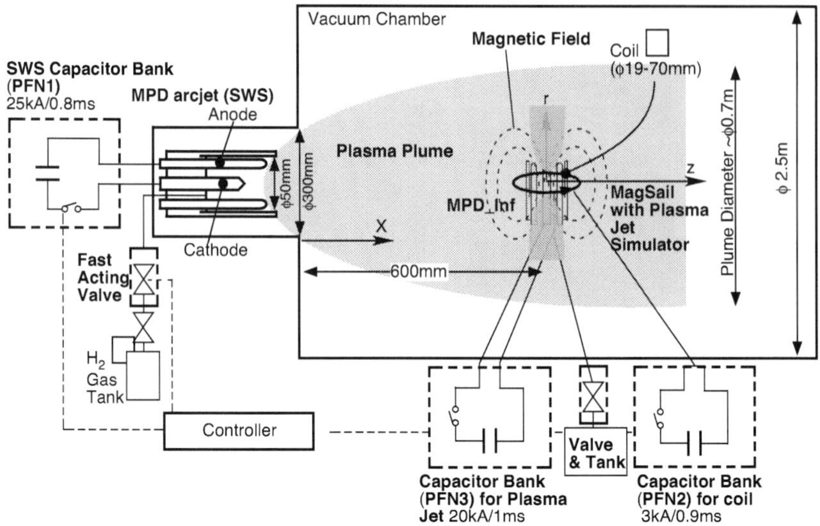

Fig. 20. Experimental setup for magnetoplasma sail (MPS).

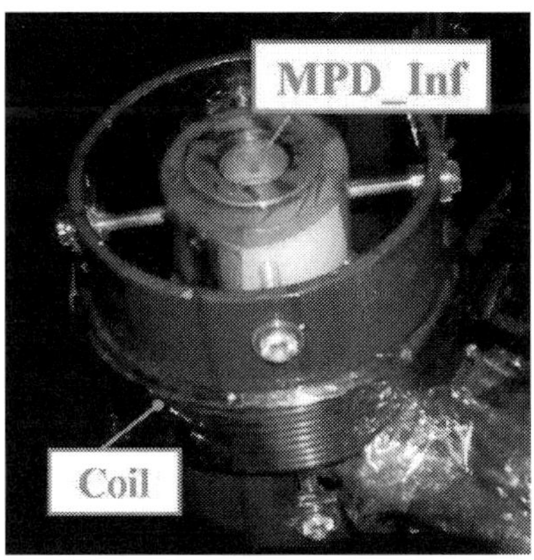

Fig. 21. Miniature MPS spacecraft (20-mm diameter MPD is installed inside 75-mm-diameter coil).

distances from the coil center to the magnetospheric boundary are 119.3 mm in Fig.23(a), and 145.3 mm in Fig.23(b); these distances are estimated from each photo. Therefore, from this evaluation, the magnetic field expansion observed is 26 mm. This is our first and successful magnetic field inflation experiment in laboratory.

Currently, experimental efforts are going on: 1) to form a large MPS magnetosphere, and 2) to detect increased thrust level by MPS operation. Direct thrust measurements of the miniature MPS is very interesting, but due to facility limitation, thrust measurement so far was not successful. Our goal is to evaluate a clear thrust gain (thrust of MPS divided by

Fig. 22. Typical flow around miniature MPS during quasi-steady operation (discharge current of SWS is 8 kA (H_2;0.4 g/s), and discharge current of MPD_Inf is 4.2 kA (H_2;0.05 g/s)).

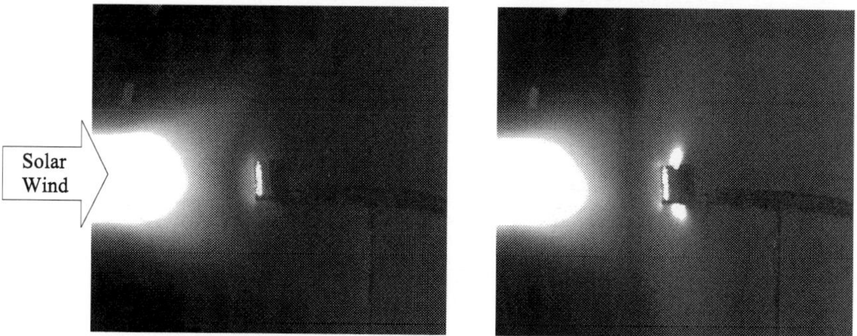

a) b)

Fig. 23. Photo of scale-model experiment of MPS; a) miniature MagSail and b) miniature MPS. (simulated solar wind is introduced to 5cm/20turn coil. SWS: u_{sw}=45 km/s, n_{sw}=1.8x10^{19} m^{-3}, H_2, 0.4 g/s. From inside the coil, a plasma jet is ejected from 2-cm diameter MPD : H_2, 0.04 g/s, 0.06 T at the surface of the MPD, and J=3.5 kA for the miniature MPS).

thrust of MagSail) that is possible by MPS. For this purpose, a higher power plasma source is going replace the small (20-mm-diameter) MPD arcjet to obtain two times larger magnetic cavity size than that of MagSail. Also, several plasma sources are prepared that offer both low velocity and high velocity plasma jet. These setups will enable thrust gain more than 200%, so that an increased thrust level might be detectable.

6. Conclusion

Starting from the basic theory of sail propulsion using the solar wind, current status of solar wind sail (MPS) research, in particular experiment using a plasma wind tunnel, is described. In the experiment, using a self-field magnetoplasmadynamic arcjet as a solar wind simulator, quasi-steady interaction between a plasma flow and a magnetic cavity is obtained for a Magsail (consisting of only a coil) and for an MPS (which accomplishes an inflated magnetic cavity by a plasma jet from inside a coil). So far, thrust measurement of these sails was conducted for only a few limited cases of Magsail (without plasma jet from spacecraft), and extension of the experimental set-up to MPS thrust measurement is still going on. During the quasi-steady interaction of about 0.8 ms, the dynamic behavior Magsail/MPS in laboratory is observed, and it was found that the magnetic cavity of Magsail is fluctuating in a frequency rage of 10-800 kHz. Thrust by Magsail is hence produced under rather turbulent plasma flow to magnetic cavity interaction.

Inspired by the ideas of MagSail and its derivatives, many new sail propulsion system are advocated. Khazanov and Akita proposed the usage of solar radiation to enhance thrust by MagSail (Khazanov et al., 2005 and Akita and Suzuki, 2004). Also, Slough proposed a way to capture the solar wind momentum by using a rotating magnetic field, and the new system is called 'the plasma magnet' (Slough, 2005). The plasma magnet is already accessed by laboratory experiment (Slough, 2007). In 2004, Janhunen proposed a completely new idea using the solar wind, which has a set of thin long wires being kept at a high positive potential so that the wires repel and deflect incident solar wind protons and it is called as electrostatic sail (Janhunen, 2004; Janhunen & Sandroos, 2007). These ideas are not so matured to proceed to a flight demonstration in space, but attractive solar wind sails that are competitive against the existing thruster technology is emerging.

7. Acknowledgment

We would like to express our acknowledgments to the members of the MPS working group (in particular, Dr. Kazuma Ueno, Mr. Yuya Oshio for providing experimental data, and Dr. Hiroyuki Nishida, Dr. Yoshihiro Kajimura, Dr. Iku Shinohara, Dr. Masaharu Matsumoto, Mr. Yasumasa Ashida, Dr. Hirotaka Otsu, Dr. Kazuhisa Fujita, and Prof. Hideyuki Usui) for their valuable activities in the analysis of MPS. This research is supported by: 1) the Grant-in-Aid for Scientific Research (A) (No.21246126) of the Japan Society for Promotion of Science, 2) the space plasma laboratory and the engineering committee of the institute of space and astronautical science in Japan Aerospace Exploration Agency (JAXA), and 3) Japan Science and Technology Agency, CREST.

8. References

Akita, D., and Suzuki, K., On the Possibility of Utilization of Radiation Pressure in Magnetic Sails, Proceedings of Symposium on Flight Mechanics and Astrodynamics, (2004), pp.1-4 (in Japanese)

Andrews, D.G., and Zubrin, R.M., Magnetic Sails and Interstellar Travel, Journal of the British Interplanetary Society, Vol.43, (1990), pp.265-272, ISSN 0007-084X

Bachynski, M.P. and Osborne, F.J.F., Laboratory Geophysics and Astrophysics, in Advances in Plasma Dynamics (Anderson, T.P. and Springer R.W. Eds), (1965). Northwestern University Press

Fujita, K., Particle Simulation of Moderately-Sized Magnetic Sails, The Journal of Space Technology and Science, Vol.20, No.2, (2004), pp.26-31, ISSN 0911-551X

Funaki, I. Kojima, H., Yamakawa, H., Nakayama, Y., and Shimizu, Y., Laboratory Experiment of Plasma Flow around Magnetic Sail, Astrophysics and Space Science, Vol.307, No.1-3, (2007), pp.63-68, ISSN 0004-640X

Funaki, I., Kimura, T., Ueno, K., Horisawa, H., Yamakawa, H., Kajimura, Y, Nakashima, H., and Shimizu, Y., Laboratory Experiment of Magnetoplasma Sail, Part 2: Magnetic

Field Inflation, Proceedings of 30th International Electric Propulsion Conference, IEPC2007-94, Florence, Sept. 2007

Frisbee, R. H., Advanced Space Propulsion for the 21st Century, Journal of Propulsion and Power, Vol.19, No.6, (2003), pp.1129-1154, ISSN 0748-4658

Giersch, L., Winglee, R.M., Slough, J., Ziemba, T., Euripides, P., Magnetic Dipole Inflation with Cascaded Arc and Applications to Mini-magnetospheric Plasma Propulsion, 39th AIAA/ASME/SAE/ASEE Joint Propulsion Conference & Exhibit, AIAA-2003-5223, Huntsville, July 2003

Janhunen, P., Electric Sails for Spacecraft Propulsion, Journal of Propulsion and Power, Vol.20, No.4, (2004), pp.763-764, ISSN 0748-4658

Janhunen, P., and Sandroos, A., Simulation Study of Solar Wind Push on a Charged Wire: Basis of Solar Wind Electric Sail Propulsion, Annales Geophysicae, Vol.25, (2007), pp.755–767, ISSN 0992-7689

Kajimura, Y., Usui, H., Ueno, K., Funaki, I., Nunami, M., Shinohara, I., Nakamura, M., and Yamakawa, H., Hybrid Particle-in-Cell Simulations of Magnetic Sail in Laboratory Experiment, Journal of Propulsion and Power, Vol.26, No.1, (2010), pp.159-166, ISSN 0748-4658

Khazanov, G., Delamere, P., Kabin, K., Linde, T. J., Fundamentals of the Plasma Sail Concept: Magnetohydrodynamic and Kinetic Studies, Journal of Propulsion and Power, Vol.21, No.5, (2005), pp.853-861, ISSN 0748-4658

Nishida, A. Ed., Magnetospheric Plasma Physics, (1982). Center for Academic Publications Japan, ISBN90-277-1345-6

Obayashi, T., Solar Terrestrial Physics (1970). Syokabo, Tokyo (in Japanese), ISBN4-7853-2405-8

Omidi, N., and Karimabadi, H., Kinetic Simulation/Modeling of Plasma Sails, 39th AIAA/ASME/SAE/ASEE Joint Propulsion Conference and Exhibit, AIAA-2003-5226, Huntsville, July 2003

Oshio, Y., Ueno, K., and Funaki, I., Fluctuation of Magnetosphere in Scale-model Experiment of Magnetic Sail, 31st International Electric Propulsion Conference, IEPC2007-94, Michigan, Sept. 2009

Slough, J., High Beta Plasma for Inflation of a Dipolar Magnetic Field as a Magnetic Sail (M2P2), Proceedings of 27th International Electric Propulsion Conference, IEPC01-202, Pasadena, Oct. 2001

Slough, J. and Giersch, L., The Plasma Magnet, 41st AIAA/ASME/SAE/ASEE Joint Propulsion Conference & Exhibit, AIAA-2005-4461, Tucson, Arizona, July 2005

Slough, J., Plasma Sail Propulsion based on the Plasma Magnet, Proceedings of 30th International Electric Propulsion Conference, IEPC2007-15, Florence, Sept. 2007

Ueno, K., Kimura, T., Ayabe, T., Funaki, I., Yamakawa, H., and Horisawa, H., Thrust Measurement of Pure Magnetic Sail, Transactions of JSASS, Space Technology Japan, Vol. 7, (2009), pp.Pb_65-Pb_69, ISSN 1884-0485

Ueno, K., Funaki, I., Kimura, T., Horisawa, H., and Yamakawa, H., Thrust Measurement of Pure Magnetic Sail using Parallelogram-pendulum Method, Journal of Propulsion and Power, Vol.25, No.2, (2009), pp.536-539, ISSN 0748-4658

Ueno, K., Oshio, Y., Funaki, I., Ayabe, T., Horisawa, H., and Yamakawa, H., Characterization of Magnetoplasma Sail in Laboratory, Proceedings of 27th

International Symposium on Space Technology and Science, ISTS-2009-b-42, Tsukuba, June 2009

Winglee, R.M., Slough, J., Ziemba, T., and Goodson, A., Mini-Magnetospheric Plasma Propulsion: Tapping the Energy of the Solar Wind for Spacecraft Propulsion, Journal of Geophysical Research, Vol.105, No.21, (2000), pp.21067-21078, ISSN0148-0227

Winglee, R.M., Ziemba, T., Euripides, P., and Slough, J., Computer Modeling of the Laboratory Testing of Mini-Magnetospheric Plasma Propulsion, Proceedings of 27th International Electric Propulsion Conference, IEPC-01-200, Pasadena, Oct. 2001

Winglee, R.M., Euripides, P., Ziemba, T., Slough, J., and Giersch, L., Simulation of Mini-Magnetospheric Plasma Propulsion (M2P2) Interacting with an External Plasma Wind, 39th AIAA/ASME/SAE/ASEE Joint Propulsion Conference & Exhibit, AIAA-2003-5224, Huntsville, July 2003

Yamakawa, H., Funaki, I., Nakayama, Y., Fujita, K., Ogawa, H., Nonaka, S., Kuninaka, H., Sawai, S., Nishida, H., Asahi, R., Otsu, H., and Nakashima, H., Magneto Plasma Sail: An Engineering Satellite Concept and its Application for Outer Planet Missions, Acta Astronautica, Vol.59, (2006), pp.777-784, ISSN 0094-5765

Ziemba, T.M., Winglee, R.M., Euripides, P., and Slough J., Parameterization of the Laboratory Performance of the Mini-Magnetospheric Plasma Propulsion (M2P2) Prototype, Proceedings of 27th International Electric Propulsion Conference, IEPC-01-201, Pasadena, Oct. 2001

Ziemba, T., Euripides, P., Winglee, R.M., Slough, J., Giersch, L., Efficient Plasma Production in Low Background Neutral Pressures with the M2P2 Prototype, 39th AIAA/ASME/SAE/ASEE Joint Propulsion Conference & Exhibit, AIAA-2003-5222, Huntsville, July 2003

Zubrin, R.M., and Andrews, D.G., Magnetic Sails and Interplanetary Travel, Journal of Spacecraft and Rockets, Vol.28, No.2, (1991), pp.197-203, ISSN 0022-4650